The Subnuclear Series • Volume 43

Proceedings of the International School of Subnuclear Physics

TOWARDS NEW MILESTONES IN OUR QUEST TO GO BEYOND THE STANDARD MODEL

THE SUBNUCLEAR SERIES

Series Editor: ANTONINO ZICHICHI, *European Physical Society, Geneva, Switzerland*

1. 1963 STRONG, ELECTROMAGNETIC, AND WEAK INTERACTIONS
2. 1964 SYMMETRIES IN ELEMENTARY PARTICLE PHYSICS
3. 1965 RECENT DEVELOPMENTS IN PARTICLE SYMMETRIES
4. 1966 STRONG AND WEAK INTERACTIONS
5. 1967 HADRONS AND THEIR INTERACTIONS
6. 1968 THEORY AND PHENOMENOLOGY IN PARTICLE PHYSICS
7. 1969 SUBNUCLEAR PHENOMENA
8. 1970 ELEMENTARY PROCESSES AT HIGH ENERGY
9. 1971 PROPERTIES OF THE FUNDAMENTAL INTERACTIONS
10. 1972 HIGHLIGHTS IN PARTICLE PHYSICS
11. 1973 LAWS OF HADRONIC MATTER
12. 1974 LEPTON AND HADRON STRUCTURE
13. 1975 NEW PHENOMENA IN SUBNUCLEAR PHYSICS
14. 1976 UNDERSTANDING THE FUNDAMENTAL CONSTITUENTS OF MATTER
15. 1977 THE WHYS OF SUBNUCLEAR PHYSICS
16. 1978 THE NEW ASPECTS OF SUBNUCLEAR PHYSICS
17. 1979 POINTLIKE STRUCTURES INSIDE AND OUTSIDE HADRONS
18. 1980 THE HIGH-ENERGY LIMIT
19. 1981 THE UNITY OF THE FUNDAMENTAL INTERACTIONS
20. 1982 GAUGE INTERACTIONS: Theory and Experiment
21. 1983 HOW FAR ARE WE FROM THE GAUGE FORCES?
22. 1984 QUARKS, LEPTONS, AND THEIR CONSTITUENTS
23. 1985 OLD AND NEW FORCES OF NATURE
24. 1986 THE SUPERWORLD I
25. 1987 THE SUPERWORLD II
26. 1988 THE SUPERWORLD III
27. 1989 THE CHALLENGING QUESTIONS
28. 1990 PHYSICS UP TO 200 TeV
29. 1991 PHYSICS AT THE HIGHEST ENERGY AND LUMINOSITY: To Understand the Origin of Mass
30. 1992 FROM SUPERSTRINGS TO THE REAL SUPERWORLD
31. 1993 FROM SUPERSYMMETRY TO THE ORIGIN OF SPACE–TIME
32. 1994 FROM SUPERSTRING TO PRESENT-DAY PHYSICS
33. 1995 VACUUM AND VACUA: The Physics of Nothing
34. 1996 EFFECTIVE THEORIES AND FUNDAMENTAL INTERACTIONS
35. 1997 HIGHLIGHTS OF SUBNUCLEAR PHYSICS: 50 Years Later
36. 1998 FROM THE PLANCK LENGTH TO THE HUBBLE RADIUS
37. 1999 BASICS AND HIGHLIGHTS IN FUNDAMENTAL PHYSICS
38. 2000 THEORY AND EXPERIMENT HEADING FOR NEW PHYSICS
39. 2001 NEW FIELDS AND STRINGS IN SUBNUCLEAR PHYSICS
40. 2002 FROM QUARKS AND GLUONS TO QUANTUM GRAVITY
41. 2003 FROM QUARKS TO BLACK HOLES: PROGRESS IN UNDERSTANDING THE LOGIC OF NATURE
42. 2004 HOW AND WHERE TO GO BEYOND THE STANDARD MODEL
43. 2005 TOWARDS NEW MILESTONES IN OUR QUEST TO GO BEYOND THE STANDARD MODEL

Volume 1 was published by W. A. Benjamin, Inc., New York; 2–8 and 11–12 by Academic Press, New York and London; 9–10 by Editrice Compositori, Bologna; 13–29 by Plenum Press, New York and London; 30–43 by World Scientific, Singapore.

The Subnuclear Series • Volume 43

Proceedings of the International School of Subnuclear Physics

TOWARDS NEW MILESTONES IN OUR QUEST TO GO BEYOND THE STANDARD MODEL

Edited by

Antonino Zichichi

European Physical Society
Geneva, Switzerland

World Scientific

NEW JERSEY • LONDON • SINGAPORE • BEIJING • SHANGHAI • HONG KONG • TAIPEI • CHENNAI

Published by

World Scientific Publishing Co. Pte. Ltd.
5 Toh Tuck Link, Singapore 596224
USA office: 27 Warren Street, Suite 401-402, Hackensack, NJ 07601
UK office: 57 Shelton Street, Covent Garden, London WC2H 9HE

British Library Cataloguing-in-Publication Data
A catalogue record for this book is available from the British Library.

The Subnuclear Series — Vol. 43
TOWARDS NEW MILESTONES IN OUR QUEST TO GO BEYOND THE STANDARD MODEL
Proceedings of the International School of Subnuclear Physics

ISBN-13 978-981-277-911-3
ISBN-10 981-277-911-6

Printed in Singapore by World Scientific Printers (S) Pte Ltd

PREFACE

During August/September 2005, a group of 76 physicists from 21 countries, working in 34 laboratories in 13 countries, met in Erice to participate in the 43rd Course of the International School of Subnuclear Physics. The countries represented by the participants were: Australia, Belarus, Canada, China, France, Germany, Greece, Hungary, Ireland, Italy, Japan, Mexico, the Netherlands, Poland, Pakistan, Russia, Spain, Switzerland, Ukraine, United Kingdom and the United States of America.

The School was sponsored by the Academies of Sciences of Estonia, Georgia, Lithuania, Russia and Ukraine; the Chinese Academy of Sciences; the Commission of the European Communities; the European Physical Society; the Italian Ministry of Education, University and Scientific Research; the Sicilian Regional Government; the Weizmann Institute of Science; the World Federation of Scientists and the World Laboratory.

The purpose of the School was to focus attention on the phenomenological and theoretical developments in Gauge Theories, as well as in global and local Supersymmetry, and in all the other sectors of Subnuclear Physics. Experimental highlights from the most relevant sources of new data were presented and discussed, including the latest news concerning future facilities, as reported in the contents.

An original feature of the School, introduced in 1996, is a series of special sessions devoted to "New Talents", with the aim of encouraging and promoting young physicists to achieve recognition at an international level. This is a serious problem in Experimental Physics, where collaborations count several hundreds of participants and it is almost impossible for young fellows to be known. Even if with much less emphasis, the problem exists also in Theoretical Physics. So we decided to offer 26 young fellows the possibility to give an open presentation of the results of their studies, followed by a discussion. Among them, 11 best "New Talents" have been selected for publication in this volume, and four were given an award. In particular, one award for an original theoretical work, one for an original experimental work, one for an original work in data analysis and one for an original work in new fields were assigned. Moreover, on the occasion of the World Year of Physics

2005, and to honour the memory of Enrico Fermi, ten "E. Fermi" Junior Grants established by the *"ENRICO FERMI" CENTRE (Rome)* have been awarded to "New Talents", of whom seven coming from developing countries.

During the organization and the running of this year's Course, I enjoyed the collaboration of my colleague and friend, Gerardus 't Hooft, who shared with me the Directorship of the Course. I would like to thank him, together with the group of invited scientists and all the people who contributed to the success of this year's Course.

I hope the reader will enjoy the book as much as the students attending the lectures and the discussion sessions, which, as every year, have been the focal point of this School's activity. Thanks to the work of the Scientific Secretaries, the discussions have been reproduced as faithfully as possible. At various stages of my work I have enjoyed the collaboration of many friends whose contributions have been extremely important for the School and are highly appreciated. I thank them most warmly. A final acknowledgement to all those in Erice, Bologna and Geneva, who have helped me on so many occasions and to whom I feel very indebted.

Antonino Zichichi
Geneva, October 2005

CONTENTS

Special Sessions for New Talents

Closing Ceremony

Experimental Signatures of Strings and Branes

I. Antoniadis[1]

Department of Physics, CERN
Theory Division
CH-1211 Geneva 23, Switzerland

Abstract

Type I string theory provides a D-brane world description of our universe and leads to two new scenaria for physics beyond the Standard Model: low string scale and split supersymmetry. Lowering the string scale in the TeV region provides a theoretical framework for solving the mass hierarchy problem and unifying all interactions. The apparent weakness of gravity can then be accounted by the existence of large internal dimensions, in the submillimeter region, and transverse to a braneworld where we must be confined. I review the main properties of this scenario and its implications for observations at both particle colliders, and in non-accelerator gravity experiments. I also review the main properties of split supersymmetry and present a concrete string realization which guarantees gauge coupling unification at the conventional scale $M_{\rm GUT} \simeq 2 \times 10^{16}$ GeV.

[1]On leave from CPHT (UMR 7644 of CNRS), Ecole Polytechnique, F-91128 Palaiseau.

1 Introduction

During the last few decades, physics beyond the Standard Model (SM) was guided from the problem of mass hierarchy. This can be formulated as the question of why gravity appears to us so weak compared to the other three known fundamental interactions corresponding to the electromagnetic, weak and strong nuclear forces. Indeed, gravitational interactions are suppressed by a very high energy scale, the Planck mass $M_P \sim 10^{19}$ GeV, associated to a length $l_P \sim 10^{-35}$ m, where they are expected to become important. In a quantum theory, the hierarchy implies a severe fine tuning of the fundamental parameters in more than 30 decimal places in order to keep the masses of elementary particles at their observed values. The reason is that quantum radiative corrections to all masses generated by the Higgs vacuum expectation value (VEV) are proportional to the ultraviolet cutoff which in the presence of gravity is fixed by the Planck mass. As a result, all masses are "attracted" to become about 10^{16} times heavier than their observed values.

Besides compositeness, there are three main theories that have been proposed and studied extensively during the last years, corresponding to different approaches of dealing with the mass hierarchy problem. (1) Low energy supersymmetry with all superparticle masses in the TeV region. Indeed, in the limit of exact supersymmetry, quadratically divergent corrections to the Higgs self-energy are exactly cancelled, while in the softly broken case, they are cutoff by the supersymmetry breaking mass splittings. (2) TeV scale strings, in which quadratic divergences are cutoff by the string scale and low energy supersymmetry is not needed. (3) Split supersymmetry, where scalar masses are heavy while fermions (gauginos and higgsinos) are light. Thus, gauge coupling unification and dark matter candidate are preserved but the mass hierarchy should be stabilized by a different way and the low energy world appears to be fine-tuned. All these ideas are experimentally testable at high-energy particle colliders and in particular at LHC. Below, I discuss their implementation in string theory.

The appropriate and most convenient framework for low energy supersymmetry and grand unification is the perturbative heterotic string. Indeed, in this theory, gravity and gauge interactions have the same origin, as massless modes of the closed heterotic string, and they are unified at the string scale M_s. As a result, the Planck mass M_P is predicted to be proportional to M_s:

$$M_P = M_s/g\,, \tag{1}$$

where g is the gauge coupling. In the simplest constructions all gauge couplings are the same at the string scale, given by the four-dimensional (4d) string coupling, and thus no grand unified group is needed for unification. In our conventions $\alpha_{\rm GUT} = g^2 \simeq 0.04$, leading to a discrepancy between the string and grand unification scale $M_{\rm GUT}$ by almost two orders of magnitude. Explaining this gap introduces in general new parameters or a new scale, and the predictive power is essentially lost. This is the main defect of this framework, which remains though an open and interesting possibility.

The other two ideas have both as natural framework of realization type I string theory with D-branes. Unlike in the heterotic string, gauge and gravitational interactions have now different origin. The latter are described again by closed strings, while the former emerge as excitations of open strings with endpoints confined on D-branes [1]. This leads to a braneworld description of our universe, which should be localized on a hypersurface, i.e. a membrane extended in p spatial dimensions, called p-brane (see Fig. 1). Closed strings propagate in all nine dimensions of string theory: in those extended along the p-brane, called parallel, as well as in the transverse ones. On the contrary, open strings are attached on the p-brane. Obviously, our p-brane world must have at least the three

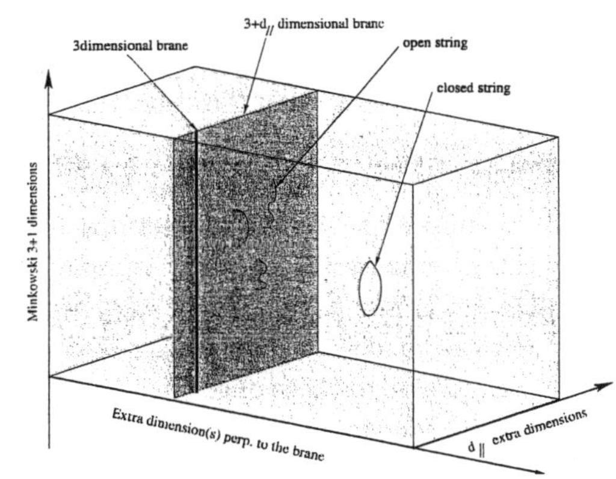

Figure 1: In the type I string framework, our Universe contains, besides the three known spatial dimensions (denoted by a single blue line), some extra dimensions ($d_{\|} = p - 3$) parallel to our world p-brane (green plane) where endpoints of open strings are confined, as well as some transverse dimensions (yellow space) where only gravity described by closed strings can propagate.

known dimensions of space. But it may contain more: the extra $d_{\|} = p - 3$ parallel dimensions must have a finite size, in order to be unobservable at present energies, and can be as large as $\text{TeV}^{-1} \sim 10^{-18}$ m [2]. On the other hand, transverse dimensions interact with us only gravitationally and experimental bounds are much weaker: their size should be less than about 0.1 mm [3]. In the following, I review the main properties and experimental signatures of low string scale [4, 5] and split supersymmetry [6, 7, 8] proposals.

2 Low string scale: Framework

In type I theory, the different origin of gauge and gravitational interactions implies that the relation between the Planck and string scales is not linear as (1) of the heterotic string. The requirement that string theory should be weakly coupled, constrain the size of all parallel dimensions to be of order of the string length, while transverse dimensions remain unrestricted. Assuming an isotropic transverse space of $n = 9 - p$ compact dimensions of common radius $R_\perp$, one finds:

$$M_P^2 = \frac{1}{g^4} M_s^{2+n} R_\perp^n \,, \qquad g_s \simeq g^2 \,. \tag{2}$$

where g_s is the string coupling. It follows that the type I string scale can be chosen hierarchically smaller than the Planck mass [9, 4] at the expense of introducing extra large transverse dimensions felt only by gravity, while keeping the string coupling small [4]. The weakness of 4d gravity compared to gauge interactions (ratio M_W/M_P) is then attributed to the largeness of the transverse space $R_\perp$ compared to the string length $l_s = M_s^{-1}$.

An important property of these models is that gravity becomes effectively $(4+n)$-dimensional with a strength comparable to those of gauge interactions at the string scale. The first relation of Eq. (2) can be understood as a consequence of the $(4+n)$-dimensional Gauss law for gravity, with

$$M_*^{(4+n)} = M_s^{2+n}/g^4 \tag{3}$$

the effective scale of gravity in $4+n$ dimensions. Taking $M_s \simeq 1$ TeV, one finds a size for the extra dimensions $R_\perp$ varying from 10^8 km, .1 mm, down to a Fermi for $n = 1, 2$, or 6 large dimensions, respectively. This shows that while $n = 1$ is excluded, $n \geq 2$ is allowed by present experimental bounds on gravitational forces [3, 10]. Thus, in these models, gravity appears to us very weak at macroscopic scales because its intensity is spread in the "hidden" extra dimensions. At distances shorter than $R_\perp$, it should deviate from Newton's law, which may be possible to explore in laboratory experiments (see Fig. 2).

The main experimental implications of TeV scale strings in particle accelerators are of three types, in correspondence with the three different sectors that are generally present:

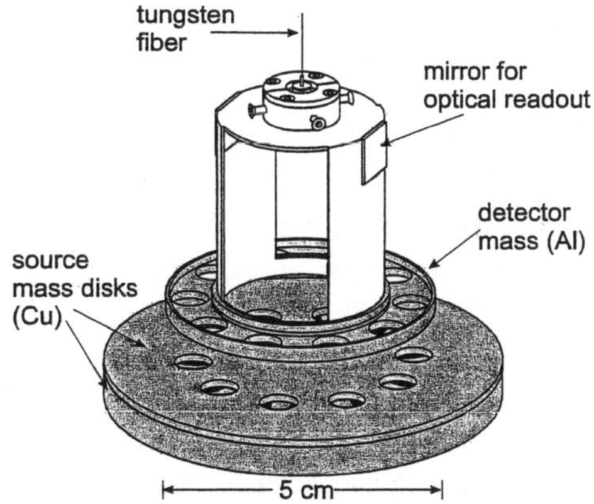

Figure 2: Torsion pendulum that tested Newton's law at 130 nm. Several sources of background noise were eliminated using appropriate devices.

(i) new compactified parallel dimensions, (ii) new extra large transverse dimensions and low scale quantum gravity, and (iii) genuine string and quantum gravity effects. On the other hand, there exist interesting implications in non accelerator table-top experiments due to the exchange of gravitons or other possible states living in the bulk.

3 Experimental implications in accelerators

3.1 World-brane extra dimensions

In this case $RM_s \gtrsim 1$, and the associated compactification scale $R_{\parallel}^{-1}$ would be the first scale of new physics that should be found increasing the beam energy [2, 11]. There are several reasons for the existence of such dimensions. It is a logical possibility, since out of the six extra dimensions of string theory only two are needed for lowering the string scale, and thus the effective p-brane of our world has in general $d_{\parallel} \equiv p - 3 \leq 4$. Moreover, they can be used to address several physical problems in braneworld models, such as obtaining different SM gauge couplings, explaining fermion mass hierarchies due to different localization points of quarks and leptons in the extra dimensions, providing calculable mechanisms of supersymmetry breaking, etc.

The main consequence is the existence of Kaluza-Klein (KK) excitations for all SM particles that propagate along the extra parallel dimensions. Their masses are given by:

$$M_m^2 = M_0^2 + \frac{m^2}{R_\parallel^2} \quad ; \quad m = 0, \pm 1, \pm 2, \ldots \tag{4}$$

where we used $d_\parallel = 1$, and M_0 is the higher dimensional mass. The zero-mode $m = 0$ is identified with the 4d state, while the higher modes have the same quantum numbers with the lowest one, except for their mass given in (4). There are two types of experimental signatures of such dimensions [11, 12, 13]: (i) virtual exchange of KK excitations, leading to deviations in cross-sections compared to the SM prediction, that can be used to extract bounds on the compactification scale; (ii) direct production of KK modes.

On general grounds, there can be two different kinds of models with qualitatively different signatures depending on the localization properties of matter fermion fields. If the latter are localized in 3d brane intersections, they do not have excitations and KK momentum is not conserved because of the breaking of translation invariance in the extra dimension(s). KK modes of gauge bosons are then singly produced giving rise to generally strong bounds on the compactification scale and new resonances that can be observed in experiments. Otherwise, they can be produced only in pairs due to the KK momentum conservation, making the bounds weaker but the resonances difficult to observe.

When the internal momentum is conserved, the interaction vertex involving KK modes has the same 4d tree-level gauge coupling. On the other hand, their couplings to localized matter have an exponential form factor suppressing the interactions of heavy modes. This form factor can be viewed as the fact that the branes intersection has a finite thickness. For instance, the coupling of the KK excitations of gauge fields $A^\mu(x,y) = \sum_m A_m^\mu \exp i\frac{my}{R_\parallel}$ to the charge density $j_\mu(x)$ of massless localized fermions is described by the effective action [14]:

$$\int d^4x \sum_m e^{-\ln 16 \frac{m^2 l_s^2}{2R_\parallel^2}} j_\mu(x)\, A_m^\mu(x)\,. \tag{5}$$

After Fourier transform in position space, it becomes:

$$\int d^4x\, dy \, \frac{1}{(2\pi \ln 16)^2} e^{-\frac{y^2 M_s^2}{2\ln 16}} j_\mu(x)\, A^\mu(x,y)\,, \tag{6}$$

from which we see that localized fermions form a Gaussian distribution of charge with a width $\sigma = \sqrt{\ln 16}\, l_s \sim 1.66\, l_s$.

To simplify the analysis, let us consider first the case $d_\parallel = 1$ where some of the gauge fields arise from an effective 4-brane, while fermions are localized states on brane intersections. Since the corresponding gauge couplings are reduced by the size of the large dimension $R_{\parallel}M_s$ compared to the others, one can account for the ratio of the weak to strong interactions strengths if the $SU(2)$ brane extends along the extra dimension, while $SU(3)$ does not. As a result, there are 3 distinct cases to study [13], denoted by (t,l,l), (t,l,t) and (t,t,l), where the three positions in the brackets correspond to the three SM gauge group factors $SU(3)\times SU(2)\times U(1)$ and those with l (longitudinal) feel the extra dimension, while those with t (transverse) do not.

In the (t,l,l) case, there are KK excitations of $SU(2)\times U(1)$ gauge bosons: $W_{\pm}^{(m)}$, $\gamma^{(m)}$ and $Z^{(m)}$. Performing a χ^2 fit of the electroweak observables, one finds that if the Higgs is a bulk state (l), $R_{\parallel}^{-1}\gtrsim 3.5$ TeV [15]. This implies that LHC can produce at most the first KK mode. Different choices for localization of matter and Higgs fields lead to bounds, lying in the range $1-5$ TeV [15].

In addition to virtual effects, KK excitations can be produced on-shell at LHC as new resonances [12] (see Fig. 3). There are two different channels, neutral Drell–Yan

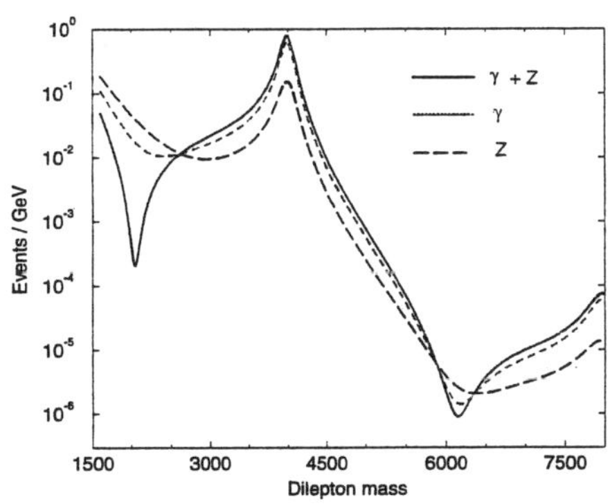

Figure 3: Production of the first KK modes of the photon and of the Z boson at LHC, decaying to electron-positron pairs. The number of expected events is plotted as a function of the energy of the pair in GeV. From highest to lowest: excitation of $\gamma+Z$, γ and Z.

processes $pp\longrightarrow l^+l^-X$ and the charged channel $l^{\pm}\nu$, corresponding to the production of the KK modes $\gamma^{(1)}$, $Z^{(1)}$ and $W_{\pm}^{(1)}$, respectively. The discovery limits are about 6 TeV, while the exclusion bounds 15 TeV. An interesting observation in the case of $\gamma^{(1)}+Z^{(1)}$ is that interferences can lead to a "dip" just before the resonance. There are some ways to distinguish the corresponding signals from other possible origin of new physics, such as models with new gauge bosons. In fact, in the (t, l, l) and (t, l, t) cases, one expects two resonances located practically at the same mass value. This property is not shared by most of other new gauge boson models. Moreover, the heights and widths of the resonances are directly related to those of SM gauge bosons in the corresponding channels.

In the (t, l, t) case, only the $SU(2)$ factor feels the extra dimension and the limits set by the KK states of $W^{\pm}$ remain the same. On the other hand, in the (t, t, l) case where only $U(1)_Y$ feels the extra dimension, the limits are weaker and the exclusion bound is around 8 TeV. In addition to these simple possibilities, brane constructions lead often to cases where part of $U(1)_Y$ is t and part is l. If $SU(2)$ is l the limits come again from $W^{\pm}$, while if it is t then it will be difficult to distinguish this case from a generic extra $U(1)'$. A good statistics would be needed to see the deviation in the tail of the resonance as being due to effects additional to those of a generic $U(1)'$ resonance. Finally, in the case of two or more parallel dimensions, the sum in the exchange of the KK modes diverges in the limit $R_{\parallel}M_s >> 1$ and needs to be regularized using the form factor (5). Cross-sections become bigger yielding stronger bounds, while resonances are closer implying that more of them could be reached by LHC.

On the other hand, if all SM particles propagate in the extra dimension (called universal)[2], KK modes can only be produced in pairs and the lower bound on the compactification scale becomes weaker, of order of 300-500 GeV. Moreover, no resonances can be observed at LHC, so that this scenario appears very similar to low energy supersymmetry. In fact, KK parity can even play the role of R-parity, implying that the lightest KK mode is stable and can be a dark matter candidate in analogy to the LSP [16].

3.2 Extra large transverse dimensions

The main experimental signal is gravitational radiation in the bulk from any physical process on the world-brane. In fact, the very existence of branes breaks translation invariance in the transverse dimensions and gravitons can be emitted from the brane into the bulk. During a collision of center of mass energy $\sqrt{s}$, there are $\sim (\sqrt{s}R_{\perp})^n$ KK excitations of gravitons with tiny masses, that can be emitted. Each of these states looks from the 4d point of view as a massive, quasi-stable, extremely weakly coupled (s/M_P^2 suppressed) particle that escapes from the detector. The total effect is a missing-energy

[2]Although interesting, this scenario seems difficult to be realized, since 4d chirality requires non-trivial action of orbifold twists with localized chiral states at the fixed points.

Table 1: Limits on $R_\perp$ in mm.

Experiment	$n = 2$	$n = 4$	$n = 6$
Collider bounds			
LEP 2	5×10^{-1}	2×10^{-8}	7×10^{-11}
Tevatron	5×10^{-1}	10^{-8}	4×10^{-11}
LHC	4×10^{-3}	6×10^{-10}	3×10^{-12}
NLC	10^{-2}	10^{-9}	6×10^{-12}
Present non-collider bounds			
SN1987A	3×10^{-4}	10^{-8}	6×10^{-10}
COMPTEL	5×10^{-5}	-	-

cross-section roughly of order:

$$\frac{(\sqrt{s}R_\perp)^n}{M_P^2} \sim \frac{1}{s}\left(\frac{\sqrt{s}}{M_s}\right)^{n+2} . \tag{7}$$

Explicit computation of these effects leads to the bounds given in Table 1. However, larger radii are allowed if one relaxes the assumption of isotropy, by taking for instance two large dimensions with different radii.

Fig. 4 shows the cross-section for graviton emission in the bulk, corresponding to the process $pp \longrightarrow jet + graviton$ at LHC, together with the SM background [17]. For a given value of M_s, the cross-section for graviton emission decreases with the number of large transverse dimensions, in contrast to the case of parallel dimensions. The reason is that gravity becomes weaker if there are more dimensions because there is more space for the gravitational field to escape. There is a particular energy and angular distribution of the produced gravitons that arise from the distribution in mass of KK states of spin-2. This can be contrasted to other sources of missing energy and might be a smoking gun for the extra dimensional nature of such a signal.

In Table 1, there are also included astrophysical and cosmological bounds. Astrophysical bounds [18, 19] arise from the requirement that the radiation of gravitons should not carry on too much of the gravitational binding energy released during core collapse of supernovae. In fact, the measurements of Kamiokande and IMB for SN1987A suggest that the main channel is neutrino fluxes. The best cosmological bound [20] is obtained from requiring that decay of bulk gravitons to photons do not generate a spike in the energy spectrum of the photon background measured by the COMPTEL instrument. Bulk gravitons are expected to be produced just before nucleosynthesis due to thermal

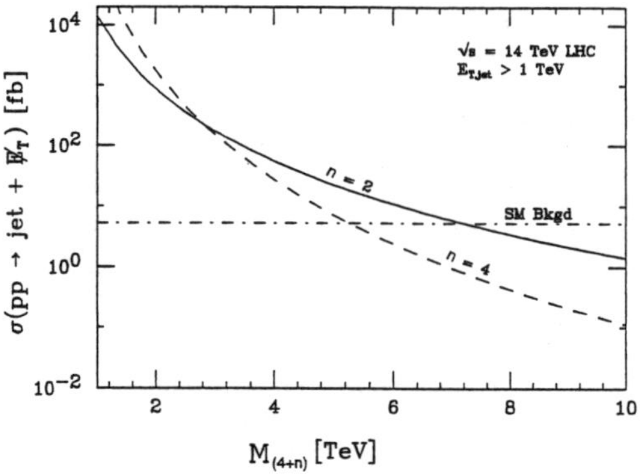

Figure 4: Missing energy due to graviton emission at LHC, as a function of the higher-dimensional gravity scale M_*, produced together with a hadronic jet. The expected cross-section is shown for $n = 2$ and $n = 4$ extra dimensions, together with the SM background.

radiation from the brane. The limits assume that the temperature was at most 1 MeV as nucleosynthesis begins, and become stronger if temperature is increased.

3.3 String effects

At low energies, the interaction of light (string) states is described by an effective field theory. Their exchange generates in particular four-fermion operators that can be used to extract independent bounds on the string scale. In analogy with the bounds on longitudinal extra dimensions, there are two cases depending on the localization properties of matter fermions. If they come from open strings with both ends on the same stack of branes, exchange of massive open string modes gives rise to dimension eight effective operators, involving four fermions and two space-time derivatives [21, 14]. The corresponding bounds on the string scale are then around 500 GeV. On the other hand, if matter fermions are localized on non-trivial brane intersections, one obtains dimension six four-fermion operators and the bounds become stronger: $M_s \gtrsim 2-3$ TeV [14, 5]. At energies higher than the string scale, new spectacular phenomena are expected to occur, related to string physics and quantum gravity effects, such as possible micro-black hole production [22]. Particle accelerators would then become the best tools for studying quantum gravity and string theory.

4 Supersymmetry in the bulk and short range forces

4.1 Sub-millimeter forces

Besides the spectacular predictions in accelerators, there are also modifications of gravitation in the sub-millimeter range, which can be tested in "table-top" experiments that measure gravity at short distances. There are three categories of such predictions:
(i) Deviations from the Newton's law $1/r^2$ behavior to $1/r^{2+n}$, which can be observable for $n = 2$ large transverse dimensions of sub-millimeter size. This case is particularly attractive on theoretical grounds because of the logarithmic sensitivity of SM couplings on the size of transverse space [23], that allows to determine the hierarchy [24].
(ii) New scalar forces in the sub-millimeter range, related to the mechanism of supersymmetry breaking, and mediated by light scalar fields φ with masses [25, 4]:

$$m_\varphi \simeq \frac{m_{susy}^2}{M_P} \simeq 10^{-4} - 10^{-6}\ \text{eV}\,, \tag{8}$$

for a supersymmetry breaking scale $m_{susy} \simeq 1 - 10$ TeV. They correspond to Compton wavelengths of 1 mm to 10 μm. m_{susy} can be either $1/R_\parallel$ if supersymmetry is broken by compactification [25], or the string scale if it is broken "maximally" on our world-brane [4]. A universal attractive scalar force is mediated by the radion modulus $\varphi \equiv M_P \ln R$, with R the radius of the longitudinal or transverse dimension(s). In the former case, the result (8) follows from the behavior of the vacuum energy density $\Lambda \sim 1/R_\parallel^4$ for large $R_\parallel$ (up to logarithmic corrections). In the latter, supersymmetry is broken primarily on the brane, and thus its transmission to the bulk is gravitationally suppressed, leading to (8). For $n = 2$, there may be an enhancement factor of the radion mass by $\ln R_\perp M_s \simeq 30$ decreasing its wavelength by an order of magnitude [24].

The coupling of the radius modulus to matter relative to gravity can be easily computed and is given by:

$$\sqrt{\alpha_\varphi} = \frac{1}{M}\frac{\partial M}{\partial \varphi}\ ; \quad \alpha_\varphi = \begin{cases} \frac{\partial \ln \Lambda_{QCD}}{\partial \ln R} \simeq \frac{1}{3} & \text{for } R_\parallel \\ \\ \frac{2n}{n+2} = 1 - 1.5 & \text{for } R_\perp \end{cases} \tag{9}$$

where M denotes a generic physical mass. In the longitudinal case, the coupling arises dominantly through the radius dependence of the QCD gauge coupling [25], while in the case of transverse dimension, it can be deduced from the rescaling of the metric which changes the string to the Einstein frame and depends slightly on the bulk dimensionality ($\alpha = 1 - 1.5$ for $n = 2 - 6$) [24]. Such a force can be tested in microgravity experiments

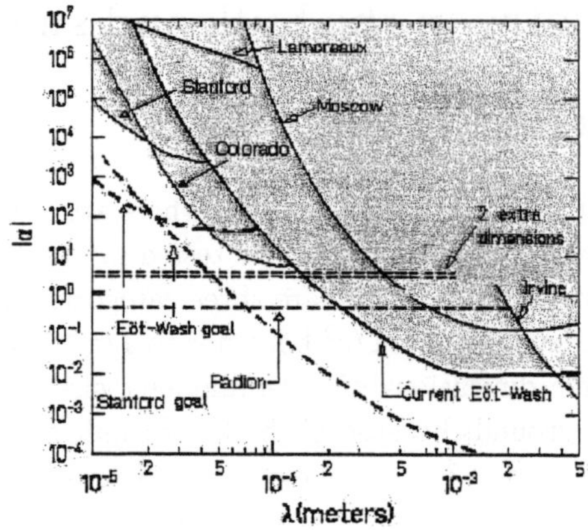

Figure 5: Present limits on non-Newtonian forces at short distances (yellow regions), as a function of their range λ and their strength relative to gravity α. The limits are compared to new forces mediated by the graviton in the case of two large extra dimensions, and by the radion.

and should be contrasted with the change of Newton's law due the presence of extra dimensions that is observable only for $n = 2$ [3, 10]. The resulting bounds from an analysis of the radion effects are [3]:

$$M_* \gtrsim 3 - 4.5\,\text{TeV} \quad \text{for} \quad n = 2 - 6\,. \tag{10}$$

In principle there can be other light moduli which couple with even larger strengths. For example the dilaton, whose VEV determines the string coupling, if it does not acquire large mass from some dynamical supersymmetric mechanism, can lead to a force of strength 2000 times bigger than gravity [26].
(iii) Non universal repulsive forces much stronger than gravity, mediated by possible abelian gauge fields in the bulk [18, 27]. Such fields acquire tiny masses of the order of M_s^2/M_P, as in (8), due to brane localized anomalies [27]. Although their gauge coupling is infinitesimally small, $g_A \sim M_s/M_P \simeq 10^{-16}$, it is still bigger that the gravitational coupling E/M_P for typical energies $E \sim 1$ GeV, and the strength of the new force would be $10^6 - 10^8$ stronger than gravity. This is an interesting region which will be soon explored in micro-gravity experiments (see Fig. 5). Note that in this case supernova constraints

impose that there should be at least four large extra dimensions in the bulk [18].

In Fig. 5 we depict the actual information from previous, present and upcoming experiments [10, 24]. The solid lines indicate the present limits from the experiments indicated. The excluded regions lie above these solid lines. Measuring gravitational strength forces at short distances is challenging. The dashed thick lines give the expected sensitivity of the various experiments, which will improve the actual limits by roughly two orders of magnitude, while the horizontal dashed lines correspond to the theoretical predictions for the graviton in the case $n = 2$ and for the radion in the transverse case. These limits are compared to those obtained from particle accelerator experiments in Table 1. Finally, in Figs. 6 and 7, we display recent improved bounds for new forces at very short distances by focusing on the right hand side of Fig. 5, near the origin [10].

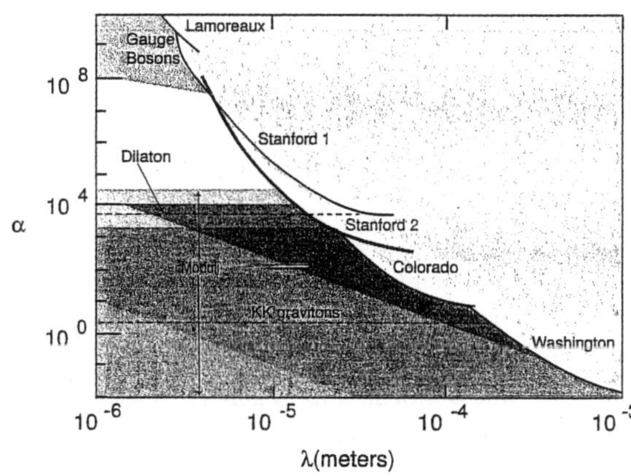

Figure 6: Bounds on non-Newtonian forces in the range 6-20 μm (see S. J. Smullin et al. in Ref. [10]).

4.2 Brane non-linear supersymmetry

When the closed string sector is supersymmetric, supersymmetry on a generic brane configuration is non-linearly realized even if the spectrum is not supersymmetric and brane fields have no superpartners. The reason is that the gravitino must couple to a conserved current locally, implying the existence of a goldstino on the brane world-volume. The goldstino is exactly massless in the infinite (transverse) volume limit and is

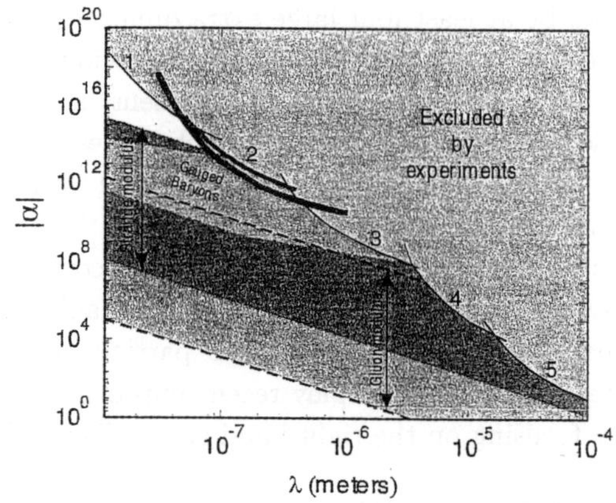

Figure 7: Bounds on non-Newtonian forces in the range around 200 nm (see R. S. Decca et al. in Ref. [10]). Curves 4 and 5 correspond to Stanford and Colorado experiments, respectively, of Fig. 6 (see also J C. Long and J. C. Price of Ref. [10]).

expected to acquire a small mass suppressed by the volume, of order (8). In the standard realization, its coupling to matter is given via the energy momentum tensor [28], while in general there are more terms invariant under non-linear supersymmetry that have been classified, up to dimension eight [29, 30].

An explicit computation was performed for a generic intersection of two brane stacks, leading to three irreducible couplings, besides the standard one [30]: two of dimension six involving the goldstino, a matter fermion and a scalar or gauge field, and one four-fermion operator of dimension eight. Their strength is set by the goldstino decay constant κ, up to model-independent numerical coefficients which are independent of the brane angles. Obviously, at low energies the dominant operators are those of dimension six. In the minimal case of (non-supersymmetric) SM, only one of these two operators may exist, that couples the goldstino χ with the Higgs H and a lepton doublet L:

$$\mathcal{L}_{\chi}^{int} = 2\kappa(D_\mu H)(LD^\mu\chi) + h.c. , \tag{11}$$

where the goldstino decay constant is given by the total brane tension

$$\frac{1}{2\,\kappa^2} = N_1 T_1 + N_2 T_2 \,; \quad T_i = \frac{M_s^4}{4\pi^2 g_i^2} , \tag{12}$$

with N_i the number of branes in each stack. It is important to notice that the effective interaction (11) conserves the total lepton number L, as long as we assign to the goldstino a total lepton number $L(\chi) = -1$ [31]. To simplify the analysis, we will consider the simplest case where (11) exists only for the first generation and L is the electron doublet [31].

The effective interaction (11) gives rise mainly to the decays $W^\pm \to e^\pm \chi$ and $Z, H \to \nu\chi$. It turns out that the invisible Z width gives the strongest limit on κ which can be translated to a bound on the string scale $M_s \gtrsim 500$ GeV, comparable to other collider bounds. This allows for the striking possibility of a Higgs boson decaying dominantly, or at least with a sizable branching ratio, via such an invisible mode, for a wide range of the parameter space (M_s, m_H), as seen in Fig. 8.

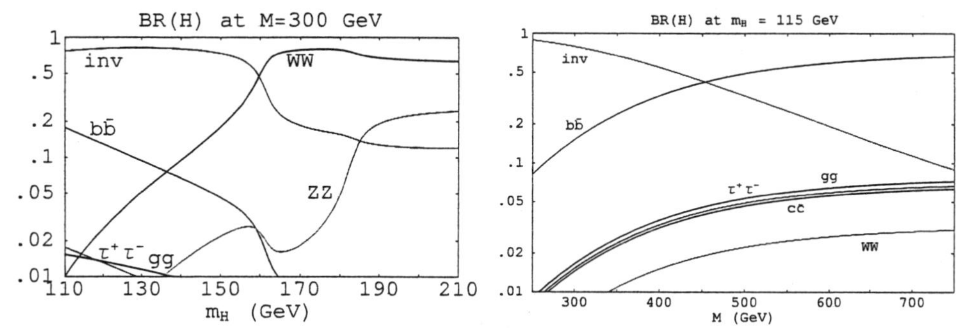

Figure 8: Higgs branching rations, as functions either of the Higgs mass m_H for a fixed value of the string scale $M_s \simeq 2M = 600$ GeV, or of $M \simeq M_s/2$ for $m_H = 115$ GeV.

5 Electroweak symmetry breaking

Non-supersymmetric TeV strings offer also a framework to realize gauge symmetry breaking radiatively. Indeed, from the effective field theory point of view, one expects quadratically divergent one-loop contributions to the masses of scalar fields. The divergences are cut off by M_s and if the corrections are negative, they can induce electroweak symmetry breaking and explain the mild hierarchy between the weak and a string scale at a few TeV, in terms of a loop factor [32]. More precisely, in the minimal case of one Higgs doublet H, the scalar potential is:

$$V = \lambda(H^\dagger H)^2 + \mu^2(H^\dagger H)\,, \tag{13}$$

where λ arises at tree-level. Moreover, in any model where the Higgs field comes from an open string with both ends fixed on the same brane stack, it is given by an appropriate truncation of a supersymmetric theory. Within the minimal spectrum of the SM, $\lambda = (g_2^2 + g'^2)/8$, with g_2 and g' the $SU(2)$ and $U(1)_Y$ gauge couplings. On the other hand, μ^2 is generated at one loop:

$$\mu^2 = -\varepsilon^2 \, g^2 \, M_s^2 \, , \tag{14}$$

where ε is a loop factor that can be estimated from a toy model computation and varies in the region $\epsilon \sim 10^{-1} - 10^{-3}$.

Indeed, consider for illustration a simple case where the whole one-loop effective potential of a scalar field can be computed. We assume for instance one extra dimension compactified on a circle of radius $R > 1$ (in string units). An interesting situation is provided by a class of models where a non-vanishing VEV for a scalar (Higgs) field ϕ results in shifting the mass of each KK excitation by a constant $a(\phi)$:

$$M_m^2 = \left(\frac{m + a(\phi)}{R} \right)^2 , \tag{15}$$

with m the KK integer momentum number. Such mass shifts arise for instance in the presence of a Wilson line, $a = q \oint \frac{dy}{2\pi} gA$, where A is the internal component of a gauge field with gauge coupling g, and q is the charge of a given state under the corresponding generator. A straightforward computation shows that the ϕ-dependent part of the one-loop effective potential is given by [33]:

$$V_{eff} = -Tr(-)^F \, \frac{R}{32 \, \pi^{3/2}} \sum_n e^{2\pi i n a} \int_0^\infty dl \; l^{3/2} \, f_s(l) \; e^{-\pi^2 n^2 R^2 l} \tag{16}$$

where $F = 0, 1$ for bosons and fermions, respectively. We have included a regulating function $f_s(l)$ which contains for example the effects of string oscillators. To understand its role we will consider the two limits $R >> 1$ and $R << 1$. In the first case only the $l \longrightarrow 0$ region contributes to the integral. This means that the effective potential receives sizable contributions only from the infrared (field theory) degrees of freedom. In this limit we would have $f_s(l) \longrightarrow 1$. For example, in the string model considered in [32]:

$$f_s(l) = \left[\frac{1}{4l} \frac{\theta_2}{\eta^3} (il + \frac{1}{2}) \right]^4 \longrightarrow 1 \qquad \text{for} \qquad l \longrightarrow 0, \tag{17}$$

and the field theory result is finite and can be explicitly computed. As a result of the Taylor expansion around $a = 0$, we are able to extract the one-loop contribution to the coefficient of the term of the potential quadratic in the Higgs field. It is given by a loop factor times the compactification scale [33]. One thus obtains $\mu^2 \sim g^2/R^2$ up to a proportionality constant which is calculable in the effective field theory. On the other hand, if we consider $R \to 0$, which by T-duality corresponds to taking the extra dimension as transverse and very large, the one-loop effective potential receives contributions from the whole tower of string oscillators as appearing in $f_s(l)$, leading to squared masses given by a loop factor times M_s^2, according to eq. (14).

More precisely, from the expression (16), one finds:

$$\varepsilon^2(R) = \frac{1}{2\pi^2}\int_0^\infty \frac{dl}{(2\,l)^{5/2}}\frac{\theta_2^4}{4\eta^{12}}\left(il + \frac{1}{2}\right) R^3 \sum_n n^2 e^{-2\pi n^2 R^2 l}\,, \tag{18}$$

which is plotted in Fig. 9. For the asymptotic value $R \to 0$ (corresponding upon T-

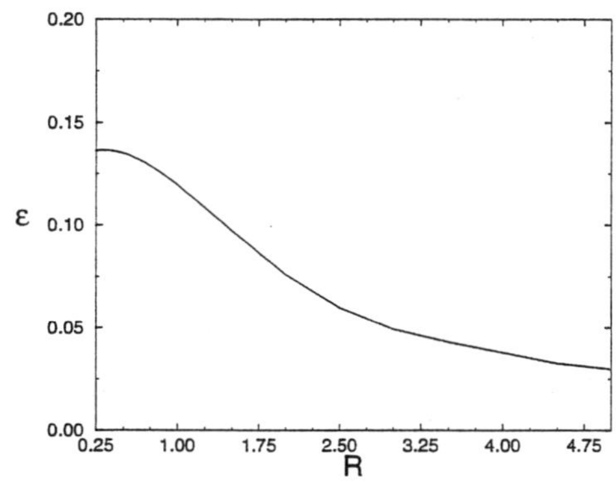

Figure 9: The coefficient ε of the one loop Higgs mass (14).

duality to a large transverse dimension of radius $1/R$), $\varepsilon(0) \simeq 0.14$, and the effective cut-off for the mass term is M_s, as can be seen from Eq. (14). At large R, $\mu^2(R)$ falls off as $1/R^2$, which is the effective cut-off in the limit $R \to \infty$, as we argued above, in agreement with field theory results in the presence of a compactified extra dimension [34, 25]. In fact, in the limit $R \to \infty$, an analytic approximation to $\varepsilon(R)$ gives:

$$\varepsilon(R) \simeq \frac{\varepsilon_\infty}{M_s R}\,, \qquad \varepsilon_\infty^2 = \frac{3\,\zeta(5)}{4\,\pi^4} \simeq 0.008\,. \tag{19}$$

The potential (13) has the usual minimum, given by the VEV of the neutral component of the Higgs doublet $v = \sqrt{-\mu^2/\lambda}$. Using the relation of v with the Z gauge boson mass, $M_Z^2 = (g_2^2 + g'^2)v^2/4$, and the expression of the quartic coupling λ, one obtains for the Higgs mass a prediction which is the MSSM value for $\tan\beta \to \infty$ and $m_A \to \infty$: $m_H = M_Z$. The tree level Higgs mass is known to receive important radiative corrections from the top-quark sector and rises to values around 120 GeV. Furthermore, from (14), one can compute M_s in terms of the Higgs mass $m_H^2 = -2\mu^2$:

$$M_s = \frac{m_H}{\sqrt{2}\, g\varepsilon}\,, \tag{20}$$

yielding naturally values in the TeV range.

6 Standard Model on D-branes

The gauge group closest to the Standard Model one can easily obtain with D-branes is $U(3) \times U(2) \times U(1)$. The first factor arises from three coincident "color" D-branes. An open string with one end on them is a triplet under $SU(3)$ and carries the same $U(1)$ charge for all three components. Thus, the $U(1)$ factor of $U(3)$ has to be identified with *gauged* baryon number. Similarly, $U(2)$ arises from two coincident "weak" D-branes and the corresponding abelian factor is identified with *gauged* weak-doublet number. Finally, an extra $U(1)$ D-brane is necessary in order to accommodate the Standard Model without breaking the baryon number [35]. In principle this $U(1)$ brane can be chosen to be independent of the other two collections with its own gauge coupling. To improve the predictability of the model, we choose to put it on top of either the color or the weak D-branes [36]. In either case, the model has two independent gauge couplings g_3 and g_2 corresponding, respectively, to the gauge groups $U(3)$ and $U(2)$. The $U(1)$ gauge coupling g_1 is equal to either g_3 or g_2.

Let us denote by Q_3, Q_2 and Q_1 the three $U(1)$ charges of $U(3) \times U(2) \times U(1)$, in a self explanatory notation. Under $SU(3) \times SU(2) \times U(1)_3 \times U(1)_2 \times U(1)_1$, the members of a family of quarks and leptons have the following quantum numbers:

$$\begin{array}{ll} Q & (\mathbf{3},\mathbf{2};1,w,0)_{1/6} \\ u^c & (\bar{\mathbf{3}},\mathbf{1};-1,0,x)_{-2/3} \\ d^c & (\bar{\mathbf{3}},\mathbf{1};-1,0,y)_{1/3} \\ L & (\mathbf{1},\mathbf{2};0,1,z)_{-1/2} \\ l^c & (\mathbf{1},\mathbf{1};0,0,1)_{1} \end{array} \tag{21}$$

The values of the $U(1)$ charges x, y, z, w will be fixed below so that they lead to the right hypercharges, shown for completeness as subscripts.

It turns out that there are two possible ways of embedding the Standard Model particle spectrum on these stacks of branes [35], which are shown pictorially in Fig. 10. The quark doublet Q corresponds necessarily to a massless excitation of an open string

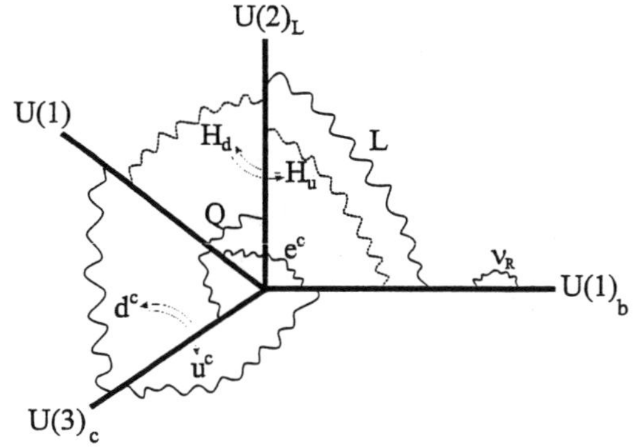

Figure 10: A minimal Standard Model embedding on D-branes.

with its two ends on the two different collections of branes (color and weak). As seen from the figure, a fourth brane stack is needed for a complete embedding, which is chosen to be a $U(1)_b$ extended in the bulk. This is welcome since one can accommodate right handed neutrinos as open string states on the bulk with sufficiently small Yukawa couplings suppressed by the large volume of the bulk [37]. The two models are obtained by an exchange of the up and down antiquarks, u^c and d^c, which correspond to open strings with one end on the color branes and the other either on the $U(1)$ brane, or on the $U(1)_b$ in the bulk. The lepton doublet L arises from an open string stretched between the weak branes and $U(1)_b$, while the antilepton l^c corresponds to a string with one end on the $U(1)$ brane and the other in the bulk. For completeness, we also show the two possible Higgs states H_u and H_d that are both necessary in order to give tree-level masses to all quarks and leptons of the heaviest generation.

The weak hypercharge Y is a linear combination of the three $U(1)$'s:

$$Y = Q_1 + \frac{1}{2}Q_2 + c_3 Q_3 \quad ; \quad c_3 = -1/3 \text{ or } 2/3\,, \tag{22}$$

where Q_N denotes the $U(1)$ generator of $U(N)$ normalized so that the fundamental representation of $SU(N)$ has unit charge. The corresponding $U(1)$ charges appearing in eq. (21) are $x = -1$ or 0, $y = 0$ or 1, $z = -1$, and $w = 1$ or -1, for $c_3 = -1/3$ or $2/3$, respectively. The hypercharge coupling g_Y is given by [3]:

$$\frac{1}{g_Y^2} = \frac{2}{g_1^2} + \frac{4c_2^2}{g_2^2} + \frac{6c_3^2}{g_3^2}. \tag{23}$$

It follows that the weak angle $\sin^2\theta_W$, is given by:

$$\sin^2\theta_W \equiv \frac{g_Y^2}{g_2^2 + g_Y^2} = \frac{1}{2 + 2g_2^2/g_1^2 + 6c_3^2 g_2^2/g_3^2}, \tag{24}$$

where g_N is the gauge coupling of $SU(N)$ and $g_1 = g_2$ or $g_1 = g_3$ at the string scale. In order to compare the theoretical predictions with the experimental value of $\sin^2\theta_W$ at M_s, we plot in Fig. 11 the corresponding curves as functions of M_s. The solid line is the

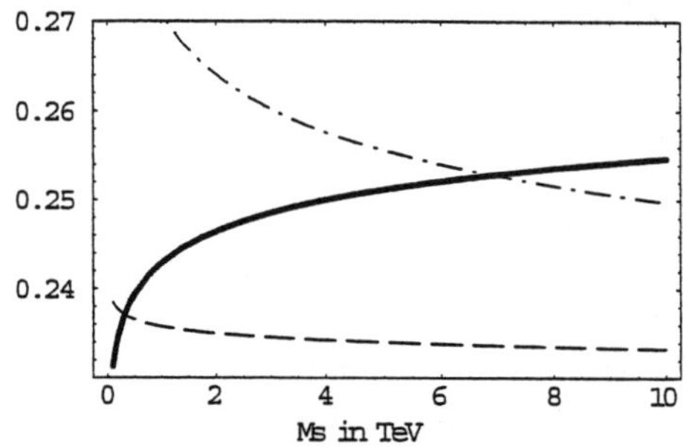

Figure 11: The experimental value of $\sin^2\theta_W$ (thick curve), and the theoretical predictions (24).

experimental curve. The dashed line is the plot of the function (24) for $g_1 = g_2$ with $c_3 = -1/3$ while the dotted-dashed line corresponds to $g_1 = g_3$ with $c_3 = 2/3$. The other two possibilities are not shown because they lead to a value of M_s which is too high to protect the hierarchy. Thus, the second case, where the $U(1)$ brane is on top of the color branes, is compatible with low energy data for $M_s \sim 6 - 8$ TeV and $g_s \simeq 0.9$.

From Eq. (24) and Fig. 11, we find the ratio of the $SU(2)$ and $SU(3)$ gauge couplings at the string scale to be $\alpha_2/\alpha_3 \sim 0.4$. This ratio can be arranged by an appropriate choice of the relevant moduli. For instance, one may choose the color and $U(1)$ branes to be D3 branes while the weak branes to be D7 branes. Then, the ratio of couplings

[3]The gauge couplings $g_{2,3}$ are determined at the tree-level by the string coupling and other moduli, like radii of longitudinal dimensions. In higher orders, they also receive string threshold corrections.

above can be explained by choosing the volume of the four compact dimensions of the seven branes to be $V_4 = 2.5$ in string units. This being larger than one is consistent with the picture above. Moreover it predicts an interesting spectrum of KK states for the Standard model, different from the naive choices that have appeared hitherto: the only Standard Model particles that have KK descendants are the W bosons as well as the hypercharge gauge boson. However, since the hypercharge is a linear combination of the three $U(1)$'s, the massive $U(1)$ KK gauge bosons do not couple to the hypercharge but to the weak doublet number.

7 Split supersymmetry: motivations

Recent precision tests of the SM, implying the absence of any deviation to a great accuracy, suggest that any new physics at a TeV needs to be fine-tuned at the per-cent level. Thus, either the underlying theory beyond the SM is very special, or our notion of naturalness should be reconsidered. The latter is also motivated from the recent evidence for the presence of a tiny non-vanishing cosmological constant that raises another more severe hierarchy problem. This raises the possibility that the same mechanism may solve both problems and casts some doubts on all previous proposals.

On the other hand, the necessity of a Dark Matter (DM) candidate and the fact that LEP data favor the unification of the three SM gauge couplings are smoking guns for the presence of new physics at high energies. Supersymmetry is then a nice candidate offering both properties. Moreover, it arises naturally in string theory, which provides a framework for incorporating the gravitational interaction in our quantum picture of the universe. It was then proposed to consider that supersymmetry might be broken at high energies without solving the gauge hierarchy problem. More precisely, making squarks and sleptons heavy does not spoil unification and the existence of a DM candidate while at the same time it gets rid of all unwanted features of the supersymmetric SM related to its complicated scalar sector. On the other hand, experimental hints to the existence of supersymmetry persist since there are still gauginos and higgsinos at the electroweak scale. This is the so-called split supersymmetry framework [6, 7].

Split supersymmetry has a natural realization in type I string theory with magnetized D9-branes, or equivalently with branes at angles [8]. We first show that the general spectrum has the required properties and then discuss the conditions for gauge coupling unification near the string scale. It turns out that equality of the two non-abelian couplings is a consequence of the correct SM spectrum for weak magnetic fields, while the value for the weak angle $\sin^2\theta_W = 3/8$ is easily obtained even in simple constructions.

Indeed, we perform a general study of SM embedding in three brane stacks and find a simple model realizing the conditions for unification [8]. We then discuss mass scales and in particular a mechanism generating light gaugino and higgsino masses in the TeV region, while scalars are superheavy, of order 10^{13} GeV [38]. Finally, we show how splitting supersymmetry reconciles toroidal models of intersecting branes with unification [39]. The gauge sector in these models arises in multiplets of extended supersymmetry while matter states are in $\mathcal{N} = 1$ representations. In general, split supersymmetry offers new possibilities for realistic string model building, that were previously unavailable because they were mainly restricted in the context of large dimensions and low string scale [2, 9, 4].

8 String realization

We start with type I string theory, or equivalently type IIB with orientifold 9-planes and D9-branes [1]. Upon compactification in four dimensions on a Calabi-Yau manifold, one gets $\mathcal{N} = 2$ supersymmetry in the bulk and $\mathcal{N} = 1$ on the branes. Moreover, various fluxes can be turned on, to stabilize part or all of the closed string moduli. We then turn on internal magnetic fields [40, 41], which, in the T-dual picture, amounts to intersecting branes [42, 43]. For generic angles, or equivalently for arbitrary magnetic fields, supersymmetry is spontaneously broken and described by effective D-terms in the four-dimensional (4d) theory [40]. In the weak field limit, $|H|\alpha' < 1$ with α' the string Regge slope, the resulting mass shifts are given by:

$$\delta M^2 = (2k+1)|qH| + 2qH\Sigma \quad ; \quad k = 0, 1, 2, \dots , \tag{25}$$

where H is the magnetic field of an abelian gauge symmetry, corresponding to a Cartan generator of the higher dimensional gauge group, on a non-contractible 2-cycle of the internal manifold. Σ is the corresponding projection of the spin operator, k is the Landau level and $q = q_L + q_R$ is the charge of the state, given by the sum of the left and right charges of the endpoints of the associated open string. We recall that the exact string mass formula has the same form as (25) with qH replaced by:

$$qH \longrightarrow \theta_L + \theta_R \quad ; \quad \theta_{L,R} = \arctan(q_{L,R}H\alpha') . \tag{26}$$

Obviously, the field theory expression (25) is reproduced in the weak field limit.

The Gauss law for the magnetic flux implies that the field H is quantized in terms of the area of the corresponding 2-cycle A:

$$H = \frac{m}{nA} , \tag{27}$$

where the integers m, n correspond to the respective magnetic and electric charges; m is the quantized flux and n is the wrapping number of the higher dimensional brane around the corresponding internal 2-cycle.

For simplicity, we consider below the case where the internal manifold is a product of three factorized tori $\prod_{I=1}^{3} T^2_{(I)}$. Then, the mass formula (25) becomes:

$$\delta M^2 = \sum_I (2k_I + 1)|qH_I| + 2qH_I\Sigma_I \,, \tag{28}$$

where Σ_I is the projection of the internal helicity along the I-th plane. For a ten-dimensional (10d) spinor, its eigenvalues are $\Sigma_I = \pm 1/2$, while for a 10d vector $\Sigma_I = \pm 1$ in one of the planes $I = I_0$ and zero in the other two ($I \neq I_0$). Thus, charged higher dimensional scalars become massive, fermions lead to chiral 4d zero modes if all $H_I \neq 0$, while the lightest scalars coming from 10d vectors have masses

$$M_0^2 = \begin{cases} |qH_1| + |qH_2| - |qH_3| \\ |qH_1| - |qH_2| + |qH_3| \\ -|qH_1| + |qH_2| + |qH_3| \end{cases} . \tag{29}$$

Note that all of them can be made positive definite, avoiding the Nielsen-Olesen instability, if all $H_I \neq 0$. Moreover, one can easily show that if a scalar mass vanishes, some supersymmetry remains unbroken [41, 42].

9 Generic spectrum

We turn on now several abelian magnetic fields H_I^a of different Cartan generators $U(1)_a$, so that the gauge group is a product of unitary factors $\prod_a U(N_a)$ with $U(N_a) = SU(N_a) \times U(1)_a$. In an appropriate T-dual representation, it amounts to consider several stacks of D6-branes intersecting in the three internal tori at angles. An open string with one end on the a-th stack has charge ± 1 under the $U(1)_a$, depending on its orientation, and is neutral with respect to all others. Using the results described above, the massless spectrum of the theory falls into three sectors [43, 41]:

1. Neutral open strings ending on the same stack, giving rise to $\mathcal{N} = 1$ gauge supermultiplets of gauge bosons and gauginos.

2. Doubled charged open strings from a single stack, with charges ± 2 under the corresponding $U(1)$, giving rise to massless fermions transforming in the antisymmetric

or symmetric representation of the associated $SU(N)$ factor. Their bosonic superpartners become massive. The multiplicities of chiral fermions are given by:

$$\begin{aligned}
\text{Antisymmetric} & \frac{1}{2}\left(\prod_I 2m_I^a\right)\left(\prod_J n_J^a + 1\right) \\
\text{Symmetric} & \frac{1}{2}\left(\prod_I 2m_I^a\right)\left(\prod_J n_J^a - 1\right)
\end{aligned} \tag{30}$$

where m_I^a, n_I^a are the integers entering in the expression of the magnetic field (27). For orbifolds or more general Calabi-Yau spaces, the above multiplicities may be further reduced by the corresponding supersymmetry projection down to $\mathcal{N} = 1$.

In the degenerate case where a magnetic field vanishes, say, along one of the tori ($m_I^a = 0$ for some I), there are no chiral fermions in $d = 4$ dimensions, but the same formula with the products extending over the other two magnetized tori gives the multiplicities of chiral fermions in $d = 6$. In this case, chirality in four dimensions may arise only when the last T^2 compactification is combined with some additional orbifold-type projection.

3. Open strings stretched between two different brane stacks, with charges ± 1 under each of the corresponding $U(1)$'s. They give rise to chiral fermions transforming in the bifundamental representation of the two associated unitary group factors. Their multiplicities, for toroidal compactifications, are given by:

$$\begin{aligned}
(N_a, N_b) &: \prod_I (m_I^a n_I^b + n_I^a m_I^b) \\
(N_a, \overline{N}_b) &: \prod_I (m_I^a n_I^b - n_I^a m_I^b)\,.
\end{aligned} \tag{31}$$

As in the previous case, when a factor in the products of the above multiplicities vanishes, there are no 4d chiral fermions, but the same formula with the product restricted over the other two magnetized tori gives the corresponding multiplicity of chiral fermions in $d = 6$.

As mentioned already above, all charged bosons are massive. Massless scalars can appear only when some supersymmetry remains unbroken. It is now clear that this framework leads to models with a tree-level spectrum realizing the idea of split supersymmetry. Embedding the Standard Model (SM) in an appropriate configuration of D-brane stacks, one obtains tree-level massless gauginos while all scalar superpartners of quarks and leptons typically get masses at the scale of the magnetic fields, whose magnitude is set by the compactification scale of the corresponding internal space. On the other hand, the condition to obtain a (tree-level) massless Higgs in the spectrum implies that supersymmetry remains unbroken in the Higgs sector, leading to a pair of massless higgsinos, as required by anomaly cancellation.

10 Gauge coupling unification

On general grounds, there are two conditions to obtain unification of SM gauge interactions, consistently with extrapolation of gauge couplings from low-energy data using the minimal supersymmetric SM spectrum. (i) Equality of the $SU(3)$ color and weak $SU(2)$ non-abelian gauge couplings and (ii) the correct prediction for the weak mixing angle $\sin^2\theta_W = 3/8$ at the grand unification (GUT) scale. On the other hand, a generic D-brane model using several stacks, as described in the framework of the previous section, does not satisfy either of the two conditions. Indeed, this framework was developed in connection to the idea of low-scale strings [4], where the concept of unification is radically different from conventional GUTs. In this section, we study precisely the general requirements for satisfying the first of the above two conditions, namely natural unification of non-abelian gauge couplings.

The 4d non-abelian gauge coupling α_{N_a} of the a-th brane stack is given by:

$$\frac{1}{\alpha_{N_a}} = \frac{V}{g_s}\prod_I |n_I^a|\sqrt{1+(H_I^a\alpha')^2}\,, \tag{32}$$

where g_s is the string coupling and V the compactification volume in string units. The presence of the wrapping numbers $|n_I^a|$ can be understood from the fact that $|n_I^a|V_I$ is the effective area of the 2-torus $T^2_{(I)}$ wrapped n_I^a times by the D9-brane, and $V=\prod_I V_I$. The additional factor in the square root follows from the non-linear Dirac-Born-Infeld (DBI) action of the abelian gauge field, $\sqrt{\det(\delta_{ij}+F_{ij}\alpha')}$, which in the case of two dimensions with $F_{ij}=\epsilon_{ij}H$, it is reduced to $\sqrt{1+(H\alpha')^2}$. Obviously, the expression (32) holds at the compactification scale, since above it gauge couplings receive important corrections and become higher dimensional. Finally, the gauge couplings of the associated abelian factors, in our convention of $U(1)$ charges, are given by

$$\alpha_{U(1)_a} = \frac{\alpha_{N_a}}{2N_a}\,. \tag{33}$$

Here, non-abelian generators are normalized according to $\mathrm{Tr}T^aT^b = \delta^{ab}/2$.

From equation (32), it follows that unification of non-abelian gauge couplings holds if (i) $\prod_I |n_I^a|$ are independent of a, and (ii) the magnetic fields are either a-independent as well, or they are much smaller than the string scale.

Condition (i) follows from eq. (30), by requiring the absence of chiral fermions transforming in the symmetric representations of the non-abelian groups, *i.e.* no chiral $SU(3)$ color sextets and no weak $SU(2)$ triplets.

Condition (ii) of weak magnetic fields is more quantitative. Allowing for 1% error in the unification condition at high scale, one should have $|H_I^a|\alpha' \lesssim 0.1$. From the quantization condition (27), this implies that the volume $V \gtrsim 10^3$ for three magnetized tori, which is rather high to keep the theory weakly coupled above the compactification scale. Indeed, eq. (32) gives a string coupling g_s of order $\mathcal{O}(10)$ for gauge couplings $\alpha_{N_a} \simeq 1/25$ at the unification scale. On the other hand, for one or two magnetized tori one obtains $V \gtrsim 10 - 10^2$, which is compatible with a string weak coupling regime ($g_s \sim 0.1 - 1$). Fortunately, this condition can be partly relaxed in some direction, by requiring the absence of chiral antiquark doublets in the spectrum. Indeed eq. (31), for open strings stretched between the strong $SU(3)$ and weak $SU(2)$ interactions brane stacks, implies the vanishing of one of the factors in the product. This leads to the equality of the ratio m_I^a/n_I^a for the two stacks and for some I, and thus, to the equality of the two corresponding magnetic fields via eq. (27).[4] As a result, the condition of perturbativity is weakened and becomes possible even in the case of three factorized magnetized tori.

The above analysis concerns the non-abelian couplings α_3 and α_2 of strong and weak interactions. The case of hypercharge is more subtle since it can be in general a linear combination of several $U(1)$'s coming from different brane stacks. In the following section, we present an explicit example with the correct prediction of the weak mixing angle. It is based on a minimal SM embedding in three brane stacks with the hypercharge being a linear combination of two abelian factors. This provides an existence proof that can be generalized in different constructions. We notice for instance that in a class of supersymmetric models with four brane stacks, the equality of the two non-abelian couplings $\alpha_2 = \alpha_3$ implies the value 3/8 for $\sin^2\theta_W$ at the unification scale [44].

[4]This argument is true only when the $U(1)$ accompanying the weak interactions brane stack participates in the hypercharge combination. Otherwise, quark anti-doublets are equivalent to quark doublets.

11 Minimal Standard Model embedding

In this section, we perform a general study of SM embedding in three brane stacks with gauge group $U(3) \times U(2) \times U(1)$ [35], and present an explicit example having realistic particle content and satisfying gauge coupling unification.

The quark and lepton doublets (Q and L) correspond to open strings stretched between the weak and the color or $U(1)$ branes, respectively. On the other hand, the u^c and d^c antiquarks can come from strings that are either stretched between the color and $U(1)$ branes, or that have both ends on the color branes and transform in the antisymmetric representation of $U(3)$ (which is an anti-triplet). There are therefore three possible models, depending on whether it is the u^c (model A), or the d^c (model B), or none of them (model C), the state coming from the antisymmetric representation of color branes. It follows that the antilepton l^c comes in a similar way from open strings with both ends either on the weak brane stack and transforming in the antisymmetric representation of $U(2)$ which is an $SU(2)$ singlet (in model A), or on the abelian brane and transforming in the "symmetric" representation of $U(1)$ (in models B and C). The three models are presented pictorially in Fig. 11.

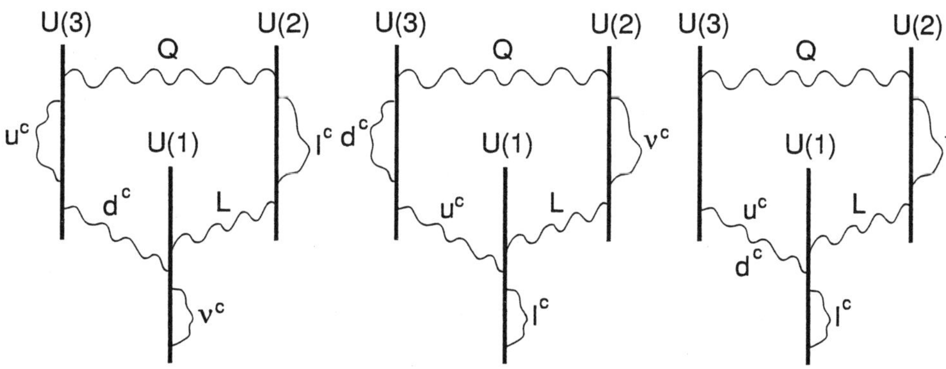

Figure 12: Pictorial representation of models A, B and C

Thus, the members of a family of quarks and leptons have the following quantum numbers:

$$\begin{array}{llll}
 & \text{Model A} & \text{Model B} & \text{Model C} \\
Q & (\mathbf{3},\mathbf{2};1,1,0)_{1/6} & (\mathbf{3},\mathbf{2};1,\varepsilon_Q,0)_{1/6} & (\mathbf{3},\mathbf{2};1,\varepsilon_Q,0)_{1/6} \\
u^c & (\bar{\mathbf{3}},\mathbf{1};2,0,0)_{-2/3} & (\bar{\mathbf{3}},\mathbf{1};-1,0,1)_{-2/3} & (\bar{\mathbf{3}},\mathbf{1};-1,0,1)_{-2/3} \\
d^c & (\bar{\mathbf{3}},\mathbf{1};-1,0,\varepsilon_d)_{1/3} & (\bar{\mathbf{3}},\mathbf{1};2,0,0)_{1/3} & (\bar{\mathbf{3}},\mathbf{1};-1,0,-1)_{1/3} \\
L & (\mathbf{1},\mathbf{2};0,-1,\varepsilon_L)_{-1/2} & (\mathbf{1},\mathbf{2};0,\varepsilon_L,1)_{-1/2} & (\mathbf{1},\mathbf{2};0,\varepsilon_L,1)_{-1/2} \\
l^c & (\mathbf{1},\mathbf{1};0,2,0)_{1} & (\mathbf{1},\mathbf{1};0,0,-2)_{1} & (\mathbf{1},\mathbf{1};0,0,-2)_{1} \\
\nu^c & (\mathbf{1},\mathbf{1};0,0,2\varepsilon_\nu)_{0} & (\mathbf{1},\mathbf{1};0,2\varepsilon_\nu,0)_{0} & (\mathbf{1},\mathbf{1};0,2\varepsilon_\nu,0)_{0}
\end{array} \tag{34}$$

where the last three digits after the semi-column in the brackets are the charges under the three abelian factors $U(1)_3 \times U(1)_2 \times U(1)$, that we will call Q_3, Q_2 and Q_1 in the following, while the subscripts denote the corresponding hypercharges. The various sign ambiguities $\varepsilon_i = \pm 1$ are due to the fact that the corresponding abelian factor does not participate in the hypercharge combination (see below). In the last lines, we also give the quantum numbers of a possible right-handed neutrino in each of the three models. These are in fact all possible ways of embedding the SM spectrum in three sets of branes.

The hypercharge combination is:

$$\begin{aligned}
\text{Model A} \quad &: \quad Y = -\frac{1}{3}Q_3 + \frac{1}{2}Q_2 \\
\text{Model B, C} \quad &: \quad Y = \ \frac{1}{6}Q_3 - \frac{1}{2}Q_1
\end{aligned} \tag{35}$$

leading to the following expressions for the weak angle:

$$\begin{aligned}
\text{Model A} \ : \ \sin^2\theta_W &= \frac{1}{2 + 2\alpha_2/3\alpha_3} = \frac{3}{8}\bigg|_{\alpha_2=\alpha_3} \\
\text{Model B, C} \ : \ \sin^2\theta_W &= \frac{1}{1 + \alpha_2/2\alpha_1 + \alpha_2/6\alpha_3} \\
&= \frac{6}{7 + 3\alpha_2/\alpha_1}\bigg|_{\alpha_2=\alpha_3}
\end{aligned} \tag{36}$$

In the second part of the above equalities, we used the unification relation $\alpha_2 = \alpha_3$, that can be naturally imposed as described in the previous section. It follows that model A admits natural gauge coupling unification of strong and weak interactions, and predicts the correct value for $\sin^2\theta_W = 3/8$ at the unification scale $M_{\rm GUT}$.

Besides the hypercharge combination, there are two additional $U(1)$'s. It is easy to check that one of the two can be identified with $B-L$. For instance, in model A choosing the signs $\varepsilon_d = \varepsilon_L = -\varepsilon_\nu = -\varepsilon_H = \varepsilon_{H'}$, it is given by:

$$B - L = -\frac{1}{6}Q_3 + \frac{1}{2}Q_2 - \frac{\varepsilon_d}{2}Q_1 \,. \tag{37}$$

Finally, the above spectrum can be easily implemented with a Higgs sector, since the Higgs field H has the same quantum numbers as the lepton doublet or its complex conjugate:

$$\begin{array}{lll} & \text{Model A} & \text{Model B, C} \\ H & (\mathbf{1},\mathbf{2};0,-1,\varepsilon_H)_{-1/2} & (\mathbf{1},\mathbf{2};0,\varepsilon_H,1)_{-1/2} \\ H' & (\mathbf{1},\mathbf{2};0,1,\varepsilon_{H'})_{1/2} & (\mathbf{1},\mathbf{2};0,\varepsilon_{H'},-1)_{1/2} \end{array} \tag{38}$$

12 Mass scales

12.1 String scale

To preserve gauge coupling unification, the compactification scale (actually the smallest, if there are several) must be of order of the unification scale $M_{\rm GUT} \simeq 10^{16}$ GeV. Above this energy, gauge interactions acquire a higher dimensional behavior. Moreover, to keep the theory weakly coupled, the string scale $M_s \equiv \alpha'^{-1/2}$ should be close to the compactification scale and therefore to $M_{\rm GUT}$. On the other hand, as we discussed above, to ensure that corrections to the unification of gauge couplings are within 1%, the magnetic fields should be weak, $|H_I^a|\alpha' \lesssim 0.1$. From the quantization condition (27), it follows that the string scale should be roughly a factor of 3 higher than the compactification scale,

$$M_s \simeq 3\, M_{\rm GUT} \,. \tag{39}$$

12.2 Scalar masses

The supersymmetry breaking scale m_0 is given by the heaviest charged scalar mass (29): $m_0^2 \sim \delta H^a \equiv \sum_{I=1}^3 \epsilon_I H_I^a$ on brane stacks, and $m_0^2 \sim \delta H^a - \delta H^b$ on brane intersections. Here, ϵ_I are signs: two positive and one negative. Thus, even for strong magnetic fields, of order of the string scale, m_0 can be much smaller and corresponds to an arbitrary parameter. Although values much lower than $M_{\rm GUT}$ require an apparent fine tuning of radii, such a tuning is technically natural since the supersymmetric point $m_0 = 0$ is radiatively stable.

All scalar masses are of the order of the supersymmetry breaking scale m_0, which is assumed to be very high in split supersymmetry, except for those coming from supersymmetric sectors, which are vanishing to lowest order, such as the higgses. The latter are expected to acquire masses from one loop corrections, proportional to m_0 but suppressed by a loop factor. Note that off diagonal elements of the 2×2 Higgs mass matrix, usually denoted by $B\mu$, should also be generated at the same order as the diagonal elements, in the absence of a Peccei-Quinn (PQ) symmetry. For high m_0, a fine tuning between $B\mu$ and the diagonal elements is then required to ensure a light Higgs.

12.3 Gaugino masses

It remains to discuss the corrections to gaugino and higgsino masses, $m_{1/2}$ and μ, which are vanishing at the tree-level. In the absence of gravity, they are both protected by an R-symmetry. Actually, higgsino masses are protected in addition by a PQ symmetry which must be broken in order to generate a $B\mu$ mixing term in the Higgs mass matrix, as we argued above. Then, a μ-term can be generated via $B\mu$, or directly using the PQ symmetry breaking, if R-symmetry is broken. Indeed, R-symmetry is in general broken in the gravitational sector by the gravitino mass $m_{3/2}$ and thus, in the presence of gravity, $m_{1/2}$ and μ are not anymore protected. Since supersymmetry breaking in the gravity sector is model dependent and brings more uncertainties, here we will assume that gravitational corrections are negligible. For instance, if supergravity breaking occurs via a Scherk-Schwarz compactification on an interval transverse to our braneworld [45], using the usual $\mathbb{Z}_2$ fermion number in the bulk, the gravitino acquires Dirac mass together with its Kaluza-Klein modes and R-symmetry remains unbroken [8]. One can therefore discuss other sources of R-symmetry breaking within only global supersymmetry.

As discussed previously, supersymmmetry breaking via internal magnetic fields is described in the 4d effective field theory by vacuum expectation values (VEVs) of D-term auxiliaries for all magnetic $U(1)$'s. In the low energy limit, one has:

$$\langle D \rangle \simeq m_0^2 \,, \tag{40}$$

and thus R-symmetry remains unbroken. However, it is broken by α'-string corrections, that modify for instance the gauge kinetic terms to the DBI form. In particular, gaugino masses can be induced by a dimension-seven effective operator which is the chiral F-term [38]:

$$F_{(0,3)} \int d^2\theta \mathcal{W}^2 \mathrm{Tr} W^2 \quad \Rightarrow \quad m_{1/2} \sim \frac{m_0^4}{M_s^3} \,, \tag{41}$$

where $\mathcal{W}$ and W denote the magnetic $U(1)$ and non-abelian gauge superfield, respectively. The coefficient $F_{(0,3)}$ is a moduli dependent function given by the topological partition function on a world-sheet with no handles and three boundaries. It is non-vanishing when the three brane stacks associated to the boundaries do not intersect at a point in any of the three internal torii. From the effective field theory point of view, it corresponds to a two-loop correction involving massive open string states. Upon a VEV $\langle\mathcal{W}\rangle = \theta\langle D\rangle$, the above F-term generates gaugino masses given in eq. (41). They are in the TeV region for scalar masses at intermediate energies, $m_0 \sim \mathcal{O}(10^{13})$ GeV.

13 Split extended supersymmetry

Implementing split supersymmetry in string theory faces a generic problem: in simple brane constructions the gauge sector comes in multiplets of extended supersymmetry [46, 47]. $\mathcal{N} = 4$ in the toroidal case, or $\mathcal{N} = 2$ is simple orbifolds. Gauginos can therefore get Dirac masses without breaking R-symmetry. Indeed, a Dirac mass [48] is induced through the dimension-five operator

$$\frac{a}{M_s}\int d^2\theta\mathcal{W}\,W^aA_a \;\Rightarrow m_D \sim a\frac{m_0^2}{M_s}\,, \tag{42}$$

where a accounts for a possible loop factor. Actually, this operator arises quite generally at one-loop level in intersecting D-brane models with a moduli-dependent coupling, determined only from the massless (topological) sector of the theory [47]. Note that this mass m_D is much higher that the Majorana induced mass of eq. (41).

It turns out that this scenario is compatible with one-loop gauge coupling unification [39]. In the energy regime between $M_{\rm GUT}$ and the electroweak scale M_W, the renormalization group equations meet three thresholds. From $M_{\rm GUT}$ to the common scalar mass m_0 all charged states contribute. Below m_0 squarks and sleptons (which do not affect unification), adjoint scalars and $2 - n_H$ higgses decouple, while below m_D the $\mathcal{N} = 2$ or $\mathcal{N} = 4$ gluinos and winos drop out. Finally, at TeV energies higgsinos decouple and we are left with the Standard Model with n_H Higgs doublets. Using $M_s \sim M_{\rm GUT}$ and varying a between $a = 1$ and $a = 1/100$, one finds realistic values for $M_{\rm GUT}$ and m_0 in both $\mathcal{N} = 4$ and $\mathcal{N} = 2$ cases. The results are summarized in Table 2.

In all cases the unification scale is high enough to avoid problems with proton decay. For the two possible cases with one light Higgs ($\mathcal{N} = 2$ or $\mathcal{N} = 4$), $M_{\rm GUT}$ is very close to the Planck scale so that there should be no need to explain the usual mismatch between these two scales.

	n_H	$M_{\rm GUT}$	m_0	m_D	$m_{1/2}$
$\mathcal{N}=2$	1	10^{18}	10^{13}	10^6	10^{-5}
	2	10^{16}	10^{13}	10^9	10^2
$\mathcal{N}=4$	1	10^{19}	10^{16}	10^{12}	10^6
	2	—	—	—	—

Table 2: *Values for the unification scale $M_{\rm GUT}$, scalar masses m_0, Dirac gaugino masses m_D, and Majorana gaugino masses $m_{1/2}$ in GeV for $\mathcal{N}=2,4$ supersymmetric gauge sector and $n_H=1,2$ light higgses.*

The low energy sector of these models contains, besides the SM, just some fermion doublets (higgsinos) and eventually two singlets (the binos from the discussion below). It therefore illustrates the fact that only these states are needed for a minimal extension of the SM consistent with unification and Dark Matter (DM) candidates, and not the full fermion spectrum of split supersymmetry [49].

Another constraint on the models is that they must provide a DM candidate. As usually in supersymmetric theories this should be the lightest neutralino. Pure higgsinos cannot be DM candidates because their mass is of Dirac type. Since DM direct detection experiments have ruled out Dirac fermions up to masses of order 50 TeV, some mixing coming from the binos is required in order to break the degeneracies of the two lightest neutralinos; the required mass difference is bigger than about 150 keV [50, 7, 39]. This translates into an upper bound on the Dirac gaugino mass of about 10^5 GeV, for the required higgsino mass splitting to be generated through the electroweak symmetry breaking mixing, which is of order m_W^2/m_D. This value compared to the values in Table 2 leads to the $\mathcal{N}=2$, $n_H=1$ case as the only possibility to accommodate it (the direct Majorana component of bino $m_{1/2}$ is negligible in this case). In fact, one needs an order of magnitude suppression of the induced Dirac mass for binos relative to the other gauginos, which is not unreasonable to assume in brane constructions.

In the other two models, the required suppression factor is much higher and the above mechanism would be very unnatural. However, since binos play no role for unification as they carry no SM charge, we could imagine a scenario where m_D vanishes identically for binos, but not for the other gauginos. For instance consider the case where Dirac masses from the operator (42) are generated by loop diagrams involving $\mathcal{N}=2$ hypermultiplets with supersymmetric masses of order $M_{\rm GUT}$ and a supersymmetry breaking splitting of order m_0. It is then possible to choose these massive states such that they carry no hypercharge, in which case binos can only have Majorana masses displayed in Table 2.

Their value in the $\mathcal{N}=4$ case corresponds to the upper bound for DM with Majorana bino mass [39]. Thus, the constraint of a viable DM candidate leaves us with two possibilities: (a) $\mathcal{N}=2$ with $n_H=1$ and Dirac masses for all gauginos and (b) $\mathcal{N}=4$ with $n_H=1$, or $\mathcal{N}=2$ with $n_H=2$ and Majorana mass binos.

Finally, the higgsinos must acquire a mass of order the electroweak scale. This can be induced by the following dimension-seven operator, generated at one loop level [38, 47]:

$$\frac{c}{M_s^3}\int d^2\theta \mathcal{W}^2\overline{D}^2\bar{H}_1\bar{H}_2 \Rightarrow \mu \sim c\frac{m_0^4}{M_s^3}\,, \tag{43}$$

where c is again a loop factor. The resulting numerical value is of the same order as $m_{1/2}$. Thus, such an operator can only give a sensible value of μ for the $\mathcal{N}=2$ $n_H=2$ model. In the other two cases, $\mathcal{N}=4$ or $\mathcal{N}=2$ with $n_H=1$, μ remains an independent parameter.

To summarize, at low energies we end up with two distinct scenarios after all massive particles are decoupled: (i) $n_H=1$ with light higgsinos (models with $\mathcal{N}=2$ and $\mathcal{N}=4$ gauge sector and $n_H=1$), and (ii) $n_H=2$ with light higgsinos and binos (model with $\mathcal{N}=2$ gauge sector and $n_H=2$). In the $n_H=1$ scenario the DM candidate is mainly higgsino, although the much heavier bino is light enough to forbid any vector couplings. The relic density reproduces the actual WMAP results for $\mu\sim 1.1$ TeV.

Acknowledgments

This work was supported in part by the European Commission under the RTN contract MRTN-CT-2004-503369, and in part by the INTAS contract 03-51-6346.

References

[1] C. Angelantonj and A. Sagnotti, *Phys. Rept.* **371**, 1 (2002) [Erratum-ibid. **376**, 339 (2003)] [arXiv:hep-th/0204089].

[2] I. Antoniadis, *Phys. Lett.* B **246** (1990) 377.

[3] C. D. Hoyle, D. J. Kapner, B. R. Heckel, E. G. Adelberger, J. H. Gundlach, U. Schmidt and H. E. Swanson, *Phys. Rev.* D **70**, 042004 (2004).

[4] N. Arkani-Hamed, S. Dimopoulos and G. R. Dvali, *Phys. Lett.* B **429**, 263 (1998) [arXiv:hep-ph/9803315]; I. Antoniadis, N. Arkani-Hamed, S. Dimopoulos and G. R. Dvali, *Phys. Lett.* B **436**, 257 (1998) [arXiv:hep-ph/9804398].

[5] For a review see e.g. I. Antoniadis, *Prepared for NATO Advanced Study Institute and EC Summer School on Progress in String, Field and Particle Theory, Cargese, Corsica, France (2002)*; and references therein.

[6] N. Arkani-Hamed and S. Dimopoulos, arXiv:hep-th/0405159;

[7] G. F. Giudice and A. Romanino, *Nucl. Phys.* B **699**, 65 (2004) [Erratum-ibid. B **706**, 65 (2005)] [arXiv:hep-ph/0406088].

[8] I. Antoniadis and S. Dimopoulos, *Nucl. Phys.* B **715**, 120 (2005).

[9] J. D. Lykken, *Phys. Rev.* D **54**, 3693 (1996) [arXiv:hep-th/9603133].

[10] J. C. Long and J. C. Price, *Comptes Rendus Physique* **4**, 337 (2003); R. S. Decca, D. Lopez, H. B. Chan, E. Fischbach, D. E. Krause and C. R. Jamell, *Phys. Rev. Lett.* **94**, 240401 (2005); S. J. Smullin, A. A. Geraci, D. M. Weld, J. Chiaverini, S. Holmes and A. Kapitulnik, arXiv:hep-ph/0508204; H. Abele, S. Haeßler and A. Westphal, in 271th WE-Heraeus-Seminar, Bad Honnef (2002).

[11] I. Antoniadis and K. Benakli, *Phys. Lett.* B **326** (1994) 69.

[12] I. Antoniadis, K. Benakli and M. Quirós, *Phys. Lett.* B **331** (1994) 313 and *Phys. Lett.* B **460** (1999) 176; P. Nath, Y. Yamada and M. Yamaguchi, *Phys. Lett.* B **466** (1999) 100 T. G. Rizzo and J. D. Wells, *Phys. Rev.* D **61** (2000) 016007; T. G. Rizzo, *Phys. Rev.* D **61** (2000) 055005; A. De Rujula, A. Donini, M. B. Gavela and S. Rigolin, *Phys. Lett.* B **482** (2000) 195;

[13] E. Accomando, I. Antoniadis and K. Benakli, *Nucl. Phys.* B **579** (2000) 3.

[14] I. Antoniadis, K. Benakli and A. Laugier, *JHEP* **0105** (2001) 044.

[15] P. Nath and M. Yamaguchi, *Phys. Rev.* D **60** (1999) 116004; *Phys. Rev.* D **60** (1999) 116006; M. Masip and A. Pomarol, *Phys. Rev.* D **60** (1999) 096005; W. J. Marciano, *Phys. Rev.* D **60** (1999) 093006; A. Strumia, *Phys. Lett.* B **466** (1999) 107; R. Casalbuoni, S. De Curtis, D. Dominici and R. Gatto, *Phys. Lett.* B **462** (1999) 48; C. D. Carone, *Phys. Rev.* D **61** (2000) 015008; A. Delgado, A. Pomarol and M. Quirós, *JHEP* **1**, 30 (2000).

[16] G. Servant and T. M. P. Tait, *Nucl. Phys.* B **650**, 391 (2003).

[17] G. F. Giudice, R. Rattazzi and J. D. Wells, *Nucl. Phys.* B **544** (1999) 3; E. A. Mirabelli, M. Perelstein and M. E. Peskin, *Phys. Rev. Lett.* **82** (1999) 2236; T. Han, J. D. Lykken and R. Zhang, *Phys. Rev.* D **59** (1999) 105006; K. Cheung and W.-Y. Keung, *Phys. Rev.* D **60** (1999) 112003; C. Balázs *et al.*, *Phys. Rev. Lett.* **83** (1999) 2112; L3 Collaboration (M. Acciarri *et al.*), *Phys. Lett.* B **464** (1999) 135 and **470**, 281 (1999): J. L. Hewett, *Phys. Rev. Lett.* **82** (1999) 4765.

[18] N. Arkani-Hamed, S. Dimopoulos and G. Dvali, *Phys. Rev.* D **59** (1999) 086004.

[19] S. Cullen and M. Perelstein, *Phys. Rev. Lett.* **83** (1999) 268; V. Barger, T. Han, C. Kao and R. J. Zhang, *Phys. Lett.* B **461** (1999) 34.

[20] K. Benakli and S. Davidson, *Phys. Rev.* D **60** (1999) 025004; L. J. Hall and D. Smith, *Phys. Rev.* D **60** (1999) 085008.

[21] S. Cullen, M. Perelstein and M. E. Peskin, *Phys. Rev.* D **62** (2000) 055012; D. Bourilkov, *Phys. Rev.* D **62** (2000) 076005; L3 Collaboration (M. Acciarri *et al.*), *Phys. Lett.* B **489** (2000) 81.

[22] S. B. Giddings and S. Thomas, *Phys. Rev.* D **65** (2002) 056010; S. Dimopoulos and G. Landsberg, *Phys. Rev. Lett.* **87**, 161602 (2001).

[23] I. Antoniadis, C. Bachas, *Phys. Lett.* B **450** (1999) 83.

[24] I. Antoniadis, K. Benakli, A. Laugier and T. Maillard, *Nucl. Phys.* B **662** (2003) 40 [arXiv:hep-ph/0211409].

[25] I. Antoniadis, S. Dimopoulos and G. Dvali, *Nucl. Phys.* B **516** (1998) 70; S. Ferrara, C. Kounnas and F. Zwirner, *Nucl. Phys.* B **429** (1994) 589.

[26] T. R. Taylor and G. Veneziano, *Phys. Lett.* B **213** (1988) 450.

[27] I. Antoniadis, E. Kiritsis and J. Rizos, *Nucl. Phys.* B **637** (2002) 92.

[28] D. V. Volkov and V. P. Akulov, *JETP Lett.* **16**, 438 (1972) and *Phys. Lett.* B **46** (1973) 109.

[29] A. Brignole, F. Feruglio and F. Zwirner, *JHEP* **9711** (1997) 001; T. E. Clark, T. Lee, S. T. Love and G. Wu, *Phys. Rev.* D **57** (1998) 5912; M. A. Luty and E. Ponton, *Phys. Rev.* D **57** (1998) 4167; I. Antoniadis, K. Benakli and A. Laugier, *Nucl. Phys.* B **631** (2002) 3.

[30] I. Antoniadis and M. Tuckmantel, *Nucl. Phys.* B **697** (2004) 3.

[31] I. Antoniadis, M. Tuckmantel and F. Zwirner, *Nucl. Phys.* B **707** (2005) 215 [arXiv:hep-ph/0410165].

[32] I. Antoniadis, K. Benakli and M. Quirós, *Nucl. Phys.* B **583** (2000) 35.

[33] I. Antoniadis, K. Benakli and M. Quiros, *New Jour. Phys.* **3** (2001) 20.

[34] I. Antoniadis, C. Muñoz and M. Quirós, *Nucl. Phys.* B **397** (1993) 515; I. Antoniadis and M. Quirós, *Phys. Lett.* B **392** (1997) 61; A. Pomarol and M. Quirós, *Phys. Lett.* B **438** (1998) 225; I. Antoniadis, S. Dimopoulos, A. Pomarol and M. Quirós, *Nucl. Phys.* B **544** (1999) 503; A. Delgado, A. Pomarol and M. Quirós, *Phys. Rev.* D **60** (1999) 095008; R. Barbieri, L. J. Hall and Y. Nomura, *Phys. Rev.* D **63** (2001) 105007.

[35] I. Antoniadis, E. Kiritsis and T. N. Tomaras, Phys. Lett. B **486** (2000) 186 [arXiv:hep-ph/0004214]; I. Antoniadis, E. Kiritsis, J. Rizos and T. N. Tomaras, Nucl. Phys. B **660** (2003) 81 [arXiv:hep-th/0210263]; R. Blumenhagen, B. Kors, D. Lust and T. Ott, Nucl. Phys. B **616**, 3 (2001) [arXiv:hep-th/0107138]; I. Antoniadis and J. Rizos, 2003 unpublished work.

[36] G. Shiu and S.-H. H. Tye, *Phys. Rev.* D **58** (1998) 106007; Z. Kakushadze and S.-H. H. Tye, *Nucl. Phys.* B **548** (1999) 180; L. E. Ibáñez, C. Muñoz and S. Rigolin, *Nucl. Phys.* B **553** (1999) 43.

[37] K. R. Dienes, E. Dudas and T. Gherghetta, *Nucl. Phys.* B **557** (1999) 25 [arXiv:hep-ph/9811428]; N. Arkani-Hamed, S. Dimopoulos, G. R. Dvali and J. March-Russell, *Phys. Rev.* D **65** (2002) 024032 [arXiv:hep-ph/9811448]; G. R. Dvali and A. Y. Smirnov, *Nucl. Phys.* B **563** (1999) 63.

[38] I. Antoniadis, K. S. Narain and T. R. Taylor, arXiv:hep-th/0507244.

[39] I. Antoniadis, A. Delgado, K. Benakli, M. Quiros and M. Tuckmantel, arXiv:hep-ph/0507192.

[40] C. Bachas, arXiv:hep-th/9503030.

[41] C. Angelantonj, I. Antoniadis, E. Dudas and A. Sagnotti, Phys. Lett. B **489** (2000) 223 [arXiv:hep-th/0007090].

[42] M. Berkooz, M. R. Douglas and R. G. Leigh, Nucl. Phys. B **480** (1996) 265 [arXiv:hep-th/9606139];

[43] R. Blumenhagen, L. Goerlich, B. Kors and D. Lust, JHEP **0010** (2000) 006 [arXiv:hep-th/0007024]; G. Aldazabal, S. Franco, L. E. Ibanez, R. Rabadan and A. M. Uranga, J. Math. Phys. **42** (2001) 3103 [arXiv:hep-th/0011073].

[44] R. Blumenhagen, D. Lust and S. Stieberger, JHEP **0307** (2003) 036 [arXiv:hep-th/0305146].

[45] I. Antoniadis, E. Dudas and A. Sagnotti, Nucl. Phys. B **544** (1999) 469 [arXiv:hep-th/9807011].

[46] For a recent review, see R. Blumenhagen, M. Cvetic, P. Langacker and G. Shiu, arXiv:hep-th/0502005; and references therein.

[47] I. Antoniadis, K. Benakli, A. Delgado, M. Quirós and M. Tuckmantel, arXiv:hep-th/0601003.

[48] P. J. Fox, A. E. Nelson and N. Weiner, JHEP **0208** (2002) 035 [arXiv:hep-ph/0206096].

[49] N. Arkani-Hamed, S. Dimopoulos and S. Kachru, arXiv:hep-th/0501082.

[50] D. R. Smith and N. Weiner, Phys. Rev. D **64** (2001) 043502 [arXiv:hep-ph/0101138].

CHAIRMAN: I. ANTONIADIS

Scientific Secretaries: G. Imponente, M. Jona Lasinio

DISCUSSION I

- *Rossi*:

Are we sure that at LHC energy we can discriminate among different models?

- *Antoniadis*:

It depends on what the signal is and to what models it is related. There are some models among which you can discriminate easily. For instance, consider the case I discussed today of TeV size dimensions. I mentioned two types of models depending on the localization properties of matter fields: one in which you have resonances and another in which you can produce them only in pairs. Thus in this case it is easy to discriminate between these two types of models. However, if you see a resonance you have to decide whether this is due to an extra dimension or due to another source, such as a new gauge boson (Z'). Then, one has to do more work. One possibility to decide if there is an extra dimension is to also look for other resonances in different channels, with the same mass. For instance you can look at charged channels, such as electron neutrino, to see a resonance due to KK excitation of W.

- *Rossi*:

The extra dimensions that could be seen at LHC have a minimum size of order 10^{-19} m, and in your case the extra dimensions are of the order of 10^{-16} cm, so we should be able to see them in LHC.

- *Antoniadis*:

If they are there, yes, but the first thing that all of us hope is that there will be some signal of new physics.Then, one has to do more work and analysis to find out what is behind.

- *Cifarelli*:

Why the number of extra dimensions cannot be n=1?

- *Antoniadis*:

Because in that case you find that the size of this extra dimension must be of the order of the Earth-Sun distance and this is excluded. The less extra dimensions there are the bigger size you need. If there are more extra dimensions, you have more space to diffuse the gravitational field and therefore they can be smaller and smaller. In the case of six extra dimensions the required size is of order 10^{-13} cm. Decreasing their number they have to become bigger; in the case of two extra dimensions you arrive to the millimeter which is the experimental bound.

- *Marrani*:

Could you speak on how the different fundamental interactions are treated with the different stacks of branes. Particularly could you mention if this influences the structures of the gauge group, their compactness and the representation in which the various particles sit.

- *Antoniadis*:

The various lines presented in the transparency correspond to branes describing different interactions. So you have one brane which corresponds to SU(3) color, another brane with SU(2) weak interaction and then you may also have U(1) brane. This space is the space of extra dimensions, so every brane can in principle point in different directions in the extra dimensions. You can then accommodate the quarks and leptons as you see. For example, the quark doublets will be an open string which stretches between the strong interaction brane and the weak interaction one, because it transforms as triplet doublet. In fact, in order to obtain the U(3) group, you need three coincident branes on top of each other, and then when you have a string which has an end on U(3) this end transforms as a triplet under U(3). Similarly, a string that ends on U(2) transforms as a doublet under U(2). So in order to have a triplet doublet you need a string to be stretched between the color branes and the weak branes. Similarly one can accommodate all the others quarks and leptons as you see.

- *Marrani*:

In the case of more complicated settings of the branes, it seems to appear a sort of redundancy. Can this be ruled out by experiments or by some theoretical arguments?

-*Antoniadis*:

The simplest thing you can try, is to embed the SM as it is. The reason is that since the string scale is at low energy, we no longer need to look for a GUT group. Having a GUT group will introduce more problems, because it contains leptoquarks and other exotic particles, that if were in the TeV region would create problems. So the simplest way is to start with the SM, which has three factors so we start with three stacks of branes. You may also have partial unification which, in a clever way, would be consistent with observations.

- *Pinfold*:

The astrophysical and cosmological bounds on extra dimensions would seem to constrain the allowed "phase space" for extra dimensions at LHC. However I am sure these indirect bounds have large systematic errors. What in your opinion is the most stringent (and "believable") astrophysical/cosmological limit on extra dimensions?

- *Antoniadis*:

The limits from supernova cooling are stringent and "believable" but concern only the case of two extra dimensions. If you consider more than two large dimensions, these limits become weaker, so the limits from astrophysics and cosmology are stronger than the accelerator bounds.

- *Pinfold*:

Would the KK graviton mode decay, or the diffuse photon background give important limits for n=3?

- *Antoniadis*:

Marginally.

- *Konikowska*:

Kaluza-Klein dimensional reduction requires some particular bulk metrics, contrary to the approach in which the bulk metric is projected on the brane-world volume which is possible to realize for any form of the bulk metric. Is this a constraint on your theory?

- *Antoniadis*:

Yes, I was restricted to flat metrics. I considered both the bulk and the branes to be flat and the extra dimensions to be toroidally compactified, or at least on a locally flat metric. In the case of curved metric you can have more general phenomena like for instance the Randall-Sundrum model. On general grounds there are two ways to hide the extra dimensions, to make them unobservable: the first is by compactification and the second by localizing our world. So compactification implies that dimensions having finite size, localization is when you have branes and generally you can combine the two.

- *Nilse*:

In which type of string theory can large extra dimensions arise ? What was the mistake in the pre-1990 arguments that linked the compactification scale with the Planck scale ?

- *Antoniadis*:

In the heterotic string you cannot realize large extra dimensions consistently with perturbation theory, keeping strings weakly coupled. The reason is that in heterotic strings, gauge interactions span the whole compact space so that there are no transverse type of extra dimensions and therefore if you want the strings to be weakly coupled, all these dimensions must be of the order of the string size. That's why this possibility was missed in the case of heterotic strings. The other string theory is the type I which I described. Sometimes you hear also the name of the type II which is the closed string part of type I.

- *Csanad*:

Gravitation can be 10^{30} stronger than thought previously on scales smaller than R_perp . Is a heavy ion collision dense enough? Or maybe a neutron star?

- *Antoniadis*:

The strength you quoted refers to the TeV scale, in other words to distances of order of 10^{-16} cm which corresponds to the string length in these theories. Traditionally, gravitation interactions become strong at the Planck scale, 10^{19} GeV. However, if you have extra dimensions of sub-mm size, so that the string scale is at the TeV, then gravity becomes strong at the TeV scale, roughly 10^{30} stronger. This possibility is still consistent with present observations, because it becomes so strong at distances which are 10^{-16} cm, and since the evolution of the gravitational interaction falls off with a power law, not logarithmically as gauge interactions, it becomes weak very rapidly at larger distances.

- *Munir*:

Where do we get the 3 TeV lower limit on brane compactification?

- *Antoniadis*:

The 3 TeV limit comes from the analysis of all indirect effects of KK modes of SM interactions, such as KK excitations of the W and the photon, in electroweak observables, for instance the Z-width and the Fermi constant which are measured with very high accuracy. Using such precision tests of the SM you put bounds on new physics and in particular on extra dimensions. An analysis of the precision tests of the SM done by several authors, leads to the bound of 3 TeV. In the case "universal extra dimensions", however, the bound becomes much weaker, about an order of magnitude less.

- *Imponente*:

How to establish the number of branes ?

- *Antoniadis*:

A single brane gives you an abelian factor, so if you want to have a unitary group U(N) you need N coincident branes. Therefore for strong interactions SU(3) you need at least 3 branes and for the SM you need at least 5 or 6.

- *Gualtieri*:

Is there any reason to assume that the transverse extra dimensions are compactified on a torus? Could other manifold change the experimental signatures and supernova bounds?

- *Antoniadis*:

No, there is no other reason than simplicity. I considered the case of flat space. If you consider curved space then you may have new phenomena, for instance the type of the Randall-Sundrum localization which gives different results. By curving the space, one can deform the KK excitations. By deformation I mean simultaneously the spacing of the excitations (they are not anymore equally spaced), and their couplings. Therefore the analysis and the bounds in general change.

CHAIRMAN: I. ANTONIADIS

Scientific Secretaries: D. Konikowska, V. Kashkan

DISCUSSION II

- *Cifarelli:*

Can I ask the first question? Actually you skipped something this morning, which seems very interesting, about the electroweak symmetry breaking.

- *Antoniadis*:

I was planning to start tomorrow with this. It will take about 10 minutes.

- *Marrani:*

I would like you to talk a little bit more about the non-linear realization of supersymmetry encoded in the goldstino lagrangian. I mean the Volkov-Akulov non-linear realization of space-time plus internal symmetries of superymmetry. In that kind of approach you have to choose a superalgebra and a co-set of it and to build the corresponding lagrangian. I was wondering what kind of expression you get and what kind of superalgebra you are treating. Can you speak a bit more about this?

- *Antoniadis:*

Well, what I considered is the simple N=1 non-linear supersymmetry. So there are no internal symmetries and one has just one goldstino.

- *Marrani:*

The goldstino corresponds to the non-linearly realized supersymmetry and is living on the brane?

- *Antoniadis:*

Right.

- *Marrani:*

And it is coupled to the gravitino that instead transforms linearly and lives in the bulk which is supersymmetric in that case. Is that right?

- *Anioniadis:*

Yes, the bulk has local supersymmetry.

- *Marrani:*

But when you add that four-fermion operator, the lagrangian becomes the most general one that you can find for that kind of the co-set of N=1 supersymmetry?

- *Antoniadis*:

This analysis can be done at the level of global non-linearly realized supersymmetry. You have the (non-linear) transformation of the goldstino and you ask which is the most general invariant action. Actually there are two ways to proceed, one is by a superfield formulation and another by writing all possible invariant terms. The standard coupling found initially by Volkov-Akulov is through the matter energy-momentum tensor, but later people realized that there are more invariant terms. Finally, there was a classification of all possible coupling up to dimension 8. By the way, one more thing: in the case we considered, of two intersecting branes, the goldstino is a linear combination of two fermions from the two branes, as seen in this formula for the goldstino decay constant.

- *Zichichi:*

Could you explain what is the fundamental difference between parallel and transverse extra dimensions? If we take the 3+1 dimensions, then we add the other 6 internal geometric dimensions, so to say. Plus one to generate the gauge coupling, then the total is 11. Are these dimensions parallel in your language?

- *Antoniadis*:

No, they can be either parallel or transverse. The definition of a parallel dimension is any spatial dimension in addition to the 3-space that we see, in which some Standard Model particles propagate.

- *Zichichi:*

Propagate means what? How do they propagate? I don't understand that, what does propagation mean? They propagate in any dimension.

- *Antoniadis:*

No, in the transverse dimensions the Standard Model fields are localized. In other words they cannot propagate.

- *Zichichi:*

Why they cannot propagate? Could you give me some hint on what you mean?

- *Antoniadis:*

Localization means that all their interactions are proportional to a delta function at a particular point in the internal space.

- *Zichichi:*

Excuse me, how can you imagine an object which cannot propagate in any dimension? This is not a dimension, it's something else.

- *Antoniadis:*

As an example, propagation means that the fields depend on x (our space) and y (extra dimension). Localization means that the fields on the brane do not depend on y, they are defined only at the position of the brane, say y=y_0.

- *Zichichi:*

How do you distinguish the 2 transverse dimensions in the example you gave?

- *Antoniadis:*

You pick up a model with branes and this defines which dimensions are parallel and which are transverse.

- *Zichichi:*

And how should we think about this? If I make a Gedanken experiment of having another sheet, the two do not communicate, even gravitationally, or not? Suppose I have two worlds, one here, and one there, do they interact transversally?

- *Antoniadis:*

The graviton can always propagate because it interacts with energy, which is everywhere. On the contrary, the other 3 Standard Model interactions can be localized, as well as the corresponding charges.

- *Zichichi:*

But the expanded dimensions of the Standard Model are 3+1.

- *Antoniadis:*

Yes, but there may be additional compact ones, along which some Standard Model interactions could propagate. In order to be unobservable, their size should be less than 10^(-16) cm. However, since these dimensions are longitudinal, it means that if we dispose enough energy, we can probe them. We can have a look inside.

- *Zichichi:*

While in the transverse case we cannot do anything. No matter what?

- *Antoniadis:*

Yes, another way to say it is that for a parallel dimension there are Kaluza-Klein excitations for some Standard Model particles. I mean for those which propagate. Those that do not propagate, do not have Kaluza-Klein excitations. This is because if you have a field, that depends on an extra dimension y, and y is some compact variable, then the corresponding momentum is quantized. This extra momentum gives rise to the tower of Kaluza-Klein excitations. However, if the field does not depend on y, there is no extra momentum, and thus there are no excitations.

- *Zichichi:*

Thank you.

- *Pinfold:*

Can I ask a question related to the model of universal extra dimensions? How to contrast it with supersymmetry at the Large Hadron Collider? I mean what would be the big smoking gun signature of the model of universal extra dimensions, as opposed to supersymmetry?

- *Antoniadis:*

The main signature of the longitudinal extra dimensions is the production of Kaluza-Klein excitations. In the universal case they can be produced only in pairs and the analysis becomes more complicated. The situation is very similar to supersymmetry.

- *'t Hooft:*

You mentioned several experimental bounds on the size of the extra dimensions, and I think you didn't mention the table-top experiments saying that gravity is still, as you know it is, up to about tenths of a millimeter. But you also mentioned some experiments like LEP or other accelerator experiments giving you a limit on the size of the compactification scale of the extra dimensions. Explain a little bit more.

- *Antoniadis:*

Well, I mentioned the table-top experiments this morning. The best limit on Newton's law comes from an experiment in Washington (Adelberger et al), which has this device. More recently, there was another experiment which obtained comparable bounds, down to approximately 200 microns (Fishbach et al).

- *'t Hooft*:

This seems to me surprisingly large.

- *Antoniadis:*

This bound restricts the case of 2 extra dimensions only. Because only in that case the size of dimensions is in that range. If you compare it with accelerator experiments, the present bound from LEP is about 0.5 mm, which is weaker than the bound of 130 microns from table-top experiments.

- *'t Hooft*:

I understand these experiments, but what did LEP to pin down the scale of gravity?

- *Antoniadis*:

LEP essentially puts bounds on the effective higher-dimensional gravitational Newton constant and then uses the relation between the Planck scales and the size of extra dimensions.

- *'t Hooft*:

And how does this number come about?

- *Antoniadis*:

It comes from an analysis of missing energy events.

- *Loginov:*

In 3 dimensions every particle has a mass. What happens to this property in the extra dimensions? Do you have some sort of D-mass in the description of extra dimension particles?

- *Antoniadis:*

Any higher dimensional particle gives rise in 4 dimensions to a tower of ordinary massive particles. The zero-mode that corresponds to zero momentum in extra dimension has the same mass as the higher dimensional particle (m_0), and then you have a tower of excitations which have a mass squared in accordance to the formula m_0^2 + (n/R)^2.

- *Loginov:*

You have n = +/-1, +/-2, etc. So will it have the same value of mass for the +1 and -1? What does it mean?

- *Antoniadis:*

This "+" and "-" corresponds to the higher dimensional parity, i.e. positive or negative momentum.

- *Loginov:*

Thank you.

- *Cifarelli:*

Can you put back this transparency? Yesterday you mentioned that normally one has KK excitations of photons, W and Z, but not of the gluons. Can you comment on that? Is it just a simplicity choice or is there any idea behind it?

- *Antoniadis:*

The idea is the following. As I mentioned today in the example of embedding the Standard Model, for every factor in the Standard Model you can associate a different brane stack, which has its own coupling. This coupling is given by the string coupling divided by the square root of the compactification volume, which is parallel to the brane, in string units. So one way to have different couplings for different interactions is by having different volumes. Since we know that the strong interaction coupling is bigger than the weak interactions one, a particular choice is to assume that strong interactions correspond to a 3-brane and there is no parallel volume here, it's equality, while weak interactions propagate to another dimension, which is compact and which has a size roughly twice the string size (reproducing the known ratio of the two couplings).

- *Korthals-Altes*:

You have this model where branes intersect, does the angle matter under which they intersect, or not at all?

- *Antoniadis:*

If the angle is not 0 or 90 degrees, supersymmetry is broken. In other words, for 2 intersecting branes, if the angle is arbitrary then supersymmetry is broken on the intersection. The magnitude of this breaking is determined by the value of the angle, or equivalently by the magnetic field in string units.

- *Marrani:*

I would like to ask if you can realize gauge groups different from unitary groups or products of U(1)'s N times for N branes in a stack. You always reproduce just the gauge groups of the Standard Model, but I was wondering, for example, if the unitary symplectic groups or other kinds of gauge groups can be reproduced in this framework.

- *Antoniadis*:

Yes. The way to do it is to introduce, in addition to D-branes, which are dynamical objects, the so-called orientifold planes. These are non-dynamical branes with no associated degrees of freedom, but actually they are part of Type I string theory. They are necessary for anomaly cancellation and for the introduction of non-oriented strings. In other words, they break the orientation of the world-sheet. The orientifold planes have also charges under some generalized gauge fields (RR antisymmetric tensors), like the D-branes.

- *Marrani:*

Do you mean that you are co-setting the world-sheet of the strings, because you're introducing a sort of symmetry on the world-sheet?

- *Antoniadis:*

Yes.

- *Marrani:*

And you are gauging this symmetry acting in some gauge groups?

- *Antoniadis:*

The orientifold plane acts as a projection on the spectrum of the branes, which are located on top of it. As a result, instead of obtaining a group which is unitary, it becomes either orthogonal or symplectic, depending on how the projection acts. But if you move the branes away from the orientifold plane, you get back the unitary group.

- *Marrani:*

And you can obtain just orthogonal or unitary symplectic groups, or also exceptional groups?

- *Antoniadis:*

Exceptional groups are not possible to obtain with such constructions in perturbation theory.

- *Marrani:*

Thank you.

- *Konikowska:*

I would also like to ask a question. Regarding the experiments, e.g. the missing energy, is there a danger that one can get misleading results, due to e.g. non-trivial bulk metric and the Einstein equations not closed on the brane?

- *Antoniadis*:

Yes, for a complete analysis you must take into account also this possibility. For instance, warped spaces give Randall-Sundrum type models, and there are known signatures that you can compare.

CHAIRMAN: I. ANTONIADIS

Scientific Secretaries: H.T.Ding, G.Vartanov

DISCUSSION III

- *Marrani*:

I would like you to comment a bit more on the Nielsen-Olesen instability and in particular on the case in which we have a magnetic field in a non-abelian gauge group you told us. We should switch it only along the Cartan subalgebra of such a group; why should it be necessarily so?

- *Antoniadis:*

If you introduce more than one magnetic fields, the Dirac quantization condition becomes very difficult to be satisfied when the corresponding generations do not commute. A simple solution is that the generations commute and each magnetic field satisfies the abelian Dirac quantization. In this case, they belong in the Cartan subalgebra of the higher dimensional gauge group. The gauge fields which are charged under the magnetized generators give rise to tachyons, if you start from 6 dimensions, corresponding to the Nielsen-Olesen instability. The nice thing about starting from ten dimensions (and thus having six compactified ones) is that you can get rid of the tachyonic instability by choosing appropriately the internal magnetic fields.

- *Marrani*:

In the case in which wc compactify from 10 to 4 dimensions, this morning you told that we may have a partial SuSy unbroken on the brane. In particular you mentioned the fact that if there is a massless scalar, this would paired with the 0-mode s=1/2 fermion, where does this massless scalar comes from?

- *Antoniadis*:

If you start from 10 dimensions, the N=1 supersymmetry decomposes as four supersymmetries in 4 dimensions. The corresponding N=4 vector multiplet contains in addition to four fermions, six scalars. These six scalars are complexified into three complex ones whose masses are given precisely by these three combinations of magnetic fields. If one of the combinations vanishes, then the corresponding massless complex scalar becomes the super-partner of the massless fermionic zero mode and N=1 SuSy is restored.

- *Cifarelli*:

Could you please better specify how you can solve many MSSM problems, such as FCNC in the high SUSY breaking scale scenario?

- *Antoniadis*:

All those problems are related to the scalar sector of the MSSM, so by making the scalars super-heavy these problems are solved. But in that case you have to pay a price, you have to do a fine tuning to the Higgs mass in order to keep it light while all other scalars are heavy.

- *Munir*:

At large tan_beta in MSSM, Higgs coupling to b-type quarks are enhanced so that the dominate role should be played by b quarks in the 1 loop correction rather than the top quark, or is this a special case as can be seen on transparency no 3 ?

- *Antoniadis*:

The reason that you get corrections only from top quark is because there is no scalar partner(s-top). The theory is not supersymmetric, so there are no scalar super-partners. Thus the corrections to this relation come from the particles which have strong Yukawa couplings. The Yukawa coupling of the top_right-top_left-Higgs is the larger one and gives the dominant correction.

- *Nilse*:

In split SUSY, why are squarks and sleptons not necessary to achieve correct unification?

- *Antoniadis*:

Because they drop in all differences of the one-loop SM beta-functions. Actually, only the differences matter for gauge coupling unification. So, in these differences of one-loop beta-functions, the contribution not only of the scalar partners of quarks and leptons but also of the quarks and leptons themselves drops. The reason is that they form SU(5) representations and satisfy automatically unification. On the other hand, the gauginos contribute in differences because they transform in adjoint representations.

- *Imponente*:

Does an eventual curvature of the metric affect the length—scale of strings? Is it possible to constrain it?

- *Antoniadis*:

In principle there can be important changes. For instance, non zero curvature in the bulk implies a bulk cosmological constant. Negative cosmological constant is even compatible with bulk supersymmetry which implies stability. A possible geometry is then anti-de Sitter space. In this case, one can find completely different results from flat space and thus curvature is important.

- *Korthals-Altes*:

You introduced fluxes in the extra space dimensions. They are quantized by integers, because if I take a charge around them, the wave function should stay single-valued. I understand that the periodicity in the extra dimensions changed the integers into fractional numbers. How does this work?

- *Antoniadis*:

I didn't say that the quantization condition changed because of the periodicity, but because of the brane wrapping. In field theory you have just the quantization of flux. But if you have a brane, then on top of the magnetic flux, you can wrap the brane around the torus. Thus, you can introduce this non-trivial wrapping number "n". In the quantization condition, this wrapping number of the brane around the torus amounts to increase the effective area of the torus. This is because the flux is defined only on the brane.

- *Cifarelli:*

Is there a theoretical constraint on the long-lived gluino mass other than the experimental limits?

- *Antoniadis*:

The gluino mass must be in the TeV range because we need gluinos for unification. There is no independent theoretical reason that it has to be light.

- *Marrani:*

When you compactify extra dimensions until now you always used toroidal compactification. What happens if you consider more general and complicated manifolds on which extra dimensions are compactified? In particular, I would like to know if there are consequences on the treatment of magnetic fluxes on the branes you gave this morning?

- *Antoniadis:*

For a general compactification manifold, one has the freedom to introduce magnetic fluxes on any non-contractible 2-cycle. The Dirac quantization condition takes similar form for every such 2-cycle a with area A_a. There is a magnetic flux m_a and a brane wrapping number n_a. If you have some topological obstructions ,you must take them into account. They will give constraints on the possible quantum numbers.

- *'t hooft:*

You introduced split supersymmetry by postulating that the scalar super-partners are much heavier than the fermions ones .Do you have an explicit model in which this can be seen to happen?

- *Antoniadis*:

Tomorrow I will present precisely a string theoretical framework that provides such models.

- *Jona-Lasinio:*

Why the metric in the bulk has to be flat? Why the argument can not be generalized to an arbitrary constant metric?

- *Antoniadis:*

There are two aspects. The curvature of the metric does not affect the possibility that you can lower the string scale by introducing large extra dimensions. However, to extract the physical consequences of such constructions, you must repeat the analysis I made for the flat case. This is not in general an easy task.

CHAIRMAN: I. ANTONIADIS

Scientific Secretaries: G. Imponente, M. Jona Lasinio

DISCUSSION IV

- *O'Leary*:

Regarding the formulae for the hypercharges for the three models you gave, could you explain how you get them? Do they come from the theory or are the coefficients of the charges from fitting to get the experimental hypercharges?

- *Antoniadis*:

The hypercharges are given by their experimental values. However, the expression of the hypercharge as a linear combination of the three U(1)'s as well as the U(1) charges are parameters that are fixed from the requirement to produce the correct hypercharges. There are three possible solutions corresponding to the three modes I presented.

- *Marrani*:

The treatment and the assumptions you did this morning seem to rely to that factorized choice of the 6 dimensional compactification manifold to be (T^2)^3. What happens if you consider more general non factorized manifold, such as 6 dimensional Calabi-Yau?

- *Antoniadis*:

For a general Calabi-Yau manifold, you may find new solutions. However you have also the choice to embed the SM in a particular combination of intersecting branes locally. The reason is that CY is locally flat. Thus the spectrum remains the same and does not depend on the global properties of the CY. Similarly, the conditions we imposed for unification translate into the same restrictions on the angles of the branes that we obtained in the flat case. In fact, small magnetic fields implies that the angles are small and the branes are almost parallel. On the other hand, the global properties of the manifold give rise to global constraints on the number of branes (for given orientifolds). Finally, the spectrum of closed strings of course changes, since this is completely dependent on the geometry of the manifold, in contrast to the spectrum of the branes which is not affected when the manifold is locally flat.

- *t'Hooft*:

You showed that the scalar fields can be massless when you choose the fluxes in a particular way. Did you really explain how to get the mass corrections, since you assume that the scalar field can get a mass? Would SUSY be broken at that level? Are they exactly massless or do you expect corrections?

- *Antoniadis*:

Massless scalars imply that you have unbroken SUSY. If the conditions on the magnetic fluxes guarantee that you have masslesss scalars in all branes and intersections, then you have an exact SUSY globally in the theory. However usually a massless scalar implies that you have SUSY only locally, for instance at the corresponding intersection where the scalar arises. But this SUSY, preserved by some intersection, is in general broken in other intersections of different branes. In this case, in the absence of global SUSY, the corrections to the massless scalars are expected to be large and you will have to do a fine tuning in order to keep them light. The corrections to the scalar masses would be in fact proportional to the SUSY breaking scale. So if you have other scalars charged under the SM, which have high masses such as 10^13 GeV that I assumed here in split SUSY, the corrections to the Higgs mass would be of the same order and you must do a fine tuning in order to keep the Higgs mass at the electroweak scale.

- *t'Hooft:*

There are so many solutions in string theory that have a landscape of different possibilities and we will never find out exactly where in the landscape our universe is. You seem to be constructing a specific model which is one specific solution of string theory, and that's it, but the claim is that there is practically a continuum of solutions and we will never be able to find out exactly which of all these solutions is going to be our universe.

- *Antoniadis:*

A way to stabilize all moduli of string theory, such as the sizes of the extra dimensions, was found recently and is based on turning on various fluxes. These are generalizations of ordinary internal magnetic fluxes to constant field strengths of generalized gauge fields from the bulk, called Ramond-Ramond fields. These are p-forms, in other words antisymmetric tensors with several indices. All these fluxes generate a potential, in the same way that magnetic fields generate a potential that breaks parts of the SUSY. Minimizing these potentials generated by fluxes one can stabilize the moduli and there have been several examples recently, showing that in principle one can stabilize in this way all moduli of string theory. The problem now is that the values of the moduli fields that one finds from the stabilization depend on the fluxes that are discrete and quantized like the magnetic fields. So you have an enormous freedom of all possible vacua, characterized by these discrete choices of fluxes. This is the so called landscape: it leads to a picture in which the string vacuum is almost infinitely degenerate, even when restricted to vacua preserving SUSY. The landscape was used as a motivation for explaining some of the fine tuning problems in physics not by dynamical mechanisms, but using principles like the anthropic one. In my opinion, the landscape is indeed there, in the sense that there are all these vacua in string theory. But it is not something new, in the sense that previously many of us knew that there are several vacua, although only recently it has been established as a result of a minimization procedure. On the other hand, to use this landscape in order to justify an anthropic principle, is still premature in my opinion.

- *t'Hooft:*

You showed that if you choose the flux sufficiently carefully, everything is stable, but if I make a stupid choice of the fluxes, then I guess, these vacua would be unstable. Maybe, I could imagine that if you look more carefully you find many more instabilities, so that perhaps a very small subset, hopefully only one of those solutions, is completely stable. Is that a possibility or would you rule it out?

- *Antoniadis:*

Yes, this is a possibility.

- *'t Hooft:*

Because the alternative is indeed the anthropic principle that brings of course to the other notion of the cosmological constant where we still haven't got a complete answer as to why it is as small as it has to be. Some people say "well, it is just the anthropic principle", which professor Zichichi calls 'intelligent design', that within the anthropic principle everything has just been tuned for us to be sitting there.

- *Antoniadis:*

Yes, I agree.

- *Zichichi*:

Should the massless Higgs have zero-coupling, because the coupling is proportional to the Higgs mass? How do you detect it? What is the production cross section that you think to be working in a complete model, for the massless Higgs?

- *Antoniadis:*

The mass term of the Higgs potential is zero at the tree level. Then, a non zero value is generated by the radiative corrections. Actually in the models I discussed the corrections are huge and you must do a fine tuning in order to keep its mass small. On the other hand, the higgsinos, remains light, and are needed in order to obtain a dark matter candidate, which is in general a combination of the higgsinos and the binos.

- *Nilse*:

The number of string vacua is often quoted to be 10^100. Where does this estimate come from? Why not 10^1000 and not say 10^1000.000?

- *Antoniadis*:

These numbers come in particular settings in which you consider some Calabi-Yau, and turn on all possible fluxes. Then you take into account all anomaly cancellation conditions and you count how many string vacua you produce by minimizing the potential. The finite number comes from the anomaly cancellation which, in the simple case without fluxes determines the number of branes to 32. In the presence of fluxes there is a new contribution on the right hand side of the equation and pays an important role for restricting the number of possibilities. The number in principle can be calculated, but in practice people only calculate the order of magnitude. This calculation is very simplistic for the moment and we need more realistic constructions. In the early days, of the heterotic string there was already a counting based on how many self-dual lattices you can have leading to heterotic four dimensional models. The result was of the order of 10^500. However, in this counting the mechanism of fixing the moduli was not taken into account. Now the counting is done in the context of type II orientifolds, implemented with moduli stabilization by fluxes and the new number is of order 10^100.

- *Konikowska*:

I was wondering if you try to analyze some cosmological problems within your approach, e.g. cosmological constant or inflation.

- *Antoniadis*:

I don't have any clever idea about the cosmological constant. I mentioned only in the context of TeV strings, the estimate on how big is the effective cosmological constant on the brane and then I imposed SUSY in the bulk to kill the dominant correction and guarantee a scaling M_string^4. Regarding cosmology and inflation, my personal criticism to several works in this context is that they are simplistic. One assumes that first the extra dimensions are stabilized, and then one introduces inflation in the effective 4 dimensional theory. However one could leave all dimensions free and try to find new mechanisms for inflating only three of them keeping the remaining ones at a finite size. It is an interesting problem, but much more complicated.

- *Marrani*:

Could you comment the flux compactifications, in particular the mechanism of moduli stabilization, in your framework? Could the Attractor Mechanism be advocated in this case as a dynamical principle to fix some (if not all) the moduli fields?

- *Antoniadis*:

It could be. Note however that this mechanism of stabilizing moduli with fluxes works also for supersymmetric vacua. For instance, by introducing only generic magnetic fluxes on the 6D torus T^6, one can stabilize all geometric moduli in the simplest toroidal compactification on T^6, keeping N=1 SUSY globally unbroken everywhere. One can still find an enormous number of vacua which are all stable because they are supersymmetric. Then you can understand what the freedom would be, when you allow also the breaking of SUSY.

- *'t Hooft:*

What is your favourite result at LHC, LHCb for instance? More generally, what is your best chance of observing something that could confirm or check the ideas of strings?

- *Antoniadis:*

Personally, I could propose some experiments which are relevant only for detecting possible signals of a particular theory. However, the important point at present is that we don't know what is the Physics beyond the SM and therefore we would like to have a discovery machine to show us what kind of Physics it is. There have been many theoretical proposals over the last 20 years, but we do not know which one is correct, if any.

- *Imponente:*

Would you expect some signature in Cosmic Rays flux?

- *Antoniadis:*

It could be, but in Cosmic Rays it is much more difficult to decide what kind of Physics is behind, unless we are very lucky. It is like when you use precision measurements to find indirect signals of new physics. On the contrary, an accelerator like LHC can make a clear discovery.

- *Zichichi*:

It would be very interesting to know what to search for, because to make a discovery means to have an instrument able to observe something. There are so many theoretical predictions that my personal recommendation would be to ignore all of them and to build up an instrument with the highest rejection and power against differentiation of particles: time of flight, momentum analysis, as good as possible.

- *Antoniadis*:

I agree. Actually, in the experimental programs the main signals of various theories are incorporated into the data analysis. Therefore, several theoretical possibilities will be explored but it is also possible that LHC finds something completely different.

- *t' Hooft*:

Do you have any favourite candidate for dark matter?

- *Antoniadis*:

In the split SUSY, the dark matter is just like in SUSY: a linear combination of higgsinos and binos. Actually, gauginos can be made heavy but still higgsinos alone can provide dark matter. The minimum theory is the SM plus a fermion doublet. In the TeV string case there are several candidates for dark matter, there is no unique proposal. For instance, in the case of universal extra dimensions, it can be the lightest KK excitation which is stable due to KK parity conservation.

Two Topics in Strongly Coupled Field Theories*

Thomas Appelquist†

Department of Physics, Yale University, New Haven, CT 06511.

Abstract

These two lectures are devoted to two aspects of strongly coupled quantum field theories. I first describe a proposed constraint on the structure of strongly coupled, asymptotically free field theories. The constraint takes the form of an inequality limiting the number of degrees of freedom in the infrared description of a theory relative to the number of underlying, ultraviolet degrees of freedom. After introducing the general idea, I discuss the application of the inequality to supersymmetric and nonsupersymmetric gauge theories, including one chiral gauge theory. In the second lecture, I discuss some of the challenges facing the idea of dynamical electroweak symmetry breaking. I review briefly constraints from precision studies and describe the idea of walking technicolor. I then discuss a model containing the ingredients for a neutrino seesaw mechanism as well as quark and lepton electric dipole moments.

1 A Constraint on Strongly Coupled Field Theories

Strongly coupled quantum field theories play a central role in the description of nature at the most fundamental level. Quantum chromodynamics describes the observed strong interactions, and other strongly interacting theories are part of most efforts to extend the standard model. But strongly coupled field theories are notoriously difficult to analyze directly. Because of this, general constraints on their behavior can be very useful.

*Presented at the the International School of Subnuclear Physics, Erice, Sicily, August 30 - September 3, 2005

†thomas.appelquist@yale.edu

In my first lecture, I will describe a constraint on strongly coupled field theories [1] . It takes the form of an inequality limiting the number of massless degrees of freedom in the infrared description of a field theory to be no larger than the number of ultraviolet degrees of freedom. It was conjectured to apply to all asymptotically free theories (those governed by a free ultraviolet fixed point), whose infrared behavior is also governed by a fixed point, not necessarily free.

The inequality is formulated using finite temperature as a device to probe all energy scales, with the degree-of-freedom count defined using the free energy of the field theory. The zero-temperature theory of interest is characterized using the quantity f_{IR}, related to the free energy by

$$f_{IR} \equiv - \lim_{T \to 0} \frac{\mathcal{F}(T)}{T^4} \frac{90}{\pi^2} , \tag{1}$$

where T is the temperature and $\mathcal{F}$ is the conventionally defined free energy per unit volume. The limit is well defined if the theory has an infrared fixed point. For the special case of an infrared-free theory, f_{IR} is simply the number of massless bosons plus 7/4 times the number of 2-component massless Weyl fermions. The corresponding expression in the large T limit is

$$f_{UV} \equiv - \lim_{T \to \infty} \frac{\mathcal{F}(T)}{T^4} \frac{90}{\pi^2} . \tag{2}$$

Again, this limit will be well defined if the theory has an ultraviolet fixed point. For an asymptotically free theory f_{UV} counts the underlying, ultraviolet degrees of freedom in a similar way.

In terms of these quantities, the conjectured inequality for any asymptotically-free theory is

$$f_{IR} \leq f_{UV} . \tag{3}$$

This inequality has not been proven, but in Ref. [1] it was shown to agree with all known results and then used to derive new results for several strongly coupled, vector-like gauge theories. It was applied to chiral gauge theories in Ref. [2].

In field theories with weakly coupled fixed points the free energy, appearing in the definition of f, may be computed perturbatively. To zeroth order in couplings the low-temperature free energy density in three spatial dimensions is

$$\mathcal{F}_{free}(T) \simeq -\frac{\pi^2 T^4}{90}[N_B + \frac{7}{8} 2N_F], \tag{4}$$

where N_B is the number of massless (real) bosonic fields, and N_F is the number of massless (two-component) fermionic fields. I have neglected the contributions of any massive fields, which vanish exponentially as $T \to 0$. A similar expression applies in the infinite T limit, with N_B and N_F including massive as well as massless fields. These expressions are exact in the case of free fixed points and approximately correct for theories governed by weak fixed points.

1.1 SUSY Example

I will first describe a SUSY $SU(N)$ gauge theory with F flavors ("quarks" and "antiquarks") of massless fermions and associated superpartners. The theory has a free UV fixed point if the number of flavors is less than 3 times the number of colors, $F < 3N$. In this case the quantity f_{UV} may be calculated using Eq. (4) to give

$$f_{UV} = \left[2(N^2 - 1) + 4NF\right](1 + \frac{7}{8}) \ . \tag{5}$$

The analysis of Seiberg [3] suggests that the infrared behavior of this theory is alternatively described through the use of F flavors of massless magnetic quarks transforming according to the fundamental representation of a dual gauge group $SU(F-N)$, along with F^2 massless "meson" chiral superfields. This theory is infrared free provided $F \leq 3N/2$. Under these circumstances f_{IR} is:

$$f_{IR} = \left[2((F-N)^2 - 1) + 4(F-N)F + 2F^2\right](1 + \frac{7}{8}) \ . \tag{6}$$

Thus our fundamental inequality becomes

$$2[(F-N)^2 - 1] + 4(F-N)F + 2F^2 \leq 2(N^2 - 1) + 4NF \tag{7}$$

Because f_{IR} grows quadratically with the number of flavors, this inequality limits the values of F for which the low energy theory can consist of massless magnetic degrees of freedom with infrared free coupling. Remarkably, this inequality gives the bound $F \leq (3/2)N$, corresponding precisely to the boundary of the weak magnetic phase determined by the analysis of Seiberg [3]. At the boundary $F = (3/2)N$ the inequality is saturated.

1.2 Non-SUSY Example

Next I will consider the non-supersymmetric version of the same $SU(N)$ gauge theory, with F massless quarks (and antiquarks). The theory has a free UV fixed point for $F < 11N/2$. Based on real QCD we expect the $SU(F) \times SU(F)$ chiral symmetries of this theory to be realized in the Nambu-Goldstone mode—at least for small enough F/N. If we assume that this is the case, the IR theory consists of $F^2 - 1$ Nambu-Goldstone bosons. The derivative interactions of these particles are irrelevant in the infrared, and consequently this theory is described by a free IR fixed point.

At these free UV and IR fixed points we may use Eq. (4) to compute f_{IR} and f_{UV}:

$$\begin{aligned} f_{IR} &= F^2 - 1 \\ f_{UV} &= 2(N^2 - 1) + \frac{7}{8} 4NF \ , \end{aligned} \tag{8}$$

and the inequality becomes

$$F^2 - 1 \leq 2(N^2 - 1) + \frac{7}{8} 4NF \ , \tag{9}$$

or, since F must be positive

$$F \leq 4\sqrt{N^2 - \frac{16}{81}} \ . \tag{10}$$

Since N must be 2 or larger, and F and N must both be integral, this is equivalent to $F < 4N$. Remarkably, the inequality says that for the number of flavors larger than or equal to four times the number of colors, this gauge theory cannot break the full set of chiral symmetries!

This bound on the onset of the chiral phase transition ($F \leq 12$ for $SU(3)$) is well above the transitional values suggested by preliminary lattice simulations [4, 5]. It is very close, however, to the value that emerges from the use of a continuum gap equation together with the assumption that the coupling is governed by an infrared fixed point appearing in the perturbative β function [6, 7]. In fact, a combination of the ladder gap equation and the two-loop beta function give a critical value $F^{crit}/N = (100N^2 - 66)/(25N^2 - 15)(\rightarrow 4$ as $N \rightarrow \infty)$. The reliability of this result is far from clear, however, since higher order effects are not obviously small. So whether the chiral phase transition saturates the inequality in this way or takes place at a lower value of F/N remains an open question.

1.3 Proving the Inequality

A natural approach to a proof starts from the dimensionless quantity (defined generally in d spatial dimensions)

$$f(T) \equiv -\frac{\mathcal{F}}{T^{d+1}}\Omega_d \tag{11}$$

where Ω_d is a constant chosen such that the contribution to $f(T)$ from a free bosonic degree of freedom is 1. The quantities f_{IR} and f_{UV} are just the limits of this function as T approaches zero and ∞ respectively.

Differentiating f with respect to T an using the standard relations $T\partial\mathcal{F}/\partial T = -u - p$, $\mathcal{F} = -p$ where p is the pressure and u is the internal energy density, we have

$$T\frac{\partial f}{\partial T} = \Omega_d \frac{u - dp}{T^{d+1}} \equiv \Omega_d \frac{\theta}{T^{d+1}} \tag{12}$$

where θ is the trace of the energy-momentum tensor. For a conformally invariant theory θ is zero, f is constant, and f_{IR} is equal to f_{UV}. Of course the theories of interest are not conformally invariant—the lack of conformal invariance arises from a scale dependence of coupling constants through renormalization. Consequently we expect the difference between

f_{UV} and f_{IR} to arise from the renormalization group flow from the ultraviolet to the infrared. If θ is positive along this trajectory, f_{IR} will necessarily be smaller than f_{UV}, proving the inequality.

Unfortunately this is not generally the case. Consider, for example, a classically scale-invariant gauge field theory. In this case the trace of the energy-momentum tensor is given by:

$$\theta = 2\frac{\beta}{g}\mathrm{Tr}G^2 \tag{13}$$

where $\mathrm{Tr}G^2$ is the trace over gauge indices of the square of the gauge field-strength, as well as over the thermal density matrix, and β is the RG beta function. Note that, at least in perturbation theory, the thermal average of the field-strength squared is negative (magnetic fluctuations are less-well screened than electric fluctuations). Therefore a negative beta function leads to a positive θ, as we desire. This observation immediately suggests examples of *negative* θ. If the low energy theory is a gauge field theory governed by a free infrared fixed point, then the β function will be positive at weak coupling where $\mathrm{Tr}G^2$ is known to be negative. This is realized if the low energy theory is either an abelian theory with massless fermions or a non-abelian theory with matter content sufficient to render it infrared free. The SUSY $SU(N)$ theory in the weak magnetic phase, discussed in the previous section, is precisely such an example. Of course this is not a counter-example to the conjectured inequality: the fact that $f(T)$ is not monotonic does not contradict the inequality involving f_{IR} and f_{UV}.

1.4 T Dependence

I will next describe in more detail the T dependence of $f(T)$ for the supersymmetric and non-supersymmetric $SU(N)$ theories, giving the T dependence for both the $T \to \infty$ and $T \to 0$ limits, where perturbation theory may be employed.

1.4.1 SUSY $SU(N)$ Theory

For $T \to \infty$, perturbation theory in the underlying, asymptotically free "electric" theory may be used, giving [8]

$$f(T) = f_{UV} - (N^2-1)(N+3F)\frac{45g_e^2(T)}{32\pi^2} + ..., \tag{14}$$

where f_{UV} is given by Eq. (5) and where T sets the scale for the electric coupling g_e. Since the β function is negative, $g_e^2(T)$ decreases as T increases, leading to positive θ and $\partial f/\partial T$, as discussed in Section 3.

For $T \to 0$, with $F \leq 3N/2$, perturbation theory gives

$$f(T) = f_{IR} - ((F-N)^2-1)(4F-N)\frac{45g_m^2(T)}{32\pi^2} - 3F^2(F-N)\frac{45y^2(T)}{32\pi^2} + ..., \tag{15}$$

where f_{IR} is given by Eq. (6), g_m is the magnetic gauge coupling, and y is the Yukawa coupling of the magnetic theory. The g_m^2 term is obtained from the g_e^2 term in Eq. (14) by the replacement $N \to F - N$. The y^2 term is obtained by evaluating the two-loop diagrams involving the Yukawa couplings and the four-scalar couplings that arise from the superpotential of the magnetic theory.

Since the theory is infrared free for $F \leq 3N/2$, both couplings increase with T, showing that θ and $T\partial f/\partial T$ are negative for small T. Still, the inequality is satisfied.

1.4.2 Non-SUSY $SU(N)$ Theory

For $T \to \infty$, perturbation theory gives

$$f(T) = f_{UV} - 10(N^2-1)(N+5F/N)\frac{g^2(T)}{16\pi^2} + ..., \tag{16}$$

where f_{UV} is given by Eq. (8) and g is the gauge coupling. Asymptotic freedom leads to positive θ and $\partial f/\partial T$.

For $T \to 0$, corrections to the free-field behavior of the Nambu-Goldstone bosons may be computed using chiral perturbation theory. The leading correction arises at second order in $1/F_\pi^2$, where F_π is the Nambu-Goldstone decay constant, and contains a chiral logarithm. The result, for $T << F_\pi$, is [9]

$$f(T) = f_{IR} + \frac{F^2(F^2-1)}{144}\frac{T^4}{F_\pi^4}\ln\left(\frac{F_\pi}{T}\right) + \ldots, \tag{17}$$

where f_{IR} is given by Eq. (8). Thus for small T, $f(T)$ increases with T.

Interestingly, for the non-supersymmetric theory in the Nambu-Goldstone phase, the function $f(T)$ is positive-monotonic for both large T and small T, the limits in which it may be computed reliably, using perturbation theory.

1.5 Another Example - QED in $2+1$ Dimensions

For $2+1$ dimensional QED (QED_3) I will show that the inequality gives an interesting constraint on the allowed infrared phase structure. QED_3 with $2F$ charged Weyl fermions (F Dirac fermions) is believed to have a phase transition as the number of flavors is varied [10, 11]. The massless theory has a $U(2F)$ global symmetry. For large F, the screening effect of the fermions prevents the formation of a condensate and the infrared theory is expected to be conformal. For small F, on the other hand, one expects global symmetry breaking and dynamical mass generation for the fermions. An analysis of the breaking using a gap equation indicates that a parity conserving mass term is formed, corresponding to the breaking of the global $U(2F)$ symmetry to its $U(F)\times U(F)$ subgroup.

The inequality places a tight constraint on this pattern of breaking. QED_3 is free in the ultraviolet and using Eq. (11) we have [1]

$$f_{UV} = 1 + \frac{3}{4}\,4F\,, \tag{18}$$

where $4F$ counts the fermionic degrees of freedom. The breaking of the $U(2F)$ symmetry to $U(F)\times U(F)$ leads to $2F^2$ Nambu-Goldstone bosons. Since the theory does not confine, the

[1] In $(2+1)$ dimensions free bosons and fermions respectively contribute 1 and $3/4$ to f.

photon remains in the infrared spectrum so we have

$$f_{IR} = 1 + 2F^2 \; . \tag{19}$$

The inequality is satisfied only for $F \leq 3/2$ which implies that chiral symmetry breaking is excluded for all $F \geq 2$.

The critical number of flavors separating the two phases has been estimated, using the gap equation with a $1/F$ expansion of the kernel, to be in the range $3 < F_{crit} < 4$ [10, 11]. The discrepancy between this result and our inequality suggests that the gap equation overestimates F_{crit} [12].

1.6 The Bars Yankielowicz Model

Finally, I will briefly describe the application of the inequality to an interesting non-supersymmetric chiral gauge theory – the Bars Yankielowicz (BY) Model [13]. This model is based on the single gauge group $SU(N \geq 3)$ and includes fermions transforming as a symmetric tensor representation, $S = \psi_L^{\{ab\}}$, $a, b = 1, \cdots, N$; $N + 4 + p$ conjugate fundamental representations: $\bar{F}_{a,i} = \psi^c_{a,iL}$, where $i = 1, \cdots, N + 4 + p$; and p fundamental representations, $F^{a,i} = \psi_L^{a,i}$, $i = 1, \cdots, p$. The $p = 0$ theory is the basic chiral theory, free of gauge anomalies by virtue of cancellation between the antisymmetric tensor and the $N + 4$ conjugate fundamentals. The additional p pairs of fundamentals and conjugate fundamentals, in a real representation of the gauge group, lead to no gauge anomalies.

The global symmetry group is

$$G_f = SU(N + 4 + p) \times SU(p) \times U_1(1) \times U_2(1) \; . \tag{20}$$

Two $U(1)$'s are the linear combination of the original $U(1)$'s generated by $S \to e^{i\theta_S} S$, $\bar{F} \to e^{i\theta_{\bar{F}}} \bar{F}$ and $F \to e^{i\theta_F} F$ that are left invariant by instantons, namely that for which $\sum_j N_{R_j} T(R_j) Q_{R_j} = 0$, where Q_{R_j} is the $U(1)$ charge of R_j and N_{R_j} denotes the number of copies of R_j.

Thus the fermionic content of the theory is

	$[SU(N)]$	$SU(N+4+p)$	$SU(p)$	$U_1(1)$	$U_2(1)$
S	$\Box\Box$	1	1	$N+4$	$2p$
$\bar{F}$	$\overline{\Box}$	$\overline{\Box}$	1	$-(N+2)$	$-p$
F	$\Box$	1	$\Box$	$N+2$	$-(N-p)$

(21)

where the first $SU(N)$ is the gauge group, indicated by the square brackets. The theory is asymptotically free for

$$p < (9/2)N - 3 \, . \tag{22}$$

Because of asymptotic freedom, the thermodynamic free-energy may be computed in the $T \to \infty$ limit. An enumeration of the degrees of freedom leads to

$$f_{UV} = 2(N^2-1) + \frac{7}{4}[\frac{N(N+1)}{2} + (N+4)N + 2pN] \, . \tag{23}$$

1.6.1 $p = 0$

I first describe the case $p = 0$. The fermions are in complex representations of the $SU(N)$ gauge group and the global symmetry group is $G_f = SU(N+4) \times U_1(1)$. The theory is strongly coupled at low energies, so it is expected either to confine or to break some of the symmetries, consistent with global anomaly matching. I will consider here only the possibility that the theory confines with the full global symmetry group G_f unbroken has been considered in Refs. [2, 14]. All the global anomalies of the underlying theory may be matched at low energies providing that the massless spectrum is composed of gauge singlet composite fermions transforming according to the antisymmetric second-rank tensor representation of $SU(N+4)$. They are described by the composite operators $\bar{F}_{[i}S\bar{F}_{j]}$ and have charge $-N$ under the $U_1(1)$ global symmetry.

With only these massless composites in the low energy spectrum, there are no dimension-four interactions, so the composites are noninteracting in the infrared. Therefore the thermodynamic free energy may be computed in the limit $T \longrightarrow 0$. Enumerating the degrees of freedom gives

$$f_{IR}^{sym}(p=0) = \frac{7}{4}\frac{(N+4)(N+3)}{2} , \qquad (24)$$

where the superscript indicates that the full global symmetry is intact. Clearly $f_{IR}^{sym}(p = 0) < f_{UV}(p = 0)$, satisfying the inequality of Eq. (3) [2].

1.6.2 $p \geq 0$

I will next discuss the full range of p allowed by asymptotic freedom: $0 < p < (9/2)N-3$. For p near $(9/2)N-3$, an infrared stable fixed point exists, determined by the first two terms in the β function. This can be arranged by taking both N and p to infinity with the difference $(9/2)N - p$ fixed, or at finite N by continuing to nonintegral p. The infrared coupling is then weak and the theory neither confines nor breaks symmetries. The fixed point leads to an approximate, long-range conformal symmetry. As p is reduced, the screening of the long range force decreases, the coupling increases, and confinement and/or symmetry breaking set in. I will describe three strong-coupling possibilities, each consistent with global anomaly matching.

It was observed by Bars and Yankielowicz [13] that confinement without chiral symmetry breaking is consistent with global anomaly matching provided that the spectrum of the theory consists of massless composite fermions transforming under the global symmetry group as follows:

	$[SU(N)]$	$SU(N+4+p)$	$SU(p)$	$U_1(1)$	$U_2(1)$
$\bar{F}S\bar{F}$	1	$\overline{\begin{smallmatrix}\square\\\square\end{smallmatrix}}$	1	$-N$	0
$\bar{F}^+S^+F$	1	$\square$	$\square$	N	$-N$
F^+SF^+	1	1	$\overline{\square\square}$	$-N$	$2N$

(25)

The effective low-energy theory is free. In Ref. [2], the thermodynamic free energy for this phase was computed, giving

$$f_{IR}^{sym} = \frac{7}{4}[\frac{1}{2}(N+4+p)(N+3+p) + p(N+4+p) + \frac{1}{2}p(p+1)] \ . \tag{26}$$

The inequality $f_{IR}^{sym} < f_{UV}$ was then invoked to argue that this phase is possible only if p is less than a certain value (well below the the asymptotic freedom bound). For large N, the condition is $p < (15/14)^{1/2}N$.

1.7 Summary

I have described a new general constraint on the structure of strongly coupled field theories. Using the finite-temperature free energy as a device to probe the long distance and short distance structure of the theories, an inequality has been proposed [1] restricting the number of degrees of freedom in the infrared to be no greater than the number of short distance degrees of freedom. The degree of freedom count is also defined using the free energy. The inequality has been tested against all known results and used to derive new constraints on a variety of supersymmetric and nonsupersymmetric theories.

The proof of the inequality remains elusive. It is not difficult to see that outside the class of asymptotically free theories, counterexamples may be constructed. Theories governed by ultraviolet stable fixed points can lead to $f_{UV} < f_{IR}$ and it is possible to construct such examples in less than four dimensions. In four dimensions, if a gauge theory could be found that was infrared free and also contained a weak (and therefore reliable) UV fixed point, it would provide a counterexample.

Even with the restriction to four dimensional asymptotically free theories, no proof yet exists. From my discussion of SUSY theories, it is clear that the proof will not emerge via monotinicity of $f(T)$. It does seem worthwhile to continue to search for a proof. The effort will no doubt help to develop more insight into the behavior of strongly coupled quantum field theories.

2 Challenges for Dynamical Electroweak Symmetry Breaking

In this second lecture, I will review briefly the idea of walking technicolor [16], emphasizing the role of approximate conformal symmetry. These theories represent an interesting and potentially important type of strongly couple quantum field theory. I will then describe two features of a recently proposed class of extended technicolor (ETC) models [17]–[21] incorporating this mechanism. The first is the possible existence of a seesaw mechanism for neutrino masses, and the second is CP violation as manifested in the quark electric (and chromo-electric) dipole moments.

2.1 Introduction

It is possible that electroweak symmetry breaking (EWSB) is dynamical at the TeV energy scale, driven by a strongly coupled gauge interaction denoted generically as technicolor (TC) [22]. The EWSB arises from the condensation of technifermion bilinears. The generation of realistic mass scales for the charged leptons and u, d, s, c, and b quarks seems attainable in this framework, via extended technicolor, in particular with slowly running ("walking") technicolor.

It is also possible that an ETC model of the above type can yield a plausible explanation for small neutrino masses. This has been a significant challenge for dynamical EWSB models, since, as conventionally formulated, they have no very large mass scale analogous to the grand unification scale M_{GUT} that can enter the seesaw mechanism. Robert Shrock and I re-examined this problem recently [18]. We noted that in the class of ETC models [17]−[21], there is a natural suppression mechanism for Dirac neutrino masses as well as large Majorana mass terms, leading to a viable seesaw.

2.2 Walking Technicolor

Suppose that the technicolor gauge group is SU(N_{TC}). The set of technifermions could include, for example, one family, viz., $Q_L = \binom{U}{D}_L$, $L_{TC,L} = \binom{N}{E}_L$, U_R, D_R, N_R, E_R transforming according to the fundamental representation of SU(N_{TC}) and the usual representations of $G_{SM} = \mathrm{SU}(3) \times \mathrm{SU}(2)_L \times \mathrm{U}(1)_Y$ (color and TC indices are usually suppressed). To generate quark and lepton mass, TC is then embedded in an extended theory (ETC) which connects the quarks and leptons to the techni-fermions. A natural class of such theories treats ETC as "horizontal", incorporating TC along with the three families of quarks and leptons. The ETC gauge group thenbreaks sequentially to TC, producing masses for the ETC gauge bosons.

To satisfy constraints from flavor-changing neutral-current processes, the ETC vector bosons, which can mediate generation-changing transitions, must have large masses. If ETC masses arise from self-breaking of the ETC gauge symmetry, ETC be a strongly coupled, chiral gauge theory. The self-breaking can occur in stages, for example at the three stages $\Lambda_1 \sim 10^3$ TeV, $\Lambda_2 \sim 50$ TeV, and $\Lambda_3 \sim 3$ TeV, corresponding to the $N_{gen} = 3$ standard-model fermion generations. Then $N_{ETC} = N_{TC} + N_{gen}$.

A particularly attractive choice for the technicolor group is $\mathrm{SU}(2)_{TC}$, which has the appeal that it minimizes the TC contributions to the S parameter, and can yield walking behavior, allowing for realistically large quark and charged lepton masses. With $N_{gen} = 3$, the choice $N_{TC} = 2$ corresponds to $N_{ETC} = 5$. With $N_f = 8$ vectorially coupled technifermions in the fundamental representation, studies suggest that this $SU(2)_{TC}$ theory could have an (approximate) infrared fixed point (IRFP) in the confining phase with spontaneous chiral symmetry breaking, but near to the phase transition (as a function of N_f for fixed N_{TC}) below which the theory would go over into a nonabelian Coulomb phase [23]. This approximate IRFP provides the walking behavior, and can enhance the technifermion condensates that control the quark and charged lepton masses. Walking can also enhance the masses of pseudo-Nambu-Goldstone bosons. It has further been argued that walking can dynamically suppress the S parameter below even that attained by using $SU(2)_{TC}$ [24].

A rough estimate of the quark and charged lepton masses can be made by considering a one-loop diagram in which a fermion f_a emits a virtual ETC gauge boson, going to a virtual technifermion F which reabsorbs the gauge boson, producing the mass term $m_{f_a}\bar{f}_{a,L}f_{a,R}+h.c.$ with

$$m_{f_a} \sim \frac{g_{ETC}^2 \eta_a N_{TC} \Lambda_{TC}^3}{4\pi^2 M_a^2} \tag{27}$$

where $M_a \sim g_{ETC}\Lambda_a$ is the mass of the ETC gauge bosons that gain mass at scale Λ_a and g_{ETC} is the running ETC gauge coupling evaluated at this scale. In Eq. (27) η_a is a possible enhancement factor incorporating walking, which can be as large as Λ_a/f_F, where f_F is the technicolor pseudoscalar decay constant (for present purposes I can take $f_L \simeq f_Q \equiv f_F$). Recall that Λ_{TC} is determined by using the relation $m_W^2 = (g^2/4)(N_c f_Q^2 + f_L^2) \simeq (g^2/4)(N_c + 1)f_F^2$, which gives $f_F \simeq 130$ GeV. In QCD, $f_\pi = 93$ MeV and $\Lambda_{QCD} \sim 170$ MeV, so that $\Lambda_{QCD}/f_\pi \sim 2$; using this as a guide to technicolor, I infer $\Lambda_{TC} \sim 260$ GeV.

2.3 A Class of ETC Models

I will next describe a class of ETC models [20] based on the gauge group $SU(5)_{ETC}$ which commutes with the standard-model (SM) gauge group. It breaks sequentially to a residual exact $\mathrm{SU}(2)_{TC}$ technicolor gauge symmetry, naturally producing a hierarchy of charged lepton and quark masses. Thus, $\mathrm{SU}(5)_{ETC} \to \mathrm{SU}(4)_{ETC}$ at a scale Λ_1, with the first-generation standard-model fermions separating from the other four components of the original 5 or $\bar{5}$ representations; then $\mathrm{SU}(4)_{ETC} \to \mathrm{SU}(3)_{ETC} \to \mathrm{SU}(2)_{TC}$ at still lower scales with the second- and third-generation fermions separating in the same way, leaving the technifermions. The models of Ref. [20] exhibit charged-current flavor mixing, intra-family mass splittings without excessive contributions to the difference $\rho - 1$ where $\rho = m_W^2/(m_Z^2 \cos^2\theta_W)$, a dynamical origin of CP-violating phases in the quark and lepton sectors, and a potential see-saw mechanism for light neutrinos without the presence of a grand unified scale [18].

The sequential breaking of the $\mathrm{SU}(5)_{ETC}$ to $\mathrm{SU}(2)_{TC}$ is driven by the condensation of standard-model-singlet fermions which are part of the models. At the scale Λ_{TC}, technifermion condensates break the electroweak symmetry. The models do not yet yield fully realistic fermion masses and mixings, and they have a small number of unacceptable Nambu-Goldstone bosons arising from spontaneously broken $U(1)$ global symmetries. Additional interactions at energies not far above Λ_1 must be invoked to give them sufficiently large masses. Nevertheless, the models share interesting generic features that are worth studying in their own right. The bilinear fermion condensates forming at each stage of ETC breaking have generically nonzero phases, providing a natural, dynamical source of CP violation. Below the electroweak symmetry breaking scale, the effective theory consists of the standard-model interactions, mass terms for the quarks, charged leptons, and neutrinos, and a tower of higher-dimension operators generated by the underlying ETC theory.

The full gauge group of the class of models is $G = \mathrm{SU}(5)_{ETC} \times \mathrm{SU}(2)_{HC} \times G_{SM}$. An additional gauge interaction, $SU(2)_{HC}$ (HC = hypercolor), is introduced along with $SU(5)_{ETC}$ and G_{SM}. Both the $SU(2)_{HC}$ and $SU(5)_{ETC}$ interactions become strong, triggering the sequential breaking pattern. The fermion content of one representative model[18] of the class is as follows, where the numbers indicate the transformation properties under $\mathrm{SU}(5)_{ETC} \times \mathrm{SU}(2)_{HC} \times \mathrm{SU}(3)_c \times \mathrm{SU}(2)_L$ and the subscript gives the weak hypercharge. There are quarks and techniquarks in the representations $(5,1,3,2)_{1/3,L}$, $(5,1,3,1)_{4/3,R}$, and $(5,1,3,1)_{-2/3,R}$, left-handed charged leptons and neutrinos and technileptons in $(5,1,1,2)_{-1,L}$, and right-handed charged leptons and technileptons in $(5,1,1,1)_{-2,R}$, together with SM-singlet fermions $\psi_{ij,R}$ and $\zeta_R^{ij,\alpha}$ transforming as $(\overline{10},1,1,1)_{0,R}$ and $(10,2,1,1)_{0,R}$. There are also fermions $\omega^{\alpha}_{p,R}$, $p=1,2$ transforming as $(1,2,1,1)_{0,R}$. The lepton number assigned to $\psi_{ij,R}$ is taken to be $L=1$ in order that Dirac terms $\bar{n}_{i,L}\psi_{jk,R}$ conserve lepton number. The full model is a chiral gauge theory, while the $\mathrm{SU}(2)_{HC}$ and $\mathrm{SU}(2)_{TC}$ sub-sectors are vectorial.

To analyze the stages of symmetry breaking, plausible preferred condensation channels are identified using a generalized most-attractive-channel (GMAC) approach. As the energy decreases from high values down to $E \sim \Lambda_1 \sim 10^3$ TeV, the coupling α_{ETC} become sufficiently large to produce condensation in the attractive channel $(\overline{10},1,1,1)_{0,R} \times (\overline{10},1,1,1)_{0,R} \to (5,1,1,1)_0$, breaking $\mathrm{SU}(5)_{ETC} \to \mathrm{SU}(4)_{ETC}$. With respect to the unbroken $\mathrm{SU}(4)_{ETC}$, we have $(\overline{10},1,1,1)_{0,R} = (\bar{4},1,1,1)_{0,R} + (\bar{6},1,1,1)_{0,R}$; we denote the $(\bar{4},1,1,1)_{0,R}$ as $\alpha_{1i,R} \equiv \psi_{1i,R}$ for $2 \le i \le 5$ and the $(\bar{6},1,1,1)_{0,R}$ as $\xi_{ij,R} \equiv \psi_{ij,R}$ for $2 \le i,j \le 5$. The associated condensate is then

$$\langle \epsilon^{1ijk\ell} \xi^T_{ij,R} C \xi_{k\ell,R} \rangle = 8 \langle \xi^T_{23,R} C \xi_{45,R} - \xi^T_{24,R} C \xi_{35,R} + \xi^T_{25,R} C \xi_{34,R} \rangle \ . \tag{28}$$

This condensate and the resultant dynamical Majorana mass terms for the six components of ξ in eq. (28) produce a violation of total lepton number by $|\Delta L| = 2$ units.

At lower scales, depending on relative strengths of couplings, different symmetry-breaking sequences can occur. In one sequence, leading to a neutrino seesaw, the $SU(4)_{ETC}$ interaction produces a condensation at a scale $\Lambda_{BHC} \le \Lambda_1$ (BHC = broken HC), in the channel $(6,2,1,1)_{0,R} \times (6,2,1,1)_{0,R} \to (1,3,1,1)_0$ with condensate $\langle \epsilon_{1ijk\ell} \zeta_R^{ij,1\ T} C \zeta_R^{k\ell,2} \rangle + (1 \leftrightarrow 2)$. This breaks $\mathrm{SU}(2)_{HC} \to \mathrm{U}(1)_{HC}$ and gives dynamical masses $\sim \Lambda_{BHC}$ to the twelve $\zeta_R^{ij,\alpha}$ fields involved. Let $\alpha = 1,2$ correspond to $Q_{HC} = \pm 1$ under the $\mathrm{U}(1)_{HC}$. This gives dynamical masses $\sim \Lambda_{BHC}$ to the twelve $\zeta_R^{ij,\alpha}$ fields involved. At the lower scale Λ_{23} there is condensation in the $4 \times 4 \to 6$ channel with condensate $\langle \epsilon_{\alpha\beta} \zeta_R^{12,\alpha\ T} C \zeta_R^{13,\beta} \rangle$, which then breaks $\mathrm{SU}(4)_{ETC} \to \mathrm{SU}(2)_{ETC}$ and is $\mathrm{U}(1)_{HC}$-invariant. The $\mathrm{U}(1)_{HC}$ interaction naturally also forms the condensates $\langle \zeta_R^{12,\alpha\ T} C \omega_{p,R}^{\beta} \rangle$ and $\langle \zeta_R^{13,\alpha\ T} C \omega_{p,R}^{\beta} \rangle$ with $p = 1, 2$, $\alpha \neq \beta$. Finally, technifermion condensation occurs at Λ_{TC}.

2.4 Neutrino Seesaw

The neutrino masses and mixing are calculated by diagonalizing the full mass matrix of neutrino-like states. The nonzero entries in this matrix arise in two different ways: (i) directly, as dynamical masses associated with various condensates, and (ii) via loop diagrams involving dynamical mass insertions on internal fermion lines. In the effective low energy theory at scales below Λ_{TC}, the light neutrinos include the three electroweak-doublet neutrinos $\nu_{\ell,L}$ with $\ell = e, \mu, \tau$, arising from a 5 of $\mathrm{SU}(5)_{ETC}$, and two electroweak-singlet neutrinos, $\alpha_{1j,R}$, $j = 2, 3$ arising from a $\overline{10}$ of $\mathrm{SU}(5)_{ETC}$. The fact that these transform as different representations leads to a strong suppression of Dirac neutrino mass terms, as in Ref. [17], because the relevant entries require mixing of ETC gauge bosons on internal lines of these loop diagrams. The Dirac neutrino mass matrix is denoted b_{ij}, with $i = 1, 2, 3$ and $j = 2, 3$. The dynamically generated technineutrino mass terms rely on the fact that we use $\mathrm{SU}(2)_{TC}$ so that 2 and $\bar{2}$ representations are equivalent. For the given symmetry breaking sequence we find

$$|b_{23}| = |b_{32}| \propto \frac{\Lambda_{TC}^4}{\Lambda_{23}^3} \tag{29}$$

with comparable values for b_{22} and b_{33}, and much smaller values for b_{1j}, $j = 2, 3$. The relevant Majorana mass matrix r_{ij} is defined by the operator product $\alpha_{1i,R}^T C r_{ij} \alpha_{1j,R}$, $i, j = 2, 3$. We find $r_{22} = r_{33} = 0$ and $r_{23} \propto \Lambda_{BHC}^n \Lambda_{23}^n / \Lambda_1^{2n-1}$ where n has the respective values 2 and 3 in a theory with walking all the way up to Λ_1 or just up to Λ_{23}. The Majorana mass r_{23} also requires ETC gauge boson mixing and is suppressed relative to the largest ETC scale.

The electroweak-doublet neutrinos are, to a very good approximation, linear combinations of three mass eigenstates having a normal hierarchy, with

$$m(\nu_j) \propto \frac{(b_{23} \pm b_{22})^2}{r_{23}} \tag{30}$$

where $j = 3, 2$ correspond to the $+, -$. Hence, for the case of a theory with walking up to Λ_1, we have $m(\nu_3) \propto (1+y)^2 \Lambda_{TC}^8 \Lambda_1^3/(\Lambda_{23}^8 \Lambda_{BHC}^2)$, where $y = b_{22}/b_{23}$. For fixed Λ_1 and Λ_{TC}, one can envision models with scales Λ_{BHC} and Λ_{23} that yield $m(\nu_3) \simeq 0.005$ eV, in agreement with the value extracted from the experimental measurement of $|\Delta m^2_{atm}|$ assuming a hierarchical neutrino mass spectrum. The prediction $m(\nu_2)/m(\nu_3) = (b_{23}-b_{22})^2/(b_{23}+b_{22})^2$ can account for the ratio of $m(\nu_2)/m(\nu_3) \sim 0.2$ extracted from the data on $|\Delta m^2_{atm}|$ and Δm^2_{sol}. The model naturally yields a large ν_μ, ν_τ mixing angle θ_{23}, although the θ_{12} mixing angle appears to be too small.

2.5 Quark Electric Dipole Moments

The quark dipole operators of interest here [21] are related to the dimension-3 mass terms of the up-type and down-type quarks, given in general by

$$\mathcal{L}_m = -\bar{f}_{L,j} M^{(f)}_{jk} f_{R,k} + h.c. \tag{31}$$

where f label the ETC eigenstates of the $Q = 2/3$ and $Q = -1/3$ quarks, respectively and the indices j, k label generation number. The mass matrices $M^{(f)}$ can be brought to real, positive diagonal form $M^{(q)}$ by the bi-unitary transformation

$$U^{(f)}_L M^{(f)} U^{(f)\ -1}_R = M^{(q)} \ . \tag{32}$$

The interaction eigenstates f are mapped to mass eigenstates q via

$$f_\chi = U^{(f)\ -1}_\chi q_\chi \ , \quad \chi = L, R \tag{33}$$

where $q = (u, c, t)$ and $q = (d, s, b)$ for $Q = 2/3$ and $Q = -1/3$ respectively. In this way, the Cabibbo-Kobayashi-Maskawa (CKM) quark mixing matrix entering the charged current interactions is generated:

$$V = U_L^{(u)} U_L^{(d)\dagger} . \tag{34}$$

Using the conventions of [34], we write

$$U_\chi^{(f)} = P_\alpha^{(f)\chi} R_{23}(\theta_{23}^{(f)\chi}) P_\delta^{(f)\chi\,*} R_{13}(\theta_{13}^{(f)\chi}) P_\delta^{(f)\chi} R(\theta_{12}^{(f)\chi}) P_\beta^{(f)\chi} \tag{35}$$

where $R_{mn}(\theta_{mn}^{(f)\chi})$ is the rotation through $\theta_{mn}^{(f)\chi}$ in the mn subspace, $P_\alpha^{(f)\chi}$ and $P_\beta^{(f)\chi}$ are given by

$$P_a^{(f)\chi} = \text{diag}(e^{ia_1^{(f)\chi}}, e^{ia_2^{(f)\chi}}, e^{ia_3^{(f)\chi}}), \quad a = \alpha, \beta \tag{36}$$

and $P_\delta^{(f)\chi} = \text{diag}(e^{i\delta^{(f)\chi}}, 1, 1)$. The mixing angles are typically small if the off-diagonal $M_{jk}^{(f)}$'s are more suppressed than the diagonal ones.

In ETC models, the off-diagonal entries of the quark mass matrices $M^{(f)}$ arise via mixing among the ETC gauge bosons. In the model of Ref. [20], employing a relatively conjugate ETC representation for the down-type quarks, this is true also of the diagonal elements [27]. We note that in this model, $M^{(u)}$ is hermitian, so that $U_L^{(u)} = U_R^{(u)}$, while $M^{(d)}$ is a more general complex matrix. For the phases, a complete theory should allow the computation of all the observable ones [26]. Here, I will describe bounds on combinations of mixing angles and phases. The mixing angles are then bounded with the assumption that the phases are generically of order unity.

2.5.1 Quark Dipole Matrices

The magnetic and electric dipole-moment matrices $D^{(f)}$ of the quarks appear in the dimension-5 operators

$$\mathcal{L}_{DM} = \frac{1}{2}\bar{f}_L D^{(f)} \sigma_{\mu\nu} f_R F_{em}^{\mu\nu} + h.c. \tag{37}$$

Similarly, the color (chromo-) magnetic and electric dipole-moment matrices $D_c^{(f)}$ enter the operators

$$\mathcal{L}_{CDM} = \frac{1}{2}\bar{f}_L T_a D_c^{(f)} \sigma_{\mu\nu} f_R G_a^{\mu\nu} + h.c., \tag{38}$$

where T_a and $G_a^{\mu\nu}$ denote a generator, and the field-strength tensor, for color $SU(3)_c$. (I include the color $SU(3)_c$ coupling g_s in the definition of $D_c^{(f)}$, just as it is usual to include the electromagnetic coupling e in the definition of the electric dipole moment.)

In the class of ETC models being considered, the ETC gauge bosons do not carry SM quantum numbers. Hence, in the respective diagrams that produce the dipole moment matrices and color dipole moment matrices, the photon and gluon couple only to the virtual (techni)fermions. Therefore,

$$D_c^{(f)} = \frac{g_s}{eQ_f} D^{(f)} \tag{39}$$

where $g_s = g_s(\mu)$ is evaluated at the appropriate scale μ.

Transforming to the mass-eigenstate basis,

$$\bar{f}_L D^{(f)} \sigma_{\mu\nu} f_R F_{em}^{\mu\nu} + h.c. = \bar{q}_L D^{(q)} \sigma_{\mu\nu} q_R F_{em}^{\mu\nu} + h.c., \tag{40}$$

where

$$D^{(q)} = U_L^{(f)} D^{(f)} U_R^{(f)\ -1} \ . \tag{41}$$

Analogously, $D_c^{(q)} = U_L^{(f)} D_c^{(f)} U_R^{(f)\ -1}$. Both $D^{(q)}$ and $D_c^{(q)}$ are independent of $P_\beta^{(f)\chi}$, $\chi = L, R$.

Decomposing $D^{(q)}$ into hermitian and anti-hermitian parts, $D^{(q)} = D_H^{(q)} + D_{AH}^{(q)}$, where $D_{H,AH}^{(q)} = (1/2)(D^{(q)} \pm D^{(q)\ \dagger})$, the dipole operator takes the form $(1/2)[\bar{q} D_H^{(q)} \sigma_{\mu\nu} q + \bar{q} D_{AH}^{(q)} \sigma_{\mu\nu} \gamma_5 q] F_{em}^{\mu\nu}$. Then the EDM of q_j is

$$d_{q_j} = -i D_{AH,jj}^{(q)} \ . \tag{42}$$

Defining $D_{c,AH}^{(q)}$ analogously, the chromo-EDM of q_j is

$$d_{c,q_j} = -i D_{c,AH,jj}^{(q)} \ . \tag{43}$$

In Refs. [20, 21], an estimate of the mass matrix $M^{(f)}$ and dipole matrix $\tilde{D}^{(f)}$ from an underlying ETC theory was described, and it was noted how they are related in the presence of the mixing of ETC interaction eigenstates to form mass eigenstates of the fermions and gauge bosons. An important result from that analysis is the relation

$$D_{jk}^{(f)} \simeq \frac{eM_{jk}^{(f)}}{\Lambda_{jk}^2} \tag{44}$$

where each Λ_{jk} is a dimensionful parameter of order the scale above which the $(j \leftrightarrow k)$ ETC propagator becomes soft. This structure reflects the fact that the leading dipole contribution contains two additional inverse factors of the ETC scale(s) relative to the mass, and that the corresponding integral is again sensitive to physics at the ETC scales. Λ_{jk} is no greater than $\min(\Lambda_j, \Lambda_k)$, and can be less. The fact that its (j,k) dependence is nontrivial implies that $D_{jk}^{(f)}$ is not, in general, $\propto M_{jk}^{(f)}$. It is therefore not diagonalized by the transformation that diagonalizes $M^{(f)}$; this transformation yields, instead, a non-diagonal and complex form for the dipole matrix $D^{(q)}$ of Eq. (41). Thus mixing has an important effect on quark dipole moments in ETC models.

For numerical estimates of the dipole matrix, I take the ETC breaking scales to be $\Lambda_1 \simeq 10^3$ TeV, $\Lambda_2 \simeq 10^2$ TeV, and $\Lambda_3 \simeq 4$ TeV. Since the ETC interactions are strong at these scales, there is resultant uncertainty in the calculations. I focus on the contribution to each element of the dipole matrix in $D^{(q)}$ in the mass-diagonal basis of third-family physics arising at the lowest ETC scale Λ_3. Then

$$D_{jk}^{(q)} \simeq \frac{eQ_q\, m_{q_3} F_{jk,3}^{(f)}}{\Lambda_3^2} \tag{45}$$

where $m_{q_3} = m_t, m_b$ for the u, d sectors respectively, and where $F_{jk,3}^{(f)}$ is a dimensionless function of the parameters in $U_L^{(f)}$ and $U_R^{(f)}$, of $O(1)$ for generic values of these parameters. Terms involving exchange of heavier ETC vector bosons with masses Λ_1 and Λ_2 are also present but are suppressed by the propagator mass ratios Λ_3^2/Λ_j^2, $j = 1, 2$. Part of this propagator suppression may be compensated for by the property that these other terms involve fewer small mixing angle factors, and hence they are not necessarily negligible; however, the Λ_3-scale terms should provide a rough measure of the overall ETC contributions.

The experimental constraints will demand small mixing angles, so I record here the small-angle form of the function $F^{(f)}_{jk,3}$. Using Eq. (41) and Eq. (35),

$$F^{(f)}_{33,3} \simeq 1 + \ldots \,, \tag{46}$$

$$F^{(f)}_{23,3} \simeq e^{i(\alpha_2^{(f)L} - \alpha_3^{(f)L})}\theta_{23}^{(f)L} + \ldots \,, \tag{47}$$

$$F^{(f)}_{32,3} \simeq e^{i(-\alpha_2^{(f)R} + \alpha_3^{(f)R})}\theta_{23}^{(f)R} + \ldots \,, \tag{48}$$

and, for $j, k \neq 3$,

$$F^{(f)}_{jk,3} \simeq \eta_{jk} e^{i[(\alpha_j^{(f)L} - \alpha_3^{(f)L}) - (\alpha_k^{(f)R} - \alpha_3^{(f)R})]}\theta_{j3}^{(f)L}\theta_{k3}^{(f)R} + \ldots \,, \tag{49}$$

where each expression is accurate up to a real coefficient of order unity, η_{jk} can contain $\delta^{(f)x}$ phases, and ... denote higher order terms.

2.5.2 The Strong CP Problem

Before considering the phenomenology of the dimension-5 operators, with their CP violating phases, I will discuss briefly the strong CP problem within the class of ETC models being considered. Can these models lead to the necessary condition

$$|\bar{\theta}| \leq 10^{-10} \tag{50}$$

where

$$\bar{\theta} = \theta - [arg(det(M^{(u)})) + arg(det(M^{(d)}))] \,, \tag{51}$$

with θ appearing via the topological term

$$\frac{\theta g_s^2}{32\pi^2} G_{a\;\mu\nu}\tilde{G}_a^{\mu\nu} \tag{52}$$

in the QCD Lagrangian?

The quark mass matrices $M^{(f)}$ ($f = u, d$) in the effective theory below Λ_{TC} are generated by integrating out short-distance physics at scales ranging from Λ_{TC} to the highest ETC scale

Λ_1. Above Λ_1, all fermions are massless. Some of the global chiral symmetries are anomalous, and hence are broken by instantons. The $F_{\mu\nu}\tilde{F}^{\mu\nu}$ terms associated with each (nonabelian) gauge interaction may be rotated away by chiral transformations through the relevant global anomalies. In particular, this renders $\theta = 0$ for SU(3)$_c$ in the underlying theory. In the effective low energy theory, then, we have $\bar{\theta} = -arg(det(M^{(u)})) - arg(det(M^{(d)}))$. If $\bar{\theta} \neq 0$, the rotation (2) to the real diagonal mass basis will, of course, regenerate the topological term through the anomaly. In the models of Refs. [17]–[20], $M^{(u)}$ is hermitian, so $\bar{\theta}$ resides in $M^{(d)}$.

More generally, the condition $|arg(det(M^{(u)}))+arg(det(M^{(d)}))| \leq 10^{-10}$ can be rewritten by letting

$$U_\chi^{(f)} = e^{i\phi_\chi^{(f)}} \mathcal{U}_\chi^{(f)} \in \mathrm{U}(3) \tag{53}$$

where $\mathcal{U}_\chi^{(f)} \in \mathrm{SU}(3)$ and $\chi = L, R$. Then $det(U_\chi^{(f)}) = e^{i\phi_\chi^{(f)}}$, and, from Eq. (32),

$$det(M^{(f)}) = e^{i(-\phi_L^{(f)}+\phi_R^{(f)})} det(M_{diag.}^{(f)}) \; . \tag{54}$$

Hence, the necessary condition reads

$$|\sum_{f=u,d} (-\phi_L^{(f)} + \phi_R^{(f)})| \leq 10^{-10} \; . \tag{55}$$

Other phases, entering the CKM matrix or the dipole operators, enter through the unimodular matrices $\mathcal{U}_\chi^{(f)}$, and are, in this sense, distinct from the strong CP phase.

In the notation of Eq. (35),

$$det(U_\chi^{(f)}) = exp[i \sum_j (\alpha_j^{(f)\chi} + \beta_j^{(f)\chi})] \; , \quad \chi = L, R \tag{56}$$

and, in terms of these quantities, Eq. (55) reads

$$\Big|\sum_{f=u,d} \sum_j \Big[-(\alpha_j^{(f)L} + \beta_j^{(f)L}) + (\alpha_j^{(f)R} + \beta_j^{(f)R})\Big]\Big| \leq 10^{-10} \; . \tag{57}$$

(independent of $\delta^{(f)\chi}$). So in terms of these parameters, only one linear combination (a sum over flavors j) of the CP-violating phases $\alpha_j^{(f)\chi}$ and $\beta_j^{(f)\chi}$ is tightly constrained.

Whether a resolution of the strong CP problem will emerge in the class of models considered here is not clear [26]. These models include certain Nambu-Goldstone bosons, one of which can be associated with a Peccei-Quinn dynamical relaxation of $\bar{\theta}$ to zero. But, as I noted earlier, these Nambu-Goldstone bosons must be given large masses by new interactions, eliminating this approach to solving the problem. Whatever the resolution of the strong CP problem turns out to be, an important point is that the relevant phase combination, entering in Eq. (55) or equivalently Eq. (57), involves a sum over the generational phases (labelled by j), whereas other CP-violating phase combinations which contribute to the quantities considered here (cf. Eqs. (47)-(49)) involve differences of generational phases (and $\delta^{(f)\chi}$).

2.5.3 Electric and Chromoelectric Dipole Moments

The diagonal electric and chromo-electric dipole moments for the up and down quarks arising from ETC interactions provide tight constraints on certain combinations of phase differences and mixing angles. These moments derive from the diagonal elements of $D^{(q)}_{jk}$ (Eq. 45). Recall that this expression is correct to leading order in ETC scales, depending explicitly on the inverse square of only the lowest ETC scale Λ_3. Thus,

$$d_u = \frac{e\,Q_u}{g_s} d_{c,u} \simeq \frac{e\,Q_u\,m_t\,Im(F^{(u)}_{11,3})}{\Lambda_3^2} \tag{58}$$

$$d_d = \frac{e\,Q_d}{g_s} d_{c,d} \simeq \frac{e\,Q_d\,m_b\,Im(F^{(d)}_{11,3})}{\Lambda_3^2}\ , \tag{59}$$

with the small-angle expressions for the $F^{(f)}_{11,3}$ given by Eq. (49) and involving flavor-differences of phases.

CP-violating electric dipole moments such as those of the neutron and certain atoms like ^{199}Hg receive contributions from the quark EDM's and the quark color EDM's. The latter lead to CP-violation in the hadronic wave function (the CEDM enters as a correction to t-channel gluon exchange between the bound quarks). There are also contributions from the CP-violating triple-gluon operator $c_{abc}G_{a\ \lambda\mu}\tilde{G}^{\mu\nu}_{b}G^{\lambda}_{c\ \nu}$ [38], and loop-induced CP-violating

$W^+W^-\gamma$ vertices [39]. I focus here on the quark-EDM and CEDM contributions. Because the direct SM contribution to these quantities is many orders of magnitude smaller than the expected ETC contribution, I neglect the SM contribution. And because the computation of the hadronic matrix elements, involving only first-generation quarks, is rather uncertain, I neglect the RG running of the ETC contribution.

2.5.4 Experimental Bounds

The current experimental upper bound on the neutron electric dipole moment d_n is $|d_n| < 6.3 \times 10^{-26}$ e-cm [40]. In setting constraints, I assume that there are no accidental cancellations between different contributions to the experimentally observable EDM's. For an estimate of the hadronic matrix element of the quark EDM operators, $\langle n|\bar{f}\sigma_{\mu\nu}\gamma_5 f|n\rangle$, $f = u, d$, various methods yield roughly similar results, which are comparable to the static quark model relation $d_n = (1/3)(4d_d - d_u)$. Using these estimates, I infer from the above limit on $|d_n|$ that $|Im(F^{(u)}_{11,3})| \leq 1 \times 10^{-6}$ and $|Im(F^{(d)}_{11,3})| \leq 3 \times 10^{-5}$.

The same quantities enter into the quark color EDM's and can, in principle, be bounded from their contributions to d_n. Since QCD is nonperturbative at the low energies relevant here, there are uncertainties in the proportionality factors connecting the CEDM's $d_{c,f}$, $f = u, d$, to d_n (e.g., [41]), and there is the related question of what value to use for the color gauge coupling g_s in Eq. (39) at such low energies. In general, from these CEDM's one obtains limits that are comparable to the ones above on $|Im(F^{(f)}_{11,3})|$, $f = u, d$.

Stringent limits on these quantities are obtained from upper bounds on EDM's of atoms, in particular from ^{199}Hg. Experimentally, $|d_{^{199}Hg}| < 2.1 \times 10^{-28}$ e-cm [42], which is about a factor of 50 smaller than the upper limit on $|d_n|$. From this I obtain the bounds

$$|Im(F^{(u)}_{11,3})| \leq 0.3 \times 10^{-7} \tag{60}$$

$$|Im(F^{(d)}_{11,3})| \leq 0.6 \times 10^{-6} \; . \tag{61}$$

Note that in the class of models Refs. [17]–[20], $Im(F^{(u)}_{11,3})$ (49) vanishes identically. This is because $M^{(u)}$ is hermitian in these models and therefore $U^{(u)}_L = U^{(u)}_R$.

From Eq. (49), we see that Eqs. (60) and (61) constrain a product of $\theta^{(f)L}_{13}\theta^{(f)R}_{13}$ times the imaginary part of a phase factor, for $f = u, d$. If the phase differences are of order unity, then

$$|\theta^{(d)L}_{13}\theta^{(d)R}_{13}| \le 0.6 \times 10^{-6} , \tag{62}$$

with a tighter bound on $|\theta^{(u)L}_{13}\theta^{(u)R}_{13}|$ if $M^{(u)}$ is not hermitian. The bound (62) is comparable to that on the corresponding product of charged-lepton mixing angles coming from the current limit on the electron EDM [21]. It may be satisfied with $|\theta^{(d)L}_{13}| \simeq |\theta^{(d)R}_{13}| \le 0.0008$. The corresponding angle θ_{13} in the CKM matrix V has the measured value $\simeq 0.004$, only a factor of five larger. Furthermore, the CKM θ_{13} contains terms proportional to products such as $\theta^{(u)L}_{12}\theta^{(d)L}_{23}$ which could be the dominant contribution and naturally be of order 0.004. Hence, although the bound (62) and its analogue for $f = u$ do imply quite small values for the indicated products of rotation angles, they can plausibly be satisfied in ETC models that successfully predict the CKM matrix.

Finally, an s-quark (chromo-) EDM can also contribute to the ^{199}Hg EDM. This contribution is more difficult to compute since there are no valence s quarks in the nucleon. It can be roughly estimated, leading to a bound on $|Im(F^{(d)}_{22,3})|$. This bound is much weaker than the above bound on $|Im(F^{(d)}_{11,3})|$, but it could have important implications for the mixing angles $\theta^{(d)L}_{23}$ and $\theta^{(d)R}_{23}$.

2.5.5 Summary

To summarize, quark EDM's and chromo-EDM's correspond to the diagonal elements of the matrices $D^{(q)}$ and $D^{(q)}_c$. The bounds on these quantities, in particular from the measured limit on the EDM of ^{199}Hg, lead to tight constraints on mixing angles and/or CP-violating phase differences in the individual sub-sectors $q = u, d$. These derive from the contributions to the elements of $D^{(q)}$ and $D_c^{(q)}$ from physics at the lowest ETC scales Λ_3 (e.g. Eq. (45)). If the phase differences are of order unity, then the down-type mixing angles are bounded as in Eq. (62), with an even tighter bound in the up sector if $M^{(u)}$ is not hermitian. The bound (62) is comparable to that on the corresponding product of charged-lepton mixing angles coming from the current limit on the electron EDM [21], and not difficult to satisfy in the class of theories being considered.

References

[1] T. Appelquist, A. G. Cohen, and M. Schmaltz, " A new constraint on Strongly Coupled Field Theories.".

[2] T. Appelquist, A. G. Cohen, M. Schmaltz, and R. Shrock, "New Constraints on Chiral Gauge Theories.".

[3] N. Seiberg, "Electric - Magnetic Duality in Supersymmetric NonAbelian Gauge Theories," *Nucl. Phys.* **B435** (1995) 129–146.

[4] D. Chen and R. D. Mawhinney, "Dependence of QCD Hadron Masses on the Number of Dynamical Quarks," *Nucl. Phys. Proc. Suppl.* **53** (1997) 216, http://xxx.lanl.gov/abs/hep-lat/9705029hep-lat/9705029.

[5] R. D. Mawhinney, "Evidence for Pronounced Quark Loop Effects in QCD," *Nucl. Phys. Proc. Suppl.* **60A** (1998) 306, http://xxx.lanl.gov/abs/hep-lat/9705031hep-lat/9705031.

[6] T. Appelquist, J. Terning, and L. C. R. Wijewardhana, "The Zero Temperature Chiral Phase Transition in SU(N) Gauge Theories," *Phys. Rev. Lett.* **77** (1996) 1214–1217, http://xxx.lanl.gov/abs/hep-ph/9602385hep-ph/9602385.

[7] T. Appelquist, A. Ratnaweera, J. Terning, and L. C. R. Wijewardhana, "The Phase Structure of an SU(N) Gauge Theory with N(F) Flavors," *Phys. Rev.* **D58** (1998) 105017, http://xxx.lanl.gov/abs/hep-ph/9806472hep-ph/9806472.

[8] J. Grundberg, T. H. Hansson, and U. Lindstrom, "Thermodynamics of N=1 Supersymmetric QCD," http://xxx.lanl.gov/abs/hep-th/9510045hep-th/9510045.

[9] P. Gerber and H. Leutwyler, "Hadrons Below the Chiral Phase Transition," *Nucl. Phys.* **B321** (1989) 387.

[10] T. Appelquist, D. Nash, and L. C. R. Wijewardhana, "Critical Behavior in (2+1)-Dimensional QED," *Phys. Rev. Lett.* **60** (1988) 2575.

[11] T. Appelquist, J. Terning, and L. C. R. Wijewardhana, "(2+1)-Dimensional QED and a Novel Phase Transition," *Phys. Rev. Lett.* **75** (1995) 2081–2084, http://xxx.lanl.gov/abs/hep-ph/9402320hep-ph/9402320.

[12] For a recent discussion of this puzzle, see "Phase Structure of Non-Compact QED3 and the Abelian Higgs Model", (with L.C.R. Wijewardhana), hep-ph/0403250, in "Quantum Theory and Symmetries", Proceedings of the 3rd International Symposium on Quantum Theory and Symmetries, University of Cincinnati, September 1-14, 2003 (World Scientific).

[13] I. Bars and S. Yankielowicz, "Composite Quarks and Leptons as Solutions of Anomaly Constraints," *Phys. Lett.* **101B** (1981) 159.

[14] E. Eichten, R. D. Peccei, J. Preskill, and D. Zeppenfeld, "Chiral Gauge Theories in the 1/N Expansion," *Nucl. Phys.* **B268** (1986) 161.

[15] M. E. Peskin, "Chiral Symmetry and Chiral Symmetry Breaking," in *Recent Advances in Field Theory and Statistical Mechanics*, J. B. Zuber and R. Stora, eds., vol. 39 of *Les Houches Summer School Proceedings*. North-Holland, 1984.

[16] B. Holdom, *Phys. Lett. B* **150** (1985) 301; K Yamawaki, M. Bando, K. Matumoto, *Phys. Rev. Lett.* **56** (1986) 1335; T. Appelquist, D. Karabali, L.C.R. Wijewardhana, *Phys. Rev. Lett.* **57** (1986) 957; T. Appelquist and L.C.R. Wijewardhana, *Phys. Rev. D* **35** (1987) 774.

[17] T. Appelquist, J. Terning, *Phys. Rev. D* **50** (1994) 2116.

[18] T. Appelquist, R. Shrock, *Phys. Lett.* **B548** (2002) 204.

[19] T. Appelquist, R. Shrock, *Phys. Rev. Lett.* **90** (2003) 201801.

[20] T. Appelquist, M. Piai, and R. Shrock, *Phys. Rev. D* **69** (2004) 015002.

[21] T. Appelquist, M. Piai, and R. Shrock, *Phys. Lett.***B595** (2004) 442 (hep-ph/0406032);

[22] For a recent review, see C. Hill and E. Simmons, hep-ph/0203079.

[23] T. Appelquist, J. Terning and L.C.R. Wijewardhana, hep-ph/9706328, Phys. Rev. Lett. 79, 2767 (1997).

[24] D.K. Hong and Ho-Ung Yee, hep-ph/0602177 and references contained therein.

[25] T. Appelquist, N. Christensen, M. Piai, and R. Shrock, *Phys. Rev. D* **70** (2004) 093010.

[26] For recent discussions of CP violation in the ETC context, see K. Lane, hep-ph/0106328, hep-ph/0202255.

[27] In this model, the ETC mixing arises from the coupling to certain condensing SM-singlet fermions, and takes the form of two-, three-, and four-point ETC gauge-boson vertices.

[28] See, e.g., the reviews A. Buras, hep-ph/9806471; M. Neubert, hep-ph/9809377; G. Isidori, hep-ph/0401079 and references therein to the extensive literature.

[29] A recent experimental review is M. Nakao, in *Proc. Lepton-Photon Symposium* 2003, hep-ex/0312041.

[30] F. Parodi, in *Proc. Fifth International Conference on Hyperons, Charm, and Beauty Hadrons*, ed. C. Kalman et al., *Nucl. Phys. B* (Proc. Suppl.) **115** (2003) 212; M. Battaglia et al., hep-ph/0304132; M. Ciuchini et al., hep-ph/0307195; http://cern.ch/ckm-workshop. With $\bar{\eta}$ and $\bar{\rho}$ defined via the rephasing-invariant ratio $-V_{ud}V_{ub}^{*}/(V_{cd}V_{cb}^{*}) = \bar{\rho}+i\bar{\eta}$, these fits give $\bar{\rho} = 0.18 \pm 0.05$ and $\bar{\eta} = 0.34 \pm 0.03$.

[31] K. Abe et al. (Belle Collab.), hep-ex/0308038.

[32] B. Aubert et al. (BABAR Collab.), hep-ex/0403035.

[33] J. Soares, *Nucl. Phys. B* **367** (1991) 575; A. Kagan, *Phys. Rev. D* **51** (1995) 6196; A. Kagan and M. Neubert, *Phys. Rev. D* **58** (1998) 094012.

[34] See http://pdg.lbl.gov (limits are 90 % CL).

[35] See, e.g., S. Bertolini, M. Fabbrichesi and J. O. Eeg, *Rev. Mod. Phys.* **72**, 65 (2000).

[36] Some reviews of EDM's are W. Bernreuther and M. Suzuki, *Rev. Mod. Phys.* **71** (1991) 1463; S. Barr, *Int. J. Mod. Phys. A* **8** (1993) 209; I. Khriplovich and S. Lamoreaux, *CP Violation without Strangeness* (Springer, New York, 1997).

[37] Some recent reviews of experimental searches for neutron and atomic EDM's are J. Pendlebury and E. Hinds, *Nucl. Instr. Meth. A* **440** (2000) 471; E. N. Fortson, P. Sandars, and S. Barr, *Phys. Today* (A.I.P., Jun. 2003); Y. Semertzidis, hep-ex/0401016.

[38] S. Weinberg, *Phys. Rev. Lett.* **63** (1998) 2333.

[39] T. Appelquist, G.-H. Wu, *Phys. Rev. D* **51** (1995) 240.

[40] P. Harris et al., *Phys. Rev. Lett.* **82** (1999) 904; see also S. Lamoreaux and R. Golub, *Phys. Rev. D* **61** (2000) 051301.

[41] T. Falk, K. Olive, M. Pospelov, R. Roiban, *Nucl. Phys. B* **560** (1999) 3.

[42] M. Romalis, W. Griffith, J. Jacobs, and E. N. Fortson, *Phys. Rev. Lett.* **86** (2001) 2505.

CHAIRMAN: T. APPELQUIST

Scientific Secretaries: S. Munir, L.Nilse

DISCUSSION I

- *Cifarelli:*

Can you point out again where the phase transition between the two monotonic behaviors of the f(T) functions comes from?

- *Appelquist:*

The phase transition comes from the strong interactions described by the gauge theory. But we have not proven monotonicity. We conjectured that a certain inequality holds. We defined an object f(T) from the free energy F(T) by taking out an overall scale factor T^4 and an appropriate constant so that f is precisely 1 for a free bosonic degree of freedom. If you plot the free energy as a function of T^4, f(0) is proportional to the slope as T approaches 0, and f(oo) is the same as T approaches oo. The conjectured inequality is f(0)<f(oo). This inequalty is true if f(T) is monotonic, but we conjecture it to be true even if f(T) is not monotonic.

One theory that I focused on was SU(N) QCD with N_f Dirac fermions. Assuming that f(0) < f(oo), it allowed me to test whether it is possible to be in the Goldstone mode, spontaneously breaking an $SU(N_f)$ X $SU(N_f)$ down to a diagonal $SU(N_f)$. There are then N_f^{2-1} broken generators and that is the number of Goldstone bosons. For example, if $N_f = 2$, as it is for up and down quarks, then there are just three Goldstone bosons. The interesting feature is that f(0) grows quadratically with N_f, whereas the underlying degree- of- freedom count is only linear. Hence you can only be in Goldstone phase if N_f is less than a certain number (=4N for SU(N)). That's the only conclusion you can draw from the inequality. With this phase assumed, I made the plot (slide- 14). I can compute f(T) as T- >oo, using the asymptotic freedom of the theory. Since it is a free field theory, I get a negative correction which vanishes as I take T to infinity. So it appears to be monotonic in that region. At low T, using chiral perturbation theory, I find that as I increase T from 0, the initial growth is proportional to T^4 log(1/T) (The log is a typical chiral log). Again, f(t) is monotonic at low T. But we have not proven that it is monotonic for all T.

If you are in the Goldstone phase at T=0, as in real QCD with N=3 and N_f = 2, and you heat it up, you go through a phase transition. This takes you into a phase that's described by the underlying quark and gluon degrees of freedom. The details of the actual phase transition lie beyond the realm of any kind of perturbation expansion that I know of. I can do perturbation theory as T- >oo and as T- >0, but the transition itself is something that requires some kind of a strong coupling approach, perhaps using the lattice.

- *Cifarelli:*

Can one try to test the monotinicity experimentally?

- *Appelquist:*

In principal, yes. It would be certainly very interesting to see if the curve is monotonic or not. It could have some kind of a peak, going up and then down again. That's a difficult thing to try to compute, but it will certainly be very important and interesting to measure it.

- *Korthales- Altes:*

You wrote the first two terms of the high T perturbative expansion, which produces a smooth curve. However higher order terms produce a very unstable series which you can not trust. How does that bear in your calculations?

- *Appelquist:*

This series converges very slowly; that's absolutely right. The temperature you have to go to, to ensure that it actually looks like this, depends very much on the order of magnitude of the g^3, the g^4 and the g^5 terms. The coefficients in these terms are typically large, meaning that you have to go to very high temperatures in order to see this clean kind of approach towards the Stefan- Boltzman law. There is evidence from the heavy ion collisions that we are so far nowhere near these kinds of temperatures.

- *Zichichi:*

But if this intermediate behavior is totally unreliable, you cannot draw any line in this region.

- *Appelquist:*

I can draw a line at high temperature. I know this confidently since it's an asymptotically free theory. When I go to infinite temperature, the coupling gets smaller and smaller, so g^2 will dominate the g^3 coupling and the higher orders. It's a question of how far out I have to go in order to make the coupling small enough in order for g^2 term to dominate all the others. That's the question of what the numerical coefficient is in front of the next- to- leading terms.

- *Zichichi:*

But the temperature could be higher than the Planck scale?

- *Appelquist:*

No. The only scale in this theory is the QCD confinement scale, which is a few hundred MeV. What you have to do is to get well above that scale. There is no Planck scale in the problem.

- *Zichichi:*

What makes you so sure that the scale that is reasonable is, say 10GeV, being generous?

- *Appelquist:*

There is no evidence for any intrinsic 10 GeV scale in real world QCD. In high energy calculations, for example in *e+e-* collisions at LEP, the coupling constant is small enough so that these corrections are well under control and you can easily see the logarithmic decrease of the coupling constant with increasing energy. In the case of a finite temperature calculation, there are coefficients that enter the free energy or the internal energy, which turn out to be rather large. But even there, I would guess that a temperature of 10 GeV would be sufficient to insure the convergence.

- *'t Hooft:*

There are quite a large set of theories that are asymptotically free but also regular in the infrared without even going beyond perturbation expansion. You do not need supersymmetry. Did you test these models?

- *Appelquist:*

Yes. The inequality is always satisfied in these cases. For example, a very interesting theory to test it in is a QCD- like theory with the parameters tuned to be very close to the loss of asymptotic freedom. The beta function then has a fixed point at a very small value of the coupling. As you go to large momentum, the coupling vanishes. As you come to low momentum, or low temperature, it is governed by a weak infrared fixed point. So I can calculate everything for the entire range of temperatures all the way from zero to infinity. The inequality is always satisfied in these cases.

- *'t Hooft:*

There are also theories which have the Higgs mechanism at the infrared domain and asymptotic freedom in the ultraviolet. Can we combine both?

- *Appelquist:*

We can combine both.

- *t Hooft:*

There's a large class of such theories.

- *Appelquist:*

Yes, there is a large class of such theories and we have applied the inequality to all that we know of. There are nice examples in 3 space- time dimensions.

- *Zichichi:*

Why 3 dimensions?

- *Appelquist:*

Mainly because it tests the inequality. There is also wide interest in 3 (2+1) dimensional theory in the condensed matter community, because there are condensed matter systems where the motion is essentially confined to planes. For example, anti-ferromagnets exist where the particle motions are confined to a plane. Let's take a quick look at a 3- d theory. Suppose I take the simplest theory that I could imagine, just electrodynamics in two space and 1 time dimensions and see what the inequality tells me here. Three dimensional theories are especially easy to work with in the UV because they are super- renormalizable. In the free energy F, we factor out an overall factor of T^3, and an appropriate constant, so that f(T) counts bosonic degrees of freedom precisely once. The photon has one such degree of freedom in 2+1 dimensions. Instead of 7/8, the Fermi- Dirac counting factor in 2+1 dimensions is 3/4. The coupling constant in this theory has dimensions of mass. So if I use four- component fermions, f(T)- >1+ (3/4)4N_f at temperatures very large compared to the coupling constant. If I go into the infrared, I can test the inequalty. There is a natural chiral symmetry associated with the set of massless fermions, just as there is chiral symmetry associated with massless fermions in 3+1 dimensions. Even if it is broken at T=0, the photon is still there; there is no confinement. If there is one 4- component fermion, the chiral symmetry is U(2). When it breaks producing a mass, a parity conserving mass, the U(2) breaks to U(2) X U(2), producing 2 Goldstone bosons. If there are N_f fermions, the chiral symmetry is U(2N_f), and it breaks to U(N_f) X U(N_f), producing 2N_f^2 broken generators, which is the count of Goldstone bosons. The inequality in this case is very simple, it says that you can be in the broken phase only if N_f<3/2. Since N_f is integer, this allows only one 4- component fermion. Any more than one and you can no longer be in the broken phase.

Some, in the condensed matter community, are very interested in this theory, QED3, as an effective theory to describe the behavior of planar anti- ferromagnets. A good description of the behavior of this theory is in fact in an abelian gauge field coupling to fermionic degrees of freedom, and the relevant number of fermions is 2. The argument leading to this is a little involved; it has to do with the D- wave structure of the wave functions and a lot of detailed condensed matter physics that I will not talk about. There is an argument that the breaking of chiral symmetry in the effective field theory corresponds to anti- ferromagnetic ordering in the planar anti- ferromagnet, so called Neel ordering. They believe they see it experimentally and therefore they would love to see chiral symmetry breaking in this theory for N=2. The inequality indicates that this will not happen. Much more work will be required to resolve this question.

- *Munir:*

Why are the second order corrections for the T- >0 limit of SU(N) independent of the gauge coupling?

- *Appelquist:*

Because there are no gauge interactions there. The gauge particles and fermions have all been confined. So instead of seeing a gauge coupling constant that you see at high temperatures, what you have at low T is an expansion in terms of T^2/f_{Pi}^2, the decay constant of the Goldstone boson. This is very typical of low energy calculations involving Goldstone Bosons. If you are doing a momentum space calculation instead of a finite temperature calculation then you would see powers of momentum over powers of f_{Pi}.

CHAIRMAN: T. APPELQUIST

Scientific Secretaries: K. Fukushima, M. Serbyn

DISCUSSION II

- Pinfold:

Can you tell us about dark matter candidates that arise from the technicolour model you presented today?

- Appelquist:

In the class of models I was talking about, you may remember that there was a new interaction called $SU(2)_{HC}$ and there were some particles ζ and ω which were both doublets under $SU(2)_{HC}$. This is a confining theory and $SU(2)_{HC}$ is exact and unbroken. Remember that $SU(2)_{HC}$ became strong at the second lowest and then the lowest ETC scale. So the lightest of these particles which are confined has a mass of order of the lowest ETC scale, which was, let's say, 3-4 TeV. These $SU(2)_{HC}$ particles form the lightest stable $SU(2)_{HC}$-singlet bound state. These bound states seem to me to be quite reasonable candidates for dark matter with a mass of 6-8 TeV. So this kind of theory has a natural dark matter candidate. But this would be extremely weakly interacting with other matter through very heavy ETC exchange.

- Pinfold

So, there is no chance to detect such dark matter particles in direct dark matter search experiments?

- Appelquist:

It looks so because they interact so weakly that it doesn't look like you'd see them very effectively in dark matter searches.

- Zichichi:

What experiment would you like most to be performed?

- Appelquist:

I would like to see careful measurements in the B_s system. I think they will tell us a lot about flavour changing neutral currents involving relatively heavy quarks. All neutral meson systems have a light eigenstate and a heavy eigenstate and a very interesting question is: what is the mass difference between two? We can estimate that based on the exchange of the ETC boson which induces this kind of transition. Our calculation for the B_s system results in some number times 10^{-7} MeV, well above the standard model.

- Giorgi:

The measurement of the mass difference in the B_s system is related to the measurement of B_s mixing. The same class of measurements are going on in $B_d \dashrightarrow s\, e^+ e^-$ where the crossing point s_0 is the crucial parameter to be measured. Could you comment?

- *Appelquist:*

This is the one that I have been particularly interested in because, if you have B_d system, the mixing is induced by a very heavy ETC boson. And so it is correspondingly weaker.

- *Zichichi:*

We still don't understand why flavours mix. It's a mystery.

- *Appelquist:*

At least we have the possibility of understanding this mystery in the framework, as I have described. Although we are not able to do convincing calculations and get answers exactly in the real world, this model naturally has the mixing angles. You don't just produce masses for ETC gauge bosons when symmetry is sequentially broken from SU(5) down to SU(2), but you also produce mixing among these ETC gauge bosons. This naturally leads to mixing among the quarks.

- *Giorgi:*

Since the LHC experiments can measure well the diagonal terms of the mass matrix, is the flavour experiment, in your view, necessary for understanding the off-diagonal terms?

- *Appelquist:*

I think that flavour experiments are very important. The B meson studies at CDF and D0 are definitely flavour experiments. In the case of the LHC experiments, of course, before you come to flavour experiments, you can forget about extended technicolour because just at the level of technicolour there is a lot to do at the LHC. There will be much new physics at the TeV scale. So the first order of business is just to see whether technicolour is right: is the Higgs boson a light particle, or some kind of a broad resonance, and are there then other kinds of particles associated with technicolour?

- *'t Hooft:*

It seems to me that there is a deep connection between the conjecture you formulated yesterday about the function $f(T)$ and your model building attempts today. Can you elucidate this?

- *Appelquist:*

There is indeed a connection. Imagine that the inequality bound on the number of (techni) fermions allowing the Nambu-Goldstone phase is nearly saturated. Then the infrared coupling is only slightly more strong than necessary to break chiral symmetry. This is the case for $SU(2)_{TC}$ with a conventional family of technifermons. The behaviour of the running coupling constant in this theory looks as though it goes to an infrared fixed point, but it is slightly too strong to stay in the Coulomb phase. But being close to the transition means the scale of the symmetry breaking will be relatively small compared with the intrinsic parameter that describes the running of this coupling constant. Therefore, the mass scale associated with the breaking and the bound states to be formed, which will define the technicolor scale, will be small compared with the intrinsic scale, which is of order the lowest ETC scale. This idea enhances condensates and allows you to keep this ETC scale and others high enough to suppress flavour changing neutral currents, while giving large enough fermion mass. That is the main connection.

- *Munir:*

For higher order corrections to the fermion masses, where would the cut-off scale lie so that the leading order masses do not get huge corrections?

-*Appelquist:*

The ETC interactions are relatively strong, so there is no suppression from the coupling constant. But, on the other hand, every ETC propagator is suppressed by the $1/M^2_{\mathrm{ETC}}$ and the integration cut-off scale is typically low compared with that.

- *'t Hooft:*

In this *S-T* plot, one of the constraints seems to be the Higgs mass, but we know it to be more than some number like 120GeV. Should this inner ellipse not be cut-off by this information?

- *Appelquist:*

The Higgs mass varying between 114-1000 GeV is indicated in the figure, which was taken from the LEP electro-weak working group. I do not think that the current lower bound of the Higgs mass takes you outside

A. DE RUJULA

"The Longest-lasting Mystery of High-Energy Physics: The Origin of Cosmic Rays"

The Content of the Lecture can be found in:

- *"A Solution of a hoary conundrum: The Origin and properties of cosmic rays"*, A. De Rujula, Nucl. Phys. Proc. Suppl. 165:93-102, 2007;
- *"What SWIFT has taught us about X-ray flashes and long-duration gamma-ray* bursts", A. De Rujula, submitted to Chin. J. Astron. Astrophys, arXiv:0707.0283, 2007.

CHAIRMAN: A. DE RÚJULA

Scientific Secretaries: V. Begun, A. Kazantsev

DISCUSSION

- *Faivre:*

What is a typical mass and size of a cannonball?

- *De Rújula:*

Cannonballs (CBs) are made of ordinary matter. Typically they have baryon number ~ 10^50 (which is one half the baryon number of planet Mercury) and their size increases to a `coasting' radius ~ 10^14 cm, some 7 times the size of the orbit of the Earth around the Sun. As it increases to its coating radius, the CB's density, which you can derive, is that of a very low density plasma.

- *Faivre:*

By fitting the afterglow profile, for which parameters you can get estimates?

- *De Rujula:*

You get fit values for the original Lorenz factor, mass and radius of cannonballs, the density of the interstellar medium, and the observer's angle relative to the CB's direction.

- *Faivre:*

Is the "Standard model" of your talk the fireball model? Could you describe it?

- *De Rujula:*

`Fireball' is a trademark for a continuously modified paradigm. The `fire' perhaps refers to the substance: electron-positron pairs with a tiny and fine-tuned `baryon-load'. Thin shells of this fancy substance are somehow accelerated to large energies. The model relies on a subsequent acceleration of the e+e- pairs, and the production of magnetic fields, in shocks created in shell-shell collisions, and the subsequent synchrotron radiation. That is the GRB, its afterglow is generally attributed to the collision of the shells with the interstellar medium, and the same type of radiation.

- *Faivre:*

At the beginning it was considered as being isotropic or …?

- *De Rújula:*

Originally the relativistically expanding shells were assumed to be spherical. Some GRBs that would, in that case, need an absurdly large energy imposed a change: the shells are now conical sections, reducing the energy requirement, as their radiation is forward collimated by the Doppler effect. In the most popular fireball scenario, a pulse of a gamma-ray burst reflects the collision between two shells moving in the same direction, the afterglow is due to the collision of the ensemble of shells with the static interstellar medium.

For the same reason that a car accident is less damaging for two cars moving in parallel than for the two of them hitting a wall, there is more energy in the GRB than in the afterglow, more than an order of magnitude for typical parameters. The observations are the opposite of this prediction. That is one reason why the model is known to be wrong from the day of its inception. Another reason is that its predicted synchrotron spectrum disagrees with that of the GRB's gamma rays. You asked me what the model is, not why it is artificially kept alive. That I do not know, or I do not want to tell.

- *Steinberg:*

What fraction of the available energy is released in CB's?

- *De Rújula:*

The average number of significant pulses in GRBs is observed to be ~ 5. That means 5 approaching cannonballs, or a total of 10 for the two opposite jets. These objects have a mass half of that of Mercury and they typically travel with a Lorenz factor of 1000. That corresponds to an energy of 2*10^51 ergs, which is about the same energy that supernova the SN transfers to its roughly spherical non-relativistic ejecta, nobody knows how, either. The total energy of a core-collapse supernova is about half the binding energy of a neutron star: 100 times bigger, and most of it carried away by few-MeV neutrinos.

- *Steinberg:*

What is the distribution of the number of CB's?

- *De Rújula:*

From 1 to ~12 observed pulses per GRB, with an average around 4 to 6, depending on how the `significant pulses' are defined.

- *Steinberg:*

And is there any sense of what the variance of this distribution would be?

- *De Rújula:*

All these things are fairly carefully studied. I just don't consider them sufficiently important to characterize them. The process of producing the pulses is fairly chaotic. And it obviously is telling you that some unstable chaotic dynamics ejects the CBs. We do not understand unstable chaotic systems. So I didn't try to memorize this variance.

- *Pinfold:*

What is the energy spectrum of the photons emitted by a GRB within the framework of the cannonball model – is it in any way characteristic? I'm particularly interested in the high energy end of the spectrum.

- *De Rújula:*

The spectrum dN/dE results from inverse Compton scattering and is predicted very simply. It is a power law 1/E changing at E ~ 250 keV to 1/E^(2.1). The exponent and `peak' energy are central predictions. The observed spectra are precisely the same. There may be some subprocess producing higher-energy gamma rays above the 1/E^2.1 tail. The

CB, after all, is also made of protons, not only electrons, and the protons are going to collide with the interstellar medium producing pi0's that will decay into gamma rays. A proton of 1TeV colliding with matter to produce pi0's results in gamma rays having about 1/10 to 1/20 of the energy of the proton: some 50 GeV. The expected flux is small, and you do not expect even harder gamma rays . Moreover, gamma rays (or neutrinos) of even higher energy made in this beam-dump-like way would be even more collimated than the lower GRB gammas. That makes the chances of detection meager.

- *Cifarelli:*

You said that in your model you can not predict time. Since you predict burst shape vs. time, which time are you talking about?

- *De Rujula:*

In the CB model we do predict the time-shape of the individual pulses of a GRB and its dependence on energy. What nobody can predict are the intervals between pulses, which are determined by instabilities of the accretion disk, as parts of it fall onto the compact object to make the CB's. And even in the cases that are fairly well understood, such as quasars and micro-quasars, the time of emission is random. The sequence of emission is like that earthquakes –it's a chaotic unstable process, whose properties are only vaguely understood statistically. One cannot predict when the next earthquake is going to happen. If we see one pulse we cannot predict when the next one is going to happen.

- *Sartorelli:*

Do you have in mind some experiment you would like to justify your model?

- *De Rújula:*

There are a lot of observations that are currently being done. Here is one interesting one: X-rays have been observed from GRBs, having specific line energies, normally attributed to highly-ionized iron. We say that the CB itself is made of protons and electrons and is emitting hydrogen lines. But if the object is moving at a Lorenz factor of 1000, the energy of these lines is Doppler boosted, in the forward direction, by ~ 1000. As it turns out, the boosted H-lines coincide, within the large observational uncertainties, with lines of Fe at rest. So our prediction is that, since the object is decelerating in the interstellar space, the Lorenz factor is going down and the lines should have an energy that is moving down with time, in a predicted way. Such data exists. It is not very precise, but one could try to extract this signal from it. It was never done. So it is not only a question of what new observations should be made. The question is also how the existing data should be analyzed. The current analyses reflect prejudices. Mainly wrong prejudices, in my opinion.

There's another important issue. The CB's are moving in the sky in a superluminal fashion. That means that a CB *appears* to move in the sky at a speed faster than light. There is one case, GRB030329, that was so close to us that it was possible to measure the superluminal separation between its two approaching CBs, which were first seen as two peaks in the gamma rays and two `shoulders' in the afterglow. We predicted the distance between the two objects as a function of time. It is measurable with very high precision radio-telescope arrays. It was measured. It agreed with our predictions. The observers claim the contrary. Didn't I say that the current analyses mainly reflect wrong prejudices?

Finally, the satellite SWIFT, specializing on X-ray signals, should teach us a lot.

- *Imponente:*

So the afterglow is fully compatible with the time scales of the cannonball evolution, isn't it?

- *De Rújula:*

Of course. If the theory was not in agreement with the data, why would I maintain it? Afterglows are fit with a few parameters. All measured afterglows of GRBs, of which we analyzed about 40, are fit by the theory simply, nicely and without any exceptions.

- *'t Hooft:*

You said that you expect that about 1% of gravitational energy is converted into gammas. Where did you get that number? I know that other theorists try to get larger percentage even closer to 100%.

- *De Rújula:*

Efficiency is a problem for those who you refer to, their true requirement is not close to 100%, but a lot bigger. Efficiency in the production of gammas is not even an issue in the CB model. Suppose you believe our explanation of the afterglow radiation. Then you can extract the typical mass and Lorentz factor of a CB. Since it is made of ordinary matter you know how many electrons it contains, ~10^(50). You let these electrons move through a known `light target': the early supernova luminosity scattered away from the radial direction by the semi-transparent medium created by the `winds' of the pre-supernova star. You compute the inverse Compton scattering of that light by the CB's electrons. You get answers, with no ad-hoc choices of any parameter, for the number of photons in a GRB's pulse, their energy distribution, their total energy etc etc. The answers are right. Where was the `efficiency'? It is the Thompson limit of the electron-photon cross section. The jets of CBs, for a reason that I do not know, carry 1% of the available gravitational energy release. The energy in the GRB gammas or in its afterglow light is orders of magnitude smaller. In our model a reason for that is that the radiation is very highly collimated.

- *Selvi:*

Together with the high energy gammas from pi0 decays we can expect also the emission of high energy neutrinos. Do you think they can be detectable at Earth? For example with large Cherenkov detector in ice, like ice- cube.

- *De Rújula:*

No, it is already very difficult to detect gamma rays of pi0 decay. It's for obvious reasons much more difficult to detect neutrinos of pi+ and pi- decays. I don't think that large underground detectors will see a burst of neutrinos in coincidence with the GRB. In this theory it is quite unlikely. Other `theories' are mainly wishful thinking, I contend.

- *Sartorelli:*

Do you have any ideas that the flux of these neutrinos going together with gammas would be?

- *De Rújula:*

Yes, we have made the estimates . Even in the most optimistic extremes, the flux of neutrinos is marginally observable and is even more forward-collimated than the observed gamma rays of a GRB. You are lucky if you detect a GRB, but you must be much luckier to catch its neutrinos, whose beam is even narrower. So we do not expect this signal.

- *Munir:*

Is there an effective method to find out that the two GRB pulses come from superposition of two cannonballs from one supernova or two different supernovas?

- *De Rújula:*

There is no need for such a method. There are only a few million supernovas in the visible Universe per day. The fraction of the total-sky solid angle covered by these supernovas or by an observed GRB is very small. The probability to find two GRB-emitting supernovas in the same direction and at times differing by seconds is totally negligible.

- *Imponente:*

Are there collimated fluxes of other types of particles besides gammas originated from a supernova?

- *De Rújula:*

Neither GRBs nor cosmic rays nor supernovae require any physics beyond the standard model (of particle physics). So the answer to your question is only affirmative for neutrinos, which we just discussed. The cannonballs made by supernovae also result in cosmic-ray electrons, protons and nuclei, scattered by the moving CBs onto a narrow forward cone of aperture its inverse Lorentz factor. But that original collimation is soon lost as the charged particles move through the galactic magnetic fields.

Updates in Local Supersymmetry and Its Spontaneous Breaking

Gianguido Dall'Agata[‡] **and Sergio Ferrara**[‡*]

Physics Department,
Theory Unit, CERN,
CH 1211, Geneva 23,
Switzerland

INFN,
Laboratori Nazionali di Frascati,
Italy

ABSTRACT

We give a basic review of some recent developments in local supersymmetry breaking in 4-dimensional effective theories coming from compactifications of string and M-theory in the presence of non-trivial form and geometrical fluxes.

Contribution to the Proceedings of the 43rd Erice International School of Subnuclear Physics "Towards New Milestones in our Quest to go Beyond the Standard Model" (29 August–7 September 2005)

Contents

1 Introduction

One of the main properties of spontaneously broken supersymmetric field theories is that the scalar potential which determines the classical spectrum of the theory is, to a large extent, controlled by supersymmetric Ward-identities [1, 2, 3]. More importantly, for theories with $N \geq 2$ supersymmetry the low-energy scalar potential is completely fixed by the gauging [3] of some invariance group of the theory up to certain "integration constants" usually called "Fayet–Iliopoulos terms" [4]. In $D = 4$, an effective action from superstring or M-theory compactifications can be described in full generality if a certain number of local supersymmetries are preserved in the process of compactification [5]. In the following paragraphs we will consider in some detail situations where this kind of possibility is realized. If N-extended local supersymmetries are preserved, the classical vacua of the theory are then selected by the scalar potential. Supergravity relates such potential to the mass deformations of the theory. The latter are in turn connected to the gauging procedure, if the theory has at least two unbroken supersymmetries.

Spontaneously broken theories of local supersymmetry have recently received a lot of attention, because some new compactifications of superstring and M-theory can be described in a rather general way and some of the new vacua are obtained by turning on different types of fluxes of some geometrical objects of the higher-dimensional theory, such as form fluxes or fluxes of the spin connection along the directions of the "internal manifold" of compactification.

The new geometrical data of the theory can then be related to the massive deformations of the lower-dimensional supergravity and it is often possible to derive general conclusions on the mass spectrum and on the broken and unbroken symmetries of the new vacua.

This review is organized as follows. In the first part (sections 2,3 and 4), we recall the so-called flow equations. These are essentially the content of the supersymmetric Ward-identities of the theory [1, 2, 3]. The second part (sections 5 and 6) deals with the construction of the $D = 4$ supergravity action with the geometrical data of the compactification procedure, in particular of the flux compactifications, which can be described by manifolds with "exceptional" G-structure [6], as an extension of manifolds with exceptional holonomy. The most famous examples of the latter are Calabi–Yau and G_2-holonomy compactifications with $N = 1$ or $N = 2$ residual supersymmetry in four dimensions. Finally, we briefly describe the black-hole attractor mechanism, which also falls in this class, as it is obtained by a flux potential. The latter is generated by the wrapping of a form (5-form in type IIB) on a product of a 2-sphere with a homology cycle of the internal manifold (a 3-cycle for Calabi–Yau 3-folds in type IIB).

2 Mass deformations of extended supergravity

The mass terms of a 4-dimensional extended supergravity lagrangian are related to the "Yukawa" couplings of the spin 3/2 spin 1/2 sector and to the scalar potential of the theory. They can be schematically written as [3]

$$\mathcal{L}_m(\det V)^{-1} = S_{AB}\bar{\psi}^A_\mu \sigma^{\mu\nu}\psi^B_\nu + i\,N^A_I \bar{\lambda}^I \gamma^\mu \psi_{\mu A} + M^{IJ}\bar{\lambda}_I\lambda_J + \text{h.c.} - V(q). \tag{2.1}$$

Here S_{AB}, N^A_I, M^{IJ} are the fermionic "mass matrices", which are related to the fermionic (supergravity) transformation laws as follows:

$$\delta\psi_{\mu A} = \ldots + \frac{1}{2}\gamma_\mu S_{AB}\epsilon^B\,, \tag{2.2}$$
$$\delta\lambda_I = \ldots + N^A_I \epsilon_A \tag{2.3}$$

and moreover, the scalar potential $V(q)$ is given by [1, 2]

$$\delta^A_B V(q) = -3\bar{S}^{AC}S_{CB} + N^A_I N^I_B, \tag{2.4}$$

where $(N^A_I)^* = N^I_A$, $S_{AB} = S_{BA} = (S^{AB})^*$. If the full R-symmetry U(N) of N-extended supergravity is manifest, then the index A labels the N-dimensional representation of U(N) and $\epsilon_A(x)$ is the local supersymmetry parameter of the transformations (as left-handed or Weyl spinor). Its complex conjugate $\epsilon^A(x)$ is a right-handed spinor. N-extended supergravity can be formulated in such a way that the R-symmetry H is a "local symmetry of the theory"; H is, in fact, part of the holonomy group of the σ-model geometry of the scalar manifold of the spin 0 components of the supersymmetric multiplets of the theory. For example in the $N = 1$ theory, U(1) is part of the U(n) holonomy of the Hodge–Kähler manifold [5, 7] of the n chiral (Wess–Zumino) multiplets. In $N = 2$, U(2) = U(1) × SU(2) is split in the U(1) part of the holonomy of the special geometry and the SU(2) part of the holonomy of the quaternionic geometry of the hypermultiplets [8, 9, 10]. Moreover, in local $N = 1, 2$ supergravity, there are associated U(1) and SU(2) bundles [7, 11] whose curvatures are in the same cohomology class as the Kähler form and as triplets of quaternionic almost complex structures of the base manifold. In simple terms, this means that in the supergravity transformations of the gravitino

$$\delta\psi_{\mu A} = D_\mu \epsilon_A + \ldots = \partial_\mu \epsilon_A + \frac{1}{4}\omega_\mu^{ab}\gamma_{ab}\epsilon_A + \omega_{\mu A}{}^B \epsilon_B + \ldots, \tag{2.5}$$

the covariant derivative contains a term $\omega_{\mu A}{}^B$ that is the U(N) connection of the principal U(N) bundle. For symmetric spaces the U(N) connection is identified with the H connection of the G/H coset space. For $N = 1$ and $N = 2$ this is the U(1) and U(2) connection of Hodge–Kähler and quaternionic geometry, respectively. All fermions of the theory are assigned to representations of H. Local H symmetry then determines their couplings to the non-linear σ-model of the scalar manifold. For rigid supersymmetry, the H connection trivializes (becomes flat) [11]. In such a case the Hodge–Kähler structure becomes simply Kähler and the quaternionic structure becomes hyper-Kähler.

For all $N > 2$ theories, the σ-model geometry is trivial (flat) in the rigid case and becomes (at least locally) a symmetric space in the case of local supersymmetry (supergravity). The local R-symmetry H becomes part of the isotropy group of the symmetric space G/H', where $H' = H \times H_{\text{Matter}}$. Here H_{Matter} is the part of the isotropy group related to matter multiplets, which can only exist for $N = 3, 4$. In such a case H_{Matter} =SU(n) for $N = 3$ and $H_{\text{Matter}} = SO(n)$ for $N = 4$, where n is the total number of matter (vector) multiplets.

An interesting case is the $N = 8$ theory. In the local $SU(8)$ covariant formulation of $N = 8$ supergravity, the rigid symmetry of the equations of motion is $E_{7(7)}$ but the maximal rigid symmetry of the lagrangian is SL(8, $\mathbb{R}$) [12, 13]. In different (duality related) formulations, the maximal rigid symmetry of the lagrangian may be a different (non-compact) subgroup of $E_{7(7)}$ and the manifest local symmetry may be a smaller subgroup of SU(8). For example, by reducing from five to four dimensions [14] the non-compact symmetry of the lagrangian is [14, 15] $E_{6(6)} \otimes \text{SO}(1,1) \rtimes T_{27}$ and the manifest local symmetry is USp(8). By performing a Scherk–Schwarz reduction from 11 to 4 [16] (and dualizing the antisymmetric tensors) the maximal non-compact symmetry of the action is [17, 18, 19] $GL(7) \rtimes N_{42}$, where N_{42} is a nilpotent algebra of dimension 42. In this case the manifest local symmetry is Spin(7). In this formulation the spin 3/2 spectrum transforms as the 8 of SU(8), USp(8) and Spin(7) respectively, but the covariant derivative still has the full SU(8) connection.

3 Electric-magnetic duality

The geometrical data of the σ-models of the scalar degrees of freedom are encoded in the Kähler (or special Kähler) geometry for $N = 1$ and $N = 2$ theories. Moreover, for $N \geq 2$ theories, where scalar and vectors lie in the same multiplet, an additional property must hold, since the non-linear σ-model must be consistently coupled to the vector fields and to the fermions. This requirement demands the existence [20, 21] in the theory of a "flat symplectic bundle" $\mathrm{Sp}(2n_V, \mathbb{R})$ where n_V is the number of vectors in the theory, and the base manifold is the scalar manifold of which the vectors are supersymmetric partners. This state of affairs implies that the $N = 2$ Hodge–Kähler geometry of vector multiplets is "special Kähler" [22, 11]. For $N > 2$ the symmetric spaces G/H' of extended supergravity are of a very special type. In particular the group G must have a "symplectic representation" R_V to which the vector field strengths $\mathcal{F}^\Lambda$ and their dual $\mathcal{G}_\Lambda$ belong. $\mathcal{G}_\Lambda$ is defined [20] through the geometrical equation $\mathcal{G}_\Lambda = 2\frac{\delta \mathcal{L}}{\delta \mathcal{F}^\Lambda}$, where the lagrangian $\mathcal{L} = \mathcal{L}(\mathcal{F}^\Lambda, \chi, \partial\chi)$ is function of the field strength $\mathcal{F}^\Lambda$ and of other fields (spin 0, 1/2, 3/2) in the theory.

Since the representation R_V is symplectic, the coset representative $L(q)$ is a symplectic matrix

$$L = \begin{pmatrix} \mathcal{U} & \mathcal{V} \\ \mathcal{W} & \mathcal{Z} \end{pmatrix}, \quad \begin{array}{l} \mathcal{U}^T\mathcal{W}, \mathcal{V}^T\mathcal{Z} \ \ \text{symmetric} \\ \mathcal{U}^T\mathcal{Z} - \mathcal{W}^T\mathcal{V} = 1. \end{array} \tag{3.1}$$

where $\mathcal{U}, \mathcal{V}, \mathcal{W}, \mathcal{Z}$ are $n_v \times n_v$ real matrices. For lagrangians quadratic in the field strength, the normalization of the "kinetic terms" of the vector fields is given through the formula

$$\mathcal{L} = \mathcal{F}^\Lambda \wedge \mathcal{G}_\Lambda, \tag{3.2}$$

with $\mathcal{G}_\Lambda = \mathrm{Re}\, \mathcal{N}_{\Lambda\Sigma}\, \mathcal{F}^\Sigma + \frac{1}{2}\mathrm{Im}\, \mathcal{N}_{\Lambda\Sigma}\, \mathcal{F}^{-\Sigma}$ or, in complex notation :

$$\mathcal{L} = \mathrm{Im}\, \mathcal{F}^{-\Lambda}\overline{\mathcal{N}}_{\Lambda\Sigma}\mathcal{F}^{-\Sigma} = \mathrm{Im}\, \mathcal{F}^{-\Lambda} \wedge \mathcal{G}^-_\Lambda. \tag{3.3}$$

The complex symmetric matrix $\mathcal{N}_{\Lambda\Sigma} = \mathcal{N}_{\Sigma\Lambda}$ with $\mathrm{Im}\, \mathcal{N} < 0$ is related to the coset representative through [20, 21]

$$\mathcal{N} = (\mathcal{W} - i\mathcal{Z})(\mathcal{U} - i\mathcal{V})^{-1} = hf^{-1}, \tag{3.4}$$

and it is subject to a fractional transformation

$$\mathcal{N}' = (C + D\mathcal{N})(A + B\mathcal{N})^{-1} \tag{3.5}$$

under a symplectic transformation of parameters A, B, C, D (A^TC, B^TD symmetric and $A^TD - C^TB = 1$) under which the vector field strengths rotate

$$\begin{pmatrix} \mathcal{F}' \\ \mathcal{G}' \end{pmatrix} = \begin{pmatrix} A & B \\ C & D \end{pmatrix} \begin{pmatrix} \mathcal{F} \\ \mathcal{G} \end{pmatrix}. \tag{3.6}$$

Here $S = \begin{pmatrix} A & B \\ C & D \end{pmatrix}$ satisfies $S^T \Omega S = \Omega$, where $\Omega = \begin{pmatrix} 0 & 1 \\ -1 & 0 \end{pmatrix}$. For infinitesimal (continuous) transformations where $S = 1 + s$, with $s = \begin{pmatrix} a & b \\ c & -a^T \end{pmatrix}$, $b = b^T$, $c = c^T$, and a arbitrary, $\mathcal{N}$ transforms as

$$\delta \mathcal{N} = -a^T \mathcal{N} - \mathcal{N} a + c - \mathcal{N} b \mathcal{N}. \tag{3.7}$$

We observe that under an S transformation the lagrangian $\mathcal{L}$ becomes

$$\delta \mathcal{L} = \frac{1}{4} \left(\mathcal{F} c \widetilde{\mathcal{F}} + \mathcal{G} b \widetilde{\mathcal{G}} \right). \tag{3.8}$$

If $b = c = 0$, $\mathcal{L}$ is invariant; if $b = 0$, $c \neq 0$ $\mathcal{L}$ is invariant up to a total derivative, while if $b \neq 0$, $\mathcal{L}$ is not invariant. In the absence of other fields, such as antisymmetric tensors, the requirement for the gauging is that $b = 0$. If $c \neq 0$ and it is a local symmetry, i.e. $c = c(x)$, the invariance can be restored up to a total derivative, provided certain conditions, discussed in the next sections, are fulfilled. If the $\mathrm{Sp}(2n_v, \mathbb{R})$ is an element of G then $\mathcal{N}'(\phi')$ since in this case the duality rotation is an isometry of the non-linear σ-model. For example, in $N = 8$, $D = 4$ supergravity, with $G = E_{7(7)}$, the $\mathrm{Sp}(56, \mathbb{R})$ duality group contain G according to the embedding $56 \to 56$. For a coordinate transformation of G/H $\phi \to \phi'$, which belongs to G,

$$\delta q^u = \xi^A k_A^u(q), \tag{3.9}$$

where $k_A^u(q)$ are the "Killing vectors" of G, $\xi^A = \xi^A(a, b, c)$. The group G has no linear action on the gauge potentials A_μ^Λ unless $b = 0$, in which case

$$\begin{aligned} \delta A_\mu^\Lambda &= a^\Lambda{}_\Sigma A_\mu^\Sigma, \\ \delta \mathcal{N} &= c - a^T \mathcal{N} - \mathcal{N} a. \end{aligned} \tag{3.10}$$

4 The gauging of duality rotations

For a duality rotation to be gauged, in a theory described only in terms of scalar and vector fields as bosonic degrees of freedom, the duality rotations must be of the lower triangular form ($b = 0$). Moreover the matrices $a_\Sigma^\Lambda(x)$, $c_{\Lambda\Sigma}(x)$ must fulfil additional restrictions, which come from the gauging, namely:

$$\begin{aligned} a_\Sigma^\Lambda(x) &= f_{\Sigma\Delta}^\Lambda \xi^\Delta(x), \\ c_{\Lambda\Sigma}(x) &= c_{\Lambda\Sigma,\Delta} \xi^\Delta(x), \end{aligned} \tag{4.1}$$

where $f_{\Sigma\Delta}^\Lambda$ are the structure constants of the gauge group $\mathcal{G} \subset G$ and $c_{\Lambda\Sigma,\Delta}$ are "constants" that must satisfy the following properties [22]

$$c_{(\Lambda\Sigma,\Delta)} = 0, \tag{4.2}$$

$$f_{\Gamma(\Pi}^\Delta c_{\Lambda)\Delta,\Sigma} - f_{\Sigma(\Pi}^\Delta c_{\Lambda)\Delta,\Gamma} + \frac{1}{2} f_{\Gamma\Sigma}^\Delta c_{\Pi\Lambda,\Delta} = 0. \tag{4.3}$$

The last equation demands that c be a non-trivial cocycle of the Lie algebra of $\mathcal{G}$. The first equation has to be fulfilled even if $\mathcal{G}$ is abelian, in which case $f^{\Lambda}_{\Sigma\Delta} = 0$. If $c_{\Lambda\Sigma,\Delta} \neq 0$, translation isometries of the θ term ($\theta F\widetilde{F}$) are gauged, which demands the presence of Wess–Zumino terms in the action (generalized Chern–Simons terms) widely discussed in the literature [22, 15, 23, 24, 25]. The complex $n_V \times n_V$ square matrix (f,h) determines most of the other couplings in the lagrangian and transformation laws. For generic N-extended theories $f^{\Lambda} = (f^{\Lambda}_{[AB]}, f^{\Lambda}_I)$, $h_{\Lambda} = (h_{\Lambda,[AB]}, h_{\Lambda,I})$, where the f^{Λ} appear in the supersymmetric variation of the gauge vector potentials [3]

$$\delta A^{\Lambda}_{\mu} = 2 f^{\Lambda}_{[AB]} \bar{\psi}^A_{\mu} \epsilon^B + i f^{\Lambda}_I \bar{\lambda}^{IA} \gamma_{\mu} \epsilon_A + \text{h.c.} \tag{4.4}$$

and the symplectic invariant combination

$$h^T \mathcal{F}^-_{\mu\nu} - f^T \mathcal{G}^-_{\mu\nu} \tag{4.5}$$

enters the supersymmetry transformation laws of the fermions [9, 26, 3]. For $N = 2$ special geometry [22, 11], where the scalar manifold is an arbitrary special Kähler manifold, these matrices are [26, 27, 11] $f^{\Lambda}_I = (\bar{f}^{\Lambda}_{\bar{\imath}}, L^{\Lambda})$, $h_{I\Lambda} = (\bar{f}_{\bar{\imath}\Lambda}, M_{\Lambda})$, $\Lambda = 0, \ldots, n_V$ in terms of the symplectic sector $(L^{\Lambda}, M_{\Lambda})$ and their covariant derivatives ($f^{\Lambda}_i = D_i L^{\Lambda}, h_{i\Lambda} = D_i M_{\Lambda}$).

We conclude this section by giving the scalar potential in terms of the geometrical data of the gauged supergravity. For $N = 1$ theory the gravitino mass $S_{AB} = L$ is given by

$$L = W(z) \mathrm{e}^{\frac{1}{2}K(z,\bar{z})}, \tag{4.6}$$

and the fermionic shifts N^A_I are $\{N^i = 2g^{i\bar{\jmath}} \nabla_{\bar{\jmath}} \bar{L}, D^{\Lambda} = 2(\mathrm{Im}\,\mathcal{N}_{\Lambda\Sigma})^{-1} P_{\Sigma}\}$.

The potential then is [28]

$$V = 4\left(-3L\bar{L} + g^{i\bar{\jmath}} \nabla_i L \nabla_{\bar{\jmath}} \bar{L} - \frac{1}{16} \mathrm{Im}\,\mathcal{N}_{\Lambda\Sigma} D^{\Lambda} D^{\Sigma}\right), \tag{4.7}$$

where P_{Λ} is the prepotential of the Killing vector $k^i_{\Lambda} = i g^{i\bar{\jmath}} \partial_{\bar{\jmath}} P_{\Lambda}$. The Fayet–Iliopoulos terms correspond to a possible shift $P_{\Lambda} \to P_{\Lambda} + \xi_{\Lambda}$ for abelian factors of the gauge group $\mathcal{G}$.

For $N = 2$ theories the geometrical data refer to both the special Kähler manifold of vector multiplets (m_V their number; note that $n_V = m_v + 1$) with metric $g_{i\bar{\jmath}}(z, \bar{z})$ and Killing vectors $k^i_{\Lambda}(z, \bar{z})$ and to the quaternionic manifold of hypermultiplets with metric $h_{uv}(q)$ and Killing vectors $k^u_{\Lambda}(q)$. Because of the special nature of these manifolds, there are additional geometrical quantities that enter the scalar potential, namely the matrix (f,h) defined before (for special geometry) and the three-holomorphic prepotentials [10, 11] P^x_{Λ} ($x = 1, 2, 3$) for the quaternionic manifold, defined through the relation [10, 11, 29]

$$2k^u_{\Lambda} \Omega^x_{uv} = \nabla_v P^x_{\Lambda}(q), \tag{4.8}$$

$$P^x_{\Lambda}(q) = \frac{1}{n_H} D_u K_{\Lambda v} \Omega^{xuv}, \tag{4.9}$$

where Ω is the SU(2) curvature of the quaternionic manifold with holonomy group SU(2) $\times$ USp(2 n_H), n_H being the number of hypermultiplets in the theory.

The gravitino mass S_{AB} and fermionic shifts N are given by [11]

$$S_{AB} = i\frac{1}{2}P^x_\Lambda L^\Lambda \sigma^x_{AB} = i\frac{1}{2}P_{AB\Lambda}L^\Lambda\sigma^x_{AB} = i\frac{1}{2}P_{AB}, \tag{4.10}$$

$$W^{iAB} = i\nabla^i P^{AB} + \epsilon^{AB}k^i, \tag{4.11}$$

$$N^A_\alpha = 2\mathcal{U}^A_{\alpha u}k^u_\Lambda \bar{L}^\Lambda, \tag{4.12}$$

where $\mathcal{U}^A_{\alpha u}$ is the vielbein of the quaternionic manifold ($A = 1, 2$, $\alpha = 1, \ldots, 2n_H$) and k^i ($i = 1, \ldots, m_V$) are the Killing vectors of the special manifold. Finally the scalar potential, through the formulae of section 2, reads [11]

$$V(z, \bar{z}, q) = \left(g_{i\bar{\jmath}}k^i_\Lambda k^{\bar{\jmath}}_\Sigma + 4h_{uv}k^u_\Lambda k^v_\Lambda\right)\bar{L}^\Lambda L^\Sigma + g^{i\bar{\jmath}}f^\Lambda_i f^\Sigma_{\bar{\jmath}}P^x_\Lambda P^x_\Sigma - 3\bar{L}^\Lambda L^\Sigma P^x_\Lambda P^x_\Sigma. \tag{4.13}$$

The last term is the contribution of the gravitino variation, the first and third terms are the contribution of the gaugino variation, while the second term is the contribution of the hyperino variation. Note that by virtue of the identity

$$g^{i\bar{\jmath}}f^\Lambda_i f^\Sigma_{\bar{\jmath}} = -\frac{1}{2}(\mathrm{Im}\,\mathcal{N}_{\Lambda\Sigma})^{-1} - \bar{L}^\Lambda L^\Sigma = U^{\Lambda\Sigma}, \tag{4.14}$$

the first and third terms can be rewritten as (also using $P_\Lambda L^\Lambda = P_\Lambda \bar{L}^\Lambda = 0$)

$$U^{\Lambda\Sigma}\left(P_\Lambda P_\Sigma + P^x_\Lambda P^x_\Sigma\right) = |\delta\lambda^A_i|^2. \tag{4.15}$$

Fayet–Iliopoulos terms are only possible in the P^x_Λ prepotentials $P^x_\Lambda \to P^x_\Lambda + \xi^x_\Lambda$ in the absence of hypermultiplets. Examples corresponding to "flat gaugings" in the context of $N = 2$ supergravity were found in [30], then realizing the no-scale structure previously found in the context of $N = 1$ supergravity [31, 32].

5 Supersymmetric vacua of supergravity

Let us now see how the gauged supergravities described above can be obtained by compactifying string or M-theory with non-trivial fluxes.

We are used to think about these compactifications as expansions around classical vacua of string or M-theory, and for this reason we will first see how the new tool of group structures on the tangent bundle can be used in order to classify and construct these vacua. However, we will also see that, in order to obtain effective 4-dimensional theories that can be described in terms of gauged supergravities, one needs only that the compactification manifold admit some globally defined spinor, not necessarily providing a vacuum of the theory. In this case the effective potential will be generically of the runaway type, but it can still be described by the structures given previously.

Let us then review how the group structure of the tangent bundle can be used to classify and construct supersymmetric vacua of supergravity theories (we follow in part [33, 34, 35, 36, 37]). We will start with a review of the ordinary geometric compactifications and the role of the holonomy group for their classification. We will then show how this gets modified in the presence of fluxes and especially how the classification in terms of holonomy groups is replaced by the more general classification in terms of structure groups.

5.1 Geometric compactifications

When solving the supersymmetry conditions, the minimal setup that can be assumed is to set to zero all the fields of the theory but the metric tensor. When only the metric is non-vanishing, all the supersymmetry transformations are trivially satisfied, with the exception of the gravitino one (at least when the theory does not contain higher-derivative terms). The latter becomes an equation imposing constraints on the geometry of the solution.

The supersymmetry requirement imposes that there exist a spinor η which is parallel with respect to the Levi–Civita connection

$$\delta\psi_m = \nabla_m \eta = 0\,. \tag{5.1}$$

By computing the integrability of this equation one obtains a set of constraints on the solution

$$[\nabla_m, \nabla_n]\,\eta = -\frac{1}{4}\,R_{mn}{}^{pq}\gamma_{pq}\,\eta = 0\,. \tag{5.2}$$

Equation (5.2) can be interpreted as the fact that certain combinations of the tangent space generators

$$T_{mn} \equiv \frac{1}{4}\,R_{mn}{}^{pq}\gamma_{pq} \tag{5.3}$$

annihilate η as well as the fact that the curvature is constrained. The first fact implies that the holonomy of the space is generically reduced. For what concerns the second comment we can see explicitly what happens by further contracting (5.2) with one gamma matrix:

$$\gamma^n\gamma^{pq}R_{mnpq}\,\eta = \gamma^{npq}R_{m[npq]}\,\eta - 2R_{mn}\gamma^n\eta = 0\,. \tag{5.4}$$

By using that $R_{m[npq]} = 0$ by construction for the Levi-Civita connection, one obtains that the solution must be Ricci-flat: $R_{mn} = 0$.

Let us now apply this procedure to M-theory, and look for spontaneous compactifications to 4-dimensional Minkowski space. We therefore assume that the 11-dimensional metric is a product of 4-dimensional Minkowski space-time with a compact internal manifold: $\mathcal{M}_{11} = M_4 \times Y_7$. According to our previous discussion, the possible Y_7 manifolds must be special-holonomy, Ricci-flat manifolds. These spaces have been classified by

Berger, and one can see that for obtaining minimal supersymmetry in four dimensions, one can compactify M-theory on G_2-manifolds, whose holonomy is contained in the group $G_2 \subset SO(7)$. Berger's classification applies more generally to all types of solutions that can be obtained for purely geometric compactifications of any (ungauged) supergravity theory preserving some supersymmetry.

Let us now explore the geometric consequences of equation (5.1) further.

The Levi–Civita connection appearing in (5.1) takes values in the tangent space group Spin$(1, d-1)$ and actually, since it preserves the metric $\nabla_M g_{NP} = 0$, in SO$(1, d-1)$. This is the structure group of the tangent bundle for a generic Riemannian manifold, i.e. the group required to patch the tangent bundle over the manifold. For the Levi–Civita connection it coincides with the holonomy group.

When solving the supersymmmetry conditions one reduces this group, as we have seen before, because the Killing spinors, solutions of (5.1), are annihilated by some of the generators of this group. This means that in order to patch together the tangent bundle over the manifold, only a subgroup $G \subset$SO$(1, d-1)$ is needed. This fact is equivalent to the Killing spinor being a singlet of G: it does not transform under an action of its generators. Clearly, in order to have a reduction of the structure group over the whole manifold, this invariant must be globally defined. This is granted for the solutions of (5.1). The Killing spinors are parallel with respect to the Levi–Civita connection and therefore any solution of (5.1) can be transported using this connection to any other point of the manifold (at least if this is simply connected). This means that once the supersymmmetry conditions are solved in one patch, the solution can be extended globally over the manifold. Moreover, since the metric is preserved by this connection, it is clear that the norm of all the invariants is preserved and the invariants whose norm is never vanishing are globally defined.

Following this discussion, any reduced group structure, and therefore any reduced holonomy group, implies the existence of a set of singlet tensor fields (or spinors) with respect to the structure group. For instance, for the compactifications of M-theory to four dimensions, we have seen that the holonomy group of the internal manifold is reduced to G_2. This means that there is one globally defined invariant spinor on the manifold, as there is only one singlet in the decomposition of the spinor representation of SO(7) in terms of G_2 representations: $\mathbf{8} \to \mathbf{1} + \mathbf{7}$. In the same way we can see that beside the metric tensor (that is a singlet of the general SO$(1, d)$ structure group of a Riemannian manifold), there is a 3-form field that is invariant under G_2: the co-associative form Φ. This can be seen again by taking the decomposition of the generic rank-3 antisymmetric tensor field with respect to G_2: $\mathbf{35} \to \mathbf{1}+\mathbf{7}+\mathbf{27}$. It is useful to notice that this tensor (and its dual) can be obtained by contractions of the invariant spinor with the 7-dimensional gamma matrices:

$$\Phi_{mnp} = -i\,\eta^\dagger \gamma_{mnp} \eta. \tag{5.5}$$

This also implies that Φ is parallel with respect to the Levi–Civita connection $\nabla\Phi = 0$, by applying (5.1). Moreover, this condition gives the known differential conditions on the 3-form Φ to define a G_2-manifold: it is closed and co-closed:

$$d\Phi = 0, \quad d \star \Phi = 0. \tag{5.6}$$

5.2 Adding fluxes

Adding fluxes obviously changes this situation. A simple consequence is given by the backreaction of these fluxes onto the internal geometry. The Einstein equation will now read

$$R_{mn} = F_m{}^{c_1 \ldots c_{p-1}} F_{nc_1 \ldots c_{p-1}} + \ldots , \tag{5.7}$$

where $F_{m_1 \ldots m_p}$ is some p-form whose vacuum expectation value (vev) is assumed to be different from zero. On the internal sector this generically implies that the space is no longer Ricci-flat. But we can say more. If we again look for supersymmetric configurations, the gravitino susy law tells us that there must exist a non-trivial spinor η which is covariantly constant with respect to a certain connection $\mathcal{D}$, which now contains also the flux information. Integrability of the gravitino supersymmetry equation now no longer implies a definite restriction of the holonomy group of the Levi–Civita connection and the manifold is not generically Ricci-flat anymore, not even with respect to the generalized connection $\mathcal{D}$.

However, we can still characterize the solutions in terms of the structure group of the tangent bundle, by using the properties of the new connection defined by $\mathcal{D}$. For a general value of the fluxes, this connection does not lie in Spin$(1, d-1)$, as not all the terms in the gravitino susy rule can be rewritten in terms of Levi–Civita-plus-torsion terms. Actually, it does not generically preserve the metric, defining the reduction of the structure group to SO$(1, d-1)$,

$$\mathcal{D}_M g_{NP} = Q_{MNP} \neq 0, \tag{5.8}$$

and the generic decomposition of the connection will contain an explicit dependence on these terms $\mathcal{D}_M = \nabla_M + \tau_M{}^{NP}\gamma_{NP} + \widetilde{Q}_M$, where τ is the contorsion tensor. As a generic consequence the spinors that solve the supersymmetry equations are no longer globally defined.

This *does not* imply that the solution does not preserve supersymmetry anymore. In order to preserve supersymmetry one just needs to solve the supersymmetry preserving conditions on every patch of the manifold, but the solutions need not be globally non-vanishing. A similar phenomenon appears when one is looking for solutions of the Killing vector equation on a manifold. Consider for instance $S^2 \equiv \frac{SO(3)}{SO(2)}$. This manifold has a local SO(3) symmetry group. This implies that in every patch one can define 3 non-vanishing vector fields that generate SO(3). At the same time, parallel transport of these fields changes their norm, as they are not parallel to the Levi–Civita connection and can therefore vanish at some point, as they actually do. Nonetheless, the group of isometries is SO(3) at each point on the manifold. The same phenomenon takes place for the supersymmetry equations and the Killing spinors, solving the supersymmetry conditions. In this case N spinor fields η satisfying $\mathcal{D}\eta = 0$ define an N-supersymmetric background, even if some of the η vanish at some point. However, for the special cases where $Q = 0$, the connection $\mathcal{D}$ lies in Spin$(1, d-1)$, and solutions to $\mathcal{D}\eta = 0$ can be parallel-transported using this connection and therefore become globally defined. In the first case, the structure group is reduced only locally, and one can use the techniques in [38] to classify the solutions. In the second case, the structure group is globally reduced and the intrinsic torsion completely specifies the supersymmetric solutions.

Before discussing how this classification can be achieved, let us give an example of the previous discussion for the spheres S^n. The spheres can be used in supergravity solutions to compactify some of the dimensions, preserving all supersymmetry. For instance, we can compactify M-theory on $AdS_4 \times S^7$, preserving all 32 supersymmetries, or 4-dimensional supergravity to $AdS_2 \times S^2$ describing the maximally supersymmetric horizon of extremal black-hole solutions. All the spheres admit a maximal number of Killing spinors satisfying the Killing equation:

$$\mathcal{D}_m \epsilon = (\nabla_m + i\Lambda\gamma_m)\,\epsilon = 0, \tag{5.9}$$

where Λ is related to the radius of the sphere and, for flux compactifications, it is also related to the expectation value of the form flux. For instance, the Fréund–Rubin ansatz leading to compactifications of M-theory on a 7-sphere assumes $F_{\mu\nu\rho\sigma} = -6\Lambda\epsilon_{\mu\nu\rho\sigma}$. Despite this, the structure group of the sphere S^n *is* not generically reduced to the identity, but for the special cases of $n = 3$ and $n = 7$. Let us consider first the case of S^2. The structure group of the tangent bundle over the manifold is SO(2) and it is not reduced by the existence of spinors solving (5.9). This happens because the $\mathcal{D}$ connection takes values in SO(3) and not simply in SO(2) and therefore the norm of these spinors is not preserved when parallel-transported by $\mathcal{D}$. Explicitly, the connection $\mathcal{D}$ can be put in the form of a standard SO(3) connection $\mathcal{D}_m = \nabla_m + \tau_m{}^{np}\gamma_{np}$, for in $d = 2$, $\gamma_m = i\epsilon_{mnp}\gamma^{np}$, with $m, n = 1, 2, 3$ and γ_3 is the matrix defining the spinor chirality. Also the vector fields that can be constructed from these spinors are therefore not globally defined and indeed they are precisely combinations of the Killing vectors defining the SO(3) isometry group.

The same relation between the γ-matrices applies in three dimensions. This now implies that the connection $\mathcal{D}$ is always in SO(3) (i.e. it preserves the metric) and therefore any solution to the sphere Killing equation can be extended globally. In this case the structure group is reduced to the identity and the manifold becomes parallelizable.

Intermediate cases exist, too. For instance S^5 has an $SU(2)$ structure, as there is one and only one globally defined vector on the manifold. Still, compactifications of type IIB supergravity on $AdS_5 \times S^5$ preserves 32 supersymmetries. The five-sphere is a

Sasaki–Einstein manifold [39]. All these manifolds have one globally defined vector.

A further example of what happens by turning on fluxes is given by string compactifications to 4 dimensions. When the internal manifold is a six-dimensional space, from the existence of a globally defined spinor we can define a 2–form $J_{mn} = \mathrm{i}\,\eta^{\dagger}\gamma_{mn}\gamma^{7}\eta$ and a 3–form $\Omega_{mnp} = \mathrm{i}\,\eta^{t}\gamma_{mnp}\,(1+\gamma^{7})\,\eta$. In the case of zero fluxes we have shown that $\nabla\eta = 0$ and therefore also $\nabla J = \nabla\Omega = 0$, i.e. J and Ω are preserved by the Levi–Civita connection. This further implies that $dJ = 0$ and $d\Omega = 0$, which means that J is the Kähler form and Ω is the holomorphic form of a Calabi–Yau 3-fold. When fluxes are turned on, we see that the internal manifold cannot be Kähler anymore, since $\nabla\eta \neq 0$ and therefore also dJ will be different from zero. The outcome is that the generic compactification in the presence of fluxes is given by a (warped) product of Minkowski space–time with a curved and non–complex internal manifold

$$\mathcal{M}_D = M_4 \times_w Y_{D-4}\,.$$

The existence of a G-structure does not a priori put any constraints on the possible holonomy groups. In particular, the failure of the holonomy of the Levi-Civita connection to reduce to $G \subset GL(n)$ is measured by the intrinsic torsion and this latter can be used to describe the G-structure. Given some G-invariant form η defining a G-structure, the derivative of η with respect to the Levi–Civita connection, $\nabla\eta$, can be decomposed into G-modules. The different types of G-structures are then specified by looking at which of these modules are present, if any. One first uses the fact that there is a connection $\nabla^{(T)}$ preserving the spinor

$$\nabla^{(T)}\eta = 0. \tag{5.10}$$

Then $\nabla^{(T)} - \nabla$ is a tensor that has values in $\Lambda^1 \otimes \Lambda^2$. Since $\Lambda^2 \cong \mathrm{so}(d) = g \oplus g^{\perp}$ where $g^{\perp}$ is the orthogonal complement of the Lie algebra g in $so(d)$, and η is invariant with respect to g, we conclude that $\nabla\eta = (\nabla - \nabla^{(T)})\eta$ can be identified with an element τ of $\Lambda^1 \otimes g^{\perp}$. Furthermore, this element is a function only of the particular G-structure, independent of the choice of $\nabla^{(T)}$ and it is in one-to-one correspondence with the intrinsic torsion. Explicitly [43], for a p-form η

$$\nabla_m \eta_{n_1 \ldots n_p} = -p\; \tau_{m\,[n_1}{}^{q}\, \eta_{|q|n_2 \ldots n_p]}\,, \tag{5.11}$$

where $\tau \in \Lambda^1 \otimes g^{\perp}$, m is the 1-form index and n, q label the 2-form $g^{\perp} \subset \Lambda^2$.

The search for supersymmetric solutions of string and supergravity theories demands the existence of spinors, which annihilate all the supersymmetry transformations. In geometrical terms, such spinors are parallel with respect to a generalized connection including the Levi–Civita connection and the fluxes contributions:

$$\nabla^{(T')}\eta = 0\,. \tag{5.12}$$

This gives us the possibility of understanding whether a certain solution preserves supersymmetry or not by analysing its group structure in terms of the intrinsic torsion. Indeed one needs its group-structure to be contained in those allowed by (5.12)

$$\nabla^{(T)} \subseteq \nabla^{(T')} . \tag{5.13}$$

It is therefore very important to express supersymmetry conditions as constraints on the intrinsic torsion and at the same time to classify the possible group-structures of the candidate solutions in terms of the irreducible components of the same intrinsic torsion. We have to stress, though, that this is still not sufficient to satisfy the equations of motion, unless one requires maximal supersymmetry. Only in certain favorable cases one can translate the extra conditions coming from such a requirement in terms of torsion classes.

Let us see once more how this works explicitly in the case of M-theory. The relevant components of the intrinsic torsion can be obtained by the product $\mathbf{7} \times \mathbf{7}$, where the first is the representation of Λ^1 and the second is the surviving representation of $g^\perp$. This latter follows from the decomposition of the adjoint of $SO(7)$ under G_2: $\mathbf{21} \to \mathbf{14} + \mathbf{7}$, where $\mathbf{14}$ is the adjoint of G_2. This produces four tensors $\tau \in \mathbf{1}+\mathbf{7}+\mathbf{14}+\mathbf{27}$ that specify completely the intrinsic torsion and hence the solutions. For computational purposes it is extremely important to observe that these components are completely determined by computing the exterior differential on the invariant forms:

$$d\Phi = X_1 \star \Phi + X_7 \wedge \Phi + X_{27}, \tag{5.14}$$

$$d \star \Phi = \frac{4}{3} X_7 \wedge \star\Phi + X_{14} \wedge \Phi. \tag{5.15}$$

A classification of the allowed tensors X_i in terms of the 4-form flux is given in [40] and the general solutions of M-theory preserving some supersymmetry are discussed in this fashion in [41, 42].

6 Data of the effective theory

In this last part of the review we will focus the discussion on the derivation and properties of the scalar potential induced by the fluxes on the effective 4-dimensional theory. We have seen how the backgrounds preserving some supersymmetry in the presence of fluxes can be classified and constructed, using the tool of the group structure of the tangent bundle. Of course, we are as well interested in effective theories coming from compactifications that do not satisfy the 10- or 11-dimensional equations of motion, but still give some supercharges in the effective theory that may be spontaneously broken. This requirement is related again to the existence of some globally defined spinors on the internal manifold (therefore implying a reduction of the group structure).

The idea is that, like the supersymmetry parameter of the previous section, all the spinor fields are reduced to effective 4-dimensional fields using these globally defined fields [44, 45]. For instance, the transverse part of the M-theory gravitino Ψ_μ can be split in the 4-dimensional part ψ_μ and the internal globally defined spinors η as $\Psi_\mu = \psi_\mu \otimes \eta + \psi_\mu^* \otimes \eta^*$. For an $N = 1$ compactification of M-theory, from the supersymmetry transformation of the 11-dimensional gravitino

$$\delta\Psi_A = \left\{ D_A[\omega] + \frac{1}{144} G_{BCDE} \left(\Gamma^{BCDE}{}_A - 8\Gamma^{CDE}\eta^B{}_A \right) \right\} \epsilon_{11}, \tag{6.1}$$

we can extract the supersymmetry transformation of the 4-dimensional field

$$\delta\psi_\mu = D_\mu \varepsilon_4 + \ldots + i e^{K/2} W \gamma_\mu \varepsilon_4^c, \tag{6.2}$$

by comparison of the various terms in the reduction after integration over the internal space. In (6.2) ϵ_4^c denotes the charge-conjugate spinor and we have emphasized only the superpotential term, neglecting in the dots the various terms with the vector fields.

The superpotential term (and in general all the shift terms) is then written as an integral over the internal space of the fluxes appearing in (6.1) and the non-vanishing contractions of the gamma matrices between the globally defined spinors. These contractions, as we saw in the previous section, describe the structure group of the internal manifold, and they are represented by globally defined forms.

6.1 Superpotentials from $p = 0, 2, 3, 4$-form fluxes

The first instance of such structure is the Gukov–Vafa–Witten superpotential for Calabi–Yau + orientifolds compactifications of type IIB string theory, with 3-form fluxes turned on [46, 47]. The superpotential reads

$$\mathcal{W} = \int G \wedge \Omega, \tag{6.3}$$

where the integral is performed over a Calabi–Yau 3-fold, $G = F_3 - \tau H_3$ is the sum of the Ramond–Ramond and Neveu–Schwarz 3-form, complexified with the axion/dilaton τ, and Ω is the holomorphic form of the Calabi–Yau. The Kähler potential describing the moduli space of the effective theory is the ordinary one for Calabi–Yau manifolds; it is the sum of the Kähler potential for the complex-structure deformations plus that for the Kähler deformations and the axion/dilaton:

$$K = K_V + K_J + K_\tau = -\log -i \int \Omega \wedge \bar{\Omega} - \log \frac{4}{3} \int J \wedge J \wedge J - \log -i(\tau - \bar{\tau}). \tag{6.4}$$

This can clearly be uplifted to F-theory, where the superpotential will have the same form as (6.3), but now G is a real 4-form and Ω is the holomorphic form of the 4-fold.

So far we discussed Calabi–Yau compactifications, where the moduli space is clearly defined and therefore also the Kähler potential of the effective theory. The first instance of a superpotential for manifolds that are no longer Calabi–Yau was obtained for the heterotic theory compactifications with fluxes preserving $N = 1$ supersymmetry in four dimensions [48] and for half-flat compactifications of type IIA supergravity [49]. For the Heterotic theory, the resulting superpotential now contains also information on the deviation of the internal manifold from the Calabi–Yau condition $dJ = 0 = d\Omega$. The expression is [50, 51]

$$\mathcal{W} = \int (H + idJ) \wedge \Omega, \tag{6.5}$$

where $H = dB + h$, h being the cohomologically non-trivial flux. It is clear that the last term in (6.5) is non-zero only for manifolds that are not Calabi–Yau, so that $dJ \neq 0$, but also generically $d\Omega \neq 0$, as can be obtained by integration by parts.

Since these papers appeared, there was a lot of study in order to understand better what the possible contributions to the effective theory potential are from the geometric part, in addition to the fluxes. We now have various derivations of the $N = 1$ superpotentials for general compactifications of type IIA/IIB and M-theory. These can be summarized by the following equations (see for instance [52, 45, 53, 54, 43]):

$$\begin{aligned} \mathcal{W}_{IIA} &= \int \mathrm{e}^{B+iJ} \wedge F - \int (H + idJ) \wedge \left(C + i\mathrm{e}^{-\phi}\mathrm{Re}\,\Omega\right) && (6.6)\\ \mathcal{W}_{IIB} &= \int \mathrm{e}^{iJ} \wedge d\Omega + \int G \wedge \Omega && (6.7)\\ \mathcal{W}_{M} &= \frac{1}{4}\int (C + i\Phi) \wedge \left[g + \frac{1}{2}d\,(C + i\Phi)\right]. && (6.8) \end{aligned}$$

The IIA and IIB expressions contain the fluxes ($F = m_0 + F_2 + F_4$, $G = G_3$) and the globally defined forms defining an SU(3) structure (J, Ω). The M-theory superpotential contains on the other hand the G_2-structure Φ. It is more difficult to give an expression for the Kähler potential governing the geometry of the moduli spaces of these compactifications. It is indeed less clear which are the light fields one should really consider in these compactifications and also when the truncation to these fields is consistent. Recently some light was shed on this issue too [45]. It has been argued that the Kähler potential for SU(3)-structure compactifications is perfectly analogous to the one for Calabi–Yau manifolds (6.4). Analogously, the M-theory Kähler potential is formally the same as the one of G_2-holonomy compactifications. It was explicitly obtained for G_2-structure manifolds, for instance in [54], and it reads

$$K = -3\log\frac{1}{7}\int \Phi \wedge \star\Phi. \tag{6.9}$$

In paper [45], a general discussion of the effective theory of a compactification of type II supergravity on a manifold with SU(3) structure, before considering any addition of orbifolds or orientifold projections, was also presented. This theory preserves $N = 2$ supersymmetry at the level of the lagrangian, and theorefore also for this type of compactification the potential can be completely specified in terms of geometrical objects through the quaternionic prepotentials P^x. These were found to be

$$\begin{aligned} P^1 &= -2\mathrm{e}^{\frac{1}{2}K_V+\phi_4}\int \mathrm{e}^{-B-iJ}\wedge d\,\mathrm{Re}\;\Omega, \\ P^2 &= -2\mathrm{e}^{\frac{1}{2}K_V+\phi_4}\int \mathrm{e}^{-B-iJ}\wedge d\,\mathrm{Im}\;\Omega, \\ P^3 &= \frac{1}{\sqrt{2}}\mathrm{e}^{2\phi_4}\int \mathrm{e}^{-B-iJ}\wedge G_{IIA} \end{aligned} \tag{6.10}$$

for type IIA compactifications, and

$$\begin{aligned} P^1 &= -2\mathrm{e}^{\frac{1}{2}K_J+\phi_4}\int \Omega\wedge d\,\mathrm{Re}\;\mathrm{e}^{-B-iJ}, \\ P^2 &= 2\mathrm{e}^{\frac{1}{2}K_J+\phi_4}\int \Omega\wedge d\,\mathrm{Im}\;\mathrm{e}^{-B-iJ}, \\ P^3 &= -\frac{1}{\sqrt{2}}\mathrm{e}^{2\phi_4}\int \Omega\wedge G_{IIB} \end{aligned} \tag{6.11}$$

for IIB compactifications, where ϕ_4 is the 4-dimensional dilaton. This result therefore includes the special cases of $N = 2$ Calabi–Yau compactifications with fluxes analysed in [55, 56, 57, 58].

6.2 The attractor potential: superpotential from 5-form fluxes

Another potential, which is also related to a holomorphic "superpotential", is the ($N = 2$) black-hole potential (for an asymptotically flat extremal black-hole) whose extrema control the formula for the black-hole entropy/area through the relation [64, 59]

$$S_{BH} = \frac{A_{\text{Hor}}}{4} = \pi V_{BH}(p, q, \phi_{fix})\Big|_{\frac{\partial V}{\partial \phi}=0}, \tag{6.12}$$

where V is given in terms of the $\mathrm{Sp}(2n, \mathbb{R})$ vector $(p^\Lambda, q_\Lambda) = \mathcal{P}$ and of the symplectic real matrix

$$\mathcal{M} = \begin{pmatrix} A & B \\ C & D \end{pmatrix}, \tag{6.13}$$

where

$$A = \mathrm{Im}\,\mathcal{N} + \mathrm{Re}\,\mathcal{N}\mathcal{N}^{-1}\mathrm{Re}\,\mathcal{N} \tag{6.14}$$

$$B = -\mathrm{Re}\,\mathcal{N}\mathrm{Im}\,\mathcal{N}^{-1} \tag{6.15}$$

$$C = -\mathrm{Im}\,\mathcal{N}^{-1}\mathrm{Re}\,\mathcal{N} \tag{6.16}$$

$$D = \mathrm{Im}\,\mathcal{N}^{-1} \tag{6.17}$$

through the formula [26, 60, 59]

$$V_{BH} = -\frac{1}{2}\mathcal{P}^T \mathcal{M} \mathcal{P}. \tag{6.18}$$

Note that $\mathcal{M}$ is symmetric and negative-definite whenever the complex symmetric matrix $\mathcal{N}$ has Im $\mathcal{N} < 0$. The matrices Im $\mathcal{N}$, Re $\mathcal{N}$ are the "normalizations" of the $F \wedge F$ and $F \wedge \widetilde{F}$ terms in the action, where F^Λ are the vector field strengths. For N-extended supergravity theories the potential V is given by [21][1]

$$V_{BH} = \frac{1}{2}|Z^{AB}|^2 + |Z_I|^2, \tag{6.19}$$

where $Z^{AB} = -Z^{BA}$ is the "central-charge" matrix of the supersymmetry algebra [61]

$$\{Q^A_\alpha, Q^B_\beta\} = \epsilon_{\alpha\beta} Z^{AB}(\phi) \tag{6.20}$$

and Z_I are the "matter" charges of the "vector multiplets". For $N = 2$, $Z^{AB} = Z\epsilon^{AB}$, $Z_I = D_i Z \equiv \left(\partial_i + \frac{1}{2}K_i\right) Z$ (recall that $\bar{D}_{\bar{\imath}} Z \equiv \left(\partial_{\bar{\imath}} - \frac{1}{2}K_{\bar{\imath}}\right) Z = 0$) and the black-hole potential simplifies to [60, 26]

$$V_{BH} = Z\bar{Z} + g^{i\bar{\jmath}} D_i Z \bar{D}_{\bar{\jmath}} \bar{Z} = |Z|^2 + |D_i Z|^2 \tag{6.21}$$

in terms of the (covariantly holomorphic) central charge [27]

$$Z = \mathrm{e}^{K/2}(X^\Lambda e_\Lambda - F_\Lambda p^\Lambda). \tag{6.22}$$

Using the special geometry identity [62, 63]

$$\mathcal{P} = 2\mathrm{Re}\,\left(-i\bar{V}Z - iD_i V g^{i\bar{\jmath}} D_{\bar{\jmath}} \bar{Z}\right) \tag{6.23}$$

in terms of the symplectic vectors [27, 26]

$$\mathcal{P} = \begin{pmatrix} m^\Lambda \\ e_\Lambda \end{pmatrix}, \quad V = \begin{pmatrix} L^\Lambda = \mathrm{e}^{K/2} X^\Lambda \\ M_\Lambda = \mathrm{e}^{K/2} F_\Lambda \end{pmatrix}, \tag{6.24}$$

the "attractor equations" $D_i V_{BH} = 0$ can be rewritten as

$$\mathcal{P} = 2\mathrm{Re}\,\left(-i\bar{V}Z + \frac{1}{2\bar{Z}} D_i V D_j Z D_k Z C_{\bar{\imath}\bar{\jmath}\bar{k}} g^{i\bar{\imath}} g^{j\bar{\jmath}} g^{k\bar{k}}\right). \tag{6.25}$$

For supersymmetric attractors ($D_i Z = 0$), this formula simply becomes [64, 65, 60]

$$\mathcal{P} = 2\mathrm{Re}\,\left(-i\bar{V}Z\right). \tag{6.26}$$

[1] An exception is $N = 6$ supergravity, where an additional singlet charge Z appears in the supergravity multiplet other than Z^{AB}.

In terms of the Calabi–Yau data Z, D_iZ are given by the "fluxes" of the IIB 5-form along the holomorphic cycles of the 3-form cohomology [26, 66]

$$Z = \int_{CY \times S^2} F_5 \wedge \Omega \tag{6.27}$$

and

$$D_iZ = \int_{CY \times S^2} F_5 \wedge D_i\Omega. \tag{6.28}$$

For supersymmetric attractors F_5 has only a (3,0) component. The 10-dimensional space is the product of $AdS_2 \times S^2 \times CY_{pq}$ where CY_{pq} is an "attractor variety" as discussed in [66]. For non-supersymmetric attractors ($D_iZ \neq 0$), F_5 has also a (2,1) component; these have recently received some attention [63, 67, 68, 69, 70].

Acknowledgments

We would like to thank R. Russo and R. Kallosh for useful discussions. The work of S.F. has been supported in part by the European Community Human Potential Program under contract MRTN-CT-2004-005104 "Constituents, fundamental forces and symmetries of the universe", in association with INFN Frascati National Laboratories and by D.O.E. grant DE-FG03-91ER40662, Task C.

References

[1] S. Ferrara and L. Maiani, *An introduction to supersymmetry breaking in extended supergravity*, CERN-TH-4232/85, *Based on lectures given at SILARG V, 5th Latin-American Symp. on Relativity and Gravitation, Bariloche, Argentina, Jan 1985*

[2] S. Cecotti, L. Girardello and M. Porrati, *Constraints On Partial Superhiggs*, Nucl. Phys. B **268**, 295 (1986).

[3] See for instance R. D'Auria and S. Ferrara, *On fermion masses, gradient flows and potential in supersymmetric theories*, JHEP **0105**, 034 (2001) [arXiv:hep-th/0103153].

[4] P. Fayet and J. Iliopoulos, *Spontaneously Broken Supergauge Symmetries And Goldstone Spinors*, Phys. Lett. B **51**, 461 (1974).

[5] B. Zumino, *Supersymmetry And Kahler Manifolds*, Phys. Lett. B **87**, 203 (1979).

[6] J. P. Gauntlett, D. Martelli, S. Pakis and D. Waldram, *G-structures and wrapped NS5-branes*, Commun. Math. Phys. **247** (2004) 421 [arXiv:hep-th/0205050].

[7] E. Witten and J. Bagger, *Quantization Of Newton's Constant In Certain Supergravity Theories*, Phys. Lett. B **115**, 202 (1982).

[8] J. Bagger and E. Witten, *Matter Couplings In N=2 Supergravity* , Nucl. Phys. B **222**, 1 (1983).

[9] A. Strominger, *Special Geometry*, Commun. Math. Phys. **133**, 163 (1990).

[10] R. D'Auria, S. Ferrara and P. Fré, *Special And Quaternionic Isometries: General Couplings In N=2 Supergravity And The Scalar Potential*, Nucl. Phys. B **359**, 705 (1991).

[11] L. Andrianopoli, M. Bertolini, A. Ceresole, R. D'Auria, S. Ferrara, P. Fré and T. Magri, *N = 2 supergravity and N = 2 super Yang-Mills theory on general scalar manifolds: Symplectic covariance, gaugings and the momentum map*, J. Geom. Phys. **23**, 111 (1997) [arXiv:hep-th/9605032].

[12] C. M. Hull and N. P. Warner, *The Structure Of The Gauged N=8 Supergravity Theories*, Nucl. Phys. B **253**, 650 (1985).

[13] F. Cordaro, P. Fré, L. Gualtieri, P. Termonia and M. Trigiante, *N = 8 gaugings revisited: An exhaustive classification*, Nucl. Phys. B **532**, 245 (1998) [arXiv:hep-th/9804056].

[14] E. Cremmer, J. Scherk and J. H. Schwarz, *Spontaneously Broken N=8 Supergravity*, Phys. Lett. B **84**, 83 (1979).

[15] L. Andrianopoli, R. D'Auria, S. Ferrara and M. A. Lledó, *Gauging of flat groups in four dimensional supergravity*, JHEP **0207**, 010 (2002) [arXiv:hep-th/0203206].
L. Andrianopoli, R. D'Auria, S. Ferrara and M. A. Lledó, *Duality and spontaneously broken supergravity in flat backgrounds*, Nucl. Phys. B **640** (2002) 63 [arXiv:hep-th/0204145].

[16] J. Scherk and J. H. Schwarz, *How To Get Masses From Extra Dimensions*, Nucl. Phys. B **153**, 61 (1979).

[17] G. Dall'Agata and S. Ferrara, *Gauged supergravity algebras from twisted tori compactifications with fluxes*, Nucl. Phys. B **717**, 223 (2005) [arXiv:hep-th/0502066].

[18] L. Andrianopoli, M. A. Lledó and M. Trigiante, JHEP **0505** (2005) 051 [arXiv:hep-th/0502083].

[19] R. D'Auria, S. Ferrara and M. Trigiante, *$E_{7(7)}$ symmetry and dual gauge algebra of M-theory on a twisted seven-torus*, Nucl. Phys. B **732**, 389 (2006) [arXiv:hep-th/0504108].

[20] M. K. Gaillard and B. Zumino, *Duality Rotations For Interacting Fields*, Nucl. Phys. B **193**, 221 (1981).

[21] L. Andrianopoli, R. D'Auria and S. Ferrara, *U-duality and central charges in various dimensions revisited*, Int. J. Mod. Phys. A **13**, 431 (1998) [arXiv:hep-th/9612105].

[22] B. de Wit, P. G. Lauwers and A. Van Proeyen, *Lagrangians Of N=2 Supergravity - Matter Systems*, Nucl. Phys. B **255**, 569 (1985).

[23] L. Andrianopoli, R. D'Auria, S. Ferrara and M. A. Lledó, *Gauged extended supergravity without cosmological constant: No-scale structure and supersymmetry breaking*, Mod. Phys. Lett. A **18**, 1001 (2003) [arXiv:hep-th/0212141].

[24] L. Andrianopoli, S. Ferrara and M. A. Lledó, *Axion gauge symmetries and generalized Chern-Simons terms in N = 1 supersymmetric theories*, JHEP **0404**, 005 (2004) [arXiv:hep-th/0402142].

[25] B. de Wit, H. Samtleben and M. Trigiante, *Magnetic charges in local field theory*, JHEP **0509**, 016 (2005) [arXiv:hep-th/0507289].

[26] A. Ceresole, R. D'Auria and S. Ferrara, *The Symplectic Structure of N=2 Supergravity and its Central Extension*, Nucl. Phys. Proc. Suppl. **46**, 67 (1996) [arXiv:hep-th/9509160].

[27] A. Ceresole, R. D'Auria, S. Ferrara and A. Van Proeyen, *Duality transformations in supersymmetric Yang-Mills theories coupled to supergravity*, Nucl. Phys. B **444**, 92 (1995) [arXiv:hep-th/9502072].

[28] E. Cremmer, S. Ferrara, L. Girardello and A. Van Proeyen, *Yang-Mills Theories With Local Supersymmetry: Lagrangian, Transformation Laws And Superhiggs Effect*, Nucl. Phys. B **212** (1983) 413.

[29] A. Ceresole, G. Dall'Agata, R. Kallosh and A. Van Proeyen, *Hypermultiplets, domain walls and supersymmetric attractors*, Phys. Rev. D **64** (2001) 104006 [arXiv:hep-th/0104056].

[30] E. Cremmer, C. Kounnas, A. Van Proeyen, J. P. Derendinger, S. Ferrara, B. de Wit and L. Girardello, *Vector Multiplets Coupled To N=2 Supergravity: Superhiggs Effect, Flat Potentials And Geometric Structure*, Nucl. Phys. B **250**, 385 (1985).

[31] E. Cremmer, S. Ferrara, C. Kounnas and D. V. Nanopoulos, *Naturally Vanishing Cosmological Constant In N=1 Supergravity*, Phys. Lett. B **133**, 61 (1983).
J. R. Ellis, A. B. Lahanas, D. V. Nanopoulos and K. Tamvakis, *No - Scale Supersymmetric Standard Model*, Phys. Lett. B **134**, 429 (1984).

[32] A. B. Lahanas and D. V. Nanopoulos, *The Road To No Scale Supergravity*, Phys. Rep. **145** (1987) 1.

[33] G. Dall'Agata, *String compactifications with fluxes*, Class. Quant. Grav. **21** (2004) S1479.

[34] J. P. Gauntlett, *Classifying supergravity solutions*, Fortsch. Phys. **53**, 468 (2005) [arXiv:hep-th/0501229].

[35] K. Behrndt, *Stabilization of moduli by fluxes*, AIP Conf. Proc. **743**, 251 (2005) [Fortsch. Phys. **53**, 861 (2005)] [arXiv:hep-th/0503129].

[36] M. Graña, *Flux compactifications in string theory: A comprehensive review*, arXiv:hep-th/0509003.

[37] D. Joyce, *Compact Manifolds with Special Holonomy.* Oxford University Press, Oxford Mathematical Monographs Editions, 2000.

[38] C. Hull, *Holonomy and symmetry in M-theory*, arXiv:hep-th/0305039.

[39] J. P. Gauntlett, D. Martelli, J. Sparks and D. Waldram, *Supersymmetric AdS(5) solutions of type IIB supergravity*, arXiv:hep-th/0510125.

[40] P. Kaste, R. Minasian and A. Tomasiello, *Supersymmetric M-theory compactifications with fluxes on seven-manifolds and G-structures*, JHEP **07** (2003) 004 [hep-th/0303127].

[41] G. Dall'Agata and N. Prezas, *$N = 1$ geometries for M-theory and type IIA strings with fluxes*, Phys. Rev. D **69** (2004) 066004 [arXiv:hep-th/0311146].

[42] K. Behrndt, M. Cvetic and T. Liu, *Classification of supersymmetric flux vacua in M theory*, [arXiv:hep-th/0512032].

[43] J. P. Gauntlett, D. Martelli and D. Waldram, *Superstrings with intrinsic torsion*, Phys. Rev. D **69** (2004) 086002 [arXiv:hep-th/0302158].

[44] M. Becker and D. Constantin, *A note on flux induced superpotentials in string theory*, JHEP **0308**, 015 (2003) [arXiv:hep-th/0210131].

[45] M. Graña, J. Louis and D. Waldram, *Hitchin functionals in $N = 2$ supergravity*, arXiv:hep-th/0505264.

[46] S. Gukov, *Solitons, superpotentials and calibrations*, Nucl. Phys. B **574**, 169 (2000) [arXiv:hep-th/9911011].

[47] S. Gukov, C. Vafa and E. Witten, *CFT's from Calabi–Yau four-folds*, Nucl. Phys. B **584**, 69 (2000) [Erratum-ibid. B **608**, 477 (2001)] [arXiv:hep-th/9906070].

[48] G. L. Cardoso, G. Curio, G. Dall'Agata, D. Lust, P. Manousselis and G. Zoupanos, *Non-Kaehler string backgrounds and their five torsion classes*, Nucl. Phys. B **652** (2003) 5 [arXiv:hep-th/0211118].

[49] S. Gurrieri, J. Louis, A. Micu and D. Waldram, *Mirror symmetry in generalized Calabi–Yau compactifications*, Nucl. Phys. B **654**, 61 (2003) [arXiv:hep-th/0211102].

[50] G. L. Cardoso, G. Curio, G. Dall'Agata and D. Lust, *BPS action and superpotential for heterotic string compactifications with fluxes*, JHEP **0310** (2003) 004 [arXiv:hep-th/0306088].

[51] K. Becker, M. Becker, K. Dasgupta and S. Prokushkin, *Properties of heterotic vacua from superpotentials*, Nucl. Phys. B **666**, 144 (2003) [arXiv:hep-th/0304001].

[52] K. Behrndt and M. Cvetic, *General $N = 1$ supersymmetric flux vacua of (massive) type IIA string theory*, Phys. Rev. Lett. **95**, 021601 (2005) [arXiv:hep-th/0403049].

[53] G. Villadoro and F. Zwirner, *$N = 1$ effective potential from dual type-IIA D6/O6 orientifolds with general fluxes*, JHEP **0506**, 047 (2005) [arXiv:hep-th/0503169].

[54] G. Dall'Agata and N. Prezas, *Scherk-Schwarz reduction of M-theory on G2-manifolds with fluxes*, JHEP **0510** (2005) 103 [arXiv:hep-th/0509052].

[55] J. Polchinski and A. Strominger, *New vacua for type II string theory*, *Phys. Lett.* **B388** (1996) 736–742 [hep-th/9510227].

[56] J. Michelson, *Compactifications of type IIB strings to four dimensions with non-trivial classical potential*, Nucl. Phys. B **495**, 127 (1997) [arXiv:hep-th/9610151].

[57] G. Dall'Agata, *Type IIB supergravity compactified on a Calabi–Yau manifold with H-fluxes*, JHEP **0111**, 005 (2001) [arXiv:hep-th/0107264].

[58] J. Louis and A. Micu, *Type II theories compactified on Calabi–Yau threefolds in the presence of background fluxes*, Nucl. Phys. B **635**, 395 (2002) [arXiv:hep-th/0202168].

[59] S. Ferrara, G. W. Gibbons and R. Kallosh, *Black holes and critical points in moduli space*, Nucl. Phys. B **500** (1997) 75 [arXiv:hep-th/9702103].

[60] S. Ferrara and R. Kallosh, *Supersymmetry and Attractors*, Phys. Rev. D **54** (1996) 1514 [arXiv:hep-th/9602136].

[61] R. Haag, J. T. Lopuszanski and M. Sohnius, *All Possible Generators Of Supersymmetries Of The S Matrix*, Nucl. Phys. B **88**, 257 (1975).

[62] S. Ferrara, M. Bodner and A. C. Cadavid, *Calabi–Yau Supermoduli Space, Field Strength Duality And Mirror Manifolds*, Phys. Lett. B **247**, 25 (1990).

[63] R. Kallosh, *New attractors*, arXiv:hep-th/0510024.

[64] S. Ferrara, R. Kallosh and A. Strominger, *N=2 extremal black holes*, Phys. Rev. D **52** (1995) 5412 [arXiv:hep-th/9508072].

[65] A. Strominger, *Macroscopic Entropy of $N = 2$ Extremal Black Holes*, Phys. Lett. B **383**, 39 (1996) [arXiv:hep-th/9602111].

[66] G. W. Moore, *Arithmetic and attractors*, arXiv:hep-th/9807087.

[67] K. Goldstein, N. Iizuka, R. P. Jena and S. P. Trivedi, *Non-supersymmetric attractors*, Phys. Rev. D **72** (2005) 124021 [arXiv:hep-th/0507096].

[68] P. K. Tripathy and S. P. Trivedi, *Non-supersymmetric attractors in string theory*, arXiv:hep-th/0511117.

[69] K. Goldstein, R. P. Jena, G. Mandal and S. P. Trivedi, *A C-function for non-supersymmetric attractors*, arXiv:hep-th/0512138.

[70] M. Alishahiha and H. Ebrahim, *Non-Supersymmetric Attractors and Entropy Function*, arXiv:hep-th/0601016.

CHAIRMAN: S. FERRARA

Scientific Secretaries: A. Marrani, S. Pacetti

DISCUSSION I

- *'t Hooft:*

You said that only group manifolds are parallelizable, that is they have vanishing Riemann tensor, but they are not torsionless. Could you elucidate this property by taking SU(2) as an example?

- *Ferrara:*

As I was saying this morning, if you compute the commutator of two covariant derivatives of an arbitrary tensor, you get the curvature term, and then you have also a term related to the torsion. Geometrically, this means that if you parallely transport the tensor along a parallelogram, it will come back rotated and also displaced. The responsible for the rotation effect is the curvature, while the displacement is due to the torsion.

In parallelizable manifolds you can always define the affine or spin connections in such a way that the Riemann tensor, which is expressed by derivatives of connection plus quadratic terms, vanishes.

Therefore, in a parallelizable manifold the tensor comes back parallel, but in general still displaced, when the parallelogram path closes. In this approach, the parallelizable manifold has vanishing Riemann tensor, and non-zero torsion.

Let me stress that in general the parallelizable manifolds are not flat at all. Indeed, in this framework the flatness condition is not given by the vanishing of the Riemann tensor, which instead implies the covariant constance of the torsion tensor.

There is a Theorem stating that all Lie group manifolds, plus the 7-sphere, are parallelizable. Actually, since the 3-sphere S^3 is also a group manifold, there exist only two parallelizable spheres: they are S^3 and S^7.

In general, the torsion tensor of a Lie group manifold may be expressed by the structure constants of the Lie algebra underlying the group itself. Roughly speaking, since the metric is related to the Cartan Killing metric of the manifold, it is also a function of the structure constants. Thus, it follows that in this case there is always a relation between the metric and torsion tensors.

The 7-sphere $S^7=SO(8)/SO(7)$ is parallelizable, though not being a group manifold. This is due to the special triality property of SO(8): this compact group has three 8-dimensional representations, i.e. the vector and two spinor ones, intertwined by the action of the γ-matrix operators.

- *Marrani:*

Could you talk a bit more on the fate of local Equivalence Principle (EP) in relativistic theories based on manifolds which are more general than Riemannian ones, such as in Einstein-Cartan case? And what about the case in which the Metric Postulate (MP) is violated?

In these cases, what do we obtain as manifold of local equivalence, i.e. as tangent space?

Might this "generalized" locally-equivalent manifolds be used to treat phenomena which are unexplained in the Minkowskian Standard Relativity, such as superluminality?

- *Ferrara:*

Clearly, when the torsion is non-vanishing, the local EP is violated. This amounts to say that in the Einstein-Cartan manifolds you don't have Minkowski space as local tangent space. Instead, you would get a sort of "generalized" Minkowski space, where the local effect of the torsion would consist in the fact that the commutator of two derivatives would correspond to another derivative.

An example of such a kind of "generalized" manifolds of local equivalence is given by the superspace. In this case, the supersymmetry algebra implies that the anticommutator of two spinor covariant derivatives is a space-time derivative. Therefore, flat superspace may be seen as a flat supermanifold with non-vanishing torsion, suitably expressed in terms of γ-matrices.

- *Marrani:*

Then, what is the fate of tetrads in the Einstein-Cartan manifolds, and in more general geometric settings?

- *Ferrara:*

Well, for simplicity's sake, let us restrict for the moment to the parallelizable manifolds. It is worth noticing that there exists another approach to such particular spaces. Indeed, due to the covariant constance of the torsion tensor, in absence of spinors you can always choose your connection to simply coincide with the Riemannian one, namely with the Levi-Civita connection expressed by the Christoffel symbols of the metric. Roughly speaking, we can get rid of the torsion, and assume the Riemannian point of view on parallelizable spaces, which then result to have constant (generally non-vanishing) curvature.

In this Riemannian torsionless approach to parallelizable manifolds, it can be shown that they have constant Riemann and Ricci tensors, with in general non-vanishing constant curvature. Clearly, the EP locally holds also for such a kind of manifolds.

In general the torsion may be considered as an extra ingredient, which doesn't change nor the EP nor the MP. Consequently, also generic Einstein-Cartan manifolds (with both non-vanishing curvature and torsion) are consistent with the local EP.

However, this is not all the story when you have spinors, because the torsion is related to the properties of local supersymmetry of the theory, and to their possible (partial) spontaneous breaking.

- *Marrani:*

What does it happen if the MP is violated?

- *Ferrara:*

The violation of such a postulate implies the non-covariant constance of the metric tensor. This yields the violation of local EP, and therefore in such a framework the tangent spaces are neither Minkowski nor "generalized" Minkowski, in the sense I mentioned before. Such a kind of geometry is clearly a more general, non-metric one, and it is called affine geometry. At the affine level, the concept of invariant distance does not make sense at all, simply because lengths cannot be measured.

Thus, I can make my previous statement more precise, by saying that the most general geometric framework where the EP locally holds is given by the torsionful Einstein-Cartan manifolds which I introduced this morning.

- *Ciufolini:*

Yes, in this case the Einstein EP would be violated and we shall find that the angular momentum orientation of a spinning particle would be different from the theoretical prediction of General Relativity. But, of course, I expect such an effect to be very small.

Is the torsion different from zero just inside matter or it can be non-vanishing also in vacuum? Could we ever hope to measure it?

For instance, we have experiments with test particles and spinning particles, namely the LAGEOS experiment and the GPB NASA experiment, and also another similar one, LARES, that is currently under study. Would a non-vanishing torsion affect the motion of a test particle in spacetime?

- *Ferrara:*

First of all, let me say that generally the torsion is not a propagating field.

Nevertheless, there are some attempts to write theories with propagating torsion; these would not have only the graviton, as a basic particle, but also other objects, as for instance non-metric "gravi-objects". They would be propagating spinning objects, more precisely they should be a mixture of spin 0, 1 and 2 and they should interact gravitationally.

However, it is very difficult to construct theories with propagating torsion. In most cases, such a torsion is simply an object that gives an additional interaction, of gravitational strength, among ordinary particles. Of course, in general this additional interaction cannot be excluded. A typical example is the four-fermions coupling term, which has a strength of order $1/M_P^2$, M_P being the Planck mass. Such a term is related to the presence of a fermionic torsion, and it gives a short-distance interaction which however, in principle, can yield a deviation from gravitation.

Moreover, there are other theories where torsion is not fermionic but instead bosonic. Also in these cases the torsion introduces a new interaction, of gravitational strength, among ordinary particles.

- *Nilse:*

Is the 1-sphere a parallelizable space?

- *Ferrara:*

Yes. Indeed, the 1-sphere S^1 is a flat space with vanishing Riemann and torsion tensors, and in general all flat torsionless spaces are trivially parallelizable.

Furthermore, the spheres have another interesting geometrical property in General Relativity: they are conformally flat spaces. This means that it is always possible to choose coordinates such that the metric is defined as a scalar function times a flat metric.

Otherwise speaking, all the spheres are flat up to a conformal rescaling.

Anyway, the parallelizability is a property of the group manifolds, and only the n-spheres S^n, with n=1, 3 and 7 are parallelizable.

As I will show in the next two lectures, the conjunction of the properties of parallelizability and conformal flatness will be important in the dimensional compactifications of supersymmetric theories, where fermionic fields play a key role. In particular, the spherical manifolds S^3 and S^7 will be of particular relevance.

- *Marrani:*

What does it happen if non-commutativity is introduced at level of coordinates? Might the non-commutativity be a source of torsion?

- *Ferrara:*

I don't think so, because, at least usually, non-commutative geometry is formulated in flat space. Moreover, the concept of torsion is invariant under diffeomorphisms, and therefore it really makes sense only when you have a theory which has coordinate covariance.

In a non-commutative setting it is very difficult, when you define a particular coordinate frame, to see what is torsion and what is something else. In order to disentangle between the torsion and the other ingredients of your geometry, you have to describe the theory in a covariant way.

At the moment, I do not know if the concept of torsion could be generalized in a non-commutative geometry.

- *Didenko:*

Could you clarify the difference between geodesic and autoparallel curves?

- *Ferrara:*

The geodesic curves minimize the distance. Instead the notion of autoparallelism may be expressed by the invariance of the velocity of a particle under parallel transport along the considered curve. In a Riemannian geometry the affine connection is expressed in terms of the Christoffel symbols of the metric, and therefore it can be easily seen that geodesic and autoparallel curves coincide.

In a more general, torsionful Einstein-Cartan geometric framework, the affine connection, which is the quantity responsible for parallel transport of tensors along the manifold, contains both the metric Christoffel symbols and a term coming from the torsion. Consequently, the features of geodesicity and autoparallelism do not coincide.

However, let me point out that, also in the Einstein-Cartan case, in order to extremize the distance we use only the metric part of the connection (given by the Christoffel symbols), while of course in the autoparallelism equation the full connection is considered.

Moreover, it can be stated that geodesics and autoparallel curves coincide also in the torsionful Einstein-Cartan case only in a particular case, namely when the completely covariant form of the torsion rank-3 tensor is totally antisymmetric in all its indices.

- *Didenko:*

In a bosonic field theory, is it possible to get rid of the torsion tensor by adjusting the stress tensor?

- *Ferrara:*

Yes, it is possible, but only if you don't have extra fields.

Let me briefly comment on the introduction of possible extra fields.

In a generic Riemannian geometry any tensor $T_{\mu\nu}{}^{\rho}=-T_{\nu\mu}{}^{\rho}$ constructed with extra fields can be interpreted as a torsion term. For instance, in string theory the $B_{\mu\nu}$ form has a field strength $H_{\mu\nu\rho}$ which can be seen as a torsion term. Other contributions to the torsion can arise from fermion bilinears, as in the Einstein-Cartan theory coupled to fermions.

In all instances, such terms modify Einstein theory, because they have gravitational self-couplings.

In the pure bosonic theory, if you have only the metric, the torsion is always zero.

CHAIRMAN: S. FERRARA

Scientific Secretaries: L. Andrianopoli, A. Santambrogio

DISCUSSION II

- *Marrani:*

In all the theories we considered so far, we always considered that the metric postulate was not violated. What happens if the metric tensor is not covariantly constant, in particular in relation to the definition of the spin connection? You mentioned briefly that such a violation of the metric postulate might arise when a connection with extra conformal invariance is introduced. Could you clarify this point? Moreover, might the violation of the metric postulatre be a new source of spontaneously broken supersymmetry from compactification, expecially in relation with non trivially conformally flat metric backgrounds, such AdS sspace? Is there any relation with the AdS/CFT correspondence?

- *Ferrara:*

The metric postulate states that the metric of space-time is covariantly constant with respect to some given affine connection Γ:

$$D_\mu g_{\nu\lambda} = 0$$

This equation contains a term with an ordinary derivative plus a term containing the affine connection. When this postulate is violated the affine connection contains, other than the Christoffel connection and the torsion term, an additional field, called the non-metricity field. It may indeed correspond to some additional degrees of freedom. As an example, let us consider a case which is related to the famous mathematician Hermann Weyl: in General Relativity, other than coordinate transformations we may have a rescaling of the metric which depends on the space-time point:

$$g'_{\mu\nu}(x) = \lambda(x) g_{\mu\nu}(x).$$

This is a scale transformation of the metric, and is called a Weyl transformation. For instance, when you couple the Yang-Mills (YM) theory, or Maxwell field, to gravitation the Maxwell lagrangian written in curved space is indeed invariant under this transformation. The term which is not invariant under Weyl rescaling is the Einstein-Hilbert action itself:

$$\int d^D x \sqrt{-g} R.$$

This is because there is a dimensionful coupling which is the Planck scale.

In a theory with Weyl invariance there are no dimensionful couplings, because Weyl invariance implies, if you go to flat space, conformal invariance. From the point of view of the theory of gravity, Weyl invariance may only be true in some particular regime, it certainly is not an invariance of our physical world, because the masses of the particles break it. However, from the mathematical point of view, you have systems which have Weyl invariance, like the electromagnetic field coupled to scalars or to spin one-half fermions (This is related to the fact that the electromagnetic and Yukawa couplings are dimensionless). So there is a part of the theory which has this symmetry. If you define now a connection with the property

$$D_\mu[\lambda(x)g_{\nu\lambda}] = \lambda(x)D_\mu g_{\nu\lambda} \ ,$$

this covariant derivative must contain a new contribution to the connection, called the Weyl connection (This is just the same as the usual requirement in gauge theories in order to get the covariance of the lagrangian under a local transformation).
So the Weyl field is a connection A transforming, under Weyl transformations, as:

$$A' = \lambda^{-1}A\lambda + \lambda^{-1}d\lambda.$$

This connection term restores the possibility of having a derivative covariant under a Weyl transformation.

However, when the Weyl connection is added, the metric postulate is no longer true, because the metric of space-time does not have this invariance: if you rescale the metric tensor, the distance among points

$$ds = \sqrt{g_{\mu\nu}dx^\mu dx^\nu}$$

gets correspondingly rescaled, as::

$$ds' = \lambda^{1/2}ds.$$

So, we have seen that the lack of invariance of the metric is related to the fact that the Weyl connection violates the metric postulate.

The hystorical fact is that Hermann Weyl introduced the Weyl connection just to explain electromagnetism, but there was a small mistake in what he did, since the Weyl rescaling is a non-compact parameter, and not a phase. So it is the gauge theory of dilatations, and not of a compact U(1) transformation, as electromagnetism is.

- *Marrani*:

Could the Weyl connection be a source of spontaneous symmetry breaking in the compactification?

- *Ferrara:*

You should make a theory which is conformally invariant, and then conformal invariance is spontaneously broken. So it could break conformal invariance, not supersymmetry.

- *Didenko:*

You have mentioned that the Riemann curvature can be treated the same way as in standard Yang-Mills theory. What the Yang-Mills action for the SO(3,1) group would correspond to, in this case?

- *Ferrara:*

That theory would have the square of the Riemann tensor in the action.
If you would like to construct the YM lagrangian as the square of the curvature, since the curvature is the Riemann tensor you would get a theory of gravity where the Riemaniann part would not be just the Einstein gravity (which is linear in the Riemann scalar) but you would replace this by the square of the Riemann tensor. Such a theory has been considered in the literature, in particular by Zee and Yang in the `60th, in order to show that the equations of motion coming from this lagrangian can have, as a particular set of solutions, the usual solutions of Einstein theory.

The problem of this action is that it is higher derivative in the metric: since the Riemann curvature is second order in the derivatives of the metric, we get an action which is fourth order in the derivatives of the metric. However, we have problems in giving a particle interpretation to an action with higher order derivatives because they have the problem of violating unitarity even at the free field theory level, in the sense that you get some extra propagating states which have negative norm.

Of course Yang had a good reason for trying to write this action, because it was the extension of the Yang-Mills theory to the gravitational case, and gravity has many resemblances with the YM case since the Riemann curvature can be interpreted as the field strength with the gauge group being the Lorentz group. However there is something which is not resemblant because there is the link between Lorentz invariance and general covariance which makes gravity more complicated than the usual YM theories (where the gauge group is now the space-time symmetry).

- *Serbyn:*

Could you clarify the notion of Killing spinors and in particular the integrability condition $[D_\mu, D_\nu]\varepsilon_\alpha = 0$?

- *Ferrara:*

The definition of a Killing spinor in a Riemanniann geometry is a spinor satisfying the equation:

$$D_\mu \varepsilon = 0.$$

In Riemanniann geometry, the integrability condition for this spinor implies that the holonomy group is reduced because it implies

$$[D_\mu, D_\nu]\varepsilon_\alpha = R_{\mu\nu\alpha}{}^{\beta}\varepsilon_\beta = 0.$$

Of course, if we have Einstein theory where the holonomy (namely the rotation of a spinor along a closed path) is an arbitrary, orthogonal transformation, this equaton has never a solution, because a spinor transforms under Lorentz transformations. The only way of getting a solution to this equation is that the holonomy group G is smaller than the orthogonal group SO(D-1,1) or, in the case of euclidean space, SO(D). G is fixed by the following requirement: the spinor representation of SO(D) must contain a singlet of G.

This is a necessary condition for this equation to have solutions.

Let me just make two examples:

a) Let us take as curved space the six dimensional Euclidean space (this case is relevant in string compactifications from 10 to 4 dimensions). In this case, the riemanniann geometry would have an holonomy group given by SO(6). The biggest group with the property of containing a singlet when SO(6) is decomposed with respect to it is SU(3). This is the case because the Lie algebra of SO(6) is the same as the one of SU(4), and the spinor representatio of SO(6) is just a vector of SU(4), a four component complex spinor. If you decompose the **4** of SU(4) under SU(3) it gives: **4** $\rightarrow$ **3** + **1**, so it contains a singlet. In this case you have a solution to the integrability condition. In fact the 6D spaces with the restricted holonomy SU(3) are called Calabi-Yau spaces.
b) A similar story is relevant in compactifications from 11 dimensions, where you replace strings with M-theory (or, if you want, eleven dimensional supergravity). To go to four dimensions now you have to compactify 7 dimensions; if you want a Killing spinor, you must again ask that its covariant derivative be zero. Now the requirement is that the group SO(7) under the holonomy group gets reduced with a

singlet. The biggest group which is contained in SO(7) and which has the property that the spinor of SO(7) contains a singlet under it is the exceptional group G_2. Indeed, the spinor represnentation of SO(7) is **8**, and under G_2 it decomposes as: **8** $\rightarrow$**7** + **1**. Spaces with this properties are called G_2-holonomy spaces, and they play an important role in M-theory compactifications, since they preserve one supersymmetry in four dimensions.

- *Grassi:*

Can you write a Lagrangian or equations of motion for the Einstein-Cartan theory, when the torsion is dynamical? There is a supergravity for this theory?

- *Ferrara:*

Although there have been attempts in the literature of introducing the spin connection as a propagating field, what I know is that in the EC theory you can introduce the vielbein V and spin connection ω as independent variables and then you can derive the expression for the spin connection by varying the Lagrangian with respect to the spin connection

$$\frac{\delta L}{\delta \omega} = 0$$

If the spin connection is not propagating, then it is actually an auxiliary field, which means that it satisfies algebraic equations of motion. In this case you can always solve algebraically these equations and if you do this in pure EC theory, this equation actually gives $\omega = \omega(V)$ which may be proven to be equivalent to the metric postulate for a Levi-civita connection.

If you have in the EC theory fermions coupled to gravity, then you can prove that this equation introduces a torsion term in the spin connection, which is proportional to a bilinear in the fermions. When you plug it back into the Lagrangian it gives a four fermions Fermi-like interaction, where however the Fermi coupling is replaced by the Planck coupling: $\frac{1}{M_P{}^2}\bar{\chi}\chi\bar{\chi}\chi$. This is the typical deviation of the minimal coupling of the fermions to General Relativity coming from EC theory. This is a very short distance interaction.

In connection to supersymmetry, we can only say that if one takes the simplest example of N=1 supergravity in four dimensions, actually in this case Deser and Zumino had shown that that theory is precisely the EC theory of the Rarita-Schwinger field. Namely, if you forget about sup[ersymmetry and you couple the Rarita-Schwinger field to gravity, not in the Einstein theory, but in the EC theory, where the spin connection is an independent variable, you get exactly a quartic coupling of the gravitino field at the same point, , with a strength proportional to the inverse of the squared planck masss, similar to the one previously described. One can prove that this coupling is precisely the one needed to make the theory invariant under local supersymmetry.

In some sense, in the N=1, D=4 case the EC theory coincides with the supergravity theory. This is however a mere accident of pure supergravity, when matter is presnt this is no longer true.

- *Konikowska:*

You have just mentioned problems encountered by the theories containing higher order gravity terms. Is it also the case for the higher dimensional theories with branes?

- *Ferrara:*

First of all, I believe that a theory with higher order curvatures terms would encounter problems in flat background (I don't know if they also encounter problems in different backgrounds like AdS), in particular problems of propagating ghosts. In the situation when you have branes in the game, I don't know precisely. However, this term encounter problems if you think at them as terms in the basic Lagrangian that defines your theory. But of course we know that these terms may be there as an effect of integrating out some massive degrees of freedom. After all, if you take string theory and you integrate out all the massive fields of the string, and you look at the string theory as an effective theory of gravity, the string theory has actually an expansion in the curvature, where the first term is the Einstein term, but then you get an infinite number of terms as an effect of the integration of the heavy modes:

$$L = R + \sum_n R^n .$$

But of course in this case you don't take this as the fundamental Lagrangian, this is an effective Lagrangian coming from the integration of the massive modes. The massless modes of the string are coming from the first term, which is just the usual Einstein sector. So, it depends very much if you consider the higher curvature terms as terms induced by integration of heavy modes, of quantum corrections, or if you consider them as the basic terms in the Lagrangian, to define your variables.

A Never Ending Story: The Nucleon Form Factors and New BaBar Time-like Results

Rinaldo Baldini Ferroli
Centro Studi e Ricerche Enrico Fermi, Roma, Italy
INFN, Laboratori Nazionali di Frascati, Italy

Abstract

Nucleon form factors have recently shown unexpected and spectacular features. For space-like $Q^2_{p\overline{p}}$ G^p_E/G^p_M is not constant at all, as believed for many decades, in spite of early farsighted predictions. For time-like $Q^2_{p\overline{p}}$ BaBar has measured with unprecedented accuracy $e^+e^- \to p\overline{p}$ from the threshold up to $Q^2_{p\overline{p}} \sim 20\ GeV^2/c^4$, finding out an unexpected cross section, with plateaux and drops, in particular confirming the one just above threshold, where evidence for a ratio $|G^p_E/G^p_M| > 1$ has also been found. The neutron time-like form factors, measured only once, and their puzzling behaviour are shortly reminded.

1 Introduction

The Nucleon is the matter building block, at least that matter we have identified in the Universe, and the Nucleon form factors (FF) are among the very first topics have been studied in subnuclear physics. In spite of many decades of experimental investigation Nucleon FF have recently shown rather unexpected and spectacular features, namely:

- in $e^-p \to e^-p$ scattering, i.e. space-like exchanged Q^2 (in the following assumed to be negative), the ratio $|G^p_E/G^p_M|$ falls down approaching zero at $Q^2 \sim 8\ GeV^2$ [1], despite that in all textbooks for about forty years it was assumed to be a constant and despite early, foresighted predictions [2,3];
- in $e^+e^- \to p\overline{p}$ annihilation , i.e. time-like (positive) exchanged Q^2, the FF have a step-like behaviour [4], despite that everybody had assumed a smoothed one;
- in particular in $e^+e^- \to p\overline{p}$ a sharp drop just above threshold [3,4] strongly suggests a narrow resonance at threshold, despite some time ago the existence of a baryonium had been dismissed;
- the $e^+e^- \to n\overline{n}$ cross section, at least according to the only measurement performed until now [5], has been found about an order of magnitude larger than expected.

  In the following the recent space-like results on Nucleon FF are summarized and the brand new time-like results, obtained by BaBar by means of Initial State Radiation (ISR), are described more in detail.

  Let remind that the FF definition is somewhat tricky: they are defined as those factors, which account for the hadronic structure, in the one photon exchange amplitude [6]. In some cases two photons exchange amplitudes are important, have to be evaluated and subtracted, affecting in a different way different observables.

2 Space-like Nucleon form factors

In the elastic electron-Nucleon scattering, i.e. space-like exchanged Q^2, magnetic $G_M(Q^2)$ and electric $G_E(Q^2)$ FF enter the spin flip and non spin flip amplitude, as defined in the Nucleon Breit frame. Because of their definition magnetic and electric FF do not interfere. In the Born approximation the scattered electron angular distribution is:

$$\frac{d\sigma(eN \to eN)}{d\cos\theta} = \left(\frac{d\sigma}{d\cos\theta}\right)_{\text{Mott}} \left[\frac{G_E(Q^2)^2 - \tau G_M(Q^2)^2}{1-\tau} - 2\tau G_M(Q^2)^2 \tan^2\left(\frac{\theta}{2}\right)\right], \qquad \tau = \frac{Q^2}{4M_N^2}. \quad (1)$$

Proton and neutron space-like magnetic form factors are quite established as well as their overall behaviour is fairly described by a dipole formula. Quark model asymptotic expectation is that the FF of a hadron, made of n constituents, should scale with Q^2 as $Q^{2(n-1)}$ [7]. Hence magnetic form factors scale according to PQCD, even if quite earlier than expected.

This is not the case of the proton space-like electric form factor. Actually it is not a straightforward measurement. As soon as $Q^2 \sim M_N^2$, because of the $\mu_N^2 \cdot Q^2/M_N^2$ factor the magnetic contribution is dominant, the electric contribution is at a few percent level and two photons exchange contribution might play an important role, overcoming the Born term. The ratio between electric and magnetic FF can also be obtained by looking to the ratio between transverse and parallel polarization of the outgoing Nucleon. This kind of measurement is not affected by the aforementioned factor and it is less sensitive to radiative corrections:

$$\frac{G_E(Q^2)}{G_M(Q^2)} = -\left(\frac{P_t}{P_l}\right)\frac{E_e + E_e^{'}}{2M_N}\tan(\theta/2), \tag{2}$$

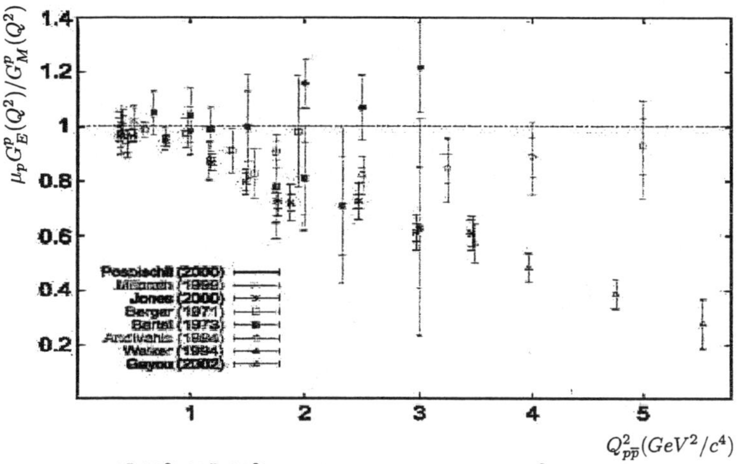

Figure 1: $G_E^p(Q^2)/G_M^p(Q^2)$ as a function of space-like Q^2, as achieved by means of the angular distribution or by means of transverse and longitudinal outgoing proton polarization.

In Fig.1, the ratio $G_E^p(Q^2)/G_M^p(Q^2)$ is plotted, as obtained, mostly at SLAC [8], according to the angular distribution by means of the Rosenbluth method, compared to what has been obtained, mostly at JLAB few years ago [1], according to the polarization ratio. The two sets of measurements disagree completely and JLAB data destroy the "scaling law" between electric and magnetic FF, considered like a dogma for many decades. As a consequence, at $Q^2_{p\bar{p}} \sim 8\ GeV^2$ a zero of $G_E^p(Q^2_{p\bar{p}})$ (followed by a negative tail reaching the value -1 asymptotically [9]) is foreseen and soon this expectation will be tested at JLAB.

Two photons contribution is difficult to evaluate, depending on the Nucleon structure too. However present evaluations [10] show that sign and order of magnitude both are in the right direction and it is very likely they are responsible of this spectacular discrepancy. A measurement by means of a positron beam will disentangle the question of two photons contribution [11].

It is worthwhile to remind that many decades ago T. Massam and A. Zichichi formulated a model of the Nucleon [2], forerunner of the present soliton models, that foresaw for the proton a fast drop with $Q^2_{p\bar{p}}$ of $G_E^p(Q^2_{p\bar{p}})/G_M^p(Q^2_{p\bar{p}})$, as shown in Fig.2. The model assumed an extended source of vector mesons with a $1/M_N$ radius and inputs of the model were essentially the Nucleon static properties and SU_3 flavour symmetry only. Adding a point-like source contribution F. Iachello, A.D. Jackson and A. Lande [3] improved this model and their prediction is in astonishing agreement with the JLAB data, as shown in Fig.2. PQCD may also predict a similar result, assuming important NLO corrections. The two calculations do not behave in the same way once extrapolated to time-like $Q^2_{p\bar{p}}$ (in a somewhat plausible but not rigorous way).

Analyticity is supposed to connect space-like FF, that are real quantities according to e.m. current conservation, and time-like FF, that have an imaginary part above the first isovector/isoscalar inelastic thresholds.

In e^+e^- annihilation into $N\overline{N}$, i.e. time-like exchanged Q^2, two FF are also defined. This time they refer to the total Nucleon helicity, as defined in the c.m. frame. Particle-antiparticle relationship relates the total helicity equal to 0 form factor to G_M and the total helicity equal to 1 form factor to G_E.

Two other FF, $F_1(Q^2)$ and $F_2(Q^2)$ (Dirac and Pauli FF), can also be defined. Their relationship with the e.m. current J^μ and the anomalous magnetic moment k is more transparent, namely: $J^\mu = F_1(Q^2)g^\mu + \frac{k}{2M}F_2(Q^2)i\sigma^{\mu\nu}q_\nu$ and

$$\begin{aligned} G_E &= F_1 + \tau F_2, \\ G_M &= F_1 + F_2. \end{aligned} \qquad (3)$$

F_1 and F_2 as well as G_E and G_M analyticity demands that exactly at threshold it is

$$G_E(4M_N^2) = G_M(4M_N^2). \qquad (4)$$

The same result is achieved if at threshold there is the S wave only.

The standard explanation of the G_E/G_M drop for space-like Q^2 is because of a negative interference between F_1 and F_2. Therefore it should become a positive interference for time-like Q^2 and an increasing ratio G_E/G_M above threshold is guessed. For time-like Q^2 F_1 and F_2 are complex functions and to get their relative phase, which has to vanish exactly at threshold, polarization tools are needed.

In principle it is possible to have experimental access to the FF below threshold by a suitable measurement of $p\bar{p} \to e^+e^-\pi^0$. Dispersion relations can also be applied to get FF in the whole unphysical region ,i.e. $0 < Q^2 < 4M_N^2$, as it is illustrated in ref. [12] in the case of G_M^p. Of course the same technique applied to G_E too, constrained by the phase measurement, will strongly improve this approach.

The neutron space-like form factor is rather small and not well known up to now.

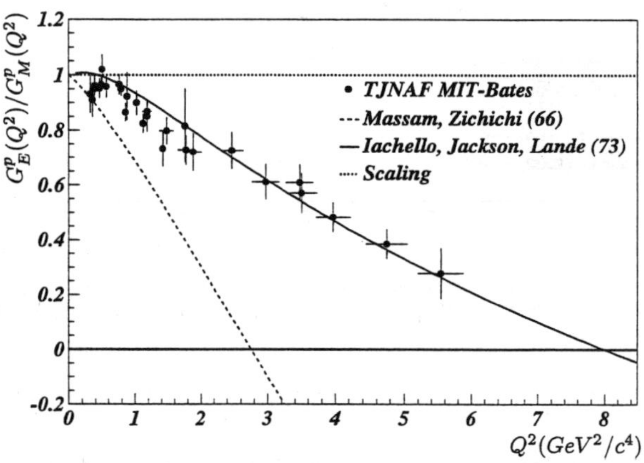

Figure 2: $G_E^p(Q^2)/G_M^p(Q^2)$, as predicted about forty years ago by T. Massam and A. Zichichi (dashed line), modelling the Nucleon as a soliton without any free parameter, and some years later by F.Iachello, A.D.Jackson, A.Lande adding a suitable point-like coupling (solid line), compared to the recent measurements by means of transverse and longitudinal outgoing proton polarization. The former dogma, the scaling law $G_E^p(Q^2)/G_M^p(Q^2) \sim 1$, is also pointed out.

3 Time-like Nucleon form factors

Let remind to the young audience of this School that in 1965 Antonino Zichichi, for the first time, identified Nucleon time-like FF as an important physics item and started up their measurement via

$$p\bar{p} \to e^+e^-.$$

At that time high energy e^+e^- storage rings were not operating yet and to get ride of the multihadronic background (about 10^6 larger than e^+e^-) he invented the preshower technique to select electrons.

Afterwords high energy e^+e^- storage rings entered the game, but for many years Nucleon FF measurements were limited from the point of view of the explored c.m. energy range as well as of the integrated luminosity. At last BaBar has recently measured [4] with unprecedented accuracy the cross section

$\sigma(e^+e^- \to p\bar{p}\gamma)$ from the threshold up to a $p\bar{p}$ c.m. total energy squared $Q^2_{p\bar{p}} \sim 20\ GeV^2/c^4$ by means of ISR, as in Fig.3. In fact it has been shown [13] that at a m_e/E_e precision level it is:

$$\frac{d\sigma}{d\Omega_\gamma dE_\gamma}(e^+e^- \to p\bar{p}\gamma) = P(s, E_\gamma, \Omega_\gamma) \cdot \sigma(Q^2_{p\bar{p}}), \tag{5}$$

where s is the e^+e^- total c.m. energy squared, E_γ and Ω_γ are energy and angles of the ISR photon in the e^+e^- c.m. frame, P is the density function for ISR emission, computed according to QED as in Fig.3, and $\sigma(Q^2_{p\bar{p}})$ is the $e^+e^- \to p\bar{p}$ cross section at the $p\bar{p}$ c.m. energy squared $Q^2_{p\bar{p}}$.

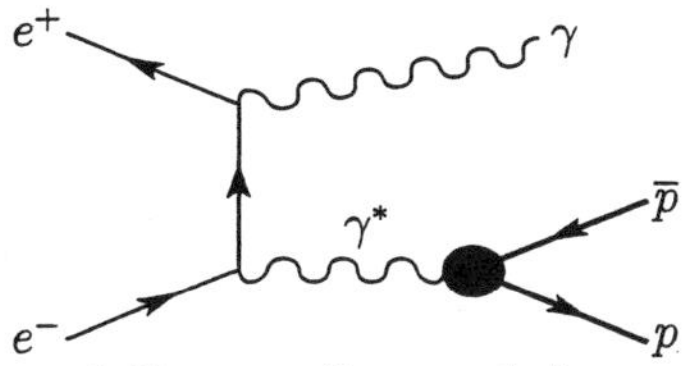

Figure 3: Feynman diagram of $e^+e^- \to p\bar{p}\gamma$.

In the Born approximation it is given by [6]:

$$\frac{d\sigma(e^+e^- \to p\bar{p})}{d\cos\theta} = \frac{\pi\alpha^2\beta C}{2Q^2_{p\bar{p}}}\Big[(1+\cos^2\theta)|G^p_M(Q^2_{p\bar{p}})|^2 + \frac{1}{\tau}\sin^2\theta|G^p_E(Q^2_{p\bar{p}})|^2\Big]. \tag{6}$$

All quantities are evaluated in the $p\bar{p}$ c.m. frame: θ is the azimuthal proton emission angle, β is the proton velocity, C is a factor introduced in the final state distorted wave approximation to take into account the Coulomb interaction [14], G^p_E and $|G^p_M|$ the analytical time-like continuation of the corresponding space-like FF. As mentioned already at $Q^2_{p\bar{p}} = 4M^2_p$ it is expected $G^p_E/G^p_M = 1$, assuming that electric, magnetic, Dirac and Pauli FF are all analytical functions with respect to $Q^2_{p\bar{p}}$, hence continuous across the $p\bar{p}$ threshold. So far this expectation has been extended to the whole $Q^2_{p\bar{p}}$ explored range [5, 15–20], lacking high statistics measurements in particular concerning G_E. Actually what is quoted is $|G^p_M|$, also because at high $Q^2_{p\bar{p}}$ its contribution is the dominant one. Concerning time-like $Q^2_{p\bar{p}}$ the first [15] and the most relevant measurements at low [16] and at high $Q^2_{p\bar{p}}$ [17], previous to BaBar, have still been performed by means of the inverse process $p\bar{p} \to e^+e^-$. Under the aforementioned hypothesis concerning $|G^p_M|$ they have shown a very steep increase approaching the threshold and a $1/Q^4_{p\bar{p}}$ overall behaviour, quite earlier than asymptotically expected according to PQCD [21] as in the space-like region.

4 BaBar results on $e^+e^- \to p\bar{p}$

In the following ISR events have been selected by asking the ISR photon is detected, to get rid of the non ISR multihadronic background. A rather large fraction of events is lost in this way, however in this case BaBar has the advantage over a conventional c.m. e^+e^- collider that the cross section can be measured even at threshold, with a $\sim 1\ MeV/c^2\ Q_{p\bar{p}}$ invariant mass resolution and with almost full $p\bar{p}$ angular coverage.

The B-factory PEP II (9 GeV/c^2 e^- colliding with 3.1 GeV/c^2 e^+) and the BaBar detector have been described in detail several times [22]. For the present purposes charged particle tracking (a 5 layer silicon vertex, SVT, and a 40 layer drift chamber, DCH) and identification (by means of an internally reflecting ring imaging Cherenkov, DIRC, of ionization in SVT and DCH and energy deposition in the calorimeter) systems are the main components. Muons are identified by means of the instrumented iron flux return.

As anticipated the ISR photon is detected and must be in the range $22^o < \theta_\gamma < 137^o$ in the lab frame, however photon information has not been used in the kinematical fit. The Monte Carlo (MC) event generator is based on the code described in [23]. Extra ISR soft photons are generated according to the structure function method [24].

At present a data sample related to 232 fb^{-1} total integrated luminosity has been analyzed. Event selection has required an energetic photon and two opposite sign charged tracks, originated from the interaction point with a polar angle within the DIRC acceptance. Radiative Bhabha events are rejected if for each track it is $0.9 < E_{\rm cal}/P_{\rm tr} < 1.1$, between the energy deposition $E_{\rm cal}$ in the calorimeter and the DCH momentum $P_{\rm tr}$. To get rid of the huge ISR $\pi^+\pi^-\gamma$, $\mu\mu\gamma$, $K^+K^-\gamma$ background both charged tracks are required to be well identified as proton candidates by means of a tight cut, leading to a loss of about 30%

of signal events. A kinematical fit is performed adding further tight cuts, that is to be compatible only with proton masses hypothesis, leading to a further loss of about 25%. The overall detection efficiency is about 18% with a mild dependence on $Q_{p\bar{p}}$. In the surviving events there is no evidence of a peak at the ρ mass or a peak at the Φ mass in a pion or kaon pair masses hypothesis and the estimated remaining contamination is negligible. Angular and energy photon distributions are consistent with the ISR expectation. Final state radiation is expected to be very small and there is no interference term, due to the different charge parity.

The most important and subtle source of background is the process $e^+e^- \to p\bar{p}\pi^0$, where easily a soft photon is lost or the two photons are merged and not disentangled by the pattern recognition. Since there are no experimental information on this process, events $e^+e^- \to p\bar{p}\pi^0$ with a fully reconstructed π^0 are identified and the contamination in the selected sample of $e^+e^- \to p\bar{p}$ candidate events is evaluated according to the MC expectation. These estimated background events are about 5% of the selected candidate events when $Q^2_{p\bar{p}} < 6\ GeV^2/c^4$, about 10% if $6 < Q^2_{p\bar{p}} < 9\ GeV^2/c^4$ and become consistent with 100% above $Q^2_{p\bar{p}} \sim 20\ GeV^2/c^4$.

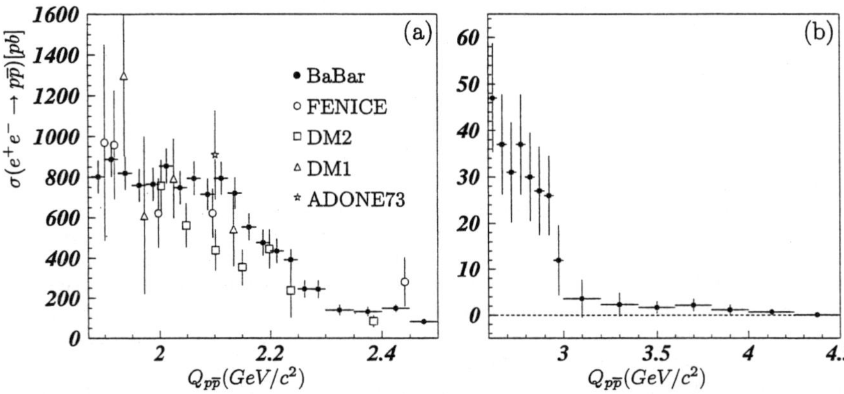

Figure 4: BaBar $e^+e^- \to p\bar{p}$ cross section in comparison with previous experiments: FENICE [5], DM2 [20], DM1 [19] and ADONE73 [18].

The ISR luminosity is calculated using the total integrated luminosity and the probability density function for ISR photon emission, as in eq.(5), taking into account the angular cuts. The luminosity, that has been integrated, depends on the $Q_{p\bar{p}}$ invariant mass bin width, varying from $\sim 0.5\, pb^{-1}$ at $Q_{p\bar{p}} \sim 2\ GeV/c^2$ up to $\sim 1\, pb^{-1}$ at $Q_{p\bar{p}} \sim 3\ GeV/c^2$ for a $10\ MeV/c^2$ bin width.

Radiative corrections have been evaluated according the structure function method. They do not include corrections due to vacuum polarization. Hence what is quoted here is the so called "dressed" cross section. The invariant mass resolution has been unfolded, however the chosen bin widths exceed the resolution.

With all these ingredients the calculated cross section $\sigma(e^+e^- \to p\bar{p})$ is shown in Fig.4, statistical and systematic errors are quoted, including the uncertainties in detection efficiency, integrated luminosity and radiative corrections. For comparison some previous measurements are also shown. The emerging cross section shape is an unexpected one: a flat plateau from the threshold up to $Q^2_{p\bar{p}} \sim 5\ GeV^2/c^4$, followed by a drop and then by a second drop at $Q^2_{p\bar{p}} \sim 8.5\ GeV^2/c^4$. At present no simple explanation has been found concerning this behaviour.

The angular distributions have been fitted, according to eq.(6) to extract the ratio $|G^p_E/G^p_M|$. The fitted ratios are reported in the Fig.5 . BaBar results are not in agreement with the APPLE [16] results at LEAR, while they are in qualitative agreement with an expectation, updated, based on dispersion relations and on the JLAB space-like $|G^p_E/G^p_M|$ measurements [25]. There is no clear evidence for an asymmetry, i.e. two photon exchange contribution, but there is not enough statistics to draw any conclusion.

As previous experiments did, $|G_M|$ is evaluated from the total cross section as shown in Fig.6, under the hypothesis $G_E = G_M$, which is at odd with the aforementioned results by the way. The asymptotic $1/(Q^2_{p\bar{p}})^2$ behaviour found in $p\bar{p} \to e^+e^-$ is confirmed as well as a very steep slope very near threshold.

The C factor in the cross section formula has a very steep slope too and is relevant only very near threshold, diverging as $1/\beta$ so that the cross section should be finite at threshold. It has been introduced to get rid of the pointlike Coulomb interaction, since the FF definition should demand that the pointlike cross section has been factorized. However this receipt concerning Coulomb interaction has been questioned [26] and a better evaluation might affect the steep threshold behaviour of the FF. Alternative interpretations are peculiar $p\overline{p}$ final state interactions, like one π exchange, as suggested in the case of J/ψ radiative decay into a pseudoscalar $p\overline{p}$ pair [27], or a narrow $p\overline{p}$ vector state below threshold as obtained by means of dispersion relations in the unphysical region [12]. It may be worthwhile to remind that no Coulomb corrections are expected in the $e^+e^- \to n\overline{n}$ process.

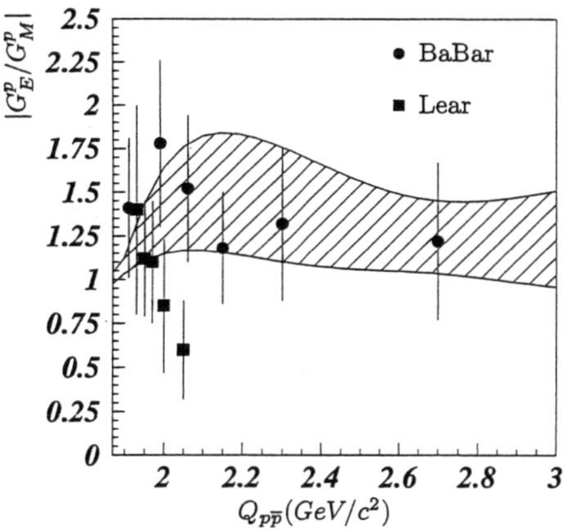

Figure 5: The measured ratio $|G_E^p/G_M^p|$: BaBar, Lear data [16] and according to analyticity [17].

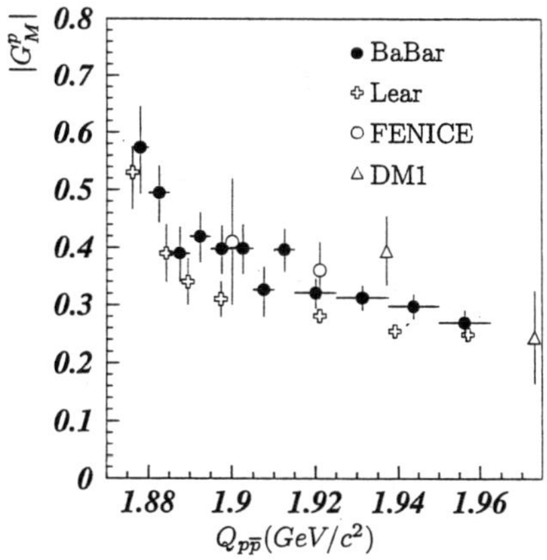

Figure 6: Proton magnetic form factor measurements near threshold, assuming $G_M^p = G_E^p$.

5 $e^+e^- \to n\overline{n}$

The $e^+e^- \to n\overline{n}$ cross section has been measured only once, 15 years ago, by the FENICE experiment at the old storage ring ADONE in Frascati [5] . The collected luminosity was very poor: $\sim 500\ nb^{-1}$ in the c.m. energy range $1.9 < Q < 2.55\ GeV$. Candidate antineutron stars were identified, testing consistently with neutron detection efficiency the presence in few cases of a candidate neutron in the opposite direction. The largest source of background were high energy cosmic neutrons and the experiment could get rid of them by looking to the timing with respect to the beam crossing. The $e^+e^- \to p\overline{p}$ cross section as well as $\sigma_{tot}(e^+e^- \to hadrons)$ and QED processes have also been measured. It has been found

$$R_N = \frac{\sigma(e^+e^- \to n\overline{n})}{\sigma(e^+e^- \to p\overline{p})} \sim 2 \tag{7}$$

as shown in Fig.7.

Assuming a leading quark in the Nucleon it is expected $R_N \sim (\frac{q_d}{q_u})^2 = 0.25$. The same result is obtained by a calculation based on a dispersive approach and PQCD asymptotic behaviour [28]. In any PQCD calculation relying on valence quarks it is quite difficult to predict R_N greater or equal to 1, while soliton models expect $R_N \geq 1$ [29] and some VMD based models predicted even $R_N >> 1$ [30].

The average angular distribution is consistent with G_E^n much lower than G_M^n as in the space-like region [5]. Taking into account the threshold constraint it would imply a vanishing magnetic form factor at threshold followed by a sharp rise, just opposite to the proton behaviour [5]: ~ 0.7 at threshold followed by a sharp drop. In a very naive picture FF are the N and $\overline{N}$ wavefunction overlap, so that concerning threshold, electric charge, neutron and proton, unbelievably ~ 0 and ~ 1 are the expected values.

The neutron magnetic form factor is shown in Fig.8, as achieved by FENICE and by DM2 also (on the basis of few candidate events because of a very poor detection efficiency or by means of the Λ form factor near threshold on the basis of U-spin invariance, i.e. times $\sim 2M_\Lambda/M_N$).

All this stress that the measurement of $e^+e^- \to n\overline{n}$ has to be redone with much higher collected luminosity, in particular in the near threshold region. Unfortunately it is a very difficult task, if not impossible, by means of ISR in BaBar. In the near future the new e^+e^- symmetric storage ring VEPP2000 in Novosibirsk should collect data up to a c.m. total energy of 2 GeV at most. However, in the near threshold region, to get rid of the background without boosting the c.m. is not a simple task too. BEPC2, the new τ/Charm Factory in Beijing and, possibly DANAE, a proposed upgrade in energy, up to 2.4 in the c.m., and in luminosity of DAΦNE in Frascati should have better chances and a wider available c.m. energy range.

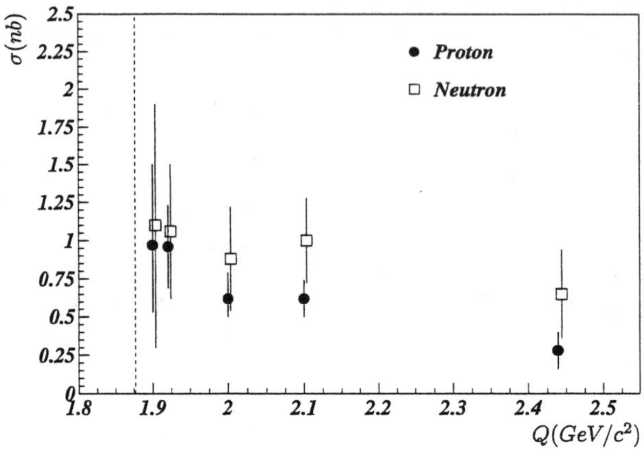

Figure 7: $e^+e^- \rightarrow n\bar{n}$ and $e^+e^- \rightarrow p\bar{p}$ cross sections, as measured by FENICE.

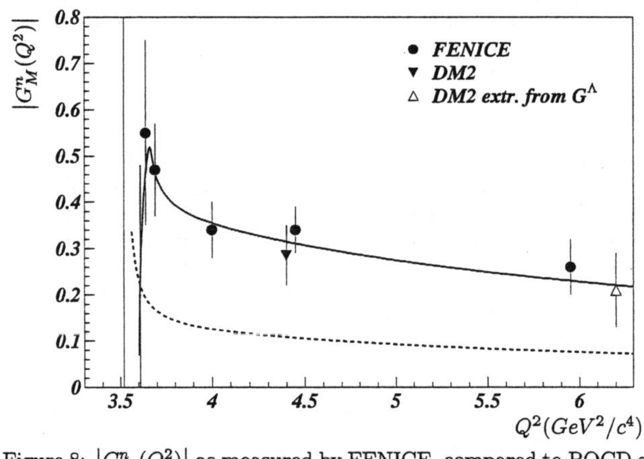

Figure 8: $|G_M^n(Q^2)|$ as measured by FENICE, compared to PQCD expectations [10] (dashed line) and as suggested by the average angular distribution (solid line).

In conclusion Nucleon FF have recently shown unexpected and spectacular features.

For space-like $Q^2_{p\bar{p}}$ the knowledge of the Nucleon has been shaken to the roots by the new measurement of G_E^p/G_M^p, which is not constant at all, as believed for many decades, in spite of early farsighted predictions.

For time-like $Q^2_{p\bar{p}}$ BaBar has measured with unprecedented accuracy, by means of ISR, the $e^+e^- \rightarrow p\bar{p}$ cross section from the threshold up to $Q^2_{p\bar{p}} \sim 20\ GeV^2/c^4$, finding out unexpected drops in the cross section at 2.2 GeV and 2.9 GeV as well as confirming a sharp drop of the magnetic form factor at threshold. Evidence for a ratio $|G_E^p/G_M^p| > 1$ has also been found just above the threshold at odd with previous measurements at LEAR, but consistent with analyticity expectation [25]. The neutron time-like FF have been poorly measured only once, nevertheless showing unexpected features.

Measurements of the proton space-like electric form factor at higher Q^2 are eagerly expected as well as new measurements of neutron and proton time-like FF, moduli and relative phase and much higher integrated luminosity, are demanded.

Acknowledgments

I warmly acknowledge the Ettore Majorana Center and his Director, Antonino Zichichi, for having the opportunity to present this still challenging topic and S. Pacetti, E. P. Solodov and A. Zallo for their inestimable and friendly collaboration.

References

[1] M.K.Jones *et al.*, Phys. Rev. Lett. **84**, 1398 (2000); O.Gayou *et al.*, Phys. Rev. Lett. **88**, 092301 (2002).

[2] T.Massam, A.Zichichi, Nuovo Cim. A **43**, 1137 (1966).

[3] F.Iachello, A.D.Jackson, A.Lande, Phys. Lett. B **43**, 191 (1973).

[4] R.Baldini representing the BaBar coll., Proc. Sci. HEP2005:285,2005. e-Print Archive: hep-ph/0512183.

[5] A.Antonelli *et al.*, Nucl. Phys. B **517**, 3 (1998).

[6] S.D.Drell, F.Zachariasen, *Electromagnetic Structure of Nucleon*, Oxford University Press (1960).

[7] S.J.Brodsky, G.R.Farrar, Phys. Rev. Lett. **31**, 1153 (1973).

[8] R.C.Walker *et al.*, Phys. Rev. D **49**, 5671 (1994).

[9] R.Baldini, C.Bini, P.Gauzzi, M.Mirazita, M.Negrini, S.Pacetti, [hep-ph/0507085].

[10] P.G.Blunden, W.Melnitchouk, J.A.Tjon, Phys. Rev. Lett. **91**, 142303 (2003).

[11] J.Arrington, Phys. Rev. C**68**, 034325 (2003).

[12] R.Baldini, S.Dubnicka, P.Gauzzi, S.Pacetti, E.Pasqualucci, Y.Srivastava, Eur. Phys. J. C **11**, 709 (1999).

[13] H.Czuz *et al.*, Eur. Phys. J. C **35**, 527 (2004).

[14] L.D.Landau, E.M.Lifshits, *Relativistic Quantum Theory*, Mir (1975).

[15] M.Conversi, T.Massam, T.Muller, A.Zichichi, Nuovo Cim. **40**, 690 (1965).

[16] G.Bardin *et al.*, Nucl. Phys. B **411**, 3 (1994).

[17] T.A. Armstrong *et al.*, Phys. Rev. Lett. **70**, 1212 (1993); M.Ambrogiani *et al.*, Phys. Rev. D **60**, 032002 (1999); M.Andreotti *et al.*, Phys. Lett. B **559**, 20 (2003).

[18] M.Castellano *et al.*, Nuovo Cim. A **14**, 1 (1973).

[19] B.Delcourt *et al.*, Phys. Lett. B **86**, 395 (1979).

[20] D.Bisello *et al.*, Nucl. Phys. B **224**, 379 (1983); Z. Phys. C **48**, 23 (1990).

[21] G.Lepage, S.Brodsky, Phys. Rev. Lett. **43**, 545 (1979).

[22] BaBar Collaboration, B.Aubert *et al.*, Nucl. Instr. And Meth. A **479**, 1 (2002).

[23] H.Czyz, J.H.Khun, Eur. Phys. J. C **18**, 497 (2001).

[24] M.Caffo, H.Czyz, E.Remiddi, Nuovo Cim. A **110**, 515 (1997); Phys. Lett. B **327**, 369 (1994).

[25] S.Pacetti, contribution to this Conference.

[26] S.Brodsky, *Nucleon05* Frascati (2005).

[27] B.S.Zou, H.C.Chiang, Phys. Rev. D **69**, 034004 (2004).

[28] H.W.Hammer, U.G.Meissner, Eur. Phys. J. A **20**, 469 (2004).

[29] J.R.Ellis, M.Karliner, New J. Phys. **4**, 18 (2002).

[30] P.Cesselli, M.Nigro, C.Voci, Proc. Physics at Lear, Erice **11**, 465 (1999).

CHAIRMAN: R. BALDINI

Scientific Secretaries: P. Bolzoni, F.M. Cardano

DISCUSSION

- *Loginov:*

I have a question about the plot on top of slide 44. That in yellow is the expectation, right? So, what is the probability to have events on the left of the plot, near threshold, above the expectation value?

- *Baldini:*

I don't know in terms of probabilities, you may look at the errors. The evidence that there is something is slight. A real evidence of something similar to the threshold effect in the form factor is in the radiative decay of J/Ψ. The other two evidences are rather marginal but they are still pointing towards the same direction.

- *Loginov*:

You have one bin which is negative, what does it mean? And on the right top corner of this plot you have a small plot with the same distribution of J/Ψ?

- *Baldini:*

I think it is $B^+ \to K^+ \ J/\Psi \to K^+$ p p. It should be a point at 3.1 which has not been put in the plot. I added this two plots but I must insist: the real evidence that something similar to the threshold effect in the form factor is really happening in an other channel is the J/Ψ radiative decay, even if the quantum numbers of the proton antiproton system are not the same. It must not be a vector meson, the charge conjugation for the proton antiproton system must be +1, because you have an additional photon. So this is not an evidence of an effect similar to that you see in the form factor, is an evidence of the same effect but with other quantum numbers.

- *Cifarelli:*

What was the FENICE apparatus for the detection of neutrons antineutrons typical events? And how can DAΦNE2 contribute to this neutron antineutron physics?

- *Baldini:*

This detector was made mostly of limited streamer tubes and scintillators so the cross you see are limited streamer tubes fired by some particles and the strips are scintillators fired by the same particles. An antineutron annihilated in the middle of the detector. On the opposite side you have just one scintillator fired that people think is the neutron going in the other direction. So if you want to detect neutron antineutron you need a detector with a good time of flight and a lot of scintillators to detect neutrons and with some kind of tracking capabilities looking for the annihilation products of antineutron.

I come to the second question. Presently in Frascati there is a discussion on what to do with DAΦNE after two years from now. For two years from now the storage ring will continue to run for KLOE, to collect up to 2 fb^{-1}, and for the FINUDA experiment, which is an experiment looking for hypernuclei. Then we must have a decision on what to do. There are two options up to now: one option is to increase the luminosity, which actually is something of the order of 10^{32} cm^{-2} s^{-1}, the other option is to increase the energy of the machine as much as possible, permitted the cost of such an operation. To increase the luminosity is an hard job if you really want to reach 10^{34} cm^{-2} s^{-1}, and increasing the luminosity of only an order of magnitude very likely you won't make much more physics than now. Conversely to increase the energy of the storage ring is something not so expensive and difficult, the luminosity is guaranteed because increasing the energy all problems that limit the luminosity will disappear and so could be very easy to have a storage ring working for instance around 2 GeV in the centre of mass with a luminosity above 10^{32} cm^{-2} s^{-1}. To make a comparison, BaBar, transforming the initial state radiation into an equivalent storage ring, has a luminosity something less then 10^{29} cm^{-2} s^{-1}, let's say 7 10^{28} cm^{-2} s^{-1}, working for five years with steps of 10 MeV (because it depends on the bin width you want to examine). So, if you have a storage ring working at 10^{32} cm^{-2} s^{-1} you will have something like 10^{3} times more than the present BaBar data. The only problem is that if you go very near threshold you have to boost somewhat the centre of mass otherwise you cannot detect proton antiproton because they don't go out from the beam pipe, but in the case of DAΦNE this will not be a problem because there is a crossing angle and just increasing a bit this angle you have enough transferred boost to detect events in this case. Very likely a decision in this direction or not will be taken within the end of the year, in October there is a workshop in Frascati just discussing about nuclear form factors. Professor Zichichi is invited to give the first talk.

- *Loginov:*

I have a question on plots and numbers on slide 59, how do you get masses and widths from the plots above?

- *Baldini:*

The masses are just the peak values you see in that plots, with their corresponding resonance widths. What is striking for me is that the masses are indeed what you expect, and this is a model independent calculation, but the widths are larger. There is one thing more: if you look at the plot on slide 60 you have the magnetic form factor for the proton and the form factor for the pion and these are peaked at the same value, but that of the pion is scaled by a factor k respect to that of the proton. This factor k is roughly 3 that is what you would expect naively according to the Vector Mesons Dominance theory (VMD): the vector part of the proton magnetic form factor should be the anomalous magnetic momentum of the proton times the π form factor. This is exactly what you get from this plot. But what is really strange is that the width is much bigger and we don't understand why. Of course once the new BaBar data will be put in this calculation, once the electric form factor will be analyzed in the same way, once for instance the phase has been measured there will be a lot of constraints for this calculation, one can test if here there is physics or there is a trick in the calculation.

- *Zichichi*:

You used the word two family resonances, I wonder why you use this statement?

- *Baldini:*

Actually the plot on slide 60 was cut but there are two peaks for the proton and the neutron.

- *Loginov:*

How did you go from q^2 to the mass?

- *Baldini:*

The mass is just the value of the peak.

- *Appelquist:*

With respect to the π form factor there is a perturbative QCD prediction for asymptotic behaviour and it seems to me interesting a comparison between experimental curve and those theoretical predictions.

- *Baldini:*

There is no prediction for the asymptotic part in this model, the model is just used to evaluate the unphysical region.

- *Appelquist:*

I'm not sure I completely believe it, but people who believe that you can do perturbative QCD predictions of the asymptotic behaviour of the π form factor are giving you a formula that says it goes like $1/q^2$ and I've seen curves that compare the data to that asymptotic prediction and it looks like the π form factor is scaling like $1/q^2$. On my understanding it's well above in normalization as perturbative correction so even at the highest q^2 is not looking perturbative if you believe those calculations.

- *Baldini:*

I believe it too, there is the same factor 2. This kind of discrepancy you can see on slide 17 (even if I agree that is not corrected to reflect the space-like region) in effect is true also for the π, which is not in this plot of course.

- *Pacetti:*

In the calculation we have only used the behaviour $1/q^2$ with a logarithmic correction but, by using a subtracted dispersion relation, we are not dependent on the asymptotic behaviour because the kernel in the integral kills all the asymptotic behaviour.

- *Zichichi*:

I want to see the old Massam, Zichichi work.

- *Baldini:*

I apologize because here I forgot to change the sign but q^2 is negative on plot on slide 8. This is an elaboration of your calculation for the ratio between electric and magnetic form factor, times the proton magnetic momentum. Let me remind to the audience that Prof. Zichichi was the first one to realize the relevance of nucleon time-like form factors. Actually he made the first measurement via proton antiproton into electron positron. He invented the preshower technique to kill the multihadronic background, which was 10^6 times higher than the electron positron signal.

- *Zichichi:*

When we made this prediction there was a sort of theoretical revolution, because the accepted behaviour was flat.

- *Baldini:*

In fact, unfortunately it decreases too much but it was a calculation without any free parameter, if you have just one fit parameter you obtain something like experimental data.

- *Zichichi:*

So these points on the plot are the experimental data now?

- *Baldini:*

Yes, people forgot about because everybody believed in the scaling law.

- *Zichichi*:

Thank you.

- *Korthals-Altes:*

What is this parameter you mentioned that reduces the slope of the curve?

- *Baldini:*

You can figure it is a free parameter which says how much hard-core is the model respect to the VMD contribution.

- *Korthals-Altes:*

What is the physical meaning of the parameter?

- *Baldini:*

In this models the coupling of vector mesons the nucleon are not point-like but with an extended source with radius of the order of the inverse of the mass of the nucleon, plus an hard-core term and the free parameter says how much hard-core there is with respect to the VMD, while in the former Massam-Zichichi model they just used vector dominance, they didn't add this term. When Massam and Zichichi made their calculation actually there was a free parameter but the value they got was just the mass of the nucleon, which is the reasonable value for the extended source.

- *'t Hooft:*

When you compared the neutron and the proton form factor you said that one expects the asymptotic value of the ratio to be ¼, can you explain this? If you take some other charges of the quark you can get something else. Why do you think that neutron and proton form factor ratio is given by the ratio of squared charges of down and up quarks?

- *Baldini:*

Because in first approximation you go from proton to neutron changing an up into a down quark, so the amplitude is proportional to this ratio. Assuming there is a leading quark actually there is a prediction by Chernyak, Zitnitsky and others which says these are the up for the proton and the down for the neutron. It is the same if you say the ratio of magnetic moments squares, as in the space-like region, which means a factor of 1/2 instead of 1/4. What I want to emphasize is that in any case if valence quarks are the most important part in an electromagnetic interaction you would expect the neutron quite lower than the proton, as in the space like region.

GRAVITOMAGNETISM AND FRAME-DRAGGING: THEORY AND MEASUREMENT

Ignazio Ciufolini

Dip. Ingegneria dell'Innovazione, Università di Lecce, Via Monteroni, 73100 Lecce, Italy and INFN Sezione di Lecce E-mail: ignazio.ciufolini@unile.it

Abstract

After a brief introduction on frame-dragging and gravitomagnetic field, including an invariant characterization of gravitomagnetism, we describe the phenomena due to spin on test gyroscopes, test particles, clocks and photons. In particular we show that when different light beams are deflected by the mass of a rotating object, with angular momentum J, and then, by gravitational lensing, observed at a far point as different images of the same source with different angular positions, there may be a significant time delay between the different images due to the spin of the deflecting body. We then discuss the time delay in the travel time of photons propagating inside a massive rotating shell. We apply this time delay, due to the spin of the shell, to the case of gravitational lensing and we show that there may be an appreciable time delay between the arrival of different images at Earth. We then consider some astrophysical configurations: a typical rotating galaxy and a typical rotating cluster, or super-cluster, of galaxies; the spin-time-delay might be large enough to be detected at Earth. This phenomenon should then be taken into account in the modeling of the time delay of different images by gravitational lensing and might be measurable in some gravitational lensing images. The spin-time-delay might give a new observable in the study of the dark matter content in rotating galaxies and clusters.
We then describe the latest results in the measurement of gravitomagnetism of Earth and Lense-Thirring effect by laser ranged satellites. Gravity Probe-B, launched by NASA on April 20 2004, will try to measure the Earth frame-dragging with accuracy of 1 % or better. A future accurate determination of

the Lense-Thirring effect, at the level of 1 % accuracy, may be provided by the LARES/WEBER-SAT experiment to measure "frame-dragging" and to give other basic tests of fundamental physics and general relativity. Here we describe the 1995-2004 measurement of Lense-Thirring effect obtained by analyzing the orbits of the two laser-ranged satellites LAGEOS and LAGEOS II; this method has provided in 2004 a direct measurement of Earth's gravitomagnetism with accuracy of the order of 10 %. We first report the measurement of the Lense-Thirring effect, obtained in 2001 over nearly 8 years of data using the nodes of the LAGEOS satellites and the perigee of LAGEOS II: it fully agrees with the previous 1998 result over a period of 4 years only. Finally, we describe the 2004 determinations of Earth's frame-dragging, using the recently released Earth's gravity field models, generated by the space mission GRACE, and analyzing about 11 years of data of the *nodes*, *only*, of the LAGEOS satellites. This new analysis agrees with our previous measurements of the Lense–Thirring effect using the LAGEOS satellites and obtained with the JGM-3 and EGM96 Earth's models. However, this new determination is much more accurate and, especially, more robust than our previous measurements. Indeed, the present analysis uses the nodal rates of the two satellites only, making no use of the perigee rate, as in our previous analyses. By using the Earth model EIGEN-GRACE02S, we obtain a relative error of the order of 4 % to 8 % of the Lense–Thirring effect due to the uncertainties in the Earth static gravity field and a total root-sum-square error budget of about 5 % to 10 %due to all the error sources. Specifically, by using EIGEN-GRACE02S, we obtain: $\mu = 0.99 \pm 0.1$. This 2004 results fully confirm and improve our previous measurements of the Earth frame-dragging: the Lense-Thirring effect exists and its experimental value is within $\sim$ 10 % of what is predicted by Einstein's theory of general relativity.

Structure of this paper

- (1) Some historical background on measurement of gravitomagnetism and gravitational field inside a rotating shell
- (2) Frame dragging, the weak field and slow motion analogy. Gravitomagnetism, an invariant characterization
- (3) Gravitomagnetic phenomena on test gyroscopes, test particles, clocks and photons
- (4) Time-delay due to spin of a central body and inside a rotating shell
- (4.1) Spin-time-delay and gravitational lensing
- (4.2) Some astrophysical sources and spin-time-delay
- (4.3) Spacetime geometry inside a rotating shell
- (4.4) Time delay inside a slowly rotating massive shell
- (4.5) Some astrophysical sources and spin-time-delay due to an external rotating shell
- (4.6) Discussion and conclusion on spin-time-delay

1 Some historical background on the measurement of gravitomagnetism and frame-dragging, and on the gravitational field inside a rotating shell

In this introduction we briefly report on some past and present experiments to measure gravitomagnetism and frame-dragging and on the problem of the gravitational field inside a rotating shell, for a more exhaustive introduction we refer to Ciufolini and Wheeler[1].

Here, we also briefly describe the milestones that led to our 2004 measurement of Lense-Thirring effect and gravitomagnetism of Earth with accuracy of about 10 % [78], see section (5.3).

In 1915 Einstein published the theory of general relativity; among the inspiring intuitions for Einstein's gravitational theory there was the Mach's idea on the origin of inertia and inertial forces [2,1]. Mach thought that centrifugal and inertial forces were the result of rotation and accelerations with respect to the masses in the universe.

Influenced by Mach ideas several investigators studied the problem of the gravitational field inside a rotating shell. In a basic paper of 1918 Thirring published a solution of Einstein field equation representing the metric inside a rotating shell to first order in $\frac{M}{R}$, mass over radius of shell, and to first order in ω, angular velocity of shell [4]. In 1966 Brill and Cohen derived the metric inside a shell with arbitrary mass and to lowest order in the angular velocity [5]. An extension of the Brill-Cohen results to higher orders on ω was then published in 1985 by Pfister and Braun [6,7].

Nevertheless, the exact solution representing the spacetime geometry inside a shell with arbitrary mass and rotating with arbitrary angular velocity is still unknown. An exact solution inside a rotating shell would give us insight on the role of the "Mach principle" in general relativity.

Indeed, the level at which general relativity satisfies the Mach's idea on the origin of inertia has been discussed in a large number of books and papers (see for example refs [2,1]). However, general relativity satisfies at least a kind of "weak manifestation" of Mach's ideas: the dragging of inertial frames. Indeed, in Einstein's gravitational theory, the concept of an inertial frame has only a local meaning, and a local inertial frame is "rotationally dragged" by mass-energy currents because moving masses influence and change the orientation of the axes of a local inertial frame, i.e. of the gyroscopes; thus, a current of mass such as the spinning Earth "drags" and changes the orientation of the gyroscopes with respect to the distant stars.

It might be surprising to know that the first experiments to detect the gravitational influence of the rotation of a mass and to measure the dragging of a gyroscope by a rotating body were performed well before the development of Einstein's general relativity [1].

In 1896 Benedikt and Immanuel Friedländer [9] tried to measure the dragging effect due to a rapidly rotating, heavy fly-wheel on a torsion balance. Immanuel Friëdlander wrote "In the same way as centrifugal force is acting on a static wheel due to the rotation of the heavy earth and the cosmos, there should, I thought, appear on accordingly smaller scale a centrifugal force action on bodies near moving heavy fly-wheels. Would this phenomenon be detectable ...". In 1904, August Föppl [10] tried to measure the dragging effect on a gyroscope due to the rotation of Earth; he reached an accuracy of about 2% of Earth angular velocity. However, the general relativistic dragging effect on a gyroscope at the surface of Earth (at a European or U.S. latitude) is about 2×10^{-10} of its rotation rate! These experiments, performed before the development of general relativity, were inspired by Mach's ideas on inertia.

In 1916, De Sitter [11] calculated the tiny shift of the perihelion of Mercury due to the rotation of the Sun, a particular case of the shift of the pericenter of an orbiting test particle due to the angular momentum of the central body, see section 2. This shift of about $-0.002''$/century is about 5×10^{-5} times smaller than the standard general relativistic Mercury precession of $\simeq 43''$/century and is too small to be measured.

In their well-known 1918 paper, Lense and Thirring [8] calculated the gravitomagnetic secular perturbations of the moons of various planets, in particular the V moon of Jupiter has a considerable gravitomagnetic secular precession, however the observations do not yet allow the separability and the measurement of this effect.

In 1959, Yilmaz [12] proposed to use polar satellites to detect the gravitomagnetic field, avoiding in such a way the effects due to the non-sphericity of Earth gravity field. In 1976, Van Patten and Everitt [13] proposed to measure the Lense-Thirring nodal precession using two drag–free, guided satellites, counter-rotating in the same polar plane. The reason to propose two counter-rotating satellites was to avoid the error associated with the determination of the inclination.

In 1959-1960 [14,15] G. E. Pugh and Leonard Schiff independently proposed an experiment using orbiting gyroscopes, this became the well-known Gravity-Probe-B experiment, or GP-B (launched on April 20, 2004). The Stanford University group has been working for more than thirty years to make and fly super-conducting gyroscopes in an Earth-orbiting satellite [16]. At an altitude of about 650 km, the axis of a gyroscope is predicted to undergo a gravitomagnetic frame dragging of about 42 milliarcsec per year. GP-B should detect the gravitomagnetism of Earth and measure it to about one percent accuracy, or less.

Several other experiments have been proposed to measure the gravitomagnetic field, for a review see Ciufolini and Wheeler 1995 [1]; here we only mention the Foucault pendulum at the South Pole [17], the ring laser gyroscopes [18] and the orbiting gradiometers. In 1980-1989 the use of gravity gradient resonant detectors orbiting Earth [19] and superconducting gravity gradiometers in polar orbit [20] was proposed to measure magnetic components of the Riemann tensor, with the accuracy needed to detect the gravitomagnetic field. Some indirect astrophysical evidence of frame dragging was obtained in 1988 by the periastron precession rate of the binary pulsar PSR 1913+16 [21].

In 1976, the LAGEOS (LAser GEOdynamics Satellite) satellite was launched by NASA.

In 1984 and 1986 we published the key idea of using the nodes of two laser ranged satellites of LAGEOS type to measure the Lense-Thirring effect [22].

In June 1985, A. Zichichi published an article to support this idea and to describe this experiment [3]).

The fundamental idea [22,24] of this experiment to measure the nodal drag on two non-polar, passive, laser-ranged satellites, called the LAGEOS III experiment, is based on two considerations:

(a) position measurements of laser–ranged satellites, of LAGEOS (1976) type (see below), are accurate enough to detect on the node of LAGEOS type satellites the very tiny effect due to the gravitomagnetic field: the Lense–Thirring precession and

(b) to "cancel out" the enormous perturbations due to the non-sphericity of Earth gravity field, we need a new satellite with inclination supplementary to that of LAGEOS, and with the other orbital parameters, a and e, nearly equal to those of LAGEOS.

The accuracy of this experiment was estimated, by several studies and papers [24], to be, *in* 1988, of the order of 10% of the Lense-Thirring effect.

In 1989, the error analysis of the LAGEOS III experiment was performed under the supervision of an official NASA committee chaired by Peter Bender; then, in 1994, J. Ries et al., published a revised error budget of the LAGEOS III experiment, with a total error of 4 % and, in 1998, with a total error of 3%-4% [see: [25]], as reported in the Supplementary Discussion of [78]. The error analysis of our 2004-measurement of the Lense-Thirring effect with the LAGEOS and LAGEOS II satellites is substantially the same as the error analysis of the LAGEOS III/LARES experiment that was carried out in a very large number of papers and studies of NASA, ASI and ESA [24] (for differences and similarities between the error analysis of the proposed LAGEOS III/LARES experiment and the one of the recent measurement of the Lense-Thirring effect with LAGEOS and LAGEOS II [78], see the Supplementary Discussion of [78] and ref. [76]. Here we just stress that *the studies of the LAGEOS III/LARES experiment were fundamental milestones in this field, indeed they analyzed in details all the main modelling errors in the nodal rates of LAGEOS type satellites.*

The idea to use the nodes of N satellites of LAGEOS type to cancel the effect of the first N-1 Earth even zonal harmonics and to measure the Lense-Thirring effect was published in 1989 [24], see also [1]. Our 2004 measurement [78] is simply the case of N=2 using two satellites only. In 1989 the error due to the even zonal harmonics was quite larger due to the much less accurate Earth gravity models available at that time and the LAGEOS II satellite was not yet launched.

In 1992, the LAGEOS II (LAser GEOdynamics Satellite 2) satellite was launched by the Italian Space Agency, ASI, and NASA.

In 1996 we published the idea to use the nodes of the two satellites LAGEOS and LAGEOS II to measure the Lense-Thirring effect, together with the corresponding formula also using the perigee of LAGEOS II [26].

Between 1995 and 2004, using the two laser-ranged satellites LAGEOS and LAGEOS II, we have performed [26–29,78,76] several measurements of the Lense-Thirring effect with accuracy ranging from about 50% to about 10%. They are described in section (5).

In 1998, the LARES experiment [25] was proposed and selected as a phase-A study by ASI, the Italian Space Agency. This space mission would allow a measurement of the Lense-Thirring effect at the 1% level and is briefly described in section (5).

In 2002, the GRACE satellites were launched by NASA and GFZ.

On April 20 2004, the Gravity Probe-B experiment was launched in order to try to measure the Earth's frame-dragging with 1% accuracy or better.

In 2004, we obtained an accurate measurement of Earth's gravitomagnetism and Lense-Thirring effect using the recently released Earth gravity field models generated by the space mission GRACE and the nodes of the LAGEOS satellites, by analyzing about 11 years of data. The accuracy of this recent measurement is of about 10 % [78] (see section 5.3), from this experiment we have concluded that the Lense-Thirring effect is in full agreement with the prediction of general relativity and its experimental value is within $\sim$ 10% of what is predicted by Einstein's general relativity [78].

This measurement reported in Nature [78] uses the nodes of the two laser ranged satellites LAGEOS and LAGEOS II in order to cancel the effect of the first even zonal harmonic coefficient, J_2, of Earth and to measure the Lense-Thirring effect, as we explicitly proposed in [26] and [32]. Furthermore, it uses the accurate model EIGEN-GRACE02S of the Earth gravity field developed by GFZ of Potsdam with the data of the GRACE satellites.

2 Frame dragging, the weak field and slow motion analogy. Gravitomagnetism, an invariant characterization

In the weak field and slow motion approximation, a formal analogy with electrodynamics has been developed by using the Einstein field equation and the geodesic equation. In geometrodynamics [2,1], in the weak field and slow motion approximation, for a stationary, localized, mass-energy distribution, the (0i) components of the Einstein field equation can be written in the Lorentz gauge: $\Delta\ h_{0i}\ \cong\ 16\ \pi\ \rho\ v^i$; formally analogous to the Maxwell-Ampère equation for the vector potential of electrodynamics, in the Coulomb gauge: $\Delta\ A^i\ =\ -\ 4\ \pi\ \rho_e\ v^i$. $\mathbf{h}$ has been called the gravitomagnetic potential. The **gravitomagnetic field** [1,30] has then been defined as: $\mathbf{H}\ =\ \nabla\ \times\ \mathbf{h}$. Furthermore, by the geodesic equation for a test particle of mass m, in the weak field and slow motion limit one has then: $m\ \frac{d^2\mathbf{x}}{dt^2}\ \cong\ m\ (\mathbf{G}\ +\ \frac{d\mathbf{x}}{dt}\ \times\ \mathbf{H})$; formally analogous to the Lorentz force. Where $\mathbf{G}\ \cong\ -\ \frac{M}{|\mathbf{x}|^2}\ \hat{\mathbf{x}}$ is the standard Newtonian acceleration and $\mathbf{H}$ is the gravitomagnetic field (see [1]).

This is the weak field and slow motion analogy of the gravitomagnetic field in geometrodynamics with the magnetic field of electrodynamics. However, a characterization of gravitomagnetism independent of any approximation has also been proposed [1,30] and is described below.

One might then call gravitomagnetism all those phenomena generated by mass-energy currents, acting as source on the right-hand side of the Einstein field equation. However, in the presence of any mass, by a boost one can always observe a mass current in the boosted frame. Therefore, a more rigorous definition of gravitomagnetism has been proposed.

This characterization of gravitomagnetism is independent on the frame and the coordinate system used, and is based on spacetime curvature invariants built using the Riemann curvature tensor (see below). It is also independent of any approximation.

In electromagnetism, in the frame in which an electric charge is at rest we only have a non-zero electric field but no magnetic field. However, if we consider an observer moving relative to the charge, in this new frame we have a magnetic field. Similarly, in general relativity, in the frame where a mass is at rest, the gravitomagnetic potential $\mathbf{h}$ is zero. However, if we consider an observer moving relative to the mass, in the local frame of the observer we have a nonzero gravitomagnetic potential $\mathbf{h}$.

Therefore, following the method to characterize curvature singularities and to classify different spacetimes [33], one should inspect the invariants of the spacetime. However, in a vacuum, as a consequence of the Einstein field equation, the Ricci curvature scalar: $R = R^{\alpha}{}_{\alpha}$ is identically equal to zero. Another scalar invariant is the Kretschmann invariant $R_{\alpha\beta\mu\nu}R^{\alpha\beta\mu\nu}$, however in the case of a metric characterized by mass and angular momentum, such as the Kerr metric, the Kretschmann invariant is a function of $\frac{M}{r^3}$ and $\frac{J}{r^4}$, with the leading term $\sim \frac{M}{r^3}$, therefore this invariant is different from zero in the presence of a mass M, whether or not there is any angular momentum.

Let us then use again the formal analogy between electromagnetism and weak field general relativity [34]. In electromagnetism to characterize the electromagnetic field one can calculate the scalar invariant $-\frac{1}{2}F_{\alpha\beta}F^{\alpha\beta} = E^2 - B^2$, which is analogous to the Kretschmann invariant $R_{\alpha\beta\mu\nu}R^{\alpha\beta\mu\nu} \sim \left(\frac{M}{r^3}\right)^2 + C\left(\frac{J}{r^4}\right)^2$, see below. However, in electrodynamics one can also construct the scalar pseudoinvariant $\frac{1}{4}F_{\alpha\beta}\,{}^*F^{\alpha\beta} = \mathbf{E}\cdot\mathbf{B}$ where * is the dual operation: ${}^*F^{\alpha\beta} = \frac{1}{2}\varepsilon^{\alpha\beta\mu\nu}F_{\mu\nu}$. We observe that if we have a charge q only, in its rest frame we have an electric field only, and the invariant $F_{\alpha\beta}\,{}^*F^{\alpha\beta}$ is zero, therefore, even in the frames where $\mathbf{B} \neq \mathbf{0}$ and $\mathbf{E} \neq \mathbf{0}$, this invariant will be zero. However, if in the rest frame we have a charge q and a magnetic dipole $\mathbf{m}$, in this frame we have in general $F_{\alpha\beta}\ {}^*F^{\alpha\beta} \neq 0$ and of course this invariant will be different from zero in any other frame.

Therefore, to characterize the spacetime geometry and curvature generated by mass-energy currents or by the intrinsic angular momentum, J, of a central body (in ref [1] is shown that in the weak field limit the angular momentum generated by mass-energy currents plays a role, in general relativity, analogous to the magnetic dipole moment generated by a loop of charge current in electromagnetism) we should look for an analogous spacetime invariant.

This invariant should therefore be built out of the dual of the Riemann tensor ${}^*R^{\alpha\beta\mu\nu} \equiv \frac{1}{2}\varepsilon^{\alpha\beta\sigma\rho}R_{\sigma\rho}{}^{\mu\nu}$, " squared" or "multiplied" by $R_{\alpha\beta\mu\nu}$. This pseudo-invariant is of the type [1,30] $\frac{1}{2}\,\varepsilon^{\alpha\beta\sigma\rho}\,R_{\sigma\rho}{}^{\mu\nu}\,R_{\alpha\beta\mu\nu}$. Because of the formal analogy with electromagnetism, and since this pseudo-invariant, ${}^*\mathbf{R}\cdot\mathbf{R}$, is built using the Levi Civita pseudo-tensor, it should change sign for time reflections: $t \to -t$, and therefore, it should be proportional to J. A list of all the possible spacetime invariants built out of the Riemann tensor and its dual is given in ref. [33].

The invariant ${}^*\mathbf{R}\cdot\mathbf{R}$ is in general relativity especially meaningful. In fact, whereas in classical electrodynamics ${}^*\mathbf{F}\cdot\mathbf{F}$ characterizes the electromagnetic field only, but not the spacetime geometry $\eta_{\alpha\beta}$, in geometrodynamics the invariant ${}^*\mathbf{R}\cdot\mathbf{R}$ characterizes the gravitational field and therefore the spacetime geometry.

Indeed, by calculating ${}^*\mathbf{R}\cdot\mathbf{R}$, the result is (for simplicity we just give here the weak field, lowest order, result; for the general, exact, expressions of $\mathbf{R}\cdot\mathbf{R}$ and ${}^*\mathbf{R}\cdot\mathbf{R}$, and related discussions see [1,30,31]):

$$ {}^*\mathbf{R}\cdot\mathbf{R} \simeq 288\,\frac{JM}{r^7} cos\theta + \cdots \tag{1} $$

whereas the Kretschmann invariant $\mathbf{R}\cdot\mathbf{R}$, for the Kerr metric, is in the weak field limit:

$$ \mathbf{R}\cdot\mathbf{R} \simeq 48\left(\frac{M^2}{r^6} - 21\,\frac{J^2}{r^8} cos^2\theta\right) + \cdots \tag{2} $$

Since the external gravitational field of a stationary black hole is determined by its mass M, charge Q and intrinsic angular momentum J, and since for the Kerr-Newman metric the invariant ${}^*\mathbf{R}\cdot\mathbf{R}$ is [31] still proportional to J, the above result is quite general in the case of black holes and is valid, asymptotically, in the weak field limit, for any quasi-stationary solution. Furthermore, the above result, that was obtained in Einstein theory, is generally valid in any metric theory of gravity (with no prior geometry) *not* necessarily described at the post-Newtonian order by the PPN formalism [1,30,31]; this can be seen in two ways. Let us first write the weak field, slow motion, expression of an asymptotically flat metric of a metric theory of gravity in the form: $g_{00} \simeq -1+2U+ \textit{higher order terms}$, $g_{ik} \simeq \delta_{ik}(1+2U)+ \textit{higher order terms}$, and $\mathbf{g} \equiv (g_{01}\,, g_{02}\,, g_{03})$. Then, the pseudo-invariant ${}^*\mathbf{R}\cdot\mathbf{R}$ can be easily calculated at the lowest order to be: ${}^*\mathbf{R}\cdot\mathbf{R} \simeq \nabla^2\left[\nabla U\cdot(\nabla\times\mathbf{g})\right]+ \textit{higher order terms}$.

In the case of a static distribution of matter and a corresponding static metric, with $\mathbf{g} = \mathbf{0}$, we then have $^*\mathbf{R} \cdot \mathbf{R} = 0$. However for a stationary distribution of matter and a corresponding stationary metric, with $\mathbf{g} \neq \mathbf{0}$, for example with $\mathbf{g} \sim J$, we have: $^*\mathbf{R} \cdot \mathbf{R} \neq 0$. In the first case, for a static distribution of matter and a static metric, with a boost with velocity $\mathbf{v}$ we have: $\mathbf{g} \sim \mathbf{v}U$, however $^*\mathbf{R} \cdot \mathbf{R}$ is, of course, still zero: $^*\mathbf{R} \cdot \mathbf{R} \sim \nabla^2 [\nabla U \cdot (\nabla \times \mathbf{v}U)] = 0$. A second argument confirms the validity of the above result in any metric theory of gravity not necessarily described by the PPN formalism. For a generic source, in any metric theory of gravity (with no prior geometry), the full expression of the scalar, $^*\mathbf{R} \cdot \mathbf{R}$ must be dependent from some of the intrinsic physical quantities characterizing the source, such as the total mass-energy of the source, its intrinsic angular momentum, its multipole mass moments... , i.e. must be dependent from some integral of the mass-energy density, ε of the mass-energy currents, $\varepsilon\, u^i, \ldots$. In particular, since $^*\mathbf{R} \cdot \mathbf{R}$ must change sign for time reflections, its full expression for a generic source, must be proportional to some odd function of the intrinsic mass-energy currents $\varepsilon\, u^i, \ldots$ (not eliminable with a change of origin or with a Lorentz transformation), characterizing the system, such as the intrinsic angular momentum of the source.

Therefore, independently from the field equations of a particular metric theory, the pseudo-invariant $^*\mathbf{R} \cdot \mathbf{R}$ may be used to determine the existence and the presence of "intrinsic" gravitomagnetism in that metric theory of gravity. Indeed, using this invariant $^*\mathbf{R} \cdot \mathbf{R} \sim \frac{JM}{r^7} cos\theta$ we can determine whether or not there is a gravitomagnetic contribution to the spacetime geometry and curvature. We just need to calculate $^*\mathbf{R} \cdot \mathbf{R}$, if it is different from zero we have a mass-energy current contribution to the spacetime curvature, if it is zero there is no mass-energy current contribution. No matter about local Lorentz trasformation or other frame and coordinate trasformations on a static background, either $^*\mathbf{R} \cdot \mathbf{R}$ is zero, as in the Schwarzschild case, or it is different from zero, as in the Kerr case. Of course a spacetime with $^*\mathbf{R} \cdot \mathbf{R} \neq 0$ is qualitatively different from a spacetime with $^*\mathbf{R} \cdot \mathbf{R} = 0$, no matter about frame and coordinate transformations.

In conclusion, we may say that gravitomagnetism [1,30,31] is that phenomenon such that the spacetime structure and curvature are determined and affected, not only by mass-energy, but also by mass-energy currents relative to other matter, i.e. mass-energy currents not generable or eliminable with a Lorentz transformation (for example the intrinsic angular momentum of a body that cannot be generated or eliminated by a Lorentz transformation). This characterization of gravitomagnetism is independent from the frame and the coordinate system used and is only based on spacetime curvature invariants.

3 Gravitomagnetic phenomena on test gyroscopes, test particles, clocks and photons

Einstein's general theory of relativity [1,2] predicts the occurrence of peculiar phenomena in the vicinity of a spinning body, caused by its rotation. The orbital period of a particle orbiting around a spinning body in the same direction as the rotation of the body, i.e. "co-rotating" with the central object, is longer than the orbital period of a particle orbiting at the same distance but in the opposite direction, "counter-rotating" with respect to the spin of the central object. The difference between co-rotating and counter-rotating orbital periods is:

$$\Delta\tau = 4\pi \frac{J}{M} \tag{3}$$

Furthermore, an orbiting particle around a spinning body has its orbital plane "dragged" around the spinning body in the same sense as the rotation of the body, and small gyroscopes that determine the axes of a local, freely falling, inertial frame, where "locally" the gravitational field is "unobservable", rotate with respect to "distant stars" due to the rotation of the body.

Thus, an external current of mass, such as the spinning Earth, "drags" and changes the orientation of gyroscopes. Indeed, a test gyroscope has a precession with respect to "an asymptotic inertial frame", in weak field, with angular velocity:

$$\dot{\Omega} = -\frac{1}{2}\mathbf{H} = \frac{[-\mathbf{J} + 3\,(\mathbf{J}\cdot\hat{\mathbf{x}})\,\hat{\mathbf{x}}]}{|\mathbf{x}|^3}, \tag{4}$$

where $\mathbf{J}$ is the angular momentum of the central object and $\mathbf{H}$ its gravitomagnetic field generated by $\mathbf{J}$ (see section 2). This is the "rotational dragging of inertial frames", or "frame dragging" ("dragging" as Einstein named it).

The whole orbital plane of a test particle is itself a kind of enormous gyroscope (for a motion under a central force) dragged by the gravitomagnetic field. Indeed, the orbit of a test particle around a central body with angular momentum $\mathbf{J}$ has a secular rate of change of the longitude of the line of the nodes (intersection between the orbital plane of the test particle and the equatorial plane of the central object), discovered by Lense-Thirring (1918) [8], in weak field:

$$\dot{\Omega}^{Lense-Thirring} = \frac{2\mathbf{J}}{[a^3\,(1-e^2)^{3/2}]}, \tag{5}$$

where a is the semimajor axis of the test particle, and e its orbital eccentricity. The pericenter of an orbiting test particle is also a kind of enormous gyroscope (for a motion under a central force $\sim 1/r^2$). Indeed, the orbit of a test particle has a secular rate of change of the mean longitude in the orbit L_0 (that is, $L_0 = n \cdot \Delta t + \tilde{\omega}$, where $n = 2\pi/P$ is the satellite' mean motion, P its orbital period, Δt the interval of time from passage of the satellite through the pericenter, $\tilde{\omega} = \Omega + \omega$ the longitude of the pericenter, and ω is the argument of the pericenter, that is the angle from the equatorial plane to the pericenter) and of the longitude of the pericenter $\dot{\tilde{\omega}}$, (defining the Runge-Lenz vector):

$$\dot{\tilde{\omega}}^{Lense-Thirring} = \frac{2J\,(\hat{\mathbf{J}} - 3\,cos\,I\,\hat{\mathbf{l}})}{[a^3(1-e^2)^{3/2}]}, \tag{6}$$

where $\hat{\mathbf{l}}$ is the orbital angular momentum, unit vector, of the test particle, and I its orbital inclination (angle between the orbital plane and the equatorial plane of the central object).

In 2004 the Lense-Thirring effect was measured with about 10% accuracy using the LAGEOS and LAGEOS II satellites [78], see section (5.2).

However, not only test particles and gyroscopes are affected by the spin of the central object but also photons and clocks. A photon co-rotating around a spinning body takes less time to return to a "fixed point" (with respect to distant stars) than a photon rotating in the opposite direction; in the Kerr metric [1] characterized by mass and angular momentum of the central object, a fixed point can be determined by constant Boyer-Lindquist spatial coordinates, i.e. by constant spatial coordinates of the weak field and slow motion metric (7) below; operationally a fixed point can be realized by a small telescope always pointing toward the same distant star, always oriented toward the center of the spinning body and at the same distance from it by using gradiometers and rockets attached to the telescope. For example, around the spinning Earth the difference between the travel-time of two pulses of electromagnetic radiation counter-propagating in the same circuit would be $\oint \frac{g_{0i}}{g_{00}}\, dx^i \sim \frac{8\pi\, J_\oplus}{r}$, or $\sim 10^{-16}$ [1,35]. Since light rays are used to synchronize clocks, the different travel-time of co-rotating and counter-rotating photons implies the impossibility of synchronization of clocks all around a closed path around a spinning body; the behavior of light rays, analyzed in this paper, and the behavior of clocks around a spinning body are intimately connected. Let us then briefly analyze the behavior of clocks around a spinning object. In several papers the "frame-dragging clock effect" around a spinning body has been estimated and space experiments have been proposed to test it [36–39,35]. We first observe that to synchronize clocks around a path in a stationary field we can use light rays or even very slowly moving clocks, so that special relativistic time-dilation is a higher order effect, always at the same distance from the central spinning body, so that mass time-dilation is equal for both clocks. Thus, when a clock, co-rotating very slowly (using rockets) around a spinning body and at a constant distance from it, returns to its starting point, it finds itself advanced relative to a clock kept there at "rest" (with respect to "distant stars", see above). Similarly a clock, counter-rotating arbitrarily slowly and at a constant distance around the spinning body, finds itself retarded relative to the clock at rest at its starting point [1,35]. For example, when a clock that co-rotates very slowly around the spinning Earth, at $\sim$ 6000 km altitude, returns to its starting point, it finds itself advanced relative to a clock kept there at "rest" (with respect to "distant stars") by $\oint \frac{g_{0i}}{g_{00}} dx^i \sim \frac{4\pi J_\oplus}{r} \sim 5\times10^{-17}$ s where $g_{0i} \sim \frac{2J_\oplus}{r^2}$ is the Earth gravitomagnetic field and $J_\oplus \cong 145\, cm^2$ is the Earth angular momentum. Similarly, a clock, that counter-rotates very slowly around the spinning Earth, finds itself retarded relative to a clock kept there at "rest" by the same amount. However, a larger clock effect, of the order of 10^{-7}, has been estimated in ref [36,37]; let us explain this apparent disagreement. The orbital period of a particle, or clock, freely co-orbiting (along a geodesic) around a spinning body is longer than the orbital period of a particle, or clock, freely counter-orbiting on the same path [36,37], see formula (7). The difference between the two orbital periods, i.e. the difference between the two times read by a clock at a fixed point (with respect to "distant stars", see above) when the two counter-rotating particles come back at this point after one revolution, is $\sim 4\pi J/M$, i.e. around the spinning Earth, is $\sim 1.4 \times 10^{-7} s$ [36–39]; this is basically the effect derived in [36,37]. Nevertheless, the difference between the time read by the two clocks when they meet again after a whole revolution is still $\sim 10^{-16}$ [35,38,39].

In Einstein's general theory of relativity, all these phenomena, on test particles, gyroscopes, photons and clocks, are the result of the rotation of the central mass.

4 Time-delay due to spin of a central body and inside a rotating shell

4.1 *Spin-time-delay and gravitational lensing*

Let us now study null geodesics around a rotating body; in particular we apply our results to the behaviour of photons. Null geodesics in the Kerr metric, also in regard to gravitational lensing and image's position, polarization and intensification, distortion and optical caustic, have been studied in several papers (see [40–42] and references therein; for gravitational lensing in a strong Schwarzschild field see [43]). However, here we derive the time delay in the arrival time of photons due to the angular momentum of the deflecting body. Using the weak gravitational field and slow motion approximation, we also derive and compare the light deflections by the angular momentum and quadrupole moment of the deflecting body.

By assuming weak gravitational field and slow motion of the source, we can write the spacetime metric at the order beyond Newtonian theory, the post-Newtonian order, in terms of small classical potentials determined by distribution and motion of mass-energy via the solution of Poisson-like equations, obtained by the weak field and slow motion limit of Einstein equation. If the source is stationary, with mass density ρ and mass-current density ρv^i, in order to study null geodesics, we thus have the metric [1,50]:

$$
\begin{aligned}
g_{00} &= -1 + 2U \\
g_{0i} &= -4V_i \\
g_{ik} &= (1 + 2U)\delta_{ik}
\end{aligned}
\tag{7}
$$

where δ_{ik} is the standard Knonecker delta and, by the Einstein equation, in the weak field and slow motion approximation, the classical potentials U and V_i satisfy the Poisson equations: $\Delta U = -4\pi\rho$, and $\Delta V_i = -4\pi\rho v_i$. V_i, or $-4V_i \equiv g_{0i}$, is the gravitomagnetic potential. As usual the Newtonian potential U for a central distribution of mass can be written as a multipole expansion; i.e. if we only include the monopole, M, and the quadrupole contributions: $U = \frac{M}{r} + \frac{1}{2}Q_{ij}\frac{x_i x_j}{r^5}$, where $Q_{ij} = \int(3x'_i x'_j - r'^2\delta_{ij})\rho(\mathbf{x}')d^3x'$ is the standard quadrupole moment tensor.

For an equilibrium ellipsoid [44,45], by assuming the outer surface to be equipotential, in spherical coordinates, we have: $U = \frac{M}{r}\left[1 - J_2(\frac{R}{r})^2 P_{20}(cos\theta)\right]$, where R is the equatorial radius of the ellipsoid and $P_{20} = \frac{1}{2}(3cos^2\theta - 1)$ is the Legendre associated function. At the lowest order in the flattening $f \equiv \frac{(R-R_p)}{R}$, where R_p is the polar radius of the ellipsoid, the quadrupole coefficient, J_2, is $J_2 = \frac{2}{3}f + O(f^2)$. If $\rho =$ constant, the quadrupole coefficient is $J_2 = \frac{2}{5}f + O(f^2)$.

The gravitomagnetic potential g_{0i}, i.e. the non-diagonal part of the metric tensor, from $\Delta V_i = -4\pi\rho v_i$, can be written:

$$
g_{0i}\,(\mathbf{X}) \cong -\,4\int \frac{\rho(\mathbf{x}')v^i(\mathbf{x}')}{|\mathbf{X} - \mathbf{x}'|}\, d^3x' \tag{8}
$$

where $\mathbf{X}$ is the position vector; far from the stationary source or and for a spheroidal rotating body with angular momentum $\mathbf{J}$, we then have:

$$
\mathbf{g}\,(\mathbf{X}) \cong \frac{-2\,\mathbf{J} \times \mathbf{X}}{|\mathbf{X}|^3} \tag{9}
$$

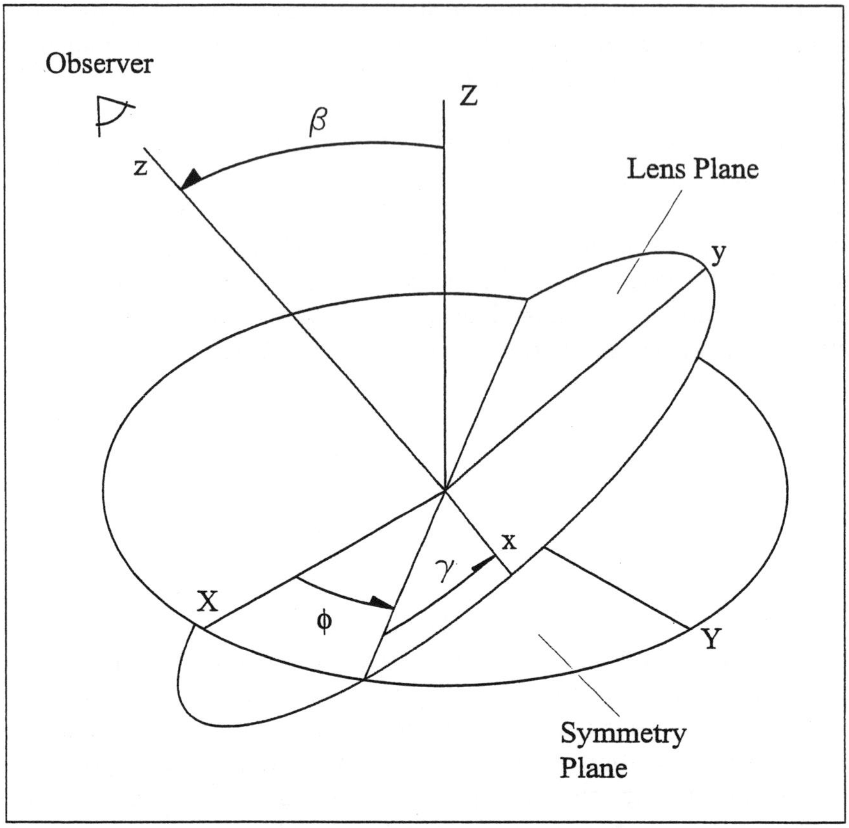

Fig. 1. Euler's angles in our notation. β is the angle between the Z axis of the body and the z axis through the observer. γ is the angle between the line of nodes (the red line) and the x axis on the lens plane, and ϕ is the angle between the line of the nodes and the X axis of the body. The origin of the coordinate systems is placed at the deflecting body.

and, when $\mathbf{J} = (0, 0, J)$, in spherical coordinates:

$$g_{0\phi} \cong - \frac{2J}{r} sin^2\theta \tag{10}$$

Let us derive time delay and deflection of electromagnetic waves due to spin and quadrupole moment of the central body. The quasi-Cartesian coordinate system (x, y, z) is the standard isotropic PPN (Parametrized Post Newtonian) system such that the coordinate z goes through the observer at Earth, while (X, Y, Z) is a coordinate system attached to the deflecting body. To relate the coordinate systems (x, y, z) and (X, Y, Z) we use the Euler's angles (ϕ, β, γ) (see Fig.1). For simplicity, we assume the deflecting body to be axially symmetric and Z, by definition, is the symmetry axis of the body. In a such case the shape of the body is invariant for rotations of ϕ and we can thus choose $\phi = 0$. The rotation from (X, Y, Z) to (x, y, z) is:

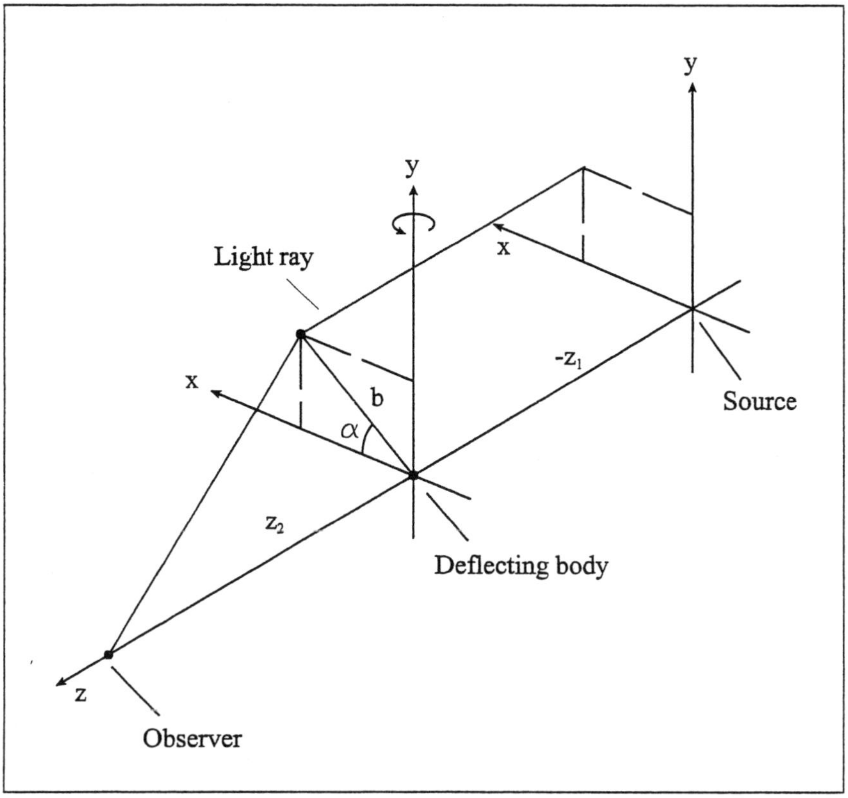

Fig. 2. Observer, source and deflecting body have the same x and y coordinates: the source is at $z = -z_1$ $(z_1 > 0)$, the observer at $z = z_2$ $(z_2 > 0)$ and the deflecting body at the origin of the coordinate system. b is the impact parameter and α is the angular position of a light ray on the lens plane. The source is very far from the deflecting body, so that we assume all the light rays from the source parallel to the optical axis (blue line).

$$\begin{pmatrix} X \\ Y \\ Z \end{pmatrix} = \begin{pmatrix} \cos\gamma & -\sin\gamma & 0 \\ \cos\beta\,\sin\gamma & \cos\beta\,\cos\gamma & -\sin\beta \\ \sin\beta\,\sin\gamma & \sin\beta\,\cos\gamma & \cos\beta \end{pmatrix} \begin{pmatrix} x \\ y \\ z \end{pmatrix} \tag{11}$$

We apply transformation (11) to the post Newtonian metric (7) and in the new coordinate system (x, y, z) we get:

$$\begin{aligned} ds^2 &= \overline{g}_{00}dt^2 + \overline{g}_{ij}dx^i\,dx^j \\ &+\tfrac{4\,J}{r^3}\left(y\cos\beta - z\cos\gamma\,\sin\beta\right)dx\,dt+ \\ &-\tfrac{4\,J}{r^3}\left(x\cos\beta - z\sin\beta\,\sin\gamma\right)dy\,dt+ \\ &+\tfrac{4\,J}{r^3}\left(x\cos\gamma\,\sin\beta - y\sin\beta\,\sin\gamma\right)dz\,dt \end{aligned} \tag{12}$$

where

$$\bar{g}_{00} = \left(-1 + 2\bar{U}\right)$$
$$\bar{g}_{ij} = \left(1 + 2\bar{U}\right)\delta_{ij}, \tag{13}$$

and:

$$\bar{U} = \frac{M}{r}\left[1 - J_2\left(\frac{R}{r}\right)^2 P_{20}\left(\frac{z\cos\beta + y\cos\gamma\sin\beta + x\sin\beta\sin\gamma}{r}\right)\right] \tag{14}$$

We can now easily derive the time delay using this metric element (12). In general, in a strong gravitational field, the time delays due to the gravitomagnetic field and to the nonsphericity of the matter distribution are nonlinearly coupled. However, in the weak field and slow motion limit, i.e. $\frac{J}{M r} \ll 1$ and $\frac{M}{r} \ll 1$, at the post-Newtonian order, we can separately analyze the two effects: the gravitomagnetic and the quadrupole moment time delays.

We have chosen the quasi-Cartesian coordinates such that the emitting and the deflecting bodies have the same x and y coordinates (see Fig. 2), but a different z coordinate: source, lens and observer are aligned. We have chosen this particularly simple configuration since here we are only interested in studying the time delay due to the gravitational field of the deflecting body (mass, quadrupole moment and gravitomagnetic time delay). However, there is an additional time delay, called geometric time delay [46, pages 143–147], due to the different geometrical path followed by different rays. Depending from the geometry of the system, this additional term can be very large and can be the main source of time delay. However, when we compare the time delay of photons that follow the same geometrical path we can neglect the geometric time delay, as in the case of two light rays with the same impact parameter but on different sides of the deflecting object. For of a small deflection angle with respect to the coordinate line $y = b\sin\alpha = constant$ and $x = b\cos\alpha = constant$ (see Fig. 2), the contribution to the travel time delay from the different path length due to the small deflection is of the order of $\sim \bar{U}^2$ [1] (to a small deflection angle of a photon path of the order of $\delta\phi \simeq \frac{4M}{r}$ corresponds a change in the total distance l traveled by the photon of the order of: $\delta l \simeq l\left(\frac{4M}{r}\right)^2$ and, depending on the geometrical configuration considered, this delay may need to be included in the total time delay. In a following paper we shall analyze higher order time delay and compare it with the gravitomagnetic time delay. Here, for simplicity, we neglect any geometrical time delay. Now, from the speed of light equal to c in a local inertial frame: $\eta_{\alpha\beta}dx^\alpha dx^\beta = 0$, we have in a general coordinate system: $ds^2 = g_{\alpha\beta}dx^\alpha dx^\beta = 0$; from this well known condition of null arc length along the world line of photons, we then have:

$$g_{00}\, dt^2 \cong -g_{33}\, dz^2 \tag{15}$$

or:

$$\frac{dz}{dt} \cong \pm\sqrt{-\frac{g_{00}}{g_{33}}} \tag{16}$$

thus, from (16), we get to first order in $\bar{U}$:

$$dt = \sqrt{\frac{1+2\bar{U}}{1-2\bar{U}}} dz \cong \left(1+2\bar{U}\right) dz \tag{17}$$

Integrating this expression from $z = -z_1$ to $z = z_2$ corresponding, respectively, to the position of source and observer, if $\bar{z} = z_1 \simeq z_2$ is much larger than the impact parameter b, we finally get:

$$\Delta t_{J_2} = 2\,\bar{z} + 4\,M \ln\left(2\frac{\bar{z}}{b}\right) + \frac{2\,M\,R^2\,J_2 \cos 2\,(\alpha+\gamma)\,\sin\beta^2}{b^2} \tag{18}$$

In the above expression the first term is the time that takes for a radio pulse to travel from the source to Earth in the absence of a central mass: $M = 0$. The second term is the Shapiro time delay and the third one is the additional delay due to the quadrupole moment, J_2, of the deflecting body.

Similarly, in regard to the time delay due to the angular moment J of the deflecting body, from the condition $ds^2 = 0$ and from $x =$ constant, $y =$ constant, solving (12) with respect to dt, we have:

$$dt = -\frac{\bar{g}_{0z}}{\bar{g}_{00}} dz \pm \frac{\sqrt{\bar{g}^2_{0z} dz^2 - \bar{g}_{00}\,\bar{g}_{zz} dz^2}}{\bar{g}_{00}} \tag{19}$$

where $\bar{g}_{0z} = \frac{4J}{r^3}\left(x\cos\gamma\,\sin\beta - y\sin\beta\,\sin\gamma\right)$. In the weak field and slow motion approximation, at the lowest order in $\bar{U}$ and $\bar{g}_{0z}$ expression (19), becomes

$$dt \simeq \bar{g}_{0z} dz \pm \left(1+2\bar{U}\right) dz \tag{20}$$

The first term in (20) is the gravitomagnetic time delay, while the other terms have already been evaluated in the previous case (18), they are just the time that takes for a photon to travel from source to observer and the Shapiro time delay by a mass. Thus, let us integrate the first term of (20) from $z = -z_1$ to

$z = z_2$ and assume that $\overline{z} = z_1 \simeq z_2 \gg b$. By setting, as before, y = constant and x = constant, we get the gravitomagnetic time delay:

$$\begin{aligned} \Delta t_J &= \lim_{\overline{z}/b \to \infty} \int_{-z_1}^{z_2} \frac{4J}{r^3} \left(x \cos\gamma \, \sin\beta - y \sin\beta \, \sin\gamma \right) dz \\ &= \frac{4J \cos(\alpha + \gamma) \, \sin\beta}{b} \end{aligned} \tag{21}$$

From (18) and (21) we see that Δt_{J_2} (the time delay due to the quadrupole moment) is of order $\sim \frac{1}{b^2}$ whereas Δt_J (the time delay due to the gravitomagnetic field) is of order $\sim \frac{1}{b}$. This shows that there is a value of the impact parameter b such that $\Delta t_J > \Delta t_{J_2}$ and, if the angular momentum of the deflecting body is large enough, the "spin time delay" may be a relevant effect

To derive the deflection of electromagnetic waves due to spin and quadrupole moment of the deflecting body, we use the geodesic equation in the weak field approximation, we then have the deflection angles due to the quadrupole moment J_2 and the deflection angles due to the angular momentum J [35]:

$$\begin{aligned} \delta_x^{J_2} &= -\frac{4M \cos\alpha}{b} - \frac{4 J_2 M \, R^2 \sin^2\beta \, \cos(3\alpha + 2\gamma)}{b^3} \\ \delta_y^{J_2} &= -\frac{4M \sin\alpha}{b} - \frac{4 J_2 M \, R^2 \sin^2\beta \, \sin(3\alpha + 2\gamma)}{b^3} \end{aligned} \tag{22}$$

$$\begin{aligned} \delta_x^{J} &= -\frac{4J \sin\beta \, \cos(2\alpha + \gamma)}{b^2} \\ \delta_y^{J} &= -\frac{4J \sin\beta \, \sin(2\alpha + \gamma)}{b^2} \end{aligned} \tag{23}$$

The first term in (22) is the standard deflection by a spherical object of mass M, whereas the second term is the additional deflection due to the quadrupole moment, J_2, of the central body.

Let us now study the possibility of determining the angular momentum J of the central deflecting body from time delay and deflection angle of different images of the source.

Let us consider three light rays emitted from a very far source, propagating parallel to the z axis and with the same impact parameter b, and let us assume that we are able to measure, determine or obtain [46] the following quantities: total time delay between the 3 rays, Δt_{12} and Δt_{13}; deflection angles δ_1, δ_2, δ_3; equatorial radius R of the deflecting body and distances from source and lens to the observer. In this way we are able to determine the angle α for each light beam and the impact parameter b, and we can write a system in which the only unknown quantities are: angular momentum, J, quadrupole moment, J_2, mass, M, and Euler's angle β and γ. Solving this system we can, in principle, determine the time delay due to the angular moment J and the other unknown quantities [35].

We have chosen a special case in which $x = y = 0$ for source, lens and observer and all the light rays have the same impact parameter. In this way, as we already remarked, we do not need to consider the relative time delays in the arrival time due to the different geometry of the paths traveled by the photons [46] and due to the difference in the Shapiro time delays by the central mass; these delays are in general much larger than spin and quadrupole moment time delays. Indeed, for other configurations in which the source is not exactly aligned with lens and observer, these effects, different geometry of the path traveled and difference in the standard Shapiro time delay, can be the main source of relative time delay. In these cases we would then need to model and remove these delays between the different images on the basis of the geometry of the system [35]. In special cases, for example if we would observe four images of the source and if the angle α of each deflected ray would differ by π – the Einstein Cross has a configuration nearly as this – we could, at least in principle, eliminate the time delay due to the quadrupole moment (see [35]) and thus determine the spin time delay.

4.2 Some astrophysical sources and spin-time-delay

Let us now calculate the time delay due to the spin of some astrophysical sources: the Sun, the lensing galaxy of the Einstein cross, Q2237+031, and a typical cluster of galaxies.

The Sun parameters are: $M_\odot = 1.477$ km, $R_\odot = 6.96 \cdot 10^5$ km, $J_{2\odot} \cong 1.7 \cdot 10^{-7}$ [1] and $a_\odot = \frac{J_\odot}{M_\odot} \cong 0.273$ km. Let us consider a co-rotating photon ($\frac{\pi}{2} \geq \alpha \geq -\frac{\pi}{2}$) and a counter-rotating one ($\frac{\pi}{2} \leq \alpha \leq \frac{3\pi}{2}$), both coming from infinity and propagating near the Sun with an impact parameter b nearly equal to the Sun equatorial radius $b \cong R = 6.96 \cdot 10^5$ km. Let us assume, for simplicity, that $\gamma = 0$ and $\beta = \frac{\pi}{2}$; the maximum gravitomagnetic and quadrupole moment time delays, according to (18) and (21), are then respectively:

$$\Delta t_{12}^{J} = \frac{8J}{b} = 1.54 \cdot 10^{-11}\, s \tag{24}$$

$$\Delta t_{12}^{J_2} = 4\frac{J_2 M R^2}{b^2} = 3.35 \cdot 10^{-12}\, s \tag{25}$$

The time delay due to the Sun spin could then, in principle, be detected using a laser-interferometer around the Sun; this would consist of a source and a detector on opposite sides of the Sun, both at a distance of about $8 \cdot 10^{10}$ km. The source, a laser, would emit, at the same time, photons toward the Sun but with slightly different angles so that they would travel on opposite sides of the Sun (i.e. photons co-rotating and counter-rotating with respect to the Sun spin). Then, by gravitational lensing, they would be focused and observed by the detector on the opposite side. Thus, according to the previous calculation, there would be a relative time delay in the arrival time of the photons due to the Sun spin. Of course all the other travel time delays should be modeled and removed from the observed delays, in particular the time delay due to the dispersion of electromagnetic waves by the solar plasma.

To derive the time delay due the lensing galaxy of the Einstein cross [47,48] we assume a simple model for rotation and shape of the central object. Details about this model can be found in [49].

The angular separation between the four images is about $0.9''$, corresponding to a radius of closest approach of about $R \simeq 650 h_{75}^{-1}$ pc, and the mass inside a shell with $R \simeq 650 h_{75}^{-1}$ pc is $\sim 1.4 \cdot 10^{10}\, h_{75}^{-1}\, M_{\odot}$ [48]. Let us assume: $J_2 \simeq 0.1$ and $J \simeq 10^{23}\, km^2\, h_{75}^{-2}$, we then have

$$\Delta t_{12}^{J} = \frac{8J}{b} \simeq 4\, min \tag{26}$$

$$\Delta t_{12}^{J_2} = 4\frac{J_2 M\, R^2}{b^2} \simeq 8\, hr \tag{27}$$

Thus, at least in principle, one could measure the time delay due to the spin of the lensing galaxy by removing the large quadrupole moment time delay using the previously described method [35]; of course one should be able to accurately enough model and remove all the other delays, due to other physical effects, from the observed time delays between the four images.

As a third example we consider the relative time delay of photons due to the spin of a typical cluster of galaxies; the precise calculations are shown elsewhere, nevertheless we give here some results. We consider a cluster of galaxies of mass $M_C \cong 10^{14} M_{\odot}$, radius $R_C \cong 5$ Mpc and angular velocity $\omega_C \cong 10^{-18} s^{-1}$; depending on the geometry of the system and on the path followed by the photons, we then find relative time delays ranging from a few minutes to several days [35].

4.3 Spacetime geometry inside a rotating shell

Let us first assume that if a body, or shell, has a steady rotation the spacetime is stationary (a spacetime is called stationary if it admits a timelike Killing vector field: ξ_t^{α}, then there exists some coordinate system in which ξ_t^{α} can be written $\xi_t^{\alpha} = (1, 0, 0, 0)$; from the Killing equation, in this coordinate system, the metric is thus time independent: $g_{\alpha\beta,0} = 0$). An external solution of this type is the stationary Kerr metric, characterized by mass and angular momentum of the central body. The well known Lense-Thirring metric [8,1] is, in Boyer-Lindquist coordinates, the weak-field and slow-motion limit of the Kerr metric. The Brill-Cohen, 1966, solution [5] describes the metric inside a shell with arbitrary mass and to lowest order in the angular velocity. It is a lowest order series expansion in the angular velocity ω of the shell on the Scharzschild background of a spherical mass shell of arbitrary mass M, valid both inside and outside the shell. The Brill-Cohen metric is:

$$ds^2 = -H(r)\,dt^2 + J(r)\,[dr^2 + r^2\,d\theta^2 + r^2\,sin^2\theta\,(d\phi - \omega(r)\,dt)^2] \tag{28}$$

where $H(r) = [(r-\alpha)/(r+\alpha)]^2$ for $r > R$; $H(r) = [(R-\alpha)/(R+\alpha)]^2$ for $r \leq R$; $J(r) = (1+\alpha/r)^4$ for $r > R$; $J(r) = (1+\alpha/R)^4$ for $r \leq R$; and where $\alpha = 2M$, and R, ω, and M are respectively the radius, angular velocity and mass of the shell (for further details on this solution see [7] and related papers in [51]).

However, in the following we shall only consider the weak field and slow motion solution of Thirring for the metric inside a rotating shell and, for simplicity, we neglect the stresses of the rotating shell (see [52,5,51] for related discussions and references). Inside a hollow, static, spherically symmetric distribution of matter, in vacuum, we have the flat metric $\eta_{\alpha\beta}$ [1]. Thus, in the weak field and slow motion limit, we assume that the metric inside a slowly rotating massive shell can be written: $g_{\alpha\beta} \cong \eta_{\alpha\beta} + h_{\alpha\beta}$, where $h_{00} = h_{ii} = 0$ and the (0i) components of the Einstein field equation then satisfy in the Lorentz gauge:

$$\Delta\, h_{0i} \cong 16\,\pi\,\rho\,v^i \tag{29}$$

with solution:

$$h_{0i}\,(\mathbf{x}) \cong -\,4\int \frac{\rho(\mathbf{x}')v^i(\mathbf{x}')}{|\mathbf{x}-\mathbf{x}'|}\,d^3x' \tag{30}$$

where ρ is the shell's mass density and ρv^i is the mass-current density.

We can then apply this result to determine the metric inside a thin shell of total mass M and radius R, rotating with small angular velocity ω, by integrating expression (30) inside the shell, where $\mathbf{v} = \omega \times \mathbf{x}$. By a rotation of the spatial axes so that $\omega = (0,0,\omega)$ and by using the mass density of a thin spherical shell, $\rho\,(\mathbf{x}') = \frac{M}{4\pi R^2}\,\delta(R-r')$ (we neglect here the stresses of the rotating shell [52,5,51]), we have:

$$\mathbf{h} = -4\int \frac{\rho(\mathbf{x}')\,(\omega\times\mathbf{x}')}{|\,\mathbf{x}-\mathbf{x}'\,|}\,r'^2\,d\Omega'\,dr' =$$

$$= -\frac{M}{\pi}\omega \times \int_0^{\pi} sin\theta'\, d\theta' \int_0^{2\pi} \frac{R\hat{\mathbf{x}}'}{|\mathbf{x} - R\hat{\mathbf{x}}'|}\, d\varphi' \tag{31}$$

The integral over $d\Omega \equiv sin\theta\, d\theta\, d\phi$, if $r < R$, is: $\int \frac{\hat{\mathbf{x}}'}{|\mathbf{x}-\hat{\mathbf{x}}'|}\, d\Omega' = \frac{4\pi}{3R}\mathbf{x}$. Therefore, for any $\mathbf{x}$ inside the shell, we have:

$$\mathbf{h} \equiv (h_{0x}, h_{0y}, h_{0z}) = -\frac{4}{3}\frac{M}{R}\omega \times \mathbf{x} =$$

$$= \left(\frac{4}{3}\frac{M}{R}\,\omega y,\ -\frac{4}{3}\frac{M}{R}\,\omega x,\ 0\right) \tag{32}$$

By substituting in the geodesic equation the components of $h_{\alpha\beta}$ inside the slowly rotating shell, we find the acceleration of a test particle inside a rotating shell, due to the spin of the shell [4,1,51]:

$$\begin{aligned} \ddot{x} &= -\frac{8}{3}\frac{M}{R}\,\omega\dot{y} + \frac{4}{15}\frac{M}{R}\,\omega^2\, x \\ \ddot{y} &= \frac{8}{3}\frac{M}{R}\,\omega\dot{x} + \frac{4}{15}\frac{M}{R}\,\omega^2\, y \\ \ddot{z} &= -\frac{8}{15}\frac{M}{R}\,\omega^2\, z \end{aligned} \tag{33}$$

where the ω^2 terms are due to the other components of $h_{\alpha\beta}$ and may be interpreted as a change in the inertial, and gravitational, mass-density of the shell due to the velocity ωR. Due to the rotation of the shell, the test particle is affected by forces *formally* similar to Coriolis and centrifugal forces. For discussions on the interpretation of the accelerations inside a rotating shell, we refer to [52,5,53,51].

Finally, we have that the axes of the local inertial frames, that is, the gyroscopes, are dragged by the rotating shell with constant angular velocity $\dot{\mathbf{\Omega}}^G$, according to [4,1]:

$$\dot{\mathbf{\Omega}}^G \cong -\frac{1}{2}\nabla \times \mathbf{h} = \frac{4}{3}\frac{M}{R}\,\omega \tag{34}$$

4.4 Time delay inside a slowly rotating massive shell

In section (4.1) we have calculated the time delay and the deflection due to the spin of a body from the rotation of the mass inside a radius r [35], however

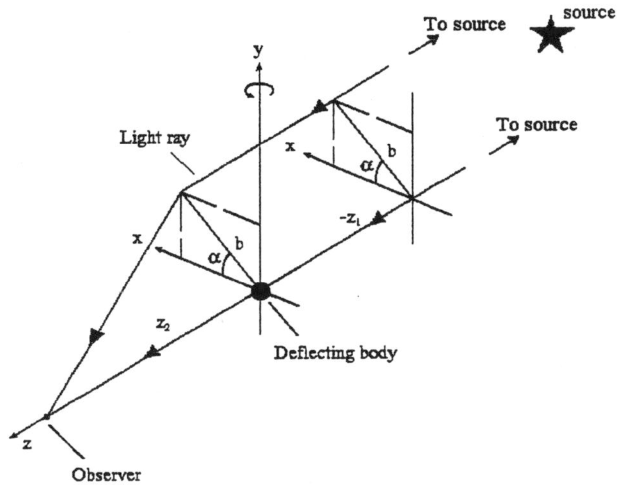

Fig. 3. Geometry of light rays, with impact parameters b and $-b$, propagating inside a rotating shell of radius R.

there is also a spin-time-delay, and an additional deflection, due to the rotation of the external mass [54].

Inside a thin shell of mass M and radius R', rotating with slow angular velocity $\omega = (0, 0, \omega)$, the $g_{0i} \cong h_{0i}$ components of the metric tensor are given by (32). Therefore, inside a rotating shell is not possible to synchronize clocks all around a closed path. Indeed, if we consider a clock co-rotating very slowly along a circular path with radius r ($r < R'$). When it comes back to its starting point it is advanced with respect to a clock kept there at rest (with respect to distant stars). The difference between the time read by the co-rotating clock and the clock at rest is given in [77]. Indicating with $\hat{\mathbf{r}}_1$ and $\hat{\mathbf{r}}_2$ the unit vectors from the center of a spherical shell to the two points on the shell where a light ray enters and leaves the sphere, respectively, we have the spin-time-delay due to the rotation of the shell [77]:

$$\Delta_{GM} = -\frac{4MR_0'}{3}\,\boldsymbol{\omega}\cdot(\hat{\mathbf{r}}_1\times\hat{\mathbf{r}}_2)\ , \tag{35}$$

A general expression of the relative spin-time-delay due the rotation of an external mass between two photons travelling inside the mass with (a) different impact parameters, b_1 and b_2, and (b) for any finite thickness of the external shell, is given in [77]

4.5 Some astrophysical sources and spin-time-delay due to an external rotating shell

Let us finally report the order of magnitude of the time delay corresponding to some astrophysical configurations.

For a lensing galaxy of the type of the lens of the "Einstein cross" [55,56], the relative time delay, due to the rotation of the external mass, of two photons travelling at a distance of $b_1 \simeq 650\,pc$ and $b_2 \simeq -650\,pc$ from the center is then [77]: $\Delta t \simeq 30\,min$.

As a further astrophysical example we consider two light rays deflected by a lensing galaxy which is inside a rotating cluster, or super-cluster of galaxies. We then calculate the amount of time delay due to the spin of the mass rotating around the deflecting galaxy. To get an order of magnitude of the time delay, we use typical super-cluster parameters [57]. By considering a galaxy in the center of the cluster and light rays with impact parameters $b_1 \simeq 15\,kpc$ and $b_2 \simeq -15\,kpc$ (of the order of the Milky Way radius), the time delay, by assuming for simplicity a constant mass density, is [77]: $\Delta t \simeq 1\,day$.

Finally, in [77] we show that if in general the lensing galaxy is not in the center of the cluster, the relative spin time delay between two photons, deflected by the galaxy, that are propagating inside a rotating cluster, or super-cluster of galaxies, may, under special conditions, be as large as several years.

4.6 Discussion and conclusion on spin-time-delay

We have derived and studied the "*spin time delay*" due to the angular momentum of a body experienced by the photons of two, or more, images of a source observed at a far point by gravitational lensing, this effect is due to the propagation of the photons in opposite directions with respect to the direction of the spin of the rotating body. We have both analyzed the spin time delay by a central rotating mass and in the case of photons propagating inside a massive rotating shell with time delay due to the rotation of the external mass.

We have also derived, in weak field and slow motion, the deflection in the path of the images due to the spin of the deflecting body. We have then compared the relative time delay of photons due to spin with the relative time delay due to the quadrupole moment of the central body.

Finally, in order to estimate the relevance of the spin time delay in some real astrophysical configurations, we have considered some possible astrophysical cases; nevertheless, these estimates are preliminary because we need to use more accurate values for angular momentum and other parameters of the considered astrophysical configurations.

We can then summarize the following conclusions of our analyses:

(a) The *spin time delay* must be taken into account in the modelling of relative time delays between images observed by gravitational lensing, in addition to other time delays such as the geometrical time delay and the delay due to the quadrupole moment of the lensing body.

(b) If other smaller time delays could be accurately enough modeled and removed from the observations, we have shown that the large relative delay due to the quadrupole moment of the lensing body could be removed, for some configurations of the images, by using special combinations of observables. With this method, we could measure the spin time delay due to the rotation of a mass.

(c) The measurement of the spin time delay might in principle be a new observable for the determination of the total mass of a rotating body, i.e. of the dark matter content of objects such as galaxies, cluster and super-clusters of galaxies [77].

(d) Depending on the geometry of the system, the relative spin time delay can be a quite large effect and may then be detected at Earth, in particular in systems with small angular separation and small relative time delay between the images such as B0218+357 [77].

5 Measurement of gravitomagnetism with laser ranged satellites

In section (5.1) we describe the proposed measurement of gravitomagnetism of Earth and Lense-Thirring effect, with relative accuracy of the order of 1%, using the satellite LARES. LARES would also perform other basic general relativistic tests. In sections (5.2) and (5.3) we report the 1995-2004 measurements of the Lense-Thirring effect obtained by analyzing the orbits of the laser ranged satellites LAGEOS and LAGEOS II, confirming the general relativistic prediction of frame-dragging with accuracy of the order of 10% in the most recent (2004) analysis [78].

5.1 LARES, LAser RElativity Satellite

The main scientific objectives of the LARES mission [25] are:

•• (1) High precision tests of Einstein's theory of general relativity, in particular:

• (1a) The accurate measurement of the Lense-Thirring effect due to the Earth's angular momentum and the high precision test of the Earth's gravitomagnetic field, with relative accuracy of the order of 1%.

• (1c) Tests of the equivalence principle and of the inverse square law for very weak-field gravity [25], improved by about four orders of magnitude at a range of a few thousands km, and high precision bounds on hypothetical very weak, long-range, gravitational forces.

• (1d) A 10^{-6} measurement (improved by about two orders of magnitude) of the PPN (Parametrized-Post-Newtonian) parameter α_1 testing the existence of preferred frames in some alternative metric theories of gravitation.

• (1e) A 10^{-3} measurement of the general relativistic perigee precession of LARES and a high precision measurement of the corresponding combination of the PPN parameters β and γ in the field of Earth. The PPN parameters β and γ test Einstein's theory of gravitation versus other metric theories of gravitation.

• Bounds on non-symmetric connections in some Einstein-Cartan theories with torsion propagating in vacuum.

• Other tests of general relativity and gravitation, such as improvements of the current limits on hypothetical spatial anisotropies of the gravitational interaction.

• (1b) A possible test, using the perigee of LARES with a specially designed orbit, of some recently proposed theories [79], based on a brane-world model, which can explain the dark-energy problem [80].

•• (2) Measurements and improved determinations in geodesy and geodynamics, in areas such as global plate tectonic, crustal deformation and determinations of variability of Earth's rotation [24,25].

The LARES (Laser RElativity Satellite) experiment is an improved version of the LAGEOS III experiment [22]. The main differences of LARES with respect to LAGEOS III are its weight and its orbital eccentricity. The new LARES satellite has been designed to be about 4 times lighter than LAGEOS, with a total weight of about 100 kg, and to be smaller than LAGEOS, with a radius of about 16 cm. The orbital eccentricity of LARES has been proposed to be 0.04 ± 0.01, whereas the proposed orbit of LAGEOS III had essentially zero eccentricity [25].

In references [22,24] and [25] is shown how, by combining the measured nodal precessions of LAGEOS, $\dot{\Omega}_{LAGEOS}$, and LARES, $\dot{\Omega}_{LARES}$, we can get a very accurate measurement of the Lense-Thirring effect, $\dot{\Omega}^{Lense-Thirring}$. The present 2004 error budget of the LARES experiment is dramatically reduced with respect to the previous estimates: present analyses show a *total statistical error in the LARES experiment of about 1%* $\dot{\Omega}^{Lense-Thirring}$*, or less,* over a 3-year period. The main improvements for this substantial reduction of the total statistical error are described in ref. [25,58].

In addition to the high precision measurement of the Lense-Thirring effect due to the Earth angular momentum, the LARES experiment will provide other important general relativistic and gravitational measurements, described above.

In the LARES phase-A study is also shown how, by measuring the LARES perigee rate, we could improve our present tests of the equivalence principle, at a range of a few thousands km, by about four orders of magnitude [25]. These tests will be realized by the use of the pericenter of LARES [59]. Indeed, a very effective way to detect a very weak, Yukawa-like, gravitational interaction is via a precise measurement of the pericenter precession; however, the precision of such measurement is inversely proportional to the orbital eccentricity, therefore the orbit of a new LAGEOS-type satellite with larger orbital eccentricity would be more effective to detect a new, hypothetical, very weak gravitational force with a range of the order of two Earth radii and would improve the present limits on such interaction by at least four orders of magnitude [60].

5.2 *The previous 1995-2001 measurements of the Lense-Thirring effect using the node of LAGEOS and the node and perigee of LAGEOS II*

5.2.1 Method

In section (3) we have seen that the whole orbital plane is dragged by the spin of the central object, this is the Lense-Thirring effect (5).

Let us now describe the 1995-2001 measurements of gravitomagnetism of Earth and Lense-Thirring effect using laser ranged satellites.

Our detection and measurement of the Lense-Thirring effect was obtained by using the satellite laser ranging data of the satellites LAGEOS (LAser GEOdynamics Satellite, NASA) and LAGEOS II (NASA and ASI, the Italian Space Agency) and the Earth gravitational field models, JGM-3 and EGM-96.

The measurement of distances has always been a fundamental issue in astronomy, engineering, and science in general. So far, laser ranging has been the most accurate technique for measuring the distances to the moon and artificial satellites [61,62]. Short-duration laser pulses are emitted from lasers on Earth, aimed at the target through a telescope, and then reflected back by optical cube-corner retroreflectors on the moon or the artificial satellite [63], such as LAGEOS. By measuring the total round-trip travel time, one can determine the distance to a retroreflector on the moon with an accuracy of about 2 cm and to the LAGEOS satellites with a few millimeters accuracy.

The LAGEOS satellites are heavy brass and aluminum satellites, of about 406 kg weight, completely passive and covered with retroreflectors, orbiting at an altitude of about 6,000 km above the surface of Earth. LAGEOS, launched in 1976 by NASA, and LAGEOS II, launched by NASA and ASI in 1992, have an essentially identical structure but they have different orbits. The semimajor axis of LAGEOS is $a \cong 12,270$ km, the period $P \cong 3.758$ hr, the eccentricity $e \cong 0.004$, and the inclination $I \cong 109.9°$. The semimajor axis of LAGEOS II is $a_{II} \cong 12,163$ km, the eccentricity $e_{II} \cong 0.014$, and the inclination $I_{II} \cong 52.65°$.

We analyzed the laser-ranging data using the principles described in [64] and adopted the IERS conventions [65] in our modeling, except that, in the 1998 analysis, we used the static and tidal EGM-96 model [66]. Error analysis of the LAGEOS orbits indicated that the EGM-96 errors can only contribute periodic root-sum-square errors of 2 to 4 mm radially, and in all three directions they do not exceed 10 to 17 mm. The initial positions and velocities of the LAGEOS satellites were adjusted for each 15-day batch of data, along with small variations in their reflectivities. Solar radiation pressure, Earth albedo, and anisotropic thermal effects were also modeled [67–70]. In modeling the thermal effects, the orientation of the satellite spin axis was obtained from [71]. Lunar, solar, and planetary perturbations were also included in the equations of motion, formulated according to Einstein's general theory of relativity with the exception of the Lense-Thirring effect, which was purposely set to zero. All of the tracking station coordinates were adjusted (accounting for tectonic motions) except for those defining the TRF terrestrial reference frame. Polar motion was also adjusted, and Earth's rotation was modeled from the very long baseline interferometry-based series SPACE96 [72]. We analyzed the orbits of the LAGEOS satellites using the orbital analysis and data reduction software GEODYN II [73].

The node and perigee of LAGEOS and LAGEOS II are dragged by the Earth's angular momentum. From the Lense-Thirring formula [22,26], we get $\dot{\Omega}_I^{Lense-Thirring} \cong 31$ mas/yr and $\dot{\Omega}_{II}^{Lense-Thirring} \cong 31.5$, where mas is a millisecond of arc. The argument of pericenter (perigee in our analysis), ω, also has a Lense-Thirring drag [1], thus, we get for LAGEOS: $\dot{\omega}_I^{Lense-Thirring} \cong 32$ mas/yr, and for LAGEOS II: $\dot{\omega}_{II}^{Lense-Thirring} \cong -57$ mas/yr [26]. The nodal precessions of LAGEOS and LAGEOS II can be determined with an accuracy of the order of 1 mas/year. Over our total observational period of about 4 years, we obtained a root mean square (RMS) of the node residuals of about 4 mas for LAGEOS and about 7 mas for LAGEOS II [29]. For the perigee, the observable quantity is the product $e \cdot a \cdot \dot{\omega}$, where e is the orbital eccentricity of the satellite. Thus, the perigee precession $\dot{\omega}$ for LAGEOS is difficult to measure because its orbital eccentricity e is $\sim 4 \times 10^{-3}$. The orbit of LAGEOS II is more eccentric, with $e \sim 0.014$, and the Lense-Thirring drag of the perigee of LAGEOS II is almost twice as large in magnitude as that of LAGEOS. Over about 4 years, we obtained a root mean square of the residuals of the LAGEOS II perigee of about 25 mas [29], whereas the total Lense-Thirring effect on the perigee is, over 4 years, $\cong -228$ mas.

To precisely quantify and measure the gravitomagnetic effects we have introduced the parameter μ that is by definition 1 in general relativity [1] and zero in Newtonian theory.

The main error in this measurement is due to the uncertainties in the Earth's even zonal harmonics and their time variations. The unmodeled orbital effects due to the harmonics of lower order are comparable to, or larger than, the Lense-Thirring effect. However, by analyzing both the JGM-3 and the EGM-96 models with their uncertainties in the even zonal harmonic coefficients, and by calculating the secular effects of these uncertainties on the orbital elements of LAGEOS and LAGEOS II, we find [26] that the main sources of error in the determination of the Lense-Thirring effect are concentrated in the first two even zonal harmonics, J_2 and J_4. We can, however, use the three observable quantities $\dot{\Omega}_I$, $\dot{\Omega}_{II}$ and $\dot{\omega}_{II}$ to determine μ [26], thereby avoiding the two largest sources of error, those arising from the uncertainties in J_2 and J_4. We do this by solving the system of the three equations for $\delta\dot{\Omega}_I$, $\delta\dot{\Omega}_{II}$ and $\delta\dot{\omega}_{II}$ in the three unknowns μ, J_2 and J_4, obtaining:

$$\delta\,\dot{\Omega}^{Exp}_{LAGEOSI} + c_1\,\delta\,\dot{\Omega}^{Exp}_{LAGEOSII} + c_2\,\delta\,\dot{\omega}^{Exp}_{LAGEOSII} =$$

$$= \mu\,(\,31\,+\,31.5\,c_1\,-\,57\,c_2)\,mas/yr\,+\,other\ errors \cong \mu\,(60.2\,mas/yr),\tag{36}$$

where $c_1 = 0.295$ and $c_2 = -0.35$. Equation (36) for μ does not depend on J_2 and J_4 nor on their uncertainties; thus, the value of μ that we obtain is unaffected by the largest errors, which are due to δJ_2 and δJ_4, and is sensitive only to the smaller errors due to δJ_{2n}, with $2n \geq 6$.

Similarly, regarding tidal, secular, and seasonal changes in the geopotential coefficients, the main effects on the nodes and perigee of LAGEOS and LAGEOS II, caused by tidal and other time variations in Earth's gravitational field [24,28], are due to changes in J_2 and J_4. However, the tidal errors in J_2 and J_4 and the errors resulting from other unmodeled, medium and long period, time variations in J_2 and J_4, including their secular and seasonal variations, are eliminated by our combination of residuals of nodes and perigee. In particular, most of the errors resulting from the 18.6- and 9.3-year tides, associated with the lunar node, are eliminated in our measurement. An extensive discussion of the various error sources that can affect our result is given in [28], only a brief discussion of the error sources is given below in section 5.2.2.

5.2.2 Results

In this section we report the 1995-2001 results of our measurements.

In [28] we measured the Lense-Thirring effect using the linear combination of the residuals of the nodes of LAGEOS and LAGEOS II and perigee of LAGEOS II according to Eq. (36) to eliminate the δJ_2 and δJ_4 errors, using the Earth gravitational model JGM-3, over a 3.1-year period.

In Fig. 4, we display the improved analysis [29] (obtained with the linear combination of the residuals of the nodes of LAGEOS and LAGEOS II and perigee of LAGEOS II according to Eq. (36)) using the more recent static and tidal Earth gravitational model EGM-96, we also refined the non-gravitational perturbations model, the total period of observations was of 4 years, longer of about 1 year with respect to the observational period corresponding to Fig. 4. We only removed four small periodic residual signals and the small observed inclination residuals. The removal of the periodic terms was achieved by a least squares fit of the residuals using a secular trend and four periodic signals with 1044-, 820-, 569-, and 365.25-day periods, corresponding to, respectively,

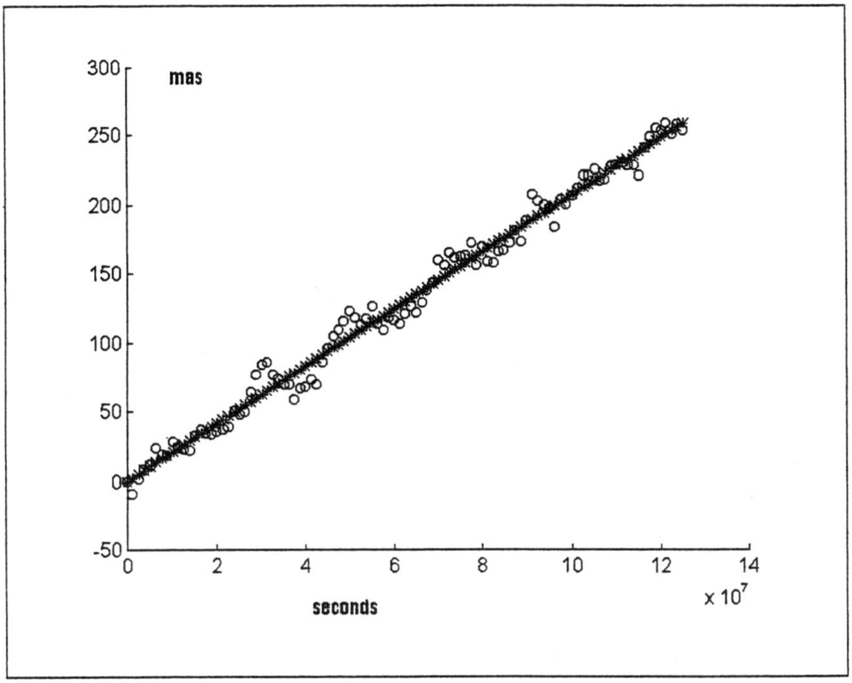

Fig. 4. Combination of the residuals of the nodes of LAGEOS and LAGEOS II and perigee of LAGEOS II according to Eq. (36), using the Earth gravitational model EGM-96, over a 4-year period. The best-fit line shown through these combined residuals has a slope of about $\mu_{Measured} \cong 1.1$.

the nodal period of LAGEOS, the perigee and nodal periods of LAGEOS II, and 1 year. The 820-day period is the period of the main odd zonal harmonics perturbations of the LAGEOS II perigee; the 1044- and 569-day periods are the periods of the main tidal orbital perturbations, with l=2 and m=1, which were not eliminated using Eq. (36). Some combinations of these frequencies correspond to the main non-gravitational perturbations of the LAGEOS II perigee. We notice that this analysis, using EGM-96 and its accurate tidal model, is substantially independent of the removed signals, whereas the previous analysis [28] was in part sensitive on the periodic terms included in the fit. In other words, our value (Fig. 4) of the secular trend does not significantly change by fitting additional periodic perturbations, and indeed, even the fit of the residuals with a secular trend only, with no periodic terms, just increases the slope by less than 10 %. Nevertheless, in this case, the root mean square of the post-fit residuals increases by about four times with respect to Fig. 4.

Our best-fit straight line of Fig. 4, using the EGM-96 gravitational model through the combined residuals of nodes and perigee, has a slope $\mu^{Measured} \cong 1.1 \pm 0.03$, where 0.03 is the standard deviation of the fit. This combined, measured, gravitomagnetic perturbation of the satellites' orbits corresponds, in a 4-year period, to about 16 m at the LAGEOS altitude, that is, about 265 mas.

The root mean square of the post-fit combined residuals corresponding to Fig. 4 is about 9 mas. Our total systematic error is estimated to be of the order of 30% to 50% of μ_{GR}.

Using the JGM-3 covariance matrix, we found the errors due to the uncertainties in the even zonal harmonics J_{2n}, with $2n \geq 6$, to be: $\delta\,\mu^{even\,zonals:J_{2n}\geq J_6} \lesssim 17\,\%$ of μ_{GR}, and using the EGM-96 covariance matrix: $\delta\,\mu^{even\,zonals:J_{2n}\geq J_6} \lesssim 13\,\%$ of μ_{GR}. The errors in the modeling of the perigee rate of LAGEOS II due to the uncertainties in the odd zonal harmonics J_{2n+1}, are, with EGM-96: $\delta\,\mu^{odd\,zonals} \lesssim 2\,\%$ of μ_{GR}. Using the EGM-96 tidal model, we estimated the effect of tidal perturbations and other variations of Earth gravitational field to be $\delta\,\mu^{tides\,+\,othervariations} \lesssim 4\,\%$ of μ_{GR}. On the basis of analyses [28,74] of the non-gravitational perturbations — in particular, those on the perigee of LAGEOS II — we found $\delta\,\mu^{non-gravitational} \lesssim 20\%$ of μ_{GR}, including uncertainties in the modeling of the satellites' reflectivities, and the error due to uncertainties in the orbital inclinations of LAGEOS and LAGEOS II was estimated to be $\delta\,\mu^{inclination} \lesssim 5\,\%$ of μ_{GR}.

Taking into account all of these error sources, we arrived at a total root-sum-square error of the order of 30% of μ_{GR}. Therefore, over an observational period of 4 years and using EGM-96, we determined $\mu^{Measured} = 1.1 \pm 0.3$, where 0.3 is the estimated total uncertainty due to all the error sources.

Based on the 1995-2001 analyses of the orbits of the laser-ranged satellites LAGEOS and LAGEOS II, we conclude that the gravitomagnetic or Lense-Thirring effect exists and its value agrees with the prediction of Einstein's theory of general relativity.

A basic concern in our analyses was to estimate the total error in our measurement of the Lense-Thirring effect. In order to support our measurement and the corresponding error analysis, we have performed a test and a preliminary blind-test simulation. Finally, below we report our latest, 2001, measurement of the Lense-Thirring effect over 7.3 years of data of LAGEOS and LAGEOS II, obtained by only modeling the radiation pressure coefficient of LAGEOS II (see Fig. 5).

This 2001 measurement fully confirms and improves our previous results and gives: $\mu \cong 1\pm0.3$ (where ±0.3 is the estimated total systematic error) and fully agrees with the general relativity prediction. It is important to observe that: (1) in the analysis corresponding to Fig. 5 we only modeled on LAGEOS II the radiation pressure coefficient of the satellite, i.e. the reflectivity coefficient, C_R, and no other parameters such as the along-track accelerations as in our previous analysis corresponding to Figures 4 (the C_R of LAGEOS II shows an apparent decay, in agreement with previous measurements [75]); (2) the

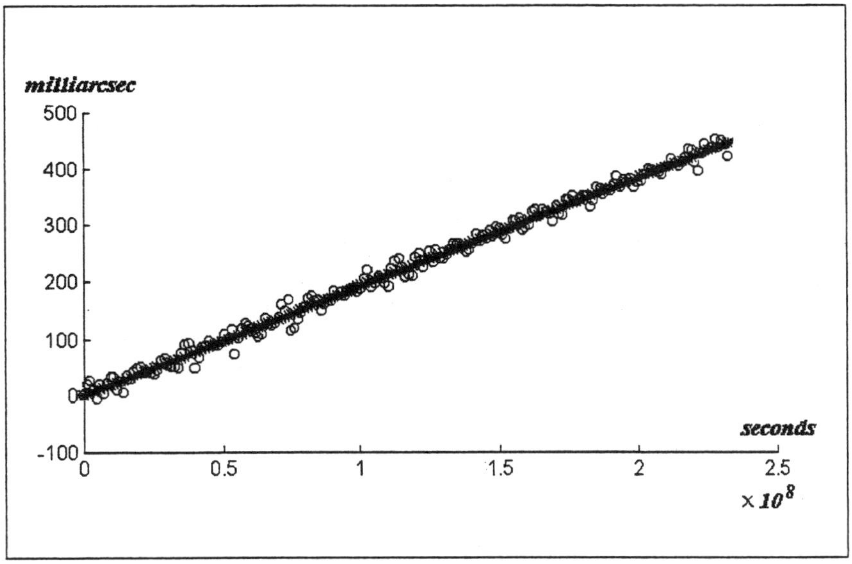

Fig. 5. Latest, 2001, measurement of the Lense-Thirring effect using LAGEOS and LAGEOS II, obtained by only modeling the radiation pressure coefficient of LAGEOS II, over nearly 8 years of data. The best-fit line shown through these combined residuals has a slope of about $\mu_{Measured} \cong 1 \pm 0.02$. The total estimated systematic error is about ± 0.3.

RMS of the residuals corresponding to Fig. 5 is about 10 milliarcsec whereas the total measured signal is about 440 milliarcsec, and finally (3) the quality of the fit and of the corresponding measurement can be further improved by further reducing the RMS of the 15-day fits (corresponding to each point of Fig. 5) with further processing of the data using GEODYN/SOLVE, thus further reducing the RMS of the final fit in Fig. 5.

In Fig. 5 we display our 2001, measurement of the Lense-Thirring effect using LAGEOS and LAGEOS II, obtained by just modeling the radiation pressure coefficient of LAGEOS II, over nearly 8 years of data, i.e. over an observational time nearly double than the longer period of our previous analyses [76].

This measurement improves our previous results and confirms the general relativistic prediction of frame-dragging. In the analysis corresponding to Fig. 5 we only modelled on LAGEOS II the radiation pressure coefficient of the satellite and no other parameters such as the along-track accelerations. The RMS of the residuals corresponding to Fig. 5 is about 10 milliarcsec whereas the total measured signal is about 440 milliarcsec; nevertheless the quality of the fit and of the corresponding measurement can be further improved by further reducing the RMS of the 15-day fits.

5.3 The recent, 2004, accurate measurements of the Lense-Thirring effect using the nodes only of the LAGEOS satellites

5.3.1 Method

The accurate measurement of the Lense–Thirring effect described in this section has been obtained, over a period of observation of about 11 years, by using the laser–ranging data of the satellites LAGEOS and LAGEOS II and the recently released Earth gravitational field model EIGEN-GRACE02S (and GGM01S) generated by the dedicated satellites GRACE [23].

The models used in this orbital analysis are described on Table 1.

In section (3) we have seen how the node and the perigee of a test-particle are dragged by the angular momentum of a central body. *However, whereas in our previous determination of the Lense–Thirring effect, described in section (5.2), we both used the nodes of LAGEOS and LAGEOS II and the perigee of LAGEOS II, in the present analyses we have only used the nodes of LAGEOS and LAGEOS II.* Indeed, the perigee of an Earth satellite such as LAGEOS II is affected by a number of perturbations whose impact in the final error budget is not easy to be assessed and this was one of our concerns (the other concern [81–83,23] was, in the EGM96 model, some favorable correlation of the errors of the Earth's spherical harmonics that might lead to some underestimated error budget). However, this concern is absent in the present analyses. Indeed, using the previous models JGM-3 and EGM96, we needed three observables and thus we also needed to use the perigee of LAGEOS II. However, with the recently released solution EIGEN-GRACE02S (and also using GGM01S) [81–83,23], thanks to the more accurate determination of the Earth's gravity field, it is just enough to eliminate the uncertainty in the quadrupole moment and thus to just use two observables, i.e. the two nodes of the LAGEOS satellites.

The nodal precessions of LAGEOS and LAGEOS II can be determined with an accuracy of the order of 1 mas/year, or less. Over our total observational period of about 11 years, we obtained a root mean square of the post–fit residuals of the nodes, combined with formula (37) below, of about 5.5 mas both with EIGEN-GRACE02S (and 6.7 mas with GGM01S).

The main error in this measurement is due to the uncertainties in the Earth's even zonal harmonics and their time variations. The un–modelled orbital effects due to the harmonics of lower degree are of order of magnitude comparable to the Lense–Thirring effect (see ref. [76]). However, by analyzing the EIGEN-GRACE02S (and GGM01S) model and its uncertainties in the even zonal harmonics, and by calculating the secular effects of these uncertainties on the orbital elements of LAGEOS and LAGEOS II, we find that the main

Table 1
Models used in the orbital analysis.

geopotential (static part)	EIGEN-GRACE02S (and GGM01S)
geopotential (tides)	Ray GOT99.2
lunisolar and planetary perturbations	JPL ephemerides DE-403
general relativistic corrections	PPN except L–T
Lense–Thirring effect	set to zero
direct solar radiation pressure	cannonball model
albedo radiation pressure	Knocke–Rubincam model
Yarkovsky–Rubincam effect	GEODYN model
spin axis evolution of LAGEOS satellites	Farinella–Vokhroulicky–Barlier model
station positions (ITRF)	ITRF2000
ocean loading	Scherneck model with GOT99.2 tides
polar motion	estimated
Earth rotation	VLBI + GPS

source of error in the determination of the Lense–Thirring effect is just due to the first even zonal harmonic, J_2 (see below).

We can however use the two observable quantities $\dot{\Omega}_I$ and $\dot{\Omega}_{II}$ to determine μ [22,24,26], thereby avoiding the largest source of error arising from the uncertainty in J_2. We do this by solving [26,32] the system of the two equations for $\delta\dot{\Omega}_I$ and $\delta\dot{\Omega}_{II}$ in the two unknowns μ and J_2, obtaining for μ:

$$\delta\,\dot{\Omega}^{Exp}_{LAGEOS\,I} + c\,\delta\,\dot{\Omega}^{Exp}_{LAGEOS\,II} =$$

$$= \mu\,(31 + c\,31.5)\,mas/yr + other\ errors \cong \mu\,(48.2\,mas/yr), \tag{37}$$

where $c = 0.545$. Equation (37) for μ does not depend on J_2 nor on its uncertainty; thus, the value of μ that we obtain is unaffected by the largest error, due to δJ_2, and is sensitive only to the smaller uncertainties due to δJ_{2n}, with $2n \geq 4$.

Similarly, regarding tidal, secular and seasonal changes in the geopotential coefficients, the main effects on the nodes of LAGEOS and LAGEOS II, caused by tidal and other time variations in Earth's gravitational field [24,25], are due to changes in J_2; e.g. by the uncertainty in the 18.6 year Lunar tide, with the period of the Moon node, the change in J_2 due to the post–glacial rebound and by the anomalous variation of the quadrupole coefficient (see below). However, the tidal errors in J_2 and the errors resulting from other un–

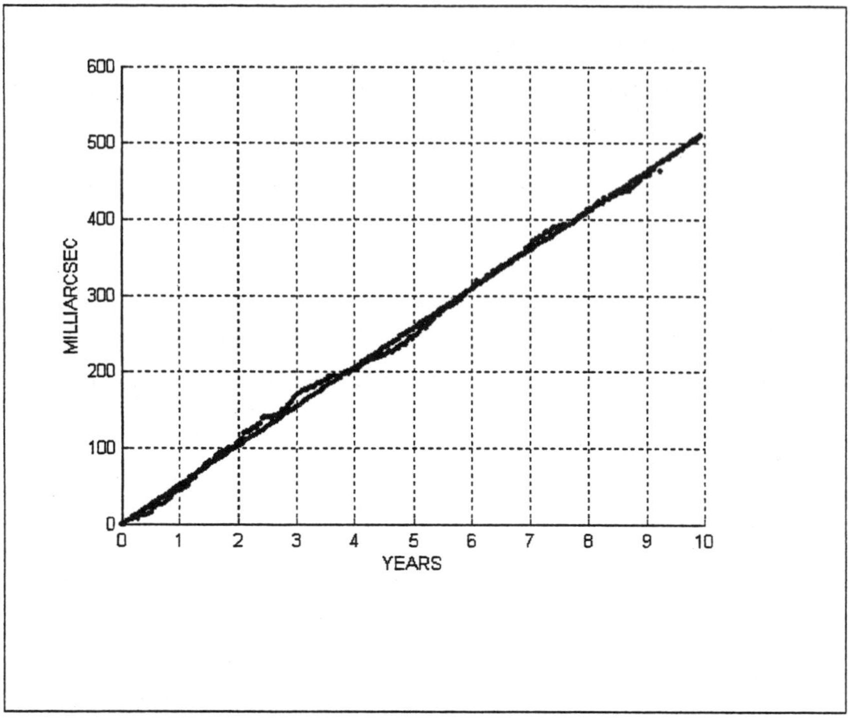

Fig. 6. Fit of the residuals of the nodes of LAGEOS and LAGEOS II, using our combination (37) and the Earths model GGM01S, with a secular trend plus ten periodical terms. The slope is $\mu \cong 1.06 \pm 0.26$, the RMS of the postfit residuals is 6.7 mas.

modelled, medium and long period, time variations in J_2, including its secular and seasonal variations, are cancelled by our combination of node residuals (37). In particular, most of the errors resulting from the 18.6 and 9.3 years tides, associated with the lunar node, are cancelled in our measurement. The various error sources that can affect the measurement of the Lense–Thirring effect using the nodes of the LAGEOS satellites have been extensively treated in a large number of papers by several authors [22,24–26,28,67–71,74]; the main error sources are treated in ref. [76].

5.3.2 Results

The orbital perturbations of a satellite may be divided in secular and periodical. Among the secular perturbations of the node of the LAGEOS satellites there are: the shift of the nodal line due to the even zonal harmonics of the Earth's gravitational field [44], the de Sitter effect and frame-dragging. The de Sitter effect has been measured with accuracy of about 7×10^{-3}, its effect is only 19.2 mas on the LAGEOS node and thus its uncertainty is negligible in the Lense–Thirring measurement, however the uncertainty in the even zonal harmonics is a crucial factor for the determination of frame-dragging since

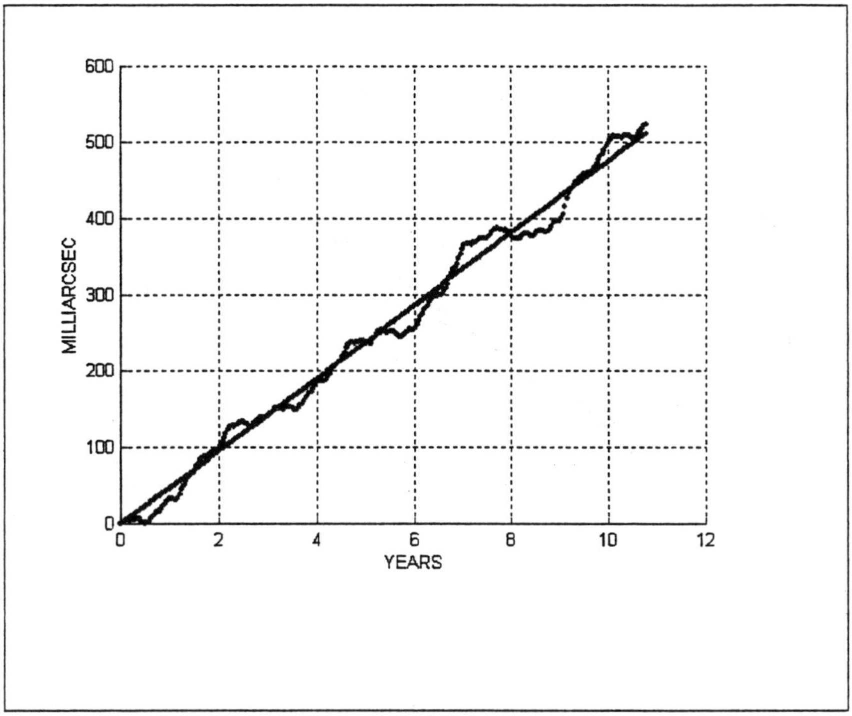

Fig. 7. Fit of the residuals of the nodes of LAGEOS and LAGEOS II, using our combination (37) and the Earths model EIGEN-GRACE02S, with a secular trend only. The slope is $\mu \cong 0.984$, the RMS of the post–fit residuals is 15.4 mas.

an error in one of the lower even zonal harmonics may be large enough to be indistinguishable from the Lense–Thirring nodal drag. This critical type of error is treated below. However, the periodical perturbations of the node of the LAGEOS satellites may also be a crucial factor in the determination of frame–dragging, particularly critical is the uncertainty in the perturbations with long period compared to the period of observation. The periodical effects with period much shorter than the observational period are averaged out. In the present determination of the Lense–Thirring effect, we have three basic factors that make negligible the error due to periodical effects in the measurement of frame–dragging and also make this error easy to be assessed in the final error budget. These basic factors are: (i) the period of the present analyses is about 11 years and thus all the periodical perturbations of the nodes are basically averaged out apart from the 18.6 years tide associated with the Moon node, however, the main effect of this tide is a change of the J_2 coefficient that is cancelled out using our combination of observables (37); (ii) since the original proposal of the LAGEOS III experiment [22], numerous researchers [22,24–26,28,67–71,74] have treated, in many papers and reports, the perturbations affecting the LAGEOS node in order to determine the Lense–Thirring effect and have concluded that the only critical perturbations on the nodes of the LAGEOS satellites are those due to the Earth's even zonal har-

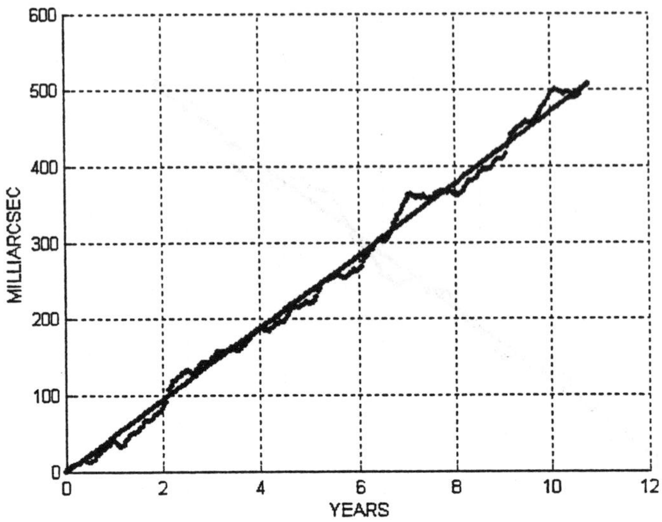

Fig. 8. Fit of the residuals of the nodes of LAGEOS and LAGEOS II, using our combination (37) and the Earths model EIGEN-GRACE02S, with a secular trend plus two periodical terms. The slope is $\mu \cong 0.97$, the RMS of the postfit residuals is 11.3 mas.

monics. However, as shown below, they contribute with an error of about 4 %, using the EIGEN-GRACE02S model, due to the high accuracy of this recent Earth's gravity field model; (iii) in regards to the periodical perturbations, in addition to a detailed treatment of the various perturbations affecting the LAGEOS node and their uncertainties given in ref. [76], a simple but very meaningful test shows that the periodic perturbations cannot introduce an error larger than about 4 % in our determination of the Lense–Thirring effect. Indeed, since the period of the gravitational and non–gravitational orbital perturbations (but not their amplitude) is very well determined, together with a secular trend we have also fitted for a number of periodic effects. We have done a number of different fits, each with different periodical perturbations, we have then compared the result in the case of the fit with a secular trend only with the various results when, together with a secular trend, we have included a different number of the main periodic perturbations. The result is that the maximum deviation of the secular trend from the case of its fit with no periodic perturbations does not exceed 2 % of Earth's frame–dragging as it is clearly displayed in figures (7), (8) and (9), (10) and corresponding captions. Of course the RMS of the fit is much smaller when we include a substantial number of periodic perturbations. Therefore, the two concerns [84], regarding really not method or the value of the Lense–Thirring effect that we had ob-

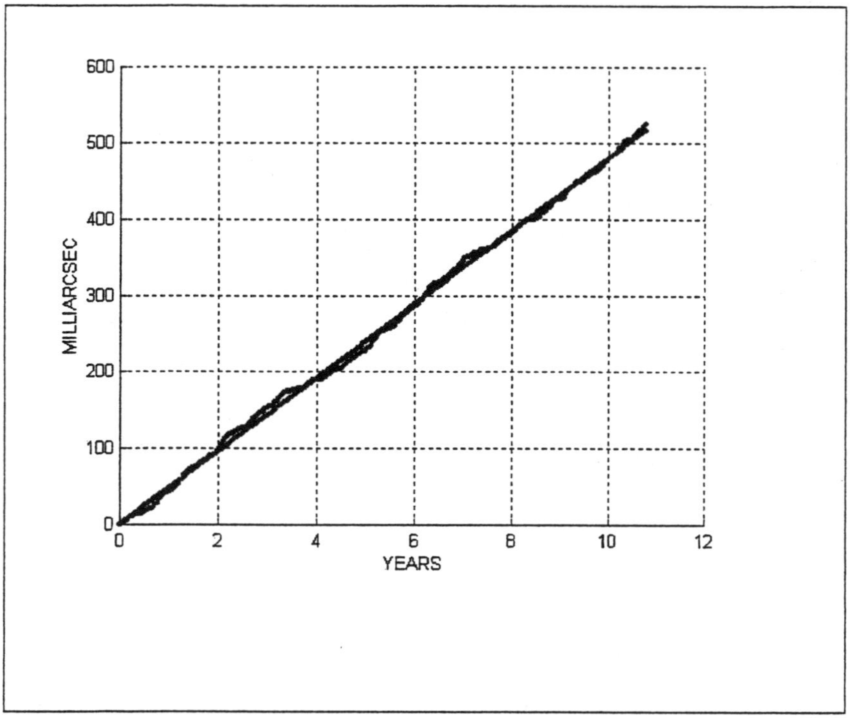

Fig. 9. Fit of the residuals of the nodes of LAGEOS and LAGEOS II, using our combination (37) and the Earths model EIGEN-GRACE02S, with a secular trend plus six periodical terms. The slope is $\mu \cong 0.994 \pm 0.1$, the RMS of the post–fit residuals is about 6 mas.

tained in our previous analyses [28–30] but our previous error budgets, cannot be applied to the present analyses, as explained below and in ref. [76]. Indeed, (i) the concern regarding the perturbations of the perigee of LAGEOS II is clearly absent in the present analyses since we use here the nodes of the LAGEOS satellites *only* and we do *not* use the perigee, (ii) the concern regarding the high correlation of some of the even zonal harmonics of the EGM96 model is also absent in the present analyses, indeed in our total error budget we summed up the absolute value of each even zonal harmonic error and, anyhow, the Earth model EIGEN-GRACE02S has low correlation between the even zonal harmonics (see below and related discussion in ref. [76].

In conclusion, in the analysis with EIGEN-GRACE02S, we have a total error budget of the order of 10 % of the Lense–Thirring effect.

The main perturbations in our determination of the Lense–Thirring effect are described and analyzed in ref. [76].

In the present analysis we have used EIGEN-GRACE02S (and GGM01S). The EIGEN-GRACE02S model is the second GRACE–only static gravity model from the GFZ group [23], derived from 110 days of GRACE only

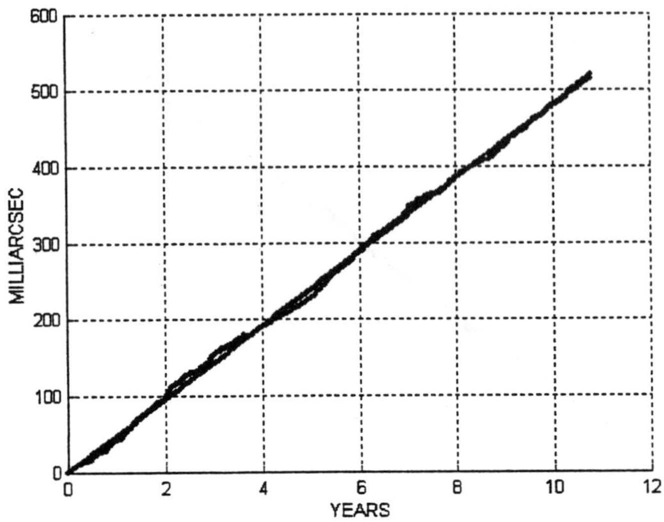

Fig. 10. Fit of the residuals of the nodes of LAGEOS and LAGEOS II, using our combination (37) and the Earths model EIGEN-GRACE02S, with a secular trend plus ten periodical terms. The slope is $\mu \cong 0.992 \pm 0.1$, the RMS of the post–fit residuals is 5.5 mas.

satellite–to–satellite observations. The model is complete to degree and order 150. However, the values of the Earth spherical harmonic coefficients may change with longer periods of observations. In particular the uncertainties in the Earth's zonal harmonics include only tentatively systematic errors in EIGEN-GRACE02S. However, our analysis is not sensitive to changes in the Earth quadrupole moment and is just affected by changes in higher zonal harmonics. Therefore, future, more accurate, determinations of the even zonal coefficients, and of their uncertainties, might lead to different values of μ and to a different total error budget. Nevertheless, it is crucial to observe that when a more accurate model using GRACE, CHAMP or GOCE will be available, it will be straightforward to, a posteriori, very accurately assess the total error of our present analyses. Indeed, one will just need to take the differences between the values of the even zonals of the EIGEN-GRACE02S model, presently used, with the corresponding values of the future, more accurate, model from GRACE, CHAMP or GOCE and just need to consider the uncertainties of these future models. Thus, one will easily re-estimate the total error in the present measurement of μ due to the uncertainties in the J_{2n} coefficients.

In figure (6) we report our determination of the Lense–Thirring effect, obtained using the Earth gravitational model GGM01S and the nodal rates of the LAGEOS and LAGEOS II satellites, over a period of about 10 years, combined according to formula (37). Corresponding to this figure, we have fitted the orbital residuals with a secular trend contemporarily with 10 main periodic effects: the slope is $\mu \cong 1.06$ with a total uncertainty, including systematic errors, estimated between about 25 % and 50 %.

Figures (7), (8), (9) and (10) are obtained using the Earth gravitational model EIGEN-GRACE02S and the nodal rates of the LAGEOS and LAGEOS II satellites, over a period of about 11 years, combined according to formula (37). Figure (7) shows our measurement of the Lense–Thirring effect using EIGEN-GRACE02S by fitting the orbital residuals with a secular trend only. Figures (8), (9), and (10) show our measurement of the Lense–Thirring effect using EIGEN-GRACE02S by fitting the orbital residuals with a secular trend together with 2, 6 and 10 periodic effects. Figures (9) and (10) represent our best measurement of the Lense–Thirring effect using the EIGEN-GRACE02S model: the slope is 0.99 μ.

In conclusion, by fitting our combined residuals with a secular trend only, using EIGEN-GRACE02S we found:

$$\mu = 0.984$$

By fitting our combined residuals with a secular trend plus 10 periodic signals, using EIGEN-GRACE02S we found:

$$\mu = 0.992 \pm 0.1. \tag{38}$$

However, the RMS of the post–fit residuals was 15.4 mas in the case (7) of the fit of a secular trend only and 5.5 mas in the case (10) of the fit of a secular trend plus 10 periodic signals.

By fitting our combined residuals with a secular trend or with a secular trend plus 10 periodic terms we practically got the same value for the Lense–Thirring effect. Furthermore, the two determinations of the Earth frame–dragging effect obtained with GGM01S and EIGENGRACE02S are practically in agreement with each other within their uncertainties ($\cong$ 25 % to 50 %for GGM01S and about $\cong$ 5 % to 10 % for GGM01S).

Our measured value of the Lense–Thirring effect corresponds to 99 % (in the fits with 6 and 10 frequencies) of the Einstein's theory prediction (using EIGEN-GRACE022S) and thus, since our experimental uncertainty is about 10 %, fully agrees with the general relativistic prediction.

We observe that the only uncertainty in our present error budget is due to the published errors of the EIGEN-GRACE02S model (and GGM01S) that might underestimate the systematic errors in the determination of the even zonal harmonics. However, when more accurate models of the Earth's gravity field will be available, it will be straightforward to evaluate the accuracy of the EIGEN-GRACE02S model (and GGM01S) we have used here (by substantially taking the difference between the corresponding coefficients). This will provide, a posteriori, a solid re-assessment of the error in the present determination of frame-dragging. Indeed, we have recently reanalyzed the LAGEOS and LAGEOS II data: (a) using the latest GRACE Earth's gravity field models, (b) with a longer period of observations and (c) using the GFZ German orbital estimator program EPOSIN, that is an orbital estimator independent of the NASA program GEODYN. These new results fully confirm our previous analysis using EIGEN-GRACE02S and GEODYN [85].

Finally, we observe that, in addition to the accurate measurement of the Lense–Thirring effect, in the orbital residuals of LAGEOS and LAGEOS II, we have observed an anomalous increase of the Earth's quadrupole coefficient J_2 since 1998. This change we observe in J_2 is in good agreement with the J_2 variation observed by Cox and Chao [86].

In conclusion, the Lense-Thirring effect exists and its experimental value: $\mu = 0.99 \pm 0.10$ *fully agrees with the prediction of general relativity [78].*

References

[1] Ciufolini I and Wheeler J A 1995 *Gravitation and Inertia* (Princeton, New Jersey: Princeton University Press)

[2] Misner C W, Thorne K S, Wheeler J A 1973 *Gravitation* (San Francisco: Freeman)

[3] Zichichi A 1985 "Da Mach ad Einstein a Wheeler: una nuova forza della Natura" *Il Tempo*. 22 Giugno 1985, 21.

[4] Thirring H 1918 *Z. Phys.* **19** 33

[5] Brill D R and Cohen J M 1966 *Phys. Rev.* **143** 1011

[6] Pfister H and Braun K 1986 *Class. Quantum Grav.* **3** 335

[7] Pfister H 1995 in *Mach's Principle. From Newton's Bucket to Quantum Gravity* ed J Barbour and H Pfister (Birkhäuser, Boston), 315

[8] Lense J and Thirring H 1918 *Phys. Z.* **19** 156. See also English translation by Mashhoon B, Hehl F W, Theiss D S 1984 *Gen. Relativ. Gravit.* **16** 711

[9] Friedländer B and I 1896 *Absolute und Relative Bewegung?* (Berlin: Simion-Verlag)

[10] Föppl A 1904 *Sitzb. Bayer. Akad. Wiss.* **34** 5; 1904 *Phys. Z.* **5**, 416; see also 1904 *Sitzb. Bayer. Akad. Wiss.* **34** 383

[11] de Sitter W 1916 *Mon. Not. Roy. Astron. Soc.* **76** 699

[12] Yilmaz H 1959 *Bull. Am. Phys. Soc.* **4** 65

[13] Van Patten R A and Everitt C W F 1976 *Phys. Rev. Lett.* **36** 629

[14] Pugh G E 1959, "Proposal for a Satellite Test of the Coriolis Prediction of General Relativity", *Weapons Systems Evaluation Group Research Memorandum No. 11* (The Pentagon, Washington)

[15] Schiff L I 1960 *Proc. Nat. Acad. Sci.* **46** 871; 1960 *Phys. Rev. Lett.* **4** 215

[16] Everitt C W F 1974, "The Gyroscope Experiment. I: General Description and Analysis of Gyroscope Performance" in *Experimental Gravitation*, ed. B. Bertotti (Academic Press, New York), 331–360

[17] Braginsky V B, Polnarev A G and Thorne K S 1984 *Phys. Rev. Lett.* **53** 863

[18] Schleich W and Scully M O 1984, in *Les Houches 1982, New Trends in Atomic Physics*, G. Grynberg and R. Stora eds., (North Holland, Amsterdam), 995

[19] Braginsky V B and Polnarev A G 1980 *Pis'ma V Zh. Eksp. & Teor. Fiz.* (USSR) **31** 444; Engl. trans. in *JETP Lett.* (USA)

[20] Mashhoon B, Paik H J and Will C M 1989 *Phys. Rev.* D **39** 2825

[21] Nordtvedt K 1988 *Int. J. Theor. Phys.* **27** 1395

[22] Ciufolini I 1986 *Phys. Rev. Lett.* **56** 278

[23] Reigber, C., Schmidt, R., Flechtner, F., Konig, R., Meyer, U., Neumayer, K. H., Schwintzer, P., and Zhu, S. Y., 2005. *J. Geodyn.* **39**, 1.

[24] Ciufolini I 1989 *Int. J. Mod. Phys. A* **4** 3083. See also: Tapley B, Ciufolini I, Ries J C, Eanes R J, Watkins M M 1989, *NASA-ASI Study on LAGEOS III*, CSR-UT publication n. CSR-89-3, Austin, Texas. See also Ciufolini I et al. 2004, *WEBER SAT/LARES Study for INFN*

[25] Ciufolini I, Paolozzi A *et al.i* 1998 *LARES phase A study for ASI*

[26] Ciufolini I 1996 *Nuovo Cimento A* **109** 1709

[27] Ciufolini I *et al.* 1996 *Nuovo Cimento A* **109** 575

[28] Ciufolini I *et al.* 1997 *Class. Quantum Grav.* **14** 2701; see also Ciufolini I *et al.* 1997 *Europhys. Lett.* **39** 35

[29] Ciufolini I, Pavlis E C *et al.* 1998 *Science* **279** 2100

[30] Ciufolini I 1994 *Class. and Quantum Grav. A* **11** 73

[31] Ciufolini I 2004 *to be published*

[32] Ciufolini, I. 2004, Proceedings of the I SIGRAV School on General Relativity and Gravitation, Frascati (Rome), September 2002 (Singapore: World Scientific).

[33] Petrov A Z 1969 *Einstein Spaces* (Oxford: Pergamon Press)

[34] Wheeler J A 1977 *Trans. N. Y. Acad. Sci., Ser.II* **38** 219

[35] Ciufolini I and Ricci F 2002 *Class. Quantum Grav.* **19** 3863

[36] Cohen J M and Mashhoon B 1993 *Phys. Lett.* A **181** 353

[37] Mashhoon, Gronwald and Theiss 1999 *Ann. Phys* **8** 135

[38] Tartaglia A 2000 Online preprint: http://babbage.sissa.it/abs/gr-qc/0001080

[39] Tartaglia A 2000 *Class. Quantum Grav.* **17** 783

[40] Pineault S and Roeder R C 1977 *Astron. J.* **212** 541

[41] Pineault S and Roeder R C 1977 *Astron. J.* **213** 548

[42] Rauch K P and Blandford R D 1994 *Astron. J.* **421** 46

[43] Virbhadra K S and Ellis G F R 2000 *Phys. Rev. D* **62** 084003

[44] Kaula W M 1966 *Theory of Satellite Geodesy* (Waltham: Blaisdell)

[45] Chandrasekhar S 1969 *Ellipsoidal Figures of Equilibrium* (New Haven: Yale Univ. Press)

[46] Schneider P, Ehlers J and Falco E 1992 *Gravitational lenses* (Berlin: Springer-Verlag)

[47] Hucra J *et al.* 1985 *Astron. J.* **90** 691

[48] Racine R 1991 *Astron. J.* **102** 454

[49] Chae, Turnshek and Khersonsky 1998 *Astrophys. J.* **495** 609

[50] Will C M 1985, *Theory and Experiment in Gravitational Physics* (Cambridge: Cambridge University Press); revised edition 1993

[51] Barbour J and Pfister H (ed) 1995 *Mach's Principle. From Newton's Bucket to Quantum Gravity* (Boston: Birkhäuser)

[52] Bass L and Pirani F A E 1955 *Phil. Mag.* **46** 850

[53] Weinberg S 1972, *Gravitation and Cosmology: Principles and Applications of the General Theory of Relativity* (New York: Wiley)

[54] Ciufolini I and Ricci F 2002 *Class. Quantum Grav.* **19** 3875

[55] Hucra J *et al.* 1985 *Astron. J.* **90** 691

[56] Chae K-H, Turnshek D A and Khersonsky V K 1998 *Astrophys. J.* **495** 609

[57] Abell G O 1975 in *in Stars and Stellar Systems, IX Galaxies and the Universe* (Chicago: Univ. of Chicago Press) 636

[58] Iorio L, Lucchesi D and Ciufolini I 2002 *Class. and Quantum Grav.* **19** 4311

[59] Iorio L, Ciufolini I and Pavlis E 2002 *Class. and Quantum Grav.* **19** 4301

[60] Nordtvedt K 1998 "LARES and Tests on New Long Range Forces", in *LARES phase A study for ASI*, I. Ciufolini, *et al.*

[61] Cohen S C and Dunn P J (ed) 1985 *LAGEOS Scientific Results, J. Geophys. Res.* **90** (B11) 9215

[62] Bender P and Goad C C 1979 *The use of satellites for geodesy and geodynamics*, in *Proceedings of the Second International Symposium on the Use of Artificial Satellites for Geodesy and Geodynamics*, Vol. II, G. Veis and E. Livieratos, eds. (National Technical University of Athens), 145

[63] Degnan J J and Pavlis E C 1994 *GPS World* **5** 62

[64] *International Earth Rotation Service (IERS) Annual Report, 1996* (Observatoire de Paris, Paris, July 1997)

[65] McCarthy D 1996 *The 1996 IERS Conventions* (Paris: Observatoire de Paris)

[66] Lemoine F G *et al.* 1998 *The Development of the Joint NASA GSFC and the National Imagery and Mapping Agency (NIMA) Geopotential Model EGM96*, NASA/TP–1998–206861

[67] Rubincam D P 1988 *J. Geophys. Res.* **93** (B11) 13803

[68] Rubincam D P 1990 *J. Geophys. Res.* **95** (B11) 4881

[69] Rubincam D P and Mallama A 1995 *J. Geophys. Res.* **100** (B10) 20285

[70] Martin C F and Rubincam D P 1996 *J. Geophys. Res.* **101** (B2) 3215

[71] Farinella P, Vokrouhlicky D V Y and Barlier F 1996 *J. Geophys. Res.* **101** (B8) 17861

[72] Gross R S 1996 *J. Geophys. Res.* **101** (B4) 8729

[73] Pavlis D E *et al.* 1997 *GEODYN II, Operations Manual*, 3

[74] D. M. Lucchesi 2002 *Plan. and Space Science*, **50**, 1067.

[75] Eanes R and Ries J 1988, Center for Space Research, University of Texas at Austin, private communication

[76] Ciufolini I, Pavlis E and Peron R. 2005 to appear in New Astronomy.

[77] Ciufolini, I., and Ricci, F. 2004 submitted to Phys. Rev. Lett.

[78] Ciufolini, I., and Pavlis, E.C. 2004, Letters to Nature, 431, 958.

[79] Dvali, G., Infrared Modification of Gravity, in: Nobel Symposium on Cosmology and String Theory, August 2003, Sigtuna, Sweden.

[80] Ciufolini, I., et al. 2004 *LARES/WEBER-SAT study for INFN.*

[81] Reigber, Ch., F. Flechtner, R. Koenig, U. Meyer, K. Neumayer, R. Schmidt, P. Schwintzer, S. Zhu 2002, GRACE Orbit and Gravity Field Recovery at GFZ Potsdam - First Experiences and Perspectives, Eos. Trans. AGU, 83(47), Fall Meet. Suppl., Abstract G12B-03.

[82] Tapley, B. D. 2002, The GRACE Mission: Status and Performance Assessment, Eos. Trans. AGU, 83(47), Fall Meet. Suppl., Abstract G12B-01.

[83] Watkins, M. M., D. Yuan, W. Bertiger, G. Kruizinga, L. Romans, S. Wu 2002, GRACE Gravity Field Results from JPL, Eos. Trans. AGU, 83(47), Fall Meet. Suppl., Abstract G12B-02.

[84] Ries J. *et al* 2004 in *Non Linear Gravitodynamics. The Lense-Thirring Effect*, World Scientific, p. 201.

[85] Ciufolini I. *et al.* 2006, to be published.

[86] Cox, C.M. and Chao, B. 2002 *Science*, **297**, 831.

CHAIRMAN: I. CIUFOLINI

Scientific Secretaries: L. Gualtieri, A. Marrani

DISCUSSION

- *t'Hooft:*

In the publication in Nature, it is indicated that the optimal inclination for these gravitating satellites is 70°. Can you give us an indication why that is so ?

- *Ciufolini:*

This is the original proposal of the LAGEOSIII experiment, which is not so much related to the result of the Nature paper. We have in orbit since 1976 the LAGEOS satellite, and its inclination is 110°. In this old proposal, we wanted to have a satellite with an inclination which is supplementary to that of LAGEOS, that is, 70°. This was for the following reason.

The largest classical contribution to the motion of the orbital plane of the satellite comes from the quadrupole moment, then there are higher multipole moments, expecially the even zonal harmonics, J_2, J_4, and so forth.

If you have a satellite with an inclination which is supplementary with respect to the first one, you have a classical precession which is equal and opposite, because the classical precession of the nodes is proportional to the cosine of the inclination angle. When you sum up these two, you get zero.

But the Lense-Thirring precession is a drag of the orbital plane. The orbital plane, which can be considered as a gyroscope in the ideal case of a perfectly spherically symmetric central body, will be dragged in the same sense of rotation of the Earth. Thus, if the inclinations are supplementary, when you combine the two measures given by LAGEOSI and LAGEOSIII, the classical precessions will disappear ; only the Lense-Tirring effects will survive, giving twice the Lense-Thirring precession.

You can put the things in a nicer way by saying that we want to build a gyroscope. If we had a perfectly spherically symmetric central body, we could just use the node of one satellite orbiting around it. But, since the Earth is not spherically symmetric, this is not possible ; nevertheless, we can still consider the bysector between the nodal lines of two satellites with supplementary inclinations.

Since the nodal lines move classically in opposite sense, the bysector, classically, would not move. Taking into account general relativity, it would be affected only by the Lense-Thirring effect.

I have not been able to launch the satellite LAGEOSIII with this special inclination, and this is why I couldn't get any results for 20 years. But in 1992 the Italian Space Agency ASI launched another satellite called LAGEOSII, and I tried very hard to convince the principal investigator of LAGEOSII to change the inclination to 70°. The principal investigator at that time was S.Zerbini of the University of Bologna, but unfortunately she wanted to have 52° of inclination because this was convenient for geodesy reasons. Therefore, I couldn't get the precise cancellation I wanted.

What we did is the following. Fortunately, then the GRACE mission, launched in 2002, produced a very accurate gravity field model of the Earth.

In particular, a good estimate of the quadrupole moment was given.

The uncertainty in the nodal precession of LAGEOSI is, in its largest part, due to the quadrupole moment.

Indeed, the uncertainty coming from the quadrupole moment and the one coming from, say, the harmonic J_{40}, are, respectively, of the order of 159% and 5% of the Lense-Thirring effect.

Then, with the satellites LAGEOSI and LAGEOSII (with 52° of inclination) I measure two quantities, namely the two nodal precessions, and therefore I have two observables. However, I only have two unkowns, the Lense-Thirring effect and the uncertainty in J_2, and consequently I can solve for the Lense-Thirring effect, and measure it.

- *Pacetti:*

Is it possible to extend the formal analogy between electromagnetism and gravitomagnetism to other phenomena as, for instance, the Faraday law, the induced current, or something like that ?

- *Ciufolini:*

This is a very interesting question. I believe the people have not completely investigated all the possibilities of extending the similarity between electromagnetism, which has of course been very well studied, and the gravitomagnetic field, which has been less studied.

But there is a basic breaking point between general relativity and electrodynamics : the equivalence principle. It is satisfied by general relativity, but there is nothing like that in electrodynamics. The inertial mass is not equal to the electric charge, while the gravitational mass is equal to the inertial mass. This makes the two theories completely different. For example, this is why there is no dipole gravitational radiation in general relativity.

Neverthless, I would like to see if some of the effects of electrodynamics could be extended to gravitomagnetism, by keeping in mind this fundamental difference.

- *Serbyn:*

The effect of precession of gyroscopes has two parts : one connected with deviation of the field from the $1/r^2$ law, and the other connected with rotation of the central body. Could you explain how we can obtain these two effects from your field H ?

- *Ciufolini:*

It does not make much sense, in my opinion, to speak of deviation from the $1/r^2$ law in general relativity: it is a more complex business, one should speak about geodesics moving in the spacetime.

For instance, some people were trying to explain the anomalous precession of Mercury of 43" per century, by simply assuming that there is a small deviation from the classical gravitational law of $1/r^2$. But they couldn't succeed to reproduce other relativistic effects.

In general relativity, if you consider the motion of a test particle in the Schwarzschild background, which is static and characterized only by mass, the plane of the orbit would stay constant. The same happens in classical mechanics : the node of the satellite would be constant if you consider only the mass. But if you consider the Lense-Thirring solution, which is the slow motion weak field approximation of the Kerr metric, there is a nondiagonal component in the metric tensor, which gives rise to a precession of the orbital plane.

Thus, no nodal precession arises from mass ; there is only nodal precession due to the angular momentum of the central object.

In addition, there is another precession, but that is due to the Schwarzschild Sun field, which is called "geodetic" or "de Sitter" precession. It is somehow similar to the Thomas precession. But what changes the nodes of the satellite is only due to the angular momentum of the Earth, plus this other effect due to the Sun. Of course, in addition to these relativistic effects, we have the huge classical nodal changes due to the Earth even zonal harmonics.

- *Serbyn:*

Why the contribution of higher harmonics is not negligible? It seems to me that higher harmonics are proportional to bigger powers of 1/r.

- *Ciufolini:*

Of course they become much smaller, but they cannot be neglected until the $J_{20\,0}$ harmonic, at least.

For instance the uncertainty on, say, J_4, would still give an uncertainty on the classical precession which is larger than the Lense-Thirring effect. This until we got the results from the GRACE mission, where the uncertainties due to the various harmonics, including J_4, are much smaller.

Probably, in order to get an uncertainty smaller than the Lense-Thirring effect, without a map of the Earth gravitational field, in 1989 we would have needed at least five or six satellites.

- *Faivre:*

At least one binary pulsar has been observed. Although you mentioned this system in your talk, you didn't show results. Is the reason that longer observational times are needed?

- *Ciufolini:*

I don't believe you can see directly the Lense-Thirring effect from any binary pulsar.

However, there is a clever argument by K. Nordvedt to measure frame dragging.

Let me recall that there are two kinds of periastron advance.

One is due to the mass of the central object, for instance the famous precession of the perihelion of Mercury, 43" per century, observed since the 19th century by Leverriero. The other is due to the angular momentum of the central object. This is the one that de Sitter was calculating, just few months after the formulation of general relativity : he calculated the shift of the perielion of Mercury due to the angular momentum of the Sun.

The problem is that we see the periastron advance of the binary pulsar which is due to the mass of the other object, but there was no chance we would measure the periastron advance due to the angular momentum of the other object. This is so far not possible.

The idea of K. Nordvedt is that since a binary pulsar is not a system of a mass plus a test particle, but it is a system of two massive objects, he considers this as a test of gravitomagnetism, because of an indirect argument.

If you have two massive objects, and you see the advance of one around the other (which has been measured very well), he concluded, on the basis of purely theoretical arguments, that this is in a sense a measurement of gravitomagnetism. His argument is based on the spacetime invariant $^*R \bullet R$, which is non-zero due to the relative angular momentum J.

- *Faivre:*

A couple of months ago there has been a direct observation of the movement of stars around the Milky Way's central black hole. Can this be used (because of the huge mass of the black hole) for frame dragging measurements?

- *Ciufolini:*

There is no clear evidence for frame dragging from any astrophysical system. For example, the 1998 paper on the observation of the Lense-Thirring effect on the accretion disk around black holes and neutron stars had an uncertainty of about 400%, due to the uncertainty in the various parameters.

- *Pinfold:*

What is the astrophysical evidence for gravitomagnetic effects from accretion disks of black holes or neutron stars?

- *Ciufolini:*

There have been a lot of papers published since 1998, by a colleague of mine, L. Stella, and other people. They get these observations, but it is very difficult to interpret them. There are some parameters which are unknown. For example there are uncertainties in the distribution of the mass. I have seen nothing like a precise measurement of gravitomagnetism, they only see some peaks in the accretion disk curve velocities, but of course it is not like a laboratory or space experiment.

- *Imponente:*

How far is the accuracy of this experiment from ruling out or confirming a non-metric space-time ?

Could ever exist an experiment able to distinguish autoparallel curves from geodesic ones? (Here or in some more extreme astrophysical enviroment)

- *Ciufolini:*

Of course, if we consider metric theories of gravity, and we make the strong assumption that we can describe all these theories with only ten parameters, then perhaps we don't need all these experiments, since we only have ten post-newtonian parameters to measure.

However, I remember that when I was in Texas I went to speak with S.Weinberg, which is a great physicist not only in particle theory but also in relativity. He told me: "I don't believe that with only 10 post-newtonian parameters you can describe any possible metric theory of gravity." On the basis of this, I wrote a chapter of my Ph.D. dissertation in which I proved that you can have in principle an infinite number of post-newtonian parameters.

Now, if however you are considering non-metric theories, which are the chances to be able to distinguish among metric and non-metric theories with experiments ? This question is somehow related with what Prof. Ferrara mentioned in his lectures.

Suppose we get, with LAGEOSIII, a precision of 1% on the measurement of Lense-Thirring effect.

Possible contributions due to the antisymmetric part of the connection cancel out from the geodesic equations, due to the symmetry of the term $u^\mu u^\nu$. But when I consider the equations expressing parallel transport of the spin S^μ, the term $u^\mu u^\nu$ is replaced by $u^\mu S^\nu$, and therefore possible contributions from the antisymmetric part of the connection could be there.

So in principle, by comparing the two experiments, with a test particle and a test gyroscope, we could test torsion effects.

But there are two kinds of problems. Firstly, these effects are incredibly small, much smaller even than the Lense-Thirring effect. Secondly, as mentioned by Prof. Ferrara, the torsion usually doesn't propagate in vacuum. Thus, we need some particular theory, such that of Rayn and Shepley with torsion propagating in vacuum, to make an experiment outside matter.

- *Marrani:*

Would it be possible to effectively take care of the frame dragging effects by introducing a four-vector on which the tetrads would depend?

- *Ciufolini:*

There are some mathematical works by B. Jantzen and by D. Bini, if I understand well your question. They may have done something similar to what you are saying, focussing on the effect of gravitomagnetism on tetrad formalism.

Experimental Status of QCD Glueballs

S. J. Lindenbaum
Physics Department, City College of New York, New York, NY 10031 USA and
Brookhaven National Laboratory, Upton, NY 11973, USA[1]

Abstract

A review of the experimental status of QCD Glueballs, and the history of major investigations and developments in the research are presented. The most important missing link in experimentally establishing QCD is the lack of consensus on the experimental establishment of at least one glueball. Therefore the lecture focused on what, if any, experimental evidence can be considered compelling for the establishment of at least one Glueball. The unique experimental evidence for the $I^G J^{PC} = 0^+2^{++}$ Glueball, has only been successfully explained in over two decades by interaction with the 2^{++} Glueball. Therefore it is resonable to conclude that the 2^{++} Glueball has been experimentally established.

1 Introduction

I had the pleasure and stimulation of presenting Lectures on Glueballs and related topics here during the ERICE International School of Subnuclear Physics courses over 2 decades ago. Some references I was able to trace for the period (1980-1985) are listed[1].

The most important missing link in experimentally establishing QCD is the lack of a scientific consensus on the experimental establishment of the existence of even one Glueball (a multi-gluon resonance state).

The Lattice Gauge Theorists have proven beyond any doubt that the existence of Glueballs are absolutely essential for QCD. In fact even long before LGT, the combination of the running coupling constant (asymptotic freedom), and color confinement convinced most of our community that the existence of Glueballs was inescapable - albeit they would likely mix with overlapping quark built states as one would expect from quantum mechanics.

Today I will consider the status of more than two decades of experimental Glueball searches. I will be focused on the question as to what, if any, experimental evidence can be considered as compelling for the establishment of at least one Glueball. Without the experimental establishment of at least one Glueball one cannot assume that QCD is consistent with experiment. Therefore if the number of Glueballs established is zero, one has to admit that QCD is at present inconsistent with experiment. If one glueball is established then this would remove this striking inconsistency.

[1]This research was supported by the U.S. Department of Energy under Contract No. DE-AC02-98CH10886.

2 Searching For Glueballs

The three most generally pursued methods of searching for glueballs are:

2.1 The Glueball Filter Reactions

- J/Ψ radiative decay which was presumed to proceed via a $\gamma + 2$. gluons decay. The hope was that this glue enriched channel would reveal Glueballs.

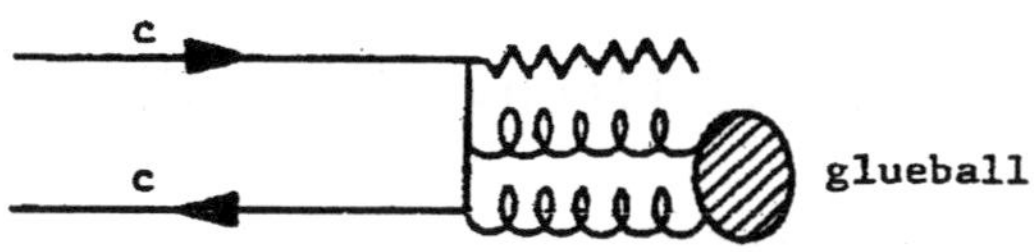

Figure 1: J/Ψ radiative decay

- Searching for $I^G J^{PC} = 0^+2^{++}$ Glueball(s) in the OZI suppressed reaction $\pi^- + p \rightarrow \phi(1020) + \phi(1020) + n$. In the $(\phi + \phi)$ cms the experiment covered the mass region of 2 to 3 GeV where the 2^{++} Glueball was theoretically expected to be.

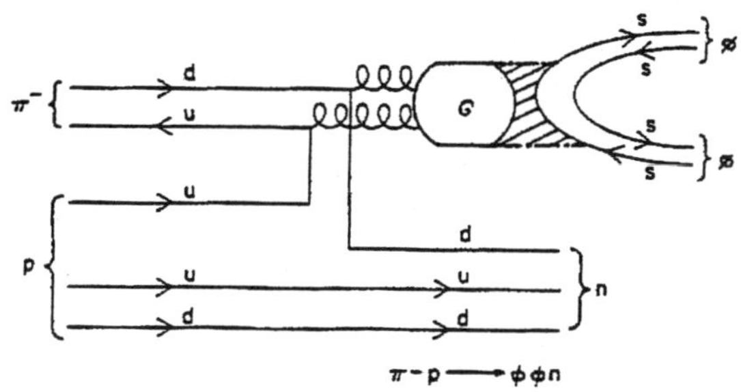

Figure 2: $\pi - p \rightarrow \phi\phi n$

My reasoning was that the two gluons in this Zweig disconnected double hairpin diagram would in the presence of a Glueball become part of a resonating system which (as expected from Quantum Mechanics) also included mixtures of overlapping or nearby quark-built states, which become rich in resonating glue. Then the Glueball and the quark-built states with the same quantum numbers that had mixed substantially with the 2^{++} Glueball would strongly selectively break down the OZI suppression and let Glueballs, and the mixed Glueball and quark-built states with the Glueball quantum numbers pass freely. At the same time strongly suppressing all states with other quantum numbers, and even those with the same quantum numbers, which had not mixed effectively. This could be an ideal justifiable filter for selecting 2^{++} Glueballs and the quark built states they mixed with substantially which become rich enough in resonating glue.

2.2 Pattern Recognition

Searching in conventional Hadron Spectroscopy for additional states beyond those expected in the nonets etc. from the quark model etc., and hoping to identify them as Glueball(s) or states, which mixed with a Glueball.

3 The J/Ψ Radiative Decay Glueball Searches

The interest in this program ran high. Any particle or structure which appeared to be different from what had previously been found was almost immediately considered a Glueball candidate. I will discuss the most popular candidates, which had an appreciable lifetime.

The Mk III research team doing J/Ψ radiative decay research at the Slac e^+e^- collider claimed to have discovered a new particle a 0^+2^{++} meson named the $\theta(1690)$ published in 1987 by R. M. Baltrusaitis[2]. It was considered new because the existing similar particle the $S^{*\prime}(1720)$ previously found by Etkin et al.[3] had 0^+0^{++} thus a spin 0 instead of 2. The $\theta(1690)$ promptly replaced the $S^{*\prime}(1720)$ in the PDG table and became an instant Glueball candidate which created quite a bit of interest. However, later the MK III Collaboration after further work concluded in 1990 that the spin of the $\theta(1690)$ was really 0 not 2 as they had claimed for many years. Therefore its Quantum Numbers were 0^+0^{++}. Thus it was not a new particle[4]. S. J. Lindenbaum and R.S. Longacre in 1992 performed an extensive coupled channel analysis including J/Ψ radiative decay which conclusively eliminated the $J = 2\ \theta(1690)$ and showed it was the first discovered $S^{*\prime}(1720)$ with spin 0[5].

The second new state considered a Glueball candidate was claimed by the MK III collaboration, R. M. Baltrusatis, et al to be a very narrow $K\bar{K}$ state called the $\xi(2230)$[6]. Ten years later it was confirmed by the BES collaboration at the Beijing electron- positron collider in J/Ψ radiative decay[7].

From the beginning this strange narrow state led to considerable interest as a Glueball candidate, but other possible explanations were offered. In the BES confirmation various decay channels were observed, which were consistent with the Glueball interpretation and appeared inconsistent with the alternatives. This led to an intensified interest in this new narrow state.

However BES became the BES II collaboration by considerably improving their detector, and with it obtaining 10 times more statistics. At the 2002 International Conference On High Energy Physics, 6 years later, they stated no clear signal for the $\xi(2230)$ had been observed in the 58 million J/Ψ data so far. All possible problems are being checked at BES II[8]. One year later they had apparently finished their checking and after an extensive analysis decided they could not find a clear signal for the $\xi(2230)$. This was reported at the

Tenth International Conf. on Hadron Spectroscopy[9]. Thus the second Glueball candidate evaporated also.

I should point out that a number of states found in conventional spectroscopy were observed in the previous experiments. The constraints imposed by the J/Ψ radiative decay would limit these. I do not expect one would observe very narrow Glueball states regardless of how they are formed. Therefore in summary, I had concluded the J/Ψ radiative decay was not an effective filter for Glueballs. However the J/Ψ decay program has contributed significantly in other areas starting with the co-discovery of the J/Ψ.

4 The OZI Suppression as a Filter for Glueballs

Now we consider the OZI suppression in the Zweig disconnected diagram (Fig. 2) as a filter for Glueball(s) and Glueball(s) mixed with quark- built states which become rich in resonating glue. For $J > 0$, an initial state composed of u, d quarks, and their anti-quarks which is connected by hard gluons, which do not resonate to a final state which is composed almost entirely of $s\bar{s}$ quarks suffers the OZI suppression.

In QCD, using Zweig diagrams the initial state would then be disconnected from the final state by 2-3 hard gluons (precocious asymptotic freedom), which provides the OZI suppression.

The thee reactions we will be considering are:

$$\begin{aligned} \pi^- + p &\rightarrow K^+K^-K^+K^-n\ (OZIallowed) && (1)\\ &\rightarrow \phi K^+K^-n\ (OZIallowed) && (2)\\ &\rightarrow \phi + \phi n\ (OZIforbidden) && (3) \end{aligned}$$

The three corresponding zweig quark line diagrams are given in Fig. 3.

Three different in time and individually set-up experiments[10, 11, 12] which increased statistics, all agreed well with each other within errors, and revealed that:

1. $\phi + \phi$ from $\pi^- p \rightarrow \phi + \phi n$ was comprised entirely of resonant states with $I = 0$, $G = +$, $J = 2$, $P = C = +$ which are the quantum numbers of the 2^{++} Glueball, and virtually nothing else was observed. The background observed was from the reaction.

2. $\pi^- p \rightarrow \phi K^+K^-n$ and was $< 13\%$ even with the wide cuts we used to avoid loss of $\phi + \phi$ events.

There was no evidence of any other background in our final analysis of reaction 1) Thus since 2) $\pi^- p \rightarrow \phi K^+K^-n$ is OZI Allowed, and $\pi^- p \rightarrow \phi + \phi n$ is OZI Forbidden.

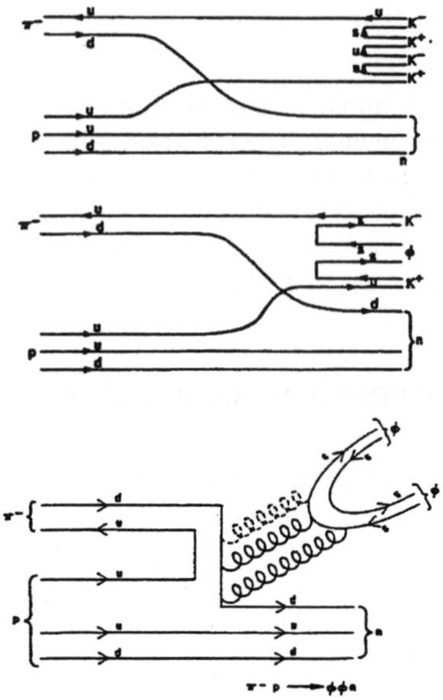

Figure3: The Zweig quark line diagrams for the three reactions investigated in this experiment. Only 2 gluons apply to the case where both ϕ are identically $\phi(1020)$

The expected OZI suppression ~100 was broken down by a factor ~800 in our $\phi + \phi$ observed range with wide cuts.

Earlier, in 1978 a low statistics (100 events) experiment by Etkin et al.[13] revealed that in $\pi^- p \rightarrow \phi + \phi n$ the OZI suppression was strongly broken down, and that the forbidden $\phi + \phi$ was an order of magnitude larger than the allowed $\phi K^+ K^-$. This indicated that the proposed Glueball filter was working. We felt that the analysis for states indicating interaction with a Glueball was important enough to warrant sufficient statistics so that very important question could be definitively answered. Our estimates were that at least 1,000 $\phi + \phi$ events, an order of magnitude higher, were required to perform a critical analysis. The 1978 experiment utilized the BNL Multi- Particle Spectrometer (MPS) user facility[14] at the AGS. It was the first generation large solid angle acceptance electronic magnetic spectrometer with on line computer analysis of the physical results. It utilized spark chamber modules read out by magnetostrictive wires, and would have required excessively long beam time runs to accumulate 1,000 events. This would interfere drastically with other users of the facility. Therefore we designed and built MPS II[15] which utilized narrow cell drift chambers and was 10 times faster for our and the user communities future needs.

4.1 First experiment with MPS II

The first $\pi^- p \to \phi + \phi n$ experiment with MPS II was performed by Etkin et al in 1982[10]. We observed 1203 $\pi^- p \to \phi + \phi n$ events.

Figure 4A shows the mass of one K^+K^- pair plotted versus the mass of the other K^+K^- pair. The black spot is due to the $\phi\phi$. The black spot occurs when both K^+K^- pairs have the $\phi(1020)$ mass. It is due to the OZI forbidden reaction $\pi^- p \to \phi + \phi n$ which exhibits a massive breakdown of the OZI suppression. The two side bands parallel to each axis are due to the reaction $\pi^- p \to \phi K^+ K^- n$ which is OZI allowed. A diagram with these characteristics was found in all our $\phi\phi$ experiments[10, 11, 12, 13].

Figure 4b part a) shows the large $\phi\phi$ signal compared to the ϕK^+K^- which is at the level of the tails. Part b) shows the large neutron peak which clearly indicates that the reaction $\pi^- p \to \phi\phi n$ has been cleanly selected.

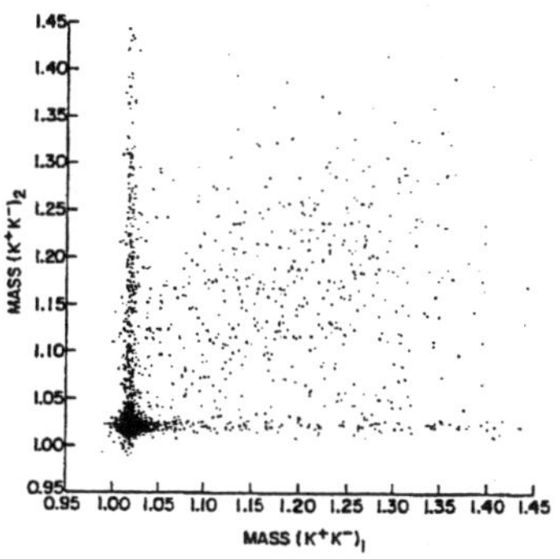

Figure 4a: The mass of one K^+K^- pair plotted vs. the mass of the other K^+K^- pair.

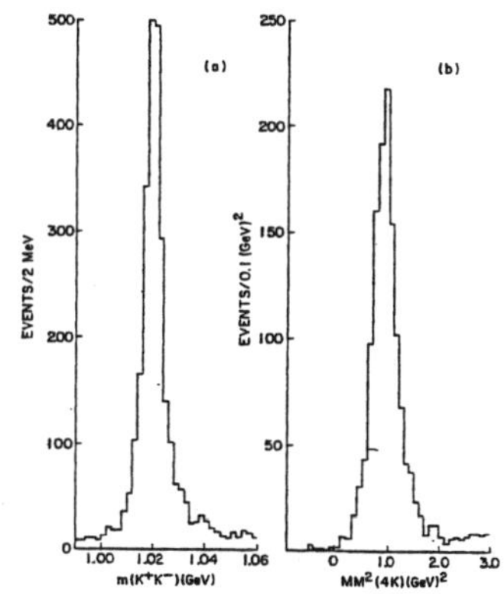

Figure 4b: (a)K^+K^- effective-mass plot for the second pair when the first pair falls in the ϕ mass band (1019.6 ± 14 MeV). (b)Missing mass squared from the $\phi\phi$ system.

In the $\to \phi\phi$ analysis there were six significant angles and their correlations due to the ϕ spin. These were two cms angles- the Gottfried-Jackson Frame angles [β(polar) and γ(azimuthal)], the polar (Θ 1, 2), and the azimuthal (α 1,2) angles for each decaying ϕ. This gave us enormous analyzing power, and enabled picking out individual partial waves from all permitted values for the quantum numbers up to $J = 4$ and $L = 3$ which were searched (52 waves). Each partial wave essentially had a detailed individual fingerprint made up of the values of the six significant angles and their correlations. Figure 5 shows various signals of six of these waves with $M = 0$, since the reaction was shown to proceed by pion exchange for which $M = 0$.

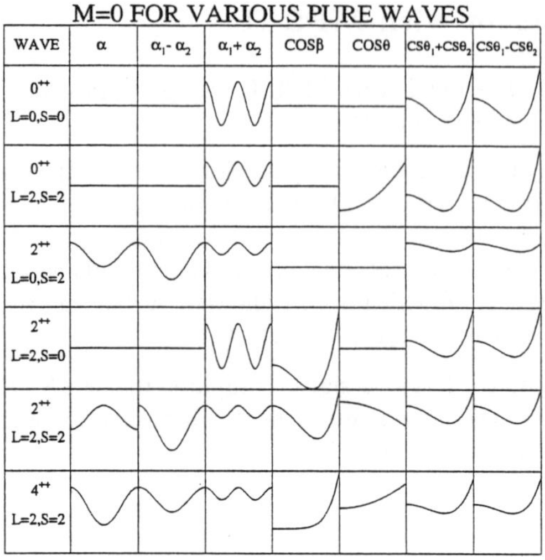

Figure 5: Above we show six out of the 114 waves tried in our $\phi\phi$ partial wave analysis. We show projections of the pure waves along: column 1—the ϕ azimuthal decay angle (α); column 2—the difference of azimuthal decay angles for ϕ_1 and ϕ_2; column 3—the sum; column 4—the cosine of the Gottfried-Jackson angle (β); column 5—the cosine of the polar ϕ decay angle (θ); column 6—the sum of the cosines of the polar decay angles of ϕ_1 and ϕ_2; column 7—the difference.

The only two significant waves found were a $J^P SLM^\eta = 2^+200^-$ S-wave and a 2^+220^- D-wave. The intensity of each wave, and their phase difference is shown in Fig. 6.

The log $(d\sigma/dt')$ plotted versus $|t'|$ was well fit by (-9.4 ± 0.7) $|t'|$ for $|t'| < 0.3$ which is expected for pion exchange . Since $M^\eta = 0^-$, it is clear that pion exchange occurred, since pion exchange is the only known exchange which has these characteristics.

Two Breit-Wigner resonances fit the data well. One resonance was rejected by $> 10\sigma$. Table I gives the details of the fit.

	$I^G J^{PC}$	Mass (GeV)	Γ_{tot}	
$g_T(2160)$	0^+2^{++}	2.16 ± 0.05	0.31 ± 0.07	$\Gamma_D/\Gamma_S \approx 0.02$
$g_T(2320)$	0^+2^{++}	2.32 ± 0.04	0.22 ± 0.07	$\Gamma_S/\Gamma_D \approx 0.04$

Table 1: Quantum numbers and parameters of the Breit-Wigner resonances fitted to the $S-$ and $D-$wave amplitudes of Fig. 3.

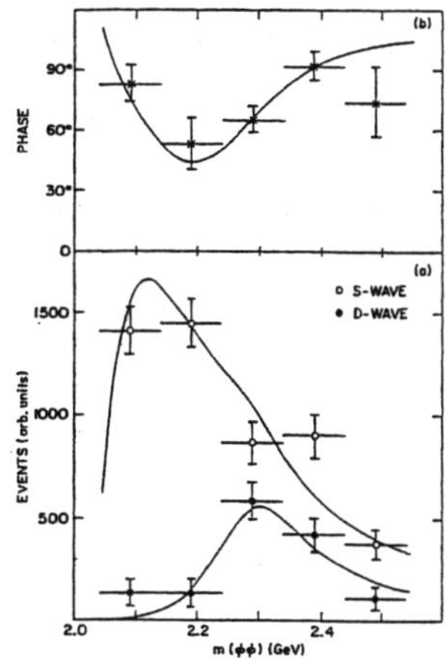

Figure 6: (a) Intensity and (b) phase difference for the two-wave fits. The curves show the fit by two Breit-Wigner resonances described in the text.

This enormous selective breakdown of the OZI suppression by only two $I = 0$, $G = +$, $J^{PC} = 2^{++}$ resonances which are the Quantum Numbers of the 2^{++} Glueball(s) was accompanied by virtually no background in the $\phi\phi$. It occurred in the expected mass region of the 2^{++} Glueball where many other states were also expected and many eventually reported, but not established due to the complexity of this region. The only other states in this mass region above 2 GeV/c that had been established in addition to the $2^{++}\phi\phi$ states are two 4^{++} states. The f_4 (2050) has $I = 0$ and $G = +$ and could in principle be detected in our experiments. The second a_4 (2040) has $I = 1$ and $G = -$ and therefore would not be detectable in our experiments. It should be noted that the total cross- section of $\pi^- p \rightarrow \phi\phi n$ is about 20 nanobarns. Thus our ability to obtain such an unusual and clean result, consisting of only 2 resonant states with the quantum numbers of the 2^{++} Glueball in a complex essentially un-analyzable region of inelastic cross-section of the order of 20 millibarns, accompanied by the massive breakdown of OZI suppression makes it clear something very unusual was occurring.

Our original explanation, that the $2^{++}\phi\phi$ resonance states were due to at least one 2^{++} Glueball mixing with quark-built states or two 2^{++} Glueballs is the only successful one within the context of QCD, and is still valid today 23 years later. We demonstrated in publications that all published alternative explanations of this 1982, and our later, two higher statistics experiments (published in 1985[11]and 1988[12]), which very clearly strengthened the case, were incorrect irrelevant or both by the following selected references, and numerous others.[16, 17, 18, 19, 20]

4.2 The Second $\pi^- p \to \phi\phi n$ Experiment.

In 1985 Etkin et al.[11] published their second experiment. The original experimental setup was to a large extent dismantled to allow other experimental use of the MPS II user facility, and the software was considerably modified. The new experimental hardware setup was made virtually identical to the first experiment. We collected an additional 2449 $\pi^- p \to \phi\phi n$ events for a total= 3652$\phi\phi$ events. We considered all partial waves up to $J = 4$ and $L = 3$ which satisfy Bose symmetry and the $\phi\phi$ system requirements which were 52 waves. With triple the statistics we were able to resolve the D-wave found in the first experiment into 2 closely overlapping D-waves as well as the previous S-wave. These results, and phase differences are shown in Fig. 7a-b respectively.

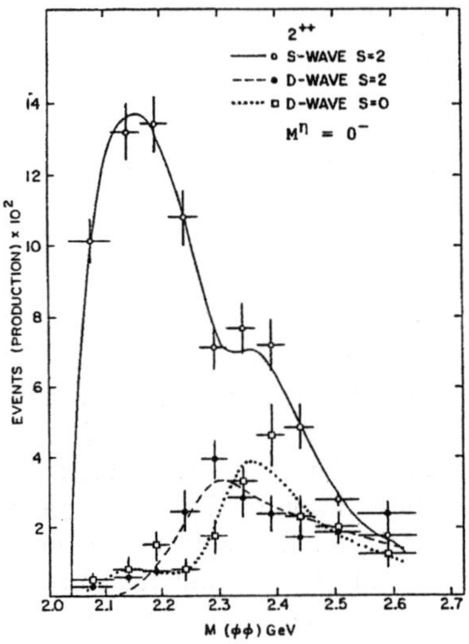

Figure 7a: Mass distribution for the three wave fits. The curves show the fit by the three K matrix poles (i.e. Breit-Wigner resonances) for second (1985) experiment.

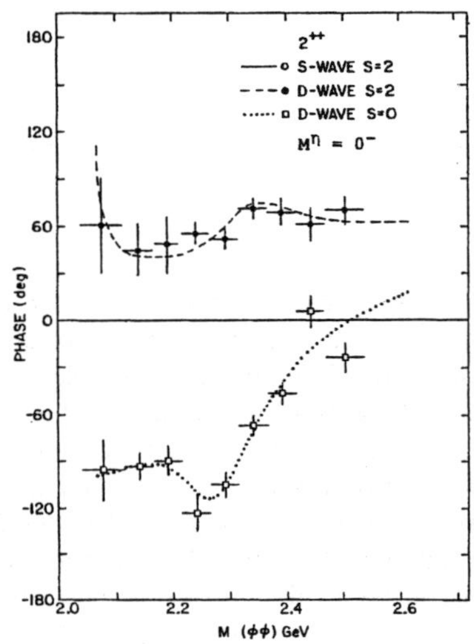

Figure 7b: Phase difference (relative to the S-wave for the three wave fits. the curves show the fit by the three K matrix poles (i.e. Breit-Wigner resonances) for second (1985) experiment.

The selection of the three waves was essentially completely determined by the angular correlations and was quite insensitive to the mass spectrum shape, acceptance etc. The good fit of the data to the angular correlations that were significant in selecting the three waves are shown in Fig 8.

All 3 waves that fit the $\phi\phi(< 1\sigma)$ had $J = 2$, $P = C = +M^\eta = 0^-$ same as previously, and $d\sigma/dt' \sim \exp(-10|t'|)$. Thus again clearly a π exchange reaction produced the $\phi\phi$. We used a K -matrix analysis to preserve unitarity. Two K-matrix pole fits were rejected by 16σ. A Good fit with 3 poles and the 3 corresponding Breit -Wigner resonances ($< 1\sigma$) was

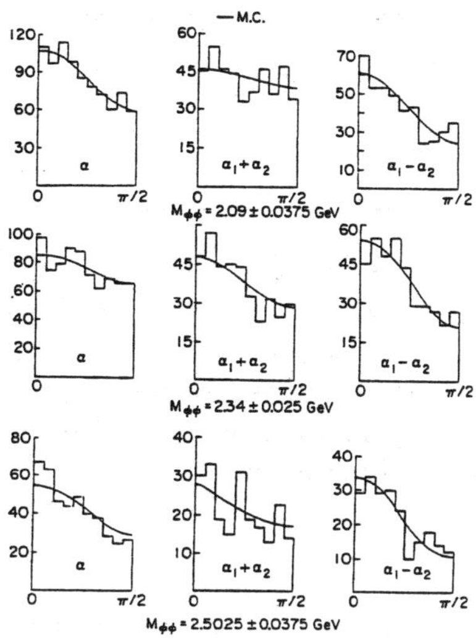

Figure 8: For second (1985) experiment angular distributions for α, $\alpha_1+\alpha_2$, $\alpha_1-\alpha_2$ in three of the ten mass bins. The $\pm$ numbers shown on the $M_{\phi\phi}$ mass values give the half bin size in which the data is grouped. The curves are the three-wave fit modified by the experimental acceptance. $\alpha(1,2)$ are the azimuthal decay angles of each ϕ_1 and ϕ_2 α_1 and α_2 (combined).

obtained. An analysis of the 13% ϕK^+K^-n background was performed. We found a 1^{--} coherent wave which allowed the phase of all three waves to be converted to approximately absolute values exhibiting clear resonance behavior Argand Diagrams. These results will be shown in the third experiment discussion.

Table II gives the details of the Breit Wigner resonances extracted from the K- matrix. Unitary effects were found to be small so therefore an analysis in terms of complex amplitude Breit-Wigner resonances would give very similar results.

4.3 The Third $\pi^-p \rightarrow \phi\phi n$ Experiment

In 1988 Etkin et al.[12] added 3006 new $\phi\phi$ events essentially doubling the statistics for a total $= 6658$. The analysis confirmed previous results within errors, and allowed us to perform a refined expanded final analysis. We searched up to $J=6$ and $L=4$ which included 114 waves more than doubling the number of waves(52) searched previously. However we found no new partial waves or resonances, therefore indicating the stability of our results. Figure 9 shows the $\phi\phi$ mass spectrum, the 2^{++} S and D wave amplitudes, and the $D-S$ phase difference. One should note that the $\phi\phi$ mass spectrum by itself does not allow one to

% data		$I^G J^{PC}$	Mass(GeV)	Γ_{tot}(GeV)	Branching ratio (%)		
					$L = 0$	$L = 2$	$L = 2$
					$S = 2$	$S = 2$	$S = 0$
					S_2	D_2	D_0
50^{+10}_{-10}	$g_T(2050)$	0^+2^{++}	$2.05^{+0.09}_{-0.05}$	$0.20^{+0.16}_{-0.05}$	98^{+2}_{-70}	0^{+50}_{-0}	2^{+25}_{-2}
20^{+20}_{-10}	$g_{T'}(2300)$	0^+2^{++}	$2.3^{+0.02}_{-0.10}$	$0.20^{+0.06}_{-0.05}$	30 ± 20	50^{+20}_{-10}	20 ± 20
30^{+10}_{-20}	$g_{T''}(2350)$	0^+2^{++}	$2.35^{+0.02}_{-0.03}$	$0.27^{+0.09}_{-0.13}$	40^{+10}_{-20}	5^{+15}_{-5}	55^{+20}_{15}

Table 2: Quantum numbers and parameters of the Breit-Wigner resonances from the three-pole K-matrix fit, for second (1985) experiment.

draw any significant conclusions. It is only the partial wave analysis and the use of angular correlations in it that leads to significant and striking results. If you just change one of the three waves selected by the fit to any other wave of the remaining 111 we used, those fits are rejected by at least 13σ. Thus the selectivity of our analysis is extremely impressive. The significant angular correlations data are compared to the three wave fit in Fig. 10 and obviously fit very well. Figure 11 shows a log (const. $\times d\sigma/dt'$) plotted vs. $|t'|$, and is well fit by $d\sigma/dt'$ = const. exp- (9.50 ± 0.10) for $|t'| < 0.3$ corresponding to pion exchange. Since $M^\eta = 0^-$ for all three waves it is clear that the three $\phi\phi$ resonances are produced by pion exchange. Only a few percent of our $\phi\phi$ at high transverse momenta are produced by other exchanges.

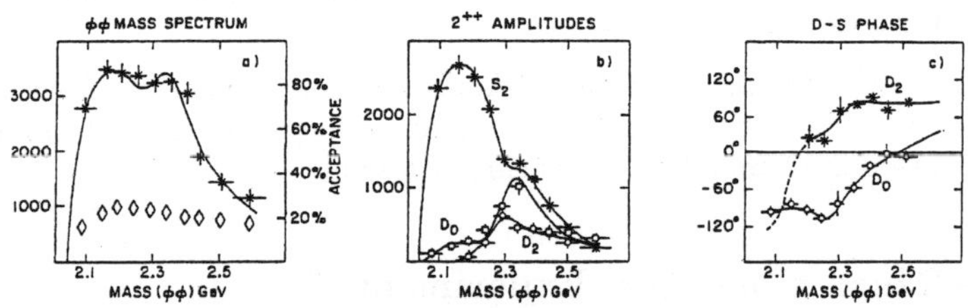

Figure 9: (a)The acceptance corrected $\phi\phi$ mass spectrum, (b) intensity and (c) phase difference for the three $J^{PC} = 2^{++}$ waves. The curves show the fit by three Breit-Wigner resonances (i.e. K-matrix poles). The resonance behavior and parameters are insensitive to the detailed shape of the mass spectrum, and are primarily determined by the six angular distributions and their correlations.

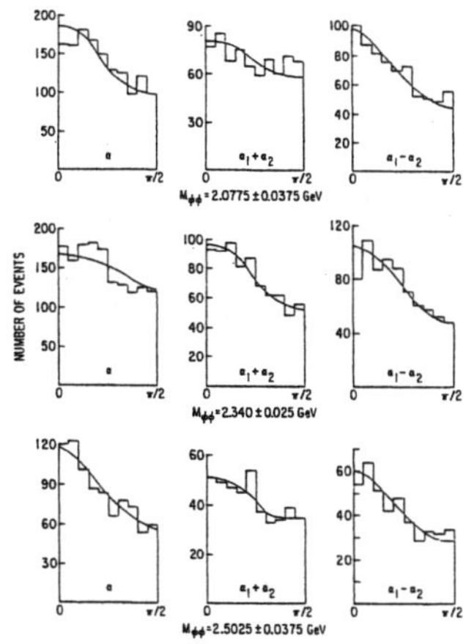

Figure 10: For third (1988) experiment the angular distributions for α, $\alpha_1 + \alpha_2$, $\alpha_1 - \alpha_2$ in three of the ten mss bins. The $\pm$ numbers shown on the $M_{\phi\phi}$ mass values give the half bin size in which the data is grouped. The curves are the three-wave fit modified by the experimental acceptance. $\alpha(1,2)$ are the azimuthal decay angles of each ϕ_1 and ϕ_2 α_1 and α_2 (combined).

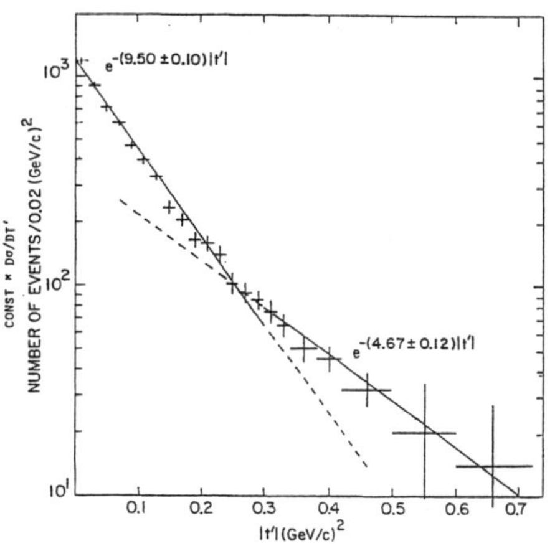

Figure 11: log(const. $\times d\sigma/dt'$) plotted versus $|t'|$

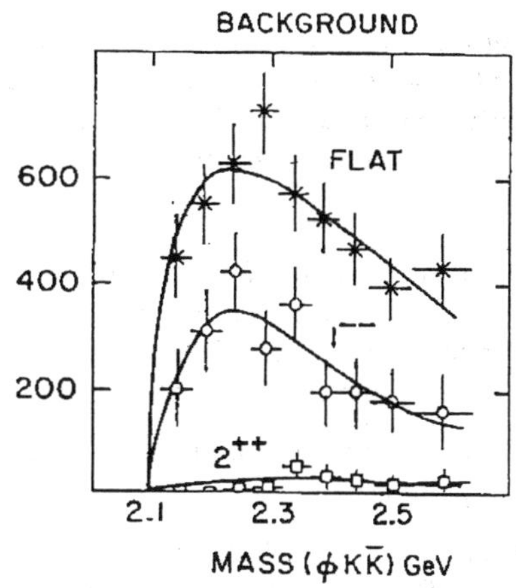

Figure 12: Intensity of the partial waves in the background reaction $\pi^- \to \phi K^+ K^- n$. The smooth curves are Breit-Wigner fits for the 1^{--} and 2^{++}, where the flat in all angular variables background fit is an exponent times the threshold factor.

We carefully analyzed the ϕK^+K^- simultaneously with the $\phi\phi$ and used the 1^{--} coherent wave within it (see Fig. 12), to approximately convert the three resonant waves phases to absolute values, and thus clearly demonstrate their resonant behavior. Errors, particularly on the phase differences were considerably reduced.

The conversion of the three resonant waves to absolute values and the corresponding classic resonant amplitude and phase behavior of each wave in the resulting Argand Diagrams is shown in Fig. 13 for the two extreme cases, and thus for the actual case in between the two extremes.

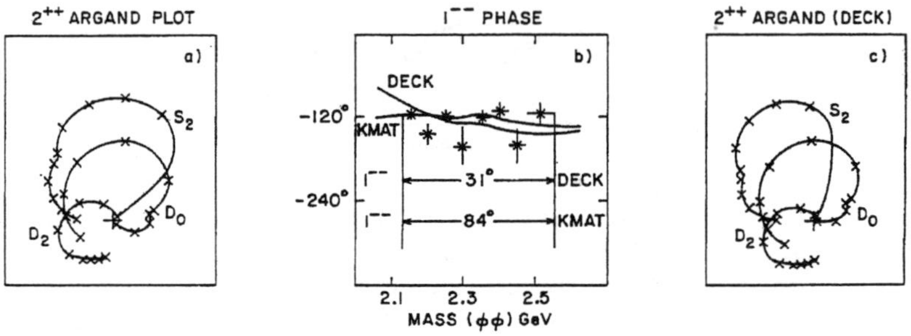

Figure 13:(a) Argand plots for the three 2^{++} waves with the absolute phase based on the $1^{--}\phi K\bar{K}$ wave being a Briet-Wigner resonance (K-matrix), while (c)is based on the $1^{--}\phi K\bar{K}$ resulting from a deck mechanism. (b) shows the 1^{--} phase (relative to the S-wave in $\phi\phi$). The curves come from the two models stated for the 1^{--} absolute phase.

We again used the K-matrix formalism to preserve unitarity. Two K-matrix pole fits were rejected by 18σ. Three K-matrix poles gave a good fit of approximately 2σ. Unitary effects were found to be small. Thus a fit which consists of three complex amplitude Breit-Wigner resonances would give consistent numbers.

Table 3 gives the parameters of the Breit-Wigner resonances corresponding to the K-matrix poles etc.

Even now the Particle Data Group (2004) still lists as established 5 resonant States for $S = C = B = 0$ in the complex mass range 2-3 Gev that we investigated.

Three of them are our g_T, $g_{T'}$ and $g_{T''}$ states. However some time ago PDG changed to a standard notation, and they have since been listed as the $f_2(2010)$, $f_2(2300)$, and $f_2(2340)$. Only one of the two other established states, the $f_4(2050)$, and the many others expected, and eventually many were reported but not established, were in principle detectable but were not seen in our experiment.

In over 2 decades the unique characteristics of our experiments has never been successfully explained within the context of QCD, except for our original explanation that they represented interactions with one or more 2^{++} glueball(s).

State	% of $\phi\phi$ data	Mass (GeV)	Width (GeV)	S_2(%)	D_2(%)	D_0(%)
g_T	45	$2.011^{+0.062}_{-0.076}$	$0.202^{+0.067}_{-0.062}$	98^{+1}_{-3}	0^{+1}	2^{+2}_{-1}
$g_{T'}$	20	2.297 ± 0.028	0.149 ± 0.041	6^{+15}_{-5}	25^{+18}_{-14}	69^{+16}_{-27}
$g_{T''}$	35	2.339 ± 0.055	$0.319^{+0.081}_{-0.069}$	37 ± 19	4^{+12}_{-4}	59^{+21}_{-19}

Table 3: Parameters of the Breit-Wigner resonances (corresponding to the K-matrix poles) and percentage of the resonances going into 2^{++} S_2, D_2, and D_0 channels. The errors come from a complete study of the χ^2 surface, for third (1988) experiment.

These characteristics are:

1. the massive breakdown of the OZI suppression by a factor $\sim$ 800, very selectively occurs only for the Quantum numbers of the 2^{++} glueball.

2. Virtually the entire mass spectrum of the $2^{++}\phi\phi$ is composed of 3 resonances with all the quantum numbers of the 2^{++} Glueball very well established in a mass region where the 2^{++} Glueball is predicted. Only two other states with $J^{PC} = 4^{++}$ have been established in this region, the $f_4(2050)$ is the only one of the two which in principle is detectable in our experiment (PDG 2004). There is little doubt many others are expected to occur. Many have been reported but it was too difficult to establish any others due to the complexity of this region.

3. Pion exchange which proceeds by annihilation of a π^- and π^+ was demonstrated to cause the $\phi\phi$ production. Thus there is a clear flavor change in the final state.

Conclusion, thus the simultaneous massive and selective 2^{++} breakdown of the OZI suppression by three 2^{++} resonances with the 2^{++} glueball quantum numbers, and the quark flavor change has only been explained by interaction of the observed resonances with 2^{++} Glueball(s).

The very extensive LGT investigation by C. J. Morningstar and M. Peardon[21] concluded that in our mass range only one 2^{++} glueball exists, with a mass of 2.40 ± 0.15 GeV. See reference[21] for other LGT investigations. Therefore we conclude that the 2^{++} glueball interacts with at least 2 other primarily quark built states which it overlaps, as would be expected from Quantum Mechanics. All 3 states would have to be rich in glue since they all break the OZI suppression.

4.4 Evidence for a Fourth State Related to the Three $\phi\phi$ States

In our most recent publication by R.S. Longacre and S.J. Lindenbaum[22], we considered whether the recently established wide $f_2(1950)$ which partially overlapped our $\phi\phi$ resonances, and showed some unexplained mini-structure on its high mass side[23] could be related to our $\phi\phi$ resonances.

We concluded by a K-matrix analysis, that by using the mini-structure to tie the two results together in a unitary preserving way, that adding a fourth K-matrix pole to our 3 wave analysis led to a good approximately 2σ overall fit, whereas a stand alone resonance fit to the data of Ref.[23]is a 6σ fit . Most of the sigma's arise from Ref.[23] not fitting the structure in their data near 2.4 GeV which overlapped with the $\phi\phi$ mass region. The results of our combined 2σ fit to our $\phi\phi$ data and Ref.[23]data are given in Figs 14 and 15.

Mass (GeV)	Width (GeV)	S_2	D_0	D_2
$2.049^{+.035}_{-.024}$	$.567^{+.064}_{-.071}$	$96\%^{+4}_{-10}$	$1\%^{+7}_{-1}$	$3\%^{+9}_{-3}$
$2.123^{+.015}_{-.033}$	$.294^{+.056}_{-.055}$	$89\%^{+7}_{-11}$	$2\%^{+13}_{-2}$	$9\%^{+10}_{-8}$
$2.340^{+.013}_{-.013}$	$.148^{+.066}_{-.032}$	$7\%^{+15}_{-7}$	$12\%^{+14}_{-10}$	$81\%^{+3}_{-14}$
$2.412^{+.028}_{-.032}$	$.362^{+.100}_{-.053}$	$45\%^{+28}_{-7}$	$21\%^{+7}_{-19}$	$34\%^{+15}_{-20}$

Table 4: Resonance Parameters for the 4 pole fit.[22]

Figure 14 is the mass spectra of the 4 waves involved, and Fig.15 shows phase differences and the argand plots. the 2σ fits in Fig. 14a-d and Fig. 15a-b are shown by the solid line and unless you are interested in the details of other fits you need only concentrate on the solid line which gives the four pole fit for the combined data . The argand diagrams (Figs. 15c and 15d) show clear resonance behavior. Table 4 gives the resonance parameters for the four pole 2σ fit. The three $\phi\phi$ resonances in this combined 4 pole fit are modified somewhat from the original values, but the differences are not significant. Reference[22] gives more detail.

The above implies it is likely that the 2^{++} state from[23] is related to the three $2^{++}\phi\phi$ states and due to the same glueball and mixing mechanism which explains the $\phi\phi$ states. This is the only state to come out of the pattern recognition program that appears relevant to the interactions with the 2^{++} Glueball. This additional 2^{++} state can be considered as an additional consistency argument for the $\phi\phi$ analyses leading to the conclusion that the 2^{++} glueball has been established.

However one can also suppose the $f_2(1950)$ is not related to our $\phi\phi$ states In that case our previous$\phi\phi$ work as described in Ref.[12] is uneffected, and the arguements we made for establishment of the 2^{++} glueball remain the same and clearly are uneffected by it.

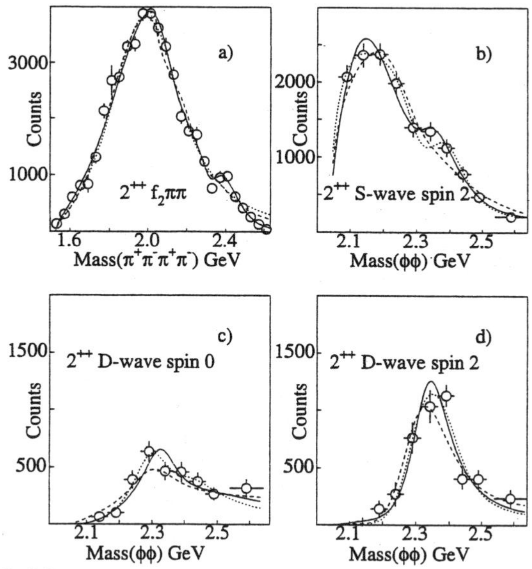

Figure 14: The $I^G J^{PC} = 0^+ 2^{++}$ mass spectra of (a) Mass $(\pi^+\pi^-\pi^+\pi^-)$ for $2^{++} f_2(1270)\pi\pi$; (b) Mass $(\phi\phi)$ for 2^{++} S-wave spin 2; (c) Mass $(\phi\phi)$ for 2^{++} D-wave spin 0; (d) Mass $(\phi\phi)$ for 2^{++} D-wave spin 2. the solid curve is the fourth resonance fit (Table I); the dashed curve is a three-resonance fit; and the dotted curve is a separate three-resonance fit to the three $\phi\phi$ waves, and a separate fit with one resonance to the $f_2(1270)\pi\pi$ wave. Note this separate fit for the $f_2(1270)\pi\pi$ is particularly poor in the region of 2.3 GeV to 2.6 GeV, whereas the four-pole (resonance) combined fit (solid line is a 2σ fit (see text).

5 Pattern Recognition of the 0^{++} Glueball

Considerable effort has been made in this area and there are obviously more states than the 0^{++} nonet can accommodate. There are ever increasing complexities to contend with. The 0^{++} glueball has the vacuum quantum numbers. Although LGT can predict the vacuum for pure glue without quark loops, it is very far from predicting the extremely complex vacuum which will likely arise if one in the future is able to do a complete enough job of including quark effects. The real vacuum can of course mix with the 0^{++} states and the 0^{++} glueball and the resultant effects are still quite unknown.

In the last Quark Model Review by PDG (2004) an attempt was made to estimate the mixing angle for the 0^{++} nonet. In order to attempt this they eliminated four established states and stated:

"The classification of the 0^{++} mesons is tentative and the mixing angle uncertain due to large uncertainties in some of the masses. The $f_0(1500)$ in the Meson Summary table is not in this table as it is hard to accommodate in the scalar nonet. The light scalars $a_0(980)$, $f_0(980)$ and $f_0(600)$ are often considered as meson-meson resonances or four quark states and are therefore not included in the table."

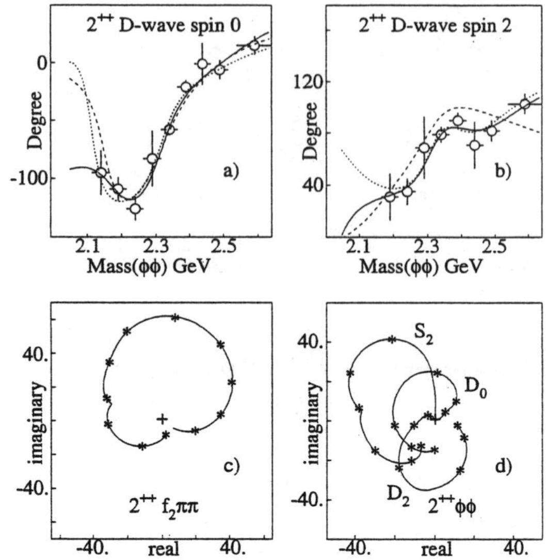

Figure 15: (a) and (b) show the phase behavior of the D-wave $\phi\phi$ amplitudes measured with respect to spin 2 S-wave for the three fits described in Fig. 14 caption and text: (a) 2^{++} spin 0 D-wave; (b) 2^{++} spin 2 D-wave, (c) and (d) show argand amplitudes which are generated by the four-pole (resonance) fit. Each $*$ is at a 0.1 GeV marking starting with 1.5 GeV for (c) the $2^{++}f_2\pi\pi$ argand amplitude, while (d) markings start at 2.1 GeV for the three $\phi\phi$ waves spin 2 S-wave (S_2), spin 0 D-wave (D_0), and spin 2 D-wave (D_2).

They then left the mixing angle column blank indicating that even with these deletions they could not determine a mixing angle.

On the other hand the 1^{--} nonet which contains the $\phi(1020)$ has a mixing angle of $\theta_{lin} = 36.0$ degrees, and is very close to ideal mixing which is 35.3 degrees. Ideal mixing decouples the $s\bar{s}$ and $(u\bar{u} + d\bar{d})$ singlet states. Thus the $\phi(1020)$ is virtually a pure $s\bar{s}$ state. Therefore the relevant pion exchange reaction $\pi^- + \pi^+ \longrightarrow \phi(1020) + \phi(1020)$ which produces the $\phi(1020) + \phi(1020)$ 2^{++} resonant states we observe, involves a clear flavor change from non strange to strange light quarks, which leads to the OZI suppression we have assumed for all $\phi\phi$ states that do not interact with a glue resonance (i.e. Glueball).

As I stated earlier the focus of this talk is to determine whether we can establish at least one Glueball experimentally.

In view of the foregoing discussion on the complexities in the 0^{++} channel, and the lack of definitive evidence for a Glueball in that channel, it is in my opinion not feasible at present to make a case for establishing a 0^{++} Glueball, and that may continue to be true for an indefinite period. For the foregoing reasons, the complexity of the 0^{++} channel, and the time restraints on this lecture, I will not treat that topic.

6 Conclusion

1. J/Ψ radiative decay groups have withdrawn their two major Glueball candidates and do not have any at present. Therefore it appears unlikely that this channel is an effective one for establishing Glueballs.

2. $\pi - p \to \phi\phi n$. Three experiments providing (1203 events in 82); (3652 total events in 85); and (6658 total events in 88) all demonstrated that the $\phi\phi$ massively broke down the OZI suppression by a factor approximately 800 very selectively, only for the quantum numbers of the 2^{++} Glueball. Virtually all of the cross-section was composed of $3\phi\phi$ resonances each of which had all of the quantum nos. of the 2^{++} Glueball established. This mass region is where LGT predicts the 2^{++} Glueball to be. These resonances have been listed as established in PDG for a long time as $f_2(2010)$, $f_2(2300)$, and $f_2(2340)$.

Our original explanation that the interaction with at least one 2^{++} Glueball caused the massive breakdown of OZI selectively, only for the resonance states resulting from the Glueball and its mixture with states having similar quantum numbers, still remains as the only viable explanation after 23 years. All published alternatives were shown by our publications to be incorrect or irrelevant or both. References were given earlier[16, 17, 18, 19, 20]. Only one other state, the recently established $f_2(1950)$ partially overlapped our states. We investigated whether it could be related, and concluded it likely is related, and due to the same Glueball and mixing mechanism which explains the $\phi\phi$ states, this analysis has been published recently by R.S. Longacre and S.J. Lindenbaum[22]. This paper also summarizes the history of the $\phi\phi$ states.

Considering the above, and the details in the lecture, and references, it is reasonable to conclude a 2^{++} Glueball has experimentally been established to exist within the context of QCD. The unusual $\phi\phi$ data is naturally explained in the context of QCD, and no other alternative explanation has emerged in over 2 decades.

References

[1] a) Hadronic Physics of $Q\bar{Q}$ Light Quark Mesons, Quark Molecules, and Glueballs, S. J. Lindenbaum, Proc. of 18th Int. School of Subnuclear Physics, Ettore Majorana, Erice, Italy July 31-Aug. 11-1980. Published in Erice Subnuclear 1980: 0509. b) QCD, Unification, and the Road to "Asymptopia", S. J. Lindenbaum, Proc. of 18th Int. School of Subnuclear Physics, Ettore Majorana, Erice, Italy July 31-Aug. 11-1980. Published in Erice Subnuclear 1980: 0575. Subnuclear 1980 : 0575. c) Evidence for Explicit Glueballs from $\pi^- p \to \phi\phi n$, S. J. Lindenbaum, Proc. of 20th Int. School of Subnuclear Physics, Erice 1982 . Published Erice Subnuclear 1982: 65. d) The Glueballs of QCD and Beyond, S.J. Lindenbaum, Proc. of 22nd Int. School of Subnuclear Physics, Ettore Majorana, Erice , Italy , 1984. Published in Erice Subnuclear 1984 : 398. e) The

Observation and Phenomenology of Glueballs , Proc. of 23rd Int. School of Subnuclear Physics, Ettore Majorana, Erice, Italy, 1985. Published Erice Subnuclear 1985:548.

[2] R. M. Baltrusaitis *et al.* Phys. Rev. D 35, 2077(1987).

[3] Etkin *et al.* Phys. Rev. D 25, 2446(1982).

[4] L. Chen *et al.* Amplitude Analysis of the $K\bar{K}$ system in $\theta(1690)$ Decay, Proc. of Rheinfeld 1990 workshop on Hadron Meson Spectroscopy (Sugort Germany Sept. 1990), SLAC PUB 5378 (1990), p 111 Hadron 91, Edited by Onieda and Peaslee, World Scientific (1991).

[5] S. J. Lindenbaum and R.S. Longacre, Phys. Lett B 274, 492(1992).

[6] R. M. Baltrusatis *et al.* Phys. Rev. Lett. 56, 107(1986)..

[7] J. Z Bai *et al.* Phys. Rev. Lett. 76, 3502(1996).

[8] Shan JIN Proc. of ICHEP 2002 pp 684-687.

[9] Feng LU AIP Conf. Proc. 717, 120. 2004

[10] A. Etkin *et al.* Phys. Rev. Lett. 49, 1620(1982).

[11] A. Etkin *et al.* Phys. Lett. B 165, 217(1985).

[12] A. Etkin *et al.* Phys. Lett. B 201, 568(1988).

[13] A. Etkin *et al.* Phys. Rev. Lett. 41, 784(1978).

[14] The BNL Multi-particle SpectrometerMPS) Facility. Invited Lecture by S. J. Lindenbaum, presented at 4th Symposium on N Anti N , Syracuse Univ. May 2-4, 1975. Published in Syracuse Sympos. 1975 VII - 19.

[15] Etkin *et al.* (presented by K. Foley), 1983 Int. Europhysics Conference on High Energy Physics, Brighton England July 20-27, 1983. Published in EPS High Energy Conference 1983:431.

[16] S.J. Lindenbaum, Phys. Lett. 131B, 221(1983).

[17] S.J. Lindenbaum, and H. J. Lipkin Phys. Lett. 149B, 407(1984).

[18] S.J. Lindenbaum , Comments Nucl. Part. Phys. 13, 285(1984).

[19] S.J. Lindenbaum and R. S. Longacre , Phys. Lett. 165B 202(1985); 234, 163(1990).

[20] S. J. Lindenbaum, *Superstrings, Supergravity And Unified Theories*, The ITCP Series in Theoretical Physics Vol. 2 (World Scientific, Singapore, 1986). pp. 548-593.

[21] C. J. Morningstar and M. Peardon , Phys. Rev. D**60**, 034509 (1999).

[22] R. S. Longacre and S. J. Lindenbaum, Phys. Rev. D**70** 094041(2004).

[23] D. Barberis et al., Phys. Lett. B 471, 440(2000).

CHAIRMAN: J. LINDENBAUM

Scientific Secretaries: I.Shreyber, M.Csanad

DISCUSSION

- *Baldini:*

You mentioned two ways to look for glueballs J/Ψ radioactive decay and double OZI rule violation), then you mentioned in the second channel the resonance to ΦΦ decay. Could you comment on J/Ψ - - > γΦΦ , why did not you mention this channel?

- *Lindenbaum:*

For a long time they said they do not see it, but eventually did see it .Just observing some ΦΦ events does not tell you anything about making a case for a glueball candidate. You need selection of a glueball quantum numbers and arguments that connects the observation to a glueball, and that was not the case.

- *Baldini:*

I remember, that they found a signal but it was not a spin=2 resonance. But if there is a glueball, they should at least see something.

- *Lindenbaum:*

The fact is that they did not have anything as a credible glueball candidate, except those I mentioned., and they have both been contradicted by MK III and BES and evaporated..

- *Pacetti:*

Could you make a comment about possible lower mass limits for the glueball and about the f0 and a0 mesons as possible candidates?

- *Lindenbaum:*

My personal view is, that so far no one has been able to establish a case for a glueball in 0++. If you go to the 0++ channel, you really need to understand the vacuum. It could be very well capable of mixing with the glueball, and quark built states that overlap the glueball. But we do not know too much about the vacuum, only what lattice groups gave us, and they did not put quarks in the theory. In the low energy 0++ channel there are all kinds of complicated objects, so quarks would be very much needed. The PDG in a 2004 listed review tried to make some analysis for the 0++ channel, they eliminated 4 particles, and they still could not reach any conclusion about what the mixing is in that channel, and concluded " The classification of the 0++ mesons is tentative"

Early in this business, many were worried about the vacuum mixing in 0++. Personally, I think that if you calculate a vacuum composed of pure glue, and you then put quarks, into the theory, and recognize that the light scalars ao(980) , fo (980), and fo (600) are often considered meson - meson resonances or four quark states it is clear this channel is very complicated. So my personal advice is, stay away from 0++, and work on 2++.

That has paid off for me up to now.

- *Zichichi:*

Understanding the vacuum is a crucial problem and there is a big effort in lattice QCD on this, but they never can answer questions. Question: why is lattice QCD in such a big trouble to understand what is going on?

When you say, we need to understand the vacuum: who cares about vacuum? Let's see, what the results are! You find that pure glueball probably does not exist, it is mixed with quarks. What did we expect?

- *Lindenbaum:*

Not only that, but we could not see a pure glueball, we can observe only quarks and objects made up from quarks. If we had only SU(3) color, as what lattice gauge people assume, you could never observe anything. We need the quarks, they need to be in your complete theory, so one has to wait for the lattice QCD groups to have that.

They cannot even tell you (do not misunderstand me, I like lattice QCD, it is the only quantitave way to calculate non- perturbative QCD) the width of the glueball spectrum. That and the mass would be a huge step forward, but there is a long way to go to understand even the pure glue theory not to mention the quark- problem again.

- *Korthals- Altes:*

You stress that the 0++ has vacuum quantum- numbers and that we should know the vacuum better. I just wonder whether that is really necessary. After all, we know from lattice simulations without quarks that 0++ is a perfectly stable state, despite the fact that it has the same quantum- numbers as the vacuum. Now imagine one couples the quarks into the theory by making them less and less massive. This renders the probability that the fluxtube like 0++ will break into several mesons. But the 2++ gets unstable through the same mechanism!

- *Lindenbaum:*

You are right, if you take the pure glue theory, you can understand the vacuum, there is no problem, to my knowledge. The problem is, that there are the quarks. If they put quarks - even if in simple form - quark loops into the gauge theory, that's going to change the vacuum considerably.

Now simple quantum mechanics tells me, that nothing stops the vacuum from mixing with the glueball, so if you do not know the 0++ vacuum, you cannot establish a 0++ glueball. But that is my personal view, and please, pay attention to all the responses, I like to throw ideas out, but I keep an open mind on responses.

The second problem with the 0++ is that that channel is hard to understand . Even if you forget about QCD, it is hard to understand, what is going on, it is hard to get a conclusion, it is a complicated channel, and it is only getting more and more complicated.

When you put all that together, I think, that no one said that they established a 0++ glueball, and I do not see how to do that. They are looking into it, they are making progress. I heard the same from J/Ψ people before for 30 years, and they are not claiming to have a glueball candidate at present.. . Maybe these things helped people to move to the spin=2 channel, but I'm not a mind- reader. However, in the present real quark- qluon hadronic world there is no experimental evidence for either vacuum mixing or instability in the 2++ channel. Therefore I do not see how you can get an appropriate LGT theory which introduces them.

Now since 23 years our results have been stable, and they are strange results so that if you never heard of QCD it appears that without something like glueballs, and an OZI filter however you name it, you will not understand our data. I think there is no other way than glueballs to understand it in the context of QCD

- *Zichichi:*

On the other hand, you should care less on theoretical predictions on vacuum. QCD without quarks is no physics. In the experiment, you have physics, in the world, you have gluons and quarks.

Whatever experiment you imagine, it will be always with quarks and gluons. Theoretical fellows tell us since 25 years, that we will have theoretical predictions on this, and then there is nothing. Even if tomorrow someone would tell me that he has calculated the pure glueballs with mass and widths with lattice gauge theory. Even if! It would be useless for physics, because they must put quarks in. Putting in quarks could change everything. So why do you care so much about understanding the vacuum. It would be of no help.

- *Lindenbaum:*

We agree on that we have to put quarks in the theory in order to understand vacuum.

- *Zichichi:*

But here you emphasized the extreme need you feel to know what you call the vacuum.

- *Lindenbaum:*

One has to be careful with the arguments if one wants to address a specific point, especially in an hour. And remember, I ended a minute earlier, , and look at the territory I covered. Of course you are correct LGT needs to put in quarks into the theory sufficiently correctly and completely enough to have a theory which can really confront the real world situation, and it, as you say could change their predictions.

By the way, when I went through the J/Ψ history and 0++ , statements I made, that was not published. Erice will publish it first. Long time ago, I made some statements on the vacuum, but that is maybe forgotten by now.

- *Imponente:*

After more that 25 years of experimental and theoretical efforts, why do you think it is so difficult to find clear glueball signatures? Is it related to "poor" accuracy of resolution of experiments or something deeper.

- *Lindenbaum:*

Let me start with this. The cross- section for π^- + p□ $\Phi\Phi$ n is approximately 20 nb. The total mess out there is enormous, at least on the order of 20 mb. Why that is, I'm not going to speculate on, because it is a very complex question. But it is a fact. You need a magnet to look for a needle in a haystack.

Now we had a magnet, which was the OZI suppression. And in fact, it probably works better in the experiment I showed you, than people ever imagined, because somehow, I believe glueballs even enhance its effects.

We have 114 waves in our analysis, and all what we can detect are those three π exchange waves, it means almost no signal in the other 111 waves. We also tried to add a fourth K- matrix pole, but nothing significant was found. There is this strange thing observed here. Now, we were lucky, but it is not pure luck, because we had made a lot of efforts to understand how we are going to pick out waves in our analysis. You have to realize what was involved here, was that we had a facility, and could do things differently than the large collaborations. We built a series of facilities, the MPS1, which did the first 100 events. We had the group which built it and ran it, but a few physicists could come along and do their own experiment approved by the PAC , and we even helped them to build the equipment, in fact we often even built it for them. They told us, what they wanted and they could do their experiment. For doing the three analyzeable glueball experiments we built MPS II which was ten times faster. That's the way you get cutting edge physics.

In my opinion, many large collaborations have a problem. I hope they will be less of a problem at LHC than they are at RHIC. The very nature of a large collaboration means that if something is off beat- out of the popular mainstream (and much of what is important in physics has come that way), many or at least some in the collaboration get nervous .I proposed long ago a large TPC magnetic spectrometer that eventually became STAR, and I hoped that they would build and operate at least one general user facility at RHIC, so that you do not need 500 people to sign the paper. One could run such a facility for a relatively small number of people who share an idea for an experiment, which may not be in the mainstream.

- *Steinberg:*

Would one expect a large (even experimentally rising) set of glueball resonances?

- *Lindenbaum:*

Yes. In this Morningstar and Peardon paper you see a large number(13) of glueball states, and they go up to approximately 4 Gev in mass .These are the sum of all combinations of PC. Multi(2or more) glueball states are implied to occur at high enough mass(beyond 4 GeV) .The implication of more states as the energy increases appears to be possible..

I think we all will be gone from this Earth before all or most that is in that paper is established experimentally- if ever.

- *Steinberg:*

How could all these states contribute to the thermodynamics, to the dynamics of the QGP seen at RHIC? Do people think along those lines, do they predict a huge variety of states which would be exponentially rising like a Hagedorn- spectrum, or do they just predict a certain distribution of states for glue- glue?

- *Lindenbaum:*

Well, there are all sorts of theoretical predictions, you may have noticed if you read the White Papers at RHIC. Everyone of them considered various theoretical approaches. STAR for example discussed several independent theories. However glueball spectra were not specifically considered. The RHIC experimental groups were under pressure to find the QGP, because RHIC was funded to find it. To their credit, they did not claim it.

- *Steinberg:*

But would the many glueball states play an important dynamical role?

- *Lindenbaum:*

I have no explicit idea. They have not been considered directly, and it is very uncertain how to do this. . Obviously the contribution from glueball states eventually appears in the observed hadronic particles and resonances, which are considered in making fits to the data , and phenomenological and theoretical interpretations.

- *'t Hooft:*

I have two small remarks. First, glueballs are closed pieces of QCD- string. According to string theory, the spinless 0++ state should be a lot lighter than 2++, so there should be good reason to look for the 0++ rather than the 2++.

Second, probably all glueball states are mixed with qq- states. But this means that in the flavor = 0 sector there are more states than in the flavor not equal 0 sectors. Is there a way to count the states? This would also be a way to look for glueballs.

- *Lindenbaum:*

This is a reasonable point of view, but the problem arises, how to identify them as glueballs.

- *'t Hooft:*

One could just count them. If you see more resonances with zero flavor quantum-numbers than resonances at the same spot from other respects, just with nonzero flavor quantum- numbers, than you know, that there is a bigger arsenal of states so some of the states can be identified as glueballs. So it is an indirect way to prove the existence of glueballs, but unfortunately, for us, experimentalists conclude, they are all mixed with two-quark states.

- *Lindenbaum:*

Right there are a lot of mechanisms that give you a large number of states there. As I said, nobody understands that channel at present.

- *'t Hooft:*

Maybe because the channel contains more bumps than the other channel.

- *Zichichi:*

Yes, very wide bumps. Could be that glueball you are talking about is the hadronic production from few hundred to seven hundred MeV. They are all there, but how you say, these are glueballs?

- *Lindenbaum:*

That's the problem. If you want my prejudice, I have seen enough of lattice gauge and enough of phenomenological things, I believe that the odds are that glueballs exist. No matter how complicated that channel becomes and vacuum that mixes with it becomes, there will be glueballs there. In that sense, I agree. But to establish them, I do not agree that you will ever convince the community that just counting them would do it.

- *'t Hooft:*

Is it possible to do a complex phase analysis?

- *Zichichi:*

The trouble is exactly this. In the low energy part, 0++, it's impossible to prove anything. They are there, but how you prove they are glueballs. Let me pick up something, which is a 10- 6 factor, but is clean, and if I can show, that this has the characteristics of being a glueball, then this is the proof. In the low energy region, the glueballs are there. I'm personally convinced, that hadron production in the few hundred MeV region, it is all coming from glueballs. But you cannot prove, so let's go to the clean channel, and this is the 2++ one.

- *Lindenbaum:*

I also tell you something else. If there was no Zweig suppression, and we were not doing the cleanest possible flavor exchange, whichis annihilating π^+- π^- into a strange quark system, and also we needed the Φ spin, to get the analysis, the mass spectrum and cms angles do not tell you enough. We would never be able to pick out those waves we picked out, if we did not have these six angles and their correlations, all significant. If the Φ was spinless, I would not have expected that experiment to have the success that it achieved.

- *Jona- Lasinio:*

What is the theoretical lifetime of a glueball? Is it possible that glueballs have never been observed due to their extremely short lifetime?

- *Lindenbaum:*

Theoretically we do not have the lifetime, but it is related of course to the width. Now I cannot see, that if you start to get a resonance, strong enough to make a glueball, how could you then get anything that is not broad, at least as broad as hadronic, and most likely even broader. But the lifetime does not matter in the observation, because what you observe are finally quark states with substantial resonating glue from mixing with the glueball , and they have the glueball quantum numbers.

- *Jona- Lasinio:*

How can you then discriminate the Φ- Φ from a glueball from another one without glueball?

- *Lindenbaum:*

Without a glueball, there is no way in QCD, to have the selectivity of 2++ glueball quantum numbers only, and have the Φ- Φ n final state consist virtually only of 3 resonant states with these quantum numbers.

Besides the states we have established there is only one more established to date in the 2- 3 GeV mass region, namely the f_4(2050), and that certainly does not have the unusual behavior that our states have..

We pick out 20 nb, which is totally clean. Nobody sees unusual sets of resonances like we do, even in the rest of hadronic physics. So there is definitely something very strange going on, even if you never heard of QCD. If you believe in QCD, and that there is a 2++ glueball state , and look at this selectivity, then you say, we found it, and our data establishes that we interacted with the 2++ glueball. .

- *Zichichi:*

The 2++ has no other reason to exist, if there are no glueballs. While in the low energy range, there are glueballs, but you can not experimentally pick them out, and convince anyone, that these are glueballs. This is what's going on since a fourth of a century.

- *Cifarelli:*

Do you think that multiquark states (provided they really exist) could be a useful tool to search for glueballs?

- *Lindenbaum:*

You would have to have an effective filter for selecting the glueball quantum numbers, and an effective way of analyzing the observed results. Offhand I do not have any idea how multiquark states would do this.

You would have to figure out a way of doing it. I do not want to say you cannot or can do that, and I'm always open to other people's ideas, I would have to see a specific proposal to draw any kind of response.

- *Loginov:*

I have a technical question related to observation of glueballs. On the M(ΦΦ) plot you have data fitted with three different fits, which correspond to different waves. You have three different sets of data points in your ΦΦ set, for every fit. Does it mean that every event in your final ΦΦ sample is in one of these subsets, i.e., you know from which wave it comes? How do you determine to which category the event goes?

- *Lindenbaum*

What we did is this. We searched for each of the waves individually first to see which waves have a signal. Then we found only the three waves that we ended up with (that includes even putting in the f2(1950) that had signals. This far exceeds what we were hoping for when we proposed this method.

But we do not have three data subsets. We end up with one set as shown in the ΦΦ mass spectrum . The points and their errors we show , which were fit for each wave come from the data analysis. If we take any of the final three of the 114 waves, and replace them with any one of the other 111 waves remaining, all the other such wave sets are rejected by 13?

- *Zichichi:*

I think this search is extremely interesting and important and currently this is the only evidence for glueballs. Other results are either weak or confusing.

D. DENISOV

"Highlights from Fermilab: Tevatron Experiments"

The Content of the Lecture can be found in:

- *"The Dzero Experiment at Fermilab: Collected Physics papers 2005"*, The Dzero Collaboration, Fermilab Preprint PUB-06-028-E, 2006.

CHAIRMAN: D. DENISOV

Scientific Secretaries: K. Hanagaki, S. Sharikov

DISCUSSION

- *Loginov:*

The deadline for the discoveries at the Tevatron is when the LHC starts taking physics data. So there are two physics "strategies": run without time-consuming upgrades if possible to get more data, as much as we can, or upgrade to take a bit less data but with better detector. Which one do you think is preferable?

- *Denisov:*

Of course it is very nontrivial to optimize the path to the discovery, because you don't know what the discovery is. There is only some general understanding what we would like to see, as I've discussed Higgs search and improvements of the top quaks measurements.

We have two teams of people, as we have two experiments at the Tevatron: D0 and CDF. And these two teams have decided to go in different ways.

D0 is planning to perform serious upgrade of their detector (so that the whole Tevatron complex will be shutdown for almost 4 months), to insert an additional layer of the silicon detector and add triggering capabilities in order to be able to handle high luminosity. CDF experiment, on the contrary, have decided to concentrate on whatever they have – their hardware capabilities and their triggering capabilities – and put all their efforts to collect as much data in current configuration as possible.

As you can understand, these two requirements are contradictory, because while one experiment would like Tevatron to be shut down, the other wants to continue collecting data. There is a lot of politics and different points of view that are expressed. At this moment the final decision is that it'll be a shut down for D0 to be upgraded.

For how long the Tevatron is going to run is also a hard question. Probably until LHC start up, but it's not a well-defined time. It's supposed to be (officially) end of 2007 but there might be some delays, as it've happened in the past with some of the large experiments and projects. We will be flexible – that means either we stop running when we collect a reasonable amount of data, enough for an interesting discovery, or just when we'll not be competitive with the LHC.

- *Munir:*

What is the reason of the substantial fall in the expected luminosity of the Tevatron to 8 fb^{-1} from the initial estimate of about 15 fb^{-1}?

- *Denisov:*

Yes, the original plan for the integrated luminosity was about 15 fb^{-1}. And if you look on the plot of the Higgs search, it's a nice number because then you can discover Higgs in all the mass range which is allowed by the Standart Model. When Tevatron started operation it became clear that there are many technical problems that prevent luminosity from reaching the planned level.

This is highly dependent on the effective production of antiprotons and also on their storage. For example, originally we had a project called "Recycler" – we were planning to recycle antiprotons after the store (since they are very expensive) and use them later. It didn't happen because of multiple technical problems. So as a result after realistically understanding what we can do, we have provided two curves, a base plan and a best-case scenario – and since then, we are on the second curve. And second curve will be, of course, better. Consider for example antiproton cooling with a beam of electrons, which was demonstrated this summer. It is very important, it makes us believe that by 2009 we will reach 8 fb^{-1}.

- *Munir:*

Have these issues been taken into account for ILC project?

- *Denisov:*

No, there perhaps won't be any clear correlation between Tevatron and ILC technically, only if you consider the finansial support to these two, because one of the options for Fermilab future is to construct ILC, and we might prefer to shut down Tevatron to save money to build the International Linear Collider.

- *Munir:*

Is there some particular effect responsible for these fluctuations in the mass of top quarks at various colliders runs and even various detectors since its discovery in 1995?

- *Denisov:*

If you carefully look at the numbers (I can show them) there is no any disagreement within errors between top mass numbers. For example. The first number that D0 provided back in 1995 was 199 GeV; but the error was ~45 GeV. Its 1 sigma, within of current measurement 173 GeV. And if you look throughout the history – you'll rarely find the differences between masses that are more than one or two sigma.

- *Dalitz:*

I was surprised to see that the CDF value for the top quark mass from their Run 1 is now settled at 173 GeV. We had a visit from Dr. Kenelli of the CDF collaboration, based at VCLA, a few months ago. She quoted 178 GeV for the analysis of Run 2 of CDF, and supposed that the value would be even larger when the analysis of Run 2 data will end.

- *Denisov:*

I have to show the plot. Look at the transparency. You can see these two numbers – 173 and 178 GeV. 178 GeV is number for the last year 2004, and not for the Run 1, as you have said, it is mainly defined by a D0 measurement, giving the error of ~4 GeV.

- *Dalitz:*

The top quark mass value from D0 was quite high originally, far above 180 GeV. Recently re-analysis of the D0 Run 1 data was completed, and I think published, by a D0 subgroup at Rochester University, although I do not yet know the result.

- *Denisov:*

Recently we published a lot of D0 Run 2 top mass measurements, and you can see that, in different channels, it is between approximately 160 – 180 GeV. And 199 for the Run 1 was mainly because there was one event with a large mass and it has affected the average. Now statistics has increased, and error is reduced. You can see that all D0 measurements are consistent with each other.

- *Loginov:*

Could you go to page 21 of your presentation?

You see there is one event with M > 1 TeV on the (t t-bar) effective mass plot. What is this event? How many events should one expect from the background in this region?

- *Denisov:*

This spectrum causes a lot of discussions at D0, and not only that region, but also region around 400 GeV, because you can see that these three points look like there is more than Standart Model prediction for the spectrum of (t t-bar) production. You can even see that the probability of resonance to be here is relatively large. And as for this specific event – it is a sign that we don't understand the backgrounds here well or, with increased amount of data, when errors will be reduced, some peak might really start to show up. At this moment we left the event on the plot, but our understanding is that it's most probably mismeasurement of some of the event parameters. It doesn't look like some kind of overlap of many events, but it looks like there might have been some extra energy measured or some other mismeasurement. And making conclusions based on a single event is risky, as you know.

- *Loginov:*

Ok, and what for this 2.6 or 2.7 sigma you were talking about during the presentation? It is quiet interesting what is a probability to have a few 2-3 sigma discrepancies when you study hundreds of different distributions.

- *Denisov:*

If you are doing 300 measurements, then probability to have a statistical fluctuation of about 3 sigma is not negligible. That's a good point. You *will* have some measurements 3 sigma away. That's why when we are talking about some real discoveries, we are talking about 5 or 6 sigma effects. Otherwise you can start making "discoveries" of 2 sigma effects every day! We are not claiming anything, but there are facts we are interested to study more carefully to see if it's only a statistical fluctuation or there is real physics behind.

- *Loginov:*

At some point probably one will have to take into account the number of channels he is looking at?

- *Denisov:*

That's a good point.

Highlights: Update from BABAR

M.A. GIORGI

INFN Sezione di Pisa and Università di Pisa Dipartimento di Fisica "Enrico Fermi"
Largo B.Pontecorvo, 3
I 56127 Pisa, Italy
E-mail: giorgi@pi.infn.it

Summary. — This paper comes 40 years after the discovery of *CP* violation [1] in neutral kaon decay in the famous experiment of Brookhaven and after the discovery by BABAR and BELLE of the indirect (2001) and the direct (2004) *CP* violation [2-5] in the b sector at PEP-II and KEKB e^+e^- asymmetric B Factories operating at the center of mass energy corresponding to the mass of the $\Upsilon(4S)$. The Unitarity Triangle is now beginning to be constrained by the measurements at B Factories of the sides, of sin2β and now also of the angles α, and γ. We are also in presence of the very intriguing results about the measurements of $\sin 2\beta$ in the time dependent analysis of decay channels via penguin loops, where $b \to (s\bar{s}s)$ and $b \to (sd\bar{d})$.

1. – Introduction

1·1. *Present* BABAR *and* BELLE *performance.* – In the last decade after 1999 the two active asymmetric B Factories, PEP-II [6] and KEKB [7], and their companion detectors, BABAR [8] and Belle [9], have produced a wealth of flavour physics results, subjecting the quark and lepton sectors of the Standard Model (SM) to a series of stringent tests, all of which have been passed. Such B Factories have in fact proved to be a unique tool for studying symmetry violations in heavy quarks and leptons. The continuing success of the two machines and experiments have demonstrated that the B sector results are fully compatible with the CKM paradigma [10, 11]. It is well known that *CP* violation occurs when an interference appears between two amplitudes and three ways are possible: *CP* violation in mixing, in decay, and in the interference between a mixing and a decay amplitude. The first one in K sector is in fact what was observed in the famous BNL experimet of 1964 [1]. The last two are well illustrated in the Fig. 1, where the mechanism

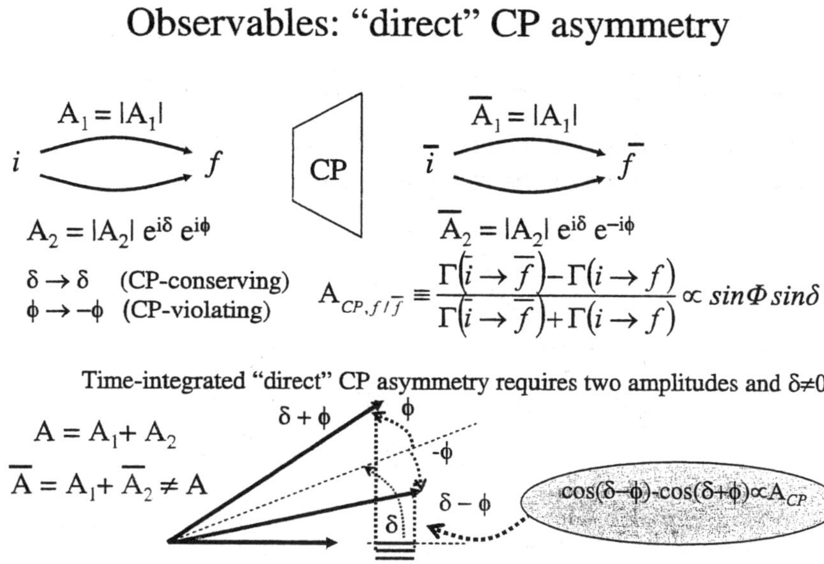

Fig. 1. – The mechanism of direct *CP* violation is based on the presence of a weak phase ϕ that shows opposite sign in the decay of B^0 or $\overline{B}^0$ and a strong phase δ that doesn't change sign.

of direct *CP* violation in the decay is pictorially explained and the violation induced by the interference between the mixing amplitude and the decay amplitude, such an interference is generating an observable time dependent asymmetry, as shown in Fig. 2.

The *CP* violation finds a natural interpretation inside the SM, although insufficient to explain the antimatter deficit in the universe, through the CKM mixing matrix of the three quark families. The presence of a complex term $i\eta$, corresponding to a violating phase, is responsible for a *CP* violating asymmetry.

Fig. 3 explains in a synthetic way the correspondence between CKM matrix elements and the sides and angles of the only Unitarity Triangle with all sides of the same order $O(\lambda^3)$, the other two triangles obtained by taking the off diagonal terms equal to zero of the unit matrix resulting from the unitarity condition $VV^\dagger = \mathrm{I}$. The sides of the Unitarity Triangle are measured with the B^0 decays as shown in the Fig. 3, while the angles can only be obtained by the measurement of the *CP* violation asymmetries. The extraction of α and β is obtained with the time dependent analysis, while γ is extracted from direct *CP* violation in the charged B decays.

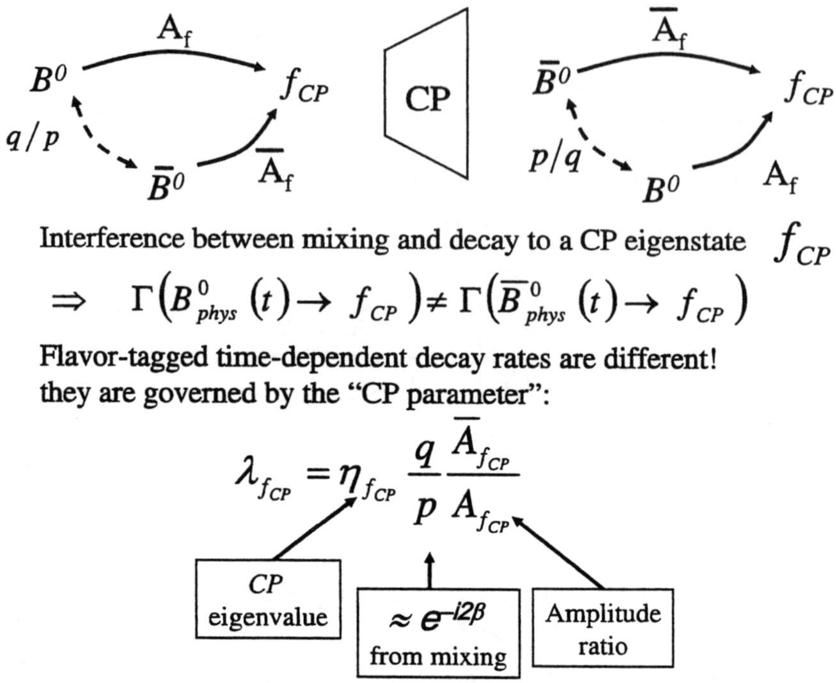

Fig. 2. – The time dependent analysis allows the measurement of the two coefficients C and S, both indicate *CP* violation: direct the first, indirect the second.

1˙2. *Detectors characteristics and analysis tools.* – Both detectors [8], [9] (see Fig. 4) are similar, they appear at first glance as the typical hermetic apparatus designed for e^+e^- colliders made of an inner Vertex Detector, a Tracking system, an Electromagnetic Calorimeter, a Solenoidal Magnet and a Muon/Hadron system. Actually they show a clear asymmetry that is a consequence of the machine energy asymmetry. The angular acceptance is infact greater in the forward direction, the direction of the high energy beam, than in the backward direction. The most relevant differences of BELLE with respect to BABAR are in the Silicon Vertex Tracker (a lower number of layers), in the Cherenkov systems (Aerogel instead of the imaging internal reflection quartz DIRC) and glass instead of Bakelite RPC for muon detection. The BABAR apparatus has shown a very high performance in the past six years, without major crisis even in the subdetector that are in principle more sensitive to radiation as the Electromagnetic Calorimeter made of Tl doped CsI crystals or the innermost subdetector, and the Silicon Vertex Tracker that is shown in Fig. 5.

1˙3. *Extracting the signal.* – In BABAR and in BELLE analyses most of the tools used to reject background and to extract the B signal are of general use in HEP (as event

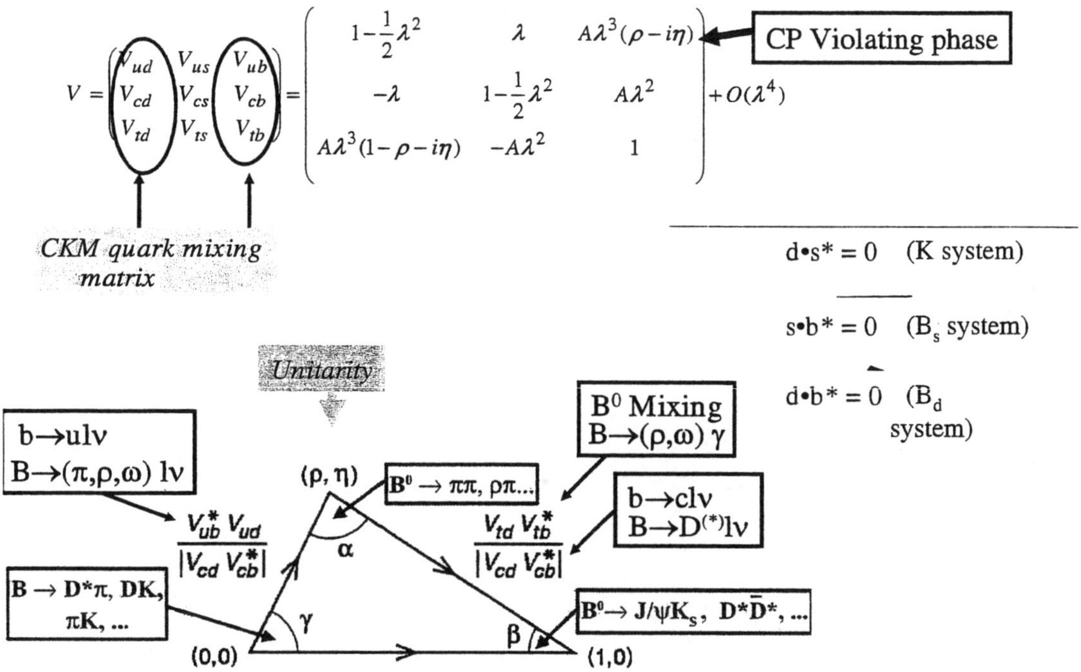

Fig. 3. – The CKM matrix in the Wolfenstein parametrization [12] and the Unitarity Triangle.

shape variables combined in some Fisher discriminants), some other instead are typical of $e^+e^- \to \Upsilon(4S) \to B\overline{B}$ experiments, as discussed in [8]. Most notably at B Factories, where the complete reconstruction of some B decay mode is needed, one can exploit the simple kinematic of the two body decay defining the variables ΔE,

$$\Delta E = E_B^* - E_{beam}^* \tag{1}$$

and the energy-substituted mass m_{ES}

$$m_{\mathrm{ES}} = \sqrt{E_{beam}^{*2} - p_B^{*2}} \tag{2}$$

where E_{beam}^* is the center-of-mass energy of the beam, and E_B^* and p_B^* are the center-of-mass energy and momentum of the reconstructed B candidate. The scatter plot of these two variables are shown in Fig. 6. Signal events will be found in the box at ΔE around 0 and m_{ES} around the B mass, $m_B = 5.28\,\mathrm{GeV}/c^2$. The typical resolutions are:

$$\sigma(m_{ES}) = 2.5\ \ \mathrm{MeV}/c^2$$

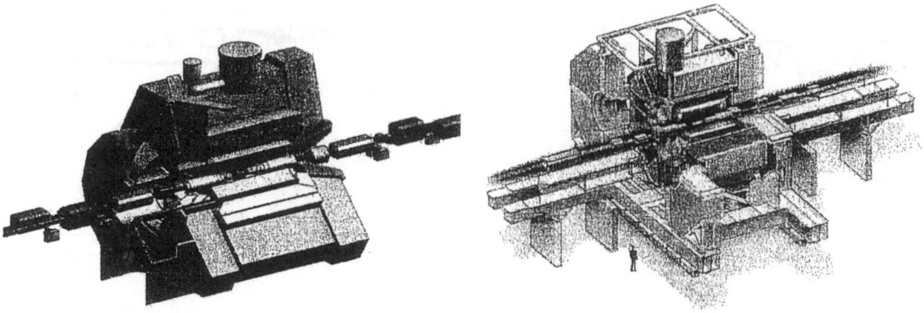

Fig. 4. – BABAR and BELLE detectors.

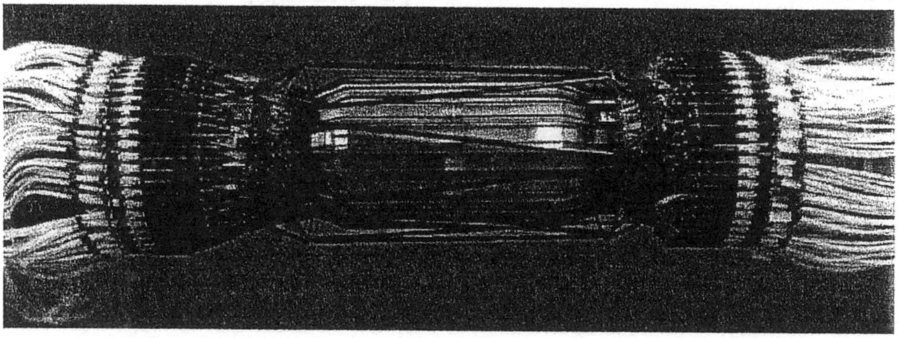

Fig. 5. – The BABAR vertex detector is built with the original shape of a lamp shade to minimize the scattering material along (i.e. the coulombian multiple scattering) the trajectories of charged particles.

that is in fact half of the beam energy spread and

$$\sigma(\Delta E) = 15 \div 30 \text{ MeV},$$

depending on the specific B decay mode. It is therefore very important for background rejection that the beam energy spread is small and under control.

2. – B Factories measurements situation in 2005

2·1. *B Factory performance.* – The two existing asymmetric B Factories have started their operations in 1999, the design peak luminosities of PEP-II and KEKB (see Table I) [13], have been exceeded in both laboratories, now PEP-II is running with a peak luminosity of $9.2 \cdot 10^{33}\,\text{cm}^{-2}\text{s}^{-1}$ and KEKB with $14.0 \cdot 10^{33}\,\text{cm}^{-2}\text{s}^{-1}$. These high luminosities have been obtained also thanks to the continuous injection of electron and

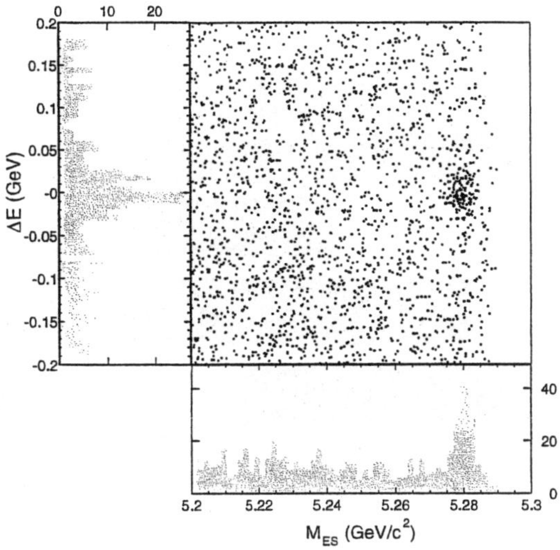

Fig. 6. – Example of ΔE vs. $m_{\rm ES}$ for $B^0 \to K^+K^-K^0$ candidates.

positron bunches that both laboratories have assumed as normal operation mode.

Parameter	PEP-II	*KEKB*
Peak Lumi.	$3.0 \cdot 10^{33}\,{\rm cm}^{-2}{\rm s}^{-1}$	$10.0 \cdot 10^{33}\,{\rm cm}^{-2}{\rm s}^{-1}$
e^- Energy	9.0 GeV	8.0 GeV
e^+ Energy	3.1 GeV	3.5 GeV

TABLE I. – *PEP-II and KEKB design parameters.*

The integrated luminosities of both B Factories doubled every 2 years as shown in Fig. 7. At present the total integrated luminosity recorded by the two experiments ($323\,{\rm fb}^{-1}$ of BABAR and the $551\,{\rm fb}^{-1}$ of BELLE) is not too far from one billion of produced $B\overline{B}$ pairs.

2·2. *Status of direct CP violation.* – In 2004 , BABAR has submitted to Phys. Rev. Lett. a paper containing the first observation of Direct *CP* violation in B decays [4].

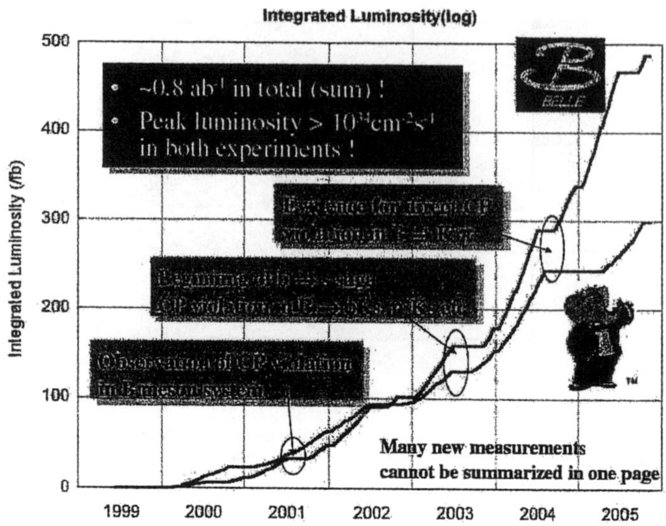

Fig. 7. – Integrated luminosities delivered by PEP-II and by KEKB.

The effect 4.2 σ was observed by measuring the asymmetry

$$A_{CP} = \frac{N(\overline{B}^0 \to K^-\pi^+) - N(B^0 \to K^+\pi^-)}{N(\overline{B}^0 \to K^-\pi^+) + N(B^0 \to K^+\pi^-)}$$

this analysis based on a sample of 227 million $B^0\overline{B}^0$ pairs allowed a clean signal of 1606 ± 51 events and has shown an asymmetry

$$A_{CP} = -0.133 \pm 0.030_{\mathbf{Stat.}} \pm 0.009_{\mathbf{Syst.}}.$$

BELLE has confirmed at ICHEP04 with a signal of 2140 ± 53 events from a sample of 274 million $B\bar{B}$ pairs, its measurement

$$A_{CP} = -0.101 \pm 0.025_{\mathbf{Stat.}} \pm 0.005_{\mathbf{Syst.}}$$

has a 3.9σ significance [14]. In Fig. 8 is shown the signal plot in the m_{ES} variable, the effect is immediately evident looking to the lower part of the figure that shows the raw asymmetry obtained before the subtraction of the background. The value after the average of BABAR and BELLE results is

$$A_{CP} = -0.114 \pm 0.020.$$

The analogous asymmetries measured in the two experiments on the same data samples for the charged B decays $B^{\pm} \to K^{\pm}\pi^{0}$ don't show any evidence of CP violation. BABAR finds [15]

$$A_{CP} = +0.06 \pm 0.06_{\text{Stat.}} \pm 0.01_{\text{Syst.}}$$

and BELLE [16]

$$A_{CP} = 0.04 \pm 0.05_{\text{Stat.}} \pm 0.02_{\text{Syst.}},$$

the average is:

$$A_{CP} = 0.049 \pm 0.040.$$

There is an evident discrepancy between the results from charged and neutral B decays, the interpretation is not straightforward and they could be not inconsistent.

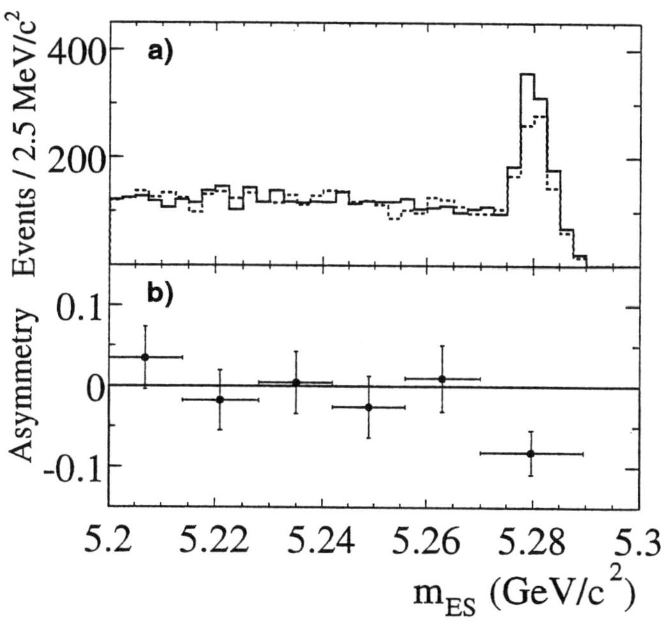

Fig. 8. – (a) Distribution of m_{ES} enhanced in $K^{+}\pi^{-}$ (solid histogram) and $K^{-}\pi^{+}$ (dashed histogram). (b) Asymmetry $A_{K\pi}$ calculated for ranges of m_{ES}. The asymmetry in the highest m_{ES} bin is somewhat diluted by the presence of background.

Starting in 2005 we have moved from the CP Violation discovery era given by the asymmetries from time dependent and time independent analyses as for example for the direct *CP* violation in $B^0 \to K^+\pi^-$, or in B^+ analysis for the extraction of γ, to a new era of precision measurements and constraints on the Unitarity Triangle.

2·3. *Update on angles.* – This summer at LP2005 there has been an update of sin2β measured with charmonium decays. In the winter 2005 the status was the following as averaged by HFAG group [17]:

$$\sin 2\beta = 0.725 \pm 0.037$$

with the contribution from $227 \cdot 10^6$ $B^0\overline{B}^0$ pairs of BABAR [18]:

$$\sin 2\beta = 0.722 \pm 0.040 \pm 0.023$$

and from BELLE [19] with $152 \cdot 10^6$ $B^0\overline{B}^0$ pairs:

$$\sin 2\phi_1 = 0.728 \pm 0.056 \pm 0.023.$$

C coefficient of the fitted asymmetry was resulting still compatible with zero, consistent with a contribution to the decay coming essentially from only one weak phase. The situation has slightly changed in the summer, due to the analysis with higher statistics ($386 \cdot 10^6$ $B^0\overline{B}^0$ pairs) from BELLE [20]:

$$\sin 2\phi_1 = 0.652 \pm 0.039 \pm 0.020.$$

C is still well compatible with zero.

The experimental value of $\cos 2\beta$, initially measured by BABAR [21] and later by BELLE [22], is excluding $\cos 2\beta < 0$ at $CL > 95\%$. These measurements remove the two-fold ambiguity on the determination of angle β as shown in Fig. 9.

An important progress has been made in the determination of α and γ, here the extraction of γ will be discussed in a more detailed way. Among various methods suggested for the extraction of γ at the $\Upsilon(4S)$ energy [23-25], where only B_d and B_u can be produced, the most promising so far is the so called Dalitz method, that is based on a fit of the Dalitz plot distribution of the events [26]. The direct *CP* violation asymmetry is generated by the interference of two decay amplitudes of charged B (B_u) leading to the same final state, but following two different routes. An example is presented in the Fig. 10. Let us use for example the following decay:

$$B^{\pm} \to DK^{\pm} = [K_S\pi^+\pi^-]K^{\pm}.$$

The Dalitz distributions to fit from data can be expressed in the following way

$$B^+ : |f(m_+^2, m_-^2) + r_+ e^{i(\delta+\gamma)} f(m_-^2, m_+^2)|^2$$

$$B^- : |f(m_-^2, m_+^2) + r_- e^{i(\delta-\gamma)} f(m_+^2, m_-^2)|^2$$

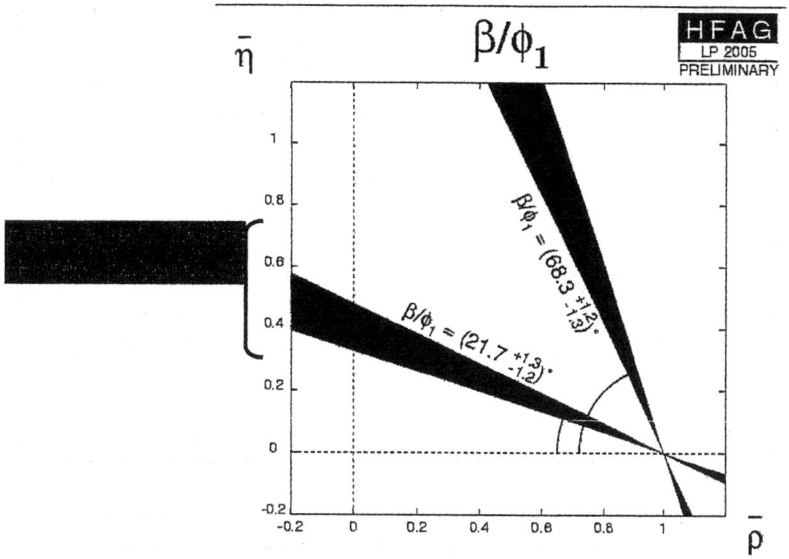

Fig. 9. – The measurement of $cos2\beta$ allows to solve the ambiguity and extract β.

where $f(m_-^2, m_+^2)$ is the D decay amplitude as a function of the Dalitz plot variables $m_\pm^2 = m^2(K_S^0\pi^\pm)$. CP is assumed to be conserved in the D decay in the formulae above. The size of the asymmetry is given by the ratio

$$r_- = \frac{|B^- \to \overline{D}^0 K^-|}{|B^- \to D^0 K^-|} \qquad r_+ = \frac{|B^+ \to D^0 K^+|}{|B^+ \to \overline{D}^0 K^+|}.$$

We expect that of course $r_+ = r_- \equiv r$ where $r \sim 0.1$ according to a simple CKM matrix elements evaluation. We keep r_+ and r_- as separate free parameters to allow a better statistical behavior.

The physics parameters r, δ and γ are quite strongly correlated and they don't allow for a reasonable estimate of the errors. BABAR has introduced a new set of variables

$$x_\pm = \Re(r_\pm e^{i(\delta\pm\gamma)}) \text{ and } y_\pm = \Im(r_\pm e^{i(\delta\pm\gamma)}),$$

which are almost uncorrelated and well behaving fitting parameters. The frequentist approach has followed in the statistical analysis with the use of "Toy Monte Carlo" to simulate pseudo experiments and extract the statistical uncertainties and the values of the physical interesting parameters r, δ, and γ from the fitted variables $x_\pm, y_\pm$. The preliminary result for BABAR [27] for the two modes D^0K and $D^{*0}K$ are presented in table II. The results from BELLE [28] with the same method are reported in table III. Inside a 2 σ interval the value of γ from BABAR is $\gamma \in [12° - 137°]$ and from BELLE is

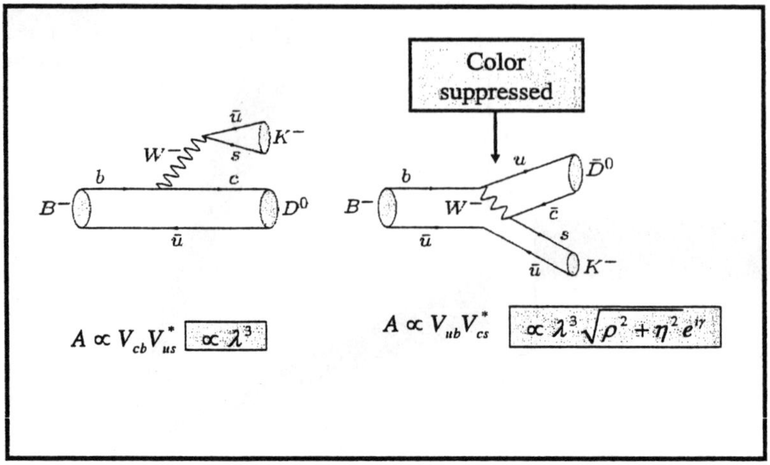

Fig. 10. – The Dalitz method allows a general fit on the Dalitz distributions of the events. Using a phenomenological description of the amplitudes of all possible resonances contributing to the D decay, a general fit allows the determination of the CP violating parameters. The choice of orthogonal variables, $x_{\pm}$ and $y_{\pm}$, as CP violating parameters is optimal to correctly determine the error on the fitted variables. A statistical (frequentist) method allows to determine the value of r, δ and γ from $x_{\pm}$ and $y_{\pm}$ fitted values.

Parameter	D^0K	$D^{*0}K$
r	$0.118 \pm 0.079 \pm 0.034^{+0.036}_{-0.034}$	$0.169 \pm 0.096^{+0.030+0.029}_{-0.028-0.026}$
$\delta(°)$	$104 \pm 45^{+17+16}_{-21-24}$	$296 \pm 41^{+14}_{-12} \pm 15$
$\gamma(°)$	$70 \pm 31^{+12+14}_{-10-11}$	

TABLE II. – BABAR *extraction of γ with Dalitz method. Errors are : statistical, systematic error due to the detector, to the $D \to K_S\pi\pi$ decay model and to non resonant contribution.*

Parameter	D^0K	$D^{*0}K$
r	$0.21 \pm 0.08 \pm 0.03 \pm 0.04$	$0.12^{+0.17}_{-0.18} \pm 0.02 \pm 0.04$
$\delta(°)$	$321 \pm 57 \pm 11 \pm 21$	-
$\gamma(°)$	$68^{+14}_{-15} \pm 13 \pm 11$	

TABLE III. – BELLE: *extraction of gamma with Dalitz method. Errors are : statistical, systematic error due to the detector, to the $D \to K_S\pi\pi$ decay model and to non resonant contribution.*

$\gamma \in [22° - 113°]$. This method needs a statistics much higher than that available so far, due to the very small value of r. However it has allowed the extraction of a preliminary value of γ.

As the extraction of α is concerned, the most sensitive methods are represented by the time dependent analyses for the charmless decays in $\pi\pi, \rho\rho$ and $\rho\pi$ [29-32]. By making use of the isospin analysis to get rid, when possible, of the strong phase due to the pollution produced by the presence of penguin amplitudes, BABAR has produced the following result [33-35]:

$$\alpha = 96^\circ \pm 10^\circ_{stat} \pm 4^\circ_{syst} \pm 11^\circ_{penguin}.$$

In the $(\overline{\rho}, \overline{\eta})$ plane the BABAR BELLE averaged result for α is reported, see Fig. 11.

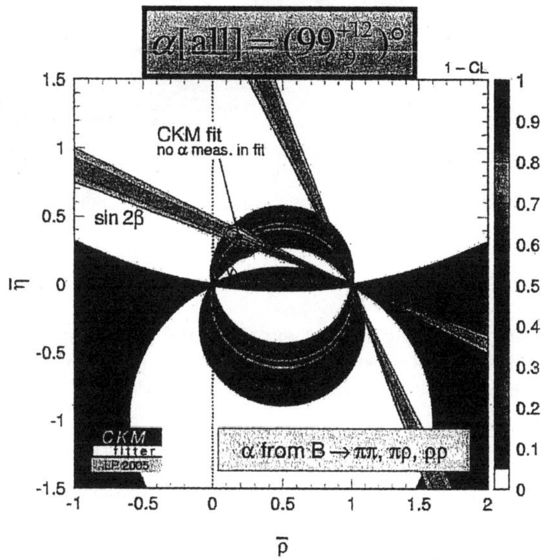

Fig. 11. – The α allowed region is presented in the $(\overline{\rho}, \overline{\eta})$ plane along with the most probable value after the combination of BABAR and BELLE results.

2·4. *CKM constraints from new measurements of sides and angles* . – The measurement of $\sin 2\beta$ has become a program of precision measurements capable to condition the general fits of CKM quantities in the $(\overline{\rho}, \overline{\eta})$ plane already in 2004. More challenging were the extractions of the other two angles of the Unitarity Triangle γ and α.

New values of $|V_{ub}|$,$|V_{cb}|$, a new value of $\sin 2\beta$ from charmonium, the measurements of α and γ have been presented last summer at LP05. They are in fact becoming the most stringent constraints of the Unitarity Triangle in the $(\overline{\rho}, \overline{\eta})$ plane, as shown in the Fig. 12. The information on $\cos 2\beta$ is included in the plot, the blue beam from B Factories is clearly constraining the region allowed for the triangle vertex and coming from the general fit.

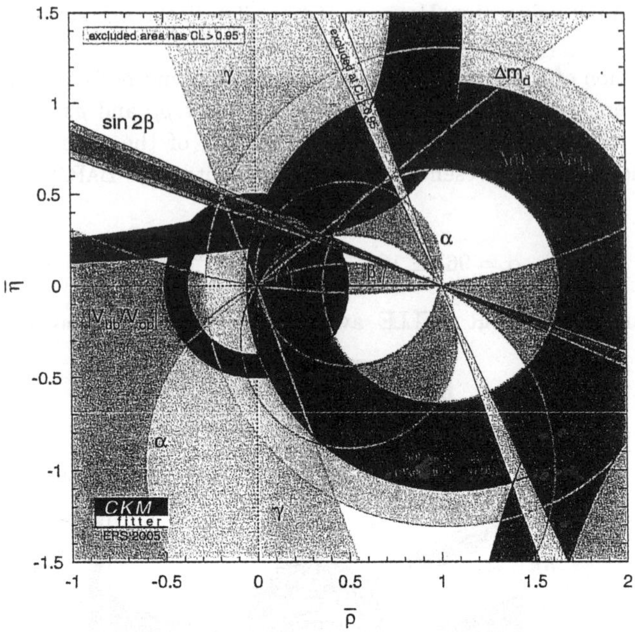

Fig. 12. – Unitarity Triangle fit using current experimental constraints. The red line around the apex of the triangle indicates the most probable region of parameters for the UT coordinates $(\overline{\rho}, \overline{\eta})$. The solution with $\cos 2\beta < 0$ is excluded at $CL > 95\%$.

A fit of UT for New Physics [36] has given 93% compatibility with SM, see Fig. 13.

2·5. *Measurements of* $\sin 2\beta$ *via decays described by penguin diagrams* . – There are however some decay channels as $b \to (s\overline{s}s)$ that can only proceed through "*penguin*" diagrams. Within the SM these channels are:

- $B^0 \to \phi K^0$
- $B^0 \to (KK)_{CP} K^0$

and the other also allowed by "*penguin*" diagrams from $b \to (d\overline{d}s)$, namely:

- $B^0 \to \eta\prime K^0$
- $B^0 \to f_0 K^0$
- $B^0 \to \pi^0 K^0$
- $B^0 \to \rho^0 K^0$
- $B^0 \to \omega K^0$

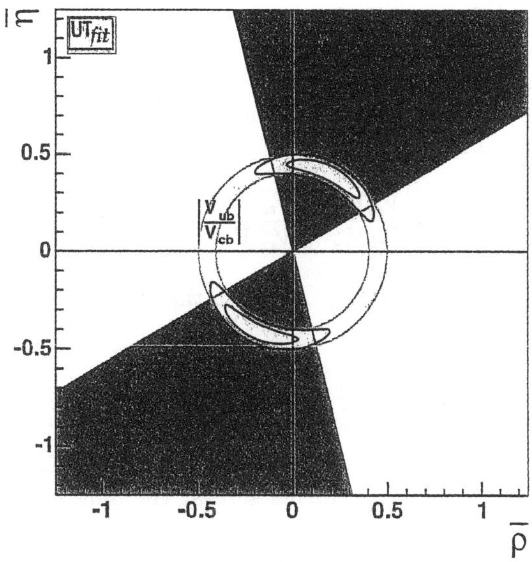

Fig. 13. – UT fit using tree-level processes, which almost unchange in presence of New Physics. The value $\bar{\eta} = 0$ is excluded at $CL > 95\%$, definitely estabilishing that CP violation is a feature of the SM.

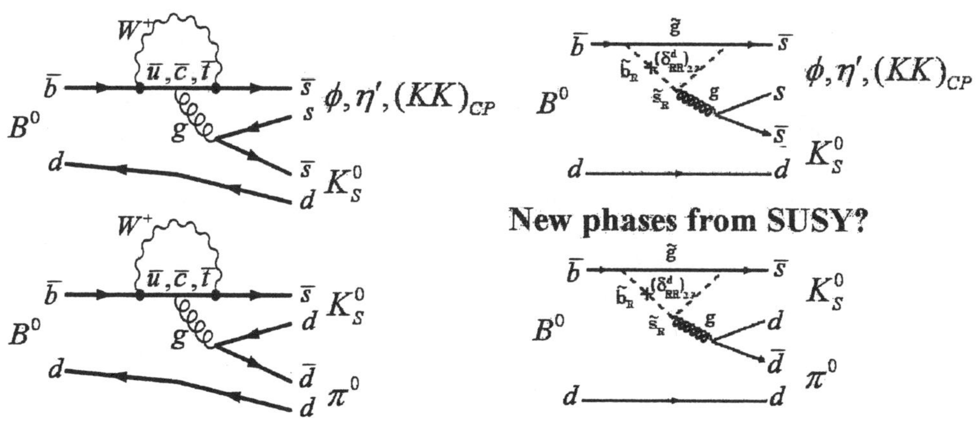

Fig. 14. – Possible New Physics particles contribution to $b \to (s\bar{s}s)$ and $b \to (sd\bar{d})$ penguin diagrams.

take contributions from the same CKM elements of charmonium channels $b \to (c\bar{c}s)$, as can be seen in the left hand part of Fig. 14. The interference B^0 Mixing - Decay -K^0 Mixing should give the same phase factor $e^{-2i\beta}$ of charmonium decays, assuming that the only relevant contribution to the mixing amplitude is coming from the heaviest fermion (top quark) in the box diagram loop

$$\lambda = \frac{V_{tb}^* V_{td}}{V_{tb} V_{td}^*} \frac{V_{tb}^* V_{ts}}{V_{tb} V_{ts}^*} \frac{V_{cs}^* V_{cd}}{V_{cs} V_{cd}^*} = e^{-2i\beta} \tag{3}$$

as can be easily checked comparing Eq. (3) with the corresponding one for the charmonium modes

$$\lambda = \frac{V_{tb}^* V_{td}}{V_{tb} V_{td}^*} \frac{V_{cs}^* V_{cb}}{V_{cs} V_{cb}^*} \frac{V_{cd}^* V_{cs}}{V_{cd} V_{cs}^*} = e^{-2i\beta}. \tag{4}$$

Using the Wolfenstein parametrization we can see that apart from V_{td} the other terms in the expressions are real matrix elements so that the only term that survives is $\frac{V_{td}}{V_{td}^*}$. If however new heavy quanta contribute to the loops, see right hand side of Fig. 14, new phases can contribute to the asymmetry and the S coefficient of the time dependent analysis could be substantially different from $\sin 2\beta$. The comparison between results from all the above B^0 decay channels and $\sin 2\beta$ from charmonium, as shown in the HFAG plots referring to the before LP05 and after LP05 Fig. 15 and Fig. 16, must be made anyway "*cum grano salis*". In addition to the diagrams with top quark insertion in the loop considered in the $\sin 2\beta$ comparison there are others contributions. For example about the cleanest channels, $B^0 \to \phi K^0$ and $B^0 \to (KK)_{CP} K^0$, we must consider a SM uncertainty due to the contribution of a similar graph with the insertion of an up quark in the loop. This uncertainty is of the order of $\lambda^2 \approx 5\%$. For the other decay channels the uncertainty could be even greater than 5% up to 10% or even 20%. These channels can receive contributions also from color suppressed *tree* diagrams $b \to (u\bar{u}s)$. Many channels have been explored so far, some of them with poor statistics, some other richer. It is clear that we are at the beginning of a very interesting season when we can start probing the SM in the flavor sector. A statistics between 20 and 50 times higher than present is needed, from 15 to 40 ab^{-1}, in order to reduce the statistical error below the theoretical uncertainties and make this comparison really possible.

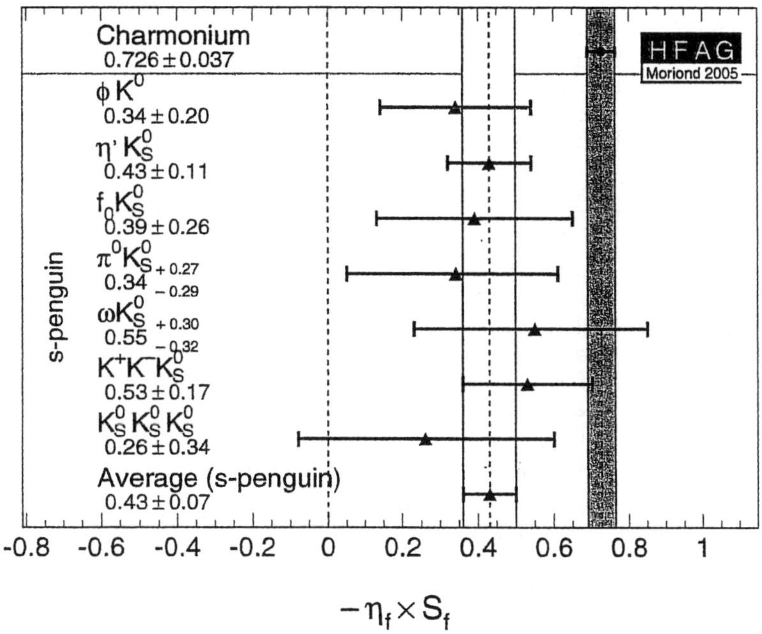

Fig. 15. – HFAG plot containing all the results about s-penguins compared with Charmonium average before LP05.

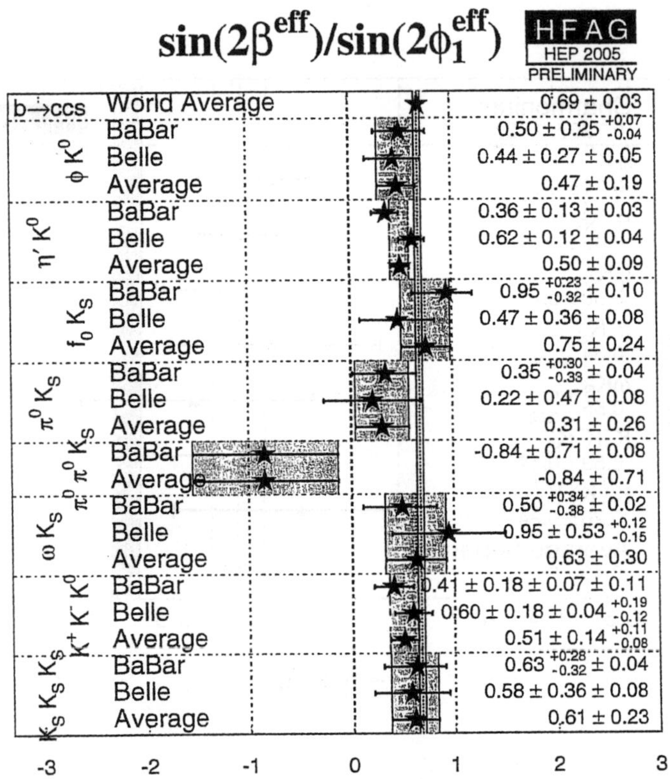

Fig. 16. – HFAG plot containing all the results about s-penguins compared with Charmonium average after LP05.

REFERENCES

[1] J. H. Christenson, J. W. Cronin, V. L. Fitch and R. Turlay, Phys. Rev. Lett. **13** (1964) 138.

[2] B. Aubert *et al.* [BABAR Collaboration], Phys. Rev. Lett. **87**, 091801 (2001)

[3] K. Abe *et al.* [Belle Collaboration], Phys. Rev. Lett. **87**, 091802 (2001)

[4] B. Aubert *et al.* [BaBar Collaboration], Phys. Rev. Lett. **93**, 131801 (2004)

[5] K. Abe *et al.* [Belle Collaboration], Phys. Rev. Lett. **93**, 021601 (2004)

[6] "PEP-II: An Asymmetric B Factory. Conceptual Design Report. June 1993," SLAC-R-418.

[7] "KEKB B-Factory Design Report", KEK Report 95-7, 1995.

[8] B. Aubert *et al.* [BABAR Collaboration], Nucl. Instrum. Meth. A **479**, 1 (2002)

[9] Nucl. Instrum. Meth. A **479**, 117 (2002).

[10] N. Cabibbo, *Phys. ReV. Lett.* **10**, 531 (1963)

[11] M. Kobayashi and T. Maskawa, *Prog. Theor. Phys.* **49**, 652 (1973)

[12] L. Wolfenstein,Phys. Rev. Lett. **51**, 1945 (1983)

[13] S. Eidelman *et al.*, Phys. Lett. B **592**, 1 (2004)

[14] Y. Chao *et al.* [Belle Collaboration], Phys. Rev. Lett. **93**, 191802 (2004)

[15] B. Aubert *et al.* [BABAR Collaboration], arXiv:hep-ex/0408081.

[16] Y. Chao *et al.* [Belle Collaboration], Phys. Rev. D **69**, 111102 (2004)

[17] K. Anikeev *et al.* [Heavy Flavor Averaging Group (HFAG)], arXiv:hep-ex/0505100.

[18] B. Aubert *et al.* [BABAR Collaboration], Phys. Rev. Lett. **94**, 161803 (2005)

[19] K. Abe *et al.* [BELLE Collaboration], Phys. Rev. D **71**, 072003 (2005) [Erratum-ibid. D **71**, 079903 (2005)]

[20] K. Abe *et al.* [Belle Collaboration], arXiv:hep-ex/0507037.

[21] B. Aubert *et al.* [BABAR Collaboration], Phys. Rev. D **71**, 032005 (2005)

[22] R. Itoh *et al.* [Belle Collaboration], Phys. Rev. Lett. **95**, 091601 (2005)

[23] M. Gronau and D. Wyler,Phys. Lett. B **265**, 172 (1991)

[24] M. Gronau and D. London.,Phys. Lett. B **253**, 483 (1991)

[25] D. Atwood, I. Dunietz and A. Soni,Phys. Rev. Lett. **78**, 3257 (1997)

[26] A. Giri, Y. Grossman, A. Soffer and J. Zupan,Phys. Rev. D **68**, 054018 (2003)

[27] B. Aubert *et al.* [BABAR Collaboration], Phys. Rev. Lett. **95**, 121802 (2005)

[28] K. Abe *et al.* [Belle Collaboration], arXiv:hep-ex/0411049.

[29] M. Gronau and D. London,Phys. Rev. Lett. **65**, 3381 (1990)

[30] Y. Grossman and H. R. Quinn,Phys. Rev. D **58**, 017504 (1998)

[31] A. E. Snyder and H. R. Quinn,Phys. Rev. D **48**, 2139 (1993)

[32] M. Gronau and J. Zupan, Phys. Rev. D **70**, 074031 (2004)

[33] B. Aubert *et al.* [BaBar Collaboration], Phys. Rev. Lett. **95**, 151803 (2005)

[34] B. Aubert *et al.* [BABAR Collaboration], arXiv:hep-ex/0408099.

[35] B. Aubert *et al.* [BABAR Collaboration], Phys. Rev. Lett. **95**, 041805 (2005)

[36] M. Bona *et al.* [UTfit Collaboration], JHEP **0603**, 080 (2006)

CHAIRMAN: M. GIORGI

Scientific Secretaries: V. Didenko, J. Mercado-Pérez

DISCUSSION

- *Hanagaki:*

The experimental error on sin(2β) measurement in the penguin mode will be smaller in future with more statistics. But can you control the theoretical uncertainty in future?

- *Giorgi:*

As I've mentioned this morning in my lecture the theoretical uncertainty in some channels is not very high. Particularly, if one looks to the channels where one has $B \rightarrow s\hat{s}s$ φKs and if φKs the uncertainty in the calculation is of the order of few percents, less than 5%. In some other channels as $B \rightarrow d\bar{d}s$ there the uncertainty is much higher. For this reason it is important to look at a some exclusive mode, namely, as I called it η'Ks and φKs. Some other modes are less sensitive. The first channel is affected by a very small theoretical uncertainty and with the second channel they are the most clean for penguin one can observe. So the discrepancies you looking for between sin(2β) of charmonium and sin(2β) of penguin is perhaps accessible. But to do this one needs nevertheless the statistics which is about of the order of 10 or more billions B^0B^0 quark pairs.

- *Preghenella:*

You said that you have a low efficiency, about 0.5% for the recoil method, and you said that you will increase this efficiency to about 1%. How will you do it?

- *Giorgi:*

First, this can be achieved for example in the designing super-B I've mentioned, since the beams are very small and there you can have very small beam size of the order 20 nm. So you can have a beam pipe which is smaller than 1 cm, at the moment we have 3 cm radius. So you can close with your tracker and increase a lot in your acceptance. Second, one can reduce the boost. At the moment the boost is 3 vs 9 and one can go down to 7 and 4, so you can have more spherical events and the efficiency of the construction is higher.

- *Faivre:*

What are the main difference detector wise between Belle and BaBar?

- *Giorgi:*

It is a very good question! There is one major difference in the flavour tagging. The "Cherenkov " used in BaBar is an imaging "Cherenkov". While the "Cherenkov" used in Belle is a conventional well performing Aerogel Cherenkov counter. A second detector is the vertex detector. The vertex detector of BaBar is lamp shade shaped and it 5 active layers instead of the 3 or 4 cylindrical layers of Belle. So its acceptance, efficiency and resolution are higher. Those are the major differencies which make the advantage of BaBar with respect to Belle. I expected a question of this kind because the statistics of Belle is much higher than statistics of the BaBar nevertheless the statistical error is of the same order. The reason is because the BaBar is more efficient and more performed. So we have much higher efficiency in the construction because of better performance of these two main detectors "Cherenkov" and vertex. In addition our reconstruction and physical analysis are based on a highly sophisticated algorithms and we make only a very moderate use of the more inefficient methods based on cuts.

- *Cifarelli:*

Can you better describe how a super-B factory could be combined with ILC project?

- *Giorgi:*

Yes, I've mentioned that we have two joined workshops between BaBar and Belle one in 2004 and second in 2005 looking for this super-B factors. The idea is to have the super B factor with very high luminosity. What about using technology of linear collider in a period where LHC is running? One interesting solution that has been proposed is to use the collision technique studied for the linear collider: sqeeze bunches at the level of few tens of nanometers and very low emittance, let them collide, the high luminosity is achieved because of the small transverse dimensions and not because of the increase of the number of electrons and positrons. It will be hard to recover these bunches and to cool them down because (due to the disruption mechanism, during the collision) they explode, so it is needed a large dynamical acceptance an in addition one has to put power making them dissipate power and cool down, thus reducing the emittance. Then get it back for further collisions. The cost of this operation is only of the order of much less then 10% of the cost of linear collider let say between 5% -10%. The detector could be built with present technology, the amount of physics one can extract from this machine is huge and there is no need to know in advance what is the mass of your particles supplied by LHC, since one looks for discrepancies, mainly in the flavour sector, with a very precise measurement looking for instance to flavour changing neutral current processes or to rare decays as lepton flavour violation in tau decays.

- *Pacetti:*

Could you make a comment about the measurement of cos (2β)?

- *Giorgi:*

It was a very difficult measurement in BaBar, because in this measurement we were looking at the interference in a k\pi system in the final state. The measurement was important to determine the sign of $\cos(2\beta)$ and therefore remove one ambiguity in the CKM plane.

- *Masina:*

What is the confidence level of the upper bound of the $\tau\rightarrow\mu\gamma$ process? Is the prospect BR<10E-10 really feasible and when it expected?

- *Giorgi:*

The bound is now less than 7 10E-8 at 90% of confidence level and this corresponds to the latest analysis of this summer. As I quoted in the project of super-B the future machine could achieve a luminosity very high as something between 10E36-10E37 and if it is 10E37, this corresponds to 150 Billions of tau events in one year. So with such a super-B factory, 10E-10 is reachable and it would not take much longer than a few years! These measurements are not difficult and very clean at BaBar.

- *Baldini:*

Can you anticipate which kind of detector you would think of for the super B-factory?

- *Giorgi:*

Yes, even though I do not have it here, I can make a sketch for you. You can have a beam of 4 GeV and another beam of 7 GeV because you intend to reduce the burst. Reducing the burst make more difficult the measurement of time-dependence because you need much higher resolution in your detector, but you have a very thin pipe (1 cm diameter) so you can have transparent pixels around. It is not like the ones used in the LHC, it is "maps", there is a new generation of maps being developed in different places: Italy, Japan and Hawaii; with these, you can have better performance. Because your background is very low, the you have a thin tracker, i.e., silicon tracker of 100 nm thickness with the same performance as we have nowadays; then a gap filled of gas (Ar) of thickness 50 cm (not the size that you have in big detectors), then crystals (for a calorimeter), then Cherenkov and at the end, something fast, perhaps scintillator plus iron. So this is roughly the kind of detector. The crucial part is the inner part, because it has to be transparent and highly performing.

- *Baldini:*

Why do you say that the 'big lambda' will be very slow with 10^{37} [$cm^{-2}s^{-1}$]?

- *Giorgi:*

Because you know that the source of background in this kind of machine is strongly related to the fact that immediately after the collision (e.g. 50 cm after) you have to bend your beams in order to have back-scattering which is the main source. And the main source is in fact going as the same power as the luminosity. However, if you instead have very small bunches in size, you may collide them in a straight line and pick them up 'far away' and put back simply for cooling, then you do not have this beam gas and the "bending" effect since they are bent far away from the detector so the concept is completely different; otherwise it would be impossible to design a detector for 10^{36} or 10^{37} [$cm^{-2}s^{-1}$] and only based on this idea you can go on.

- *Tecker:*

When you collide the small bunches, do you get already significant broader in the luminosities spectrum due to Beamsstrahlung?

- *Giorgi:*

Well, the are several effects. One is the Beamsstrahlung, which is important, and another important effect to be taken into account is the beam disruption. To cope with them you need (I am not a machine physicist) to find an equilibrium, to optimize the parameters, which are essentially, the length of the bunch and its transverse size. Therefore, what is important, is to have quite big dynamical aperture in the machine in order to recover the beams afterwards. You have a factor of 50, 100 or even bigger than that according to simulations from the machine people from Frascati and SLAC have performed.

- *Tecker:*

But your luminosity spectrum still stay narrow.

- *Giorgi:*

Yes, it is still narrow.

- *'t Hooft:*

You said that the KS measurements of CP violation is not really CP violation but T violation; now, I always see experimentalists measuring CP invariance, how is T invariance measured?

- *Giorgi:*

The measurement related to $|q/p|$ coming from mixing is essentially what is measured in the K-sector ε_K. On the other hand, ε_B is, in B sector the quantity related to the T violation which comes in the off-diagonal terms of the mass matrix. However, you can try to measure independently CPT, T, CP, if you perform de full analysis fit by measuring all the parameters. Indeed, has been performed and, so far, without assumptions, because normally in the analysis you make the assumption that $|q/p| = 1$. This is simply made in the short analysis. We have performed the full general analysis leaving free all these parameters with very high statistics, and the result so far is still compatible with the fact that we do not see discrepancy from $|q/p| = 1$. Another possibility to look for T violation is to measure the polarization of D* in the semi-leptonic decay of $B \rightarrow D^{*} + l + \nu$. It may happen to have a contribution from charged X to $l + \nu$, it is actually part of our programme. At the moment both experiments, Belle and BaBar, do not see any discrepancy from $|q/p| = 1$ and the measurement of polarisation of D* will be very important.

- *Baldini:*

Do you think that once the super B-factory is built, the energy could be tuned to have a τ/c (tau/charm) factor at 10^{37} [$cm^{-2}s^{-1}$]?

- *Giorgi:*

Well, first of all I would like to comment that a B-factory is also a tau and charm factory such that the amount of charm and tau pairs is roughly the same as Bs. In fact, BaBar has demonstrated to be able to put quite strict limit on the tau decay into $\mu\gamma$ and tau decay into μ + hadron, and also in the charm sector we have demonstrated to be able to discover new states. Therefore, in principle, is also tau and charm. One advantage that you have in a B-factory is the fact that taus and charms are quite energetic so you can measure them quite well; nevertheless, one disadvantage is the asymmetry because it makes more difficult to reconstruct in the centre of mass because you have neutrinos so the analyses are more complicated ,but in the idea of the super B-factory there is also the option of reducing the boost, i.e., the beam asymmetry and in addition the possibility of tuning the beams.

- *Faivre:*

You mentioned the discovery of new particles, but did you search for pentaquarks, included in channels not searched for by other experiments?

- *Giorgi:*

Yes, we have searched for, but we have not had any evidence or signal for any of the candidates in any of the channels.

The LHC Supercollider

Lucio Rossi

CERN – Accelerator Technology Department, Geneva
Geneva 23, CH-1211 *(lucio.rossi@cern.ch)*

Abstract : The LHC is the last of a series of projects making use of superconductivity to investigate the composition of matter in details as fine as 10^{-19} m. Based on the performance of about 1750 large superconducting magnets it presents a numbers of technological challenges, of which the most striking is may be the construction of the 1232 15 m-long dipoles that will fill 2/3 of the 27 km long tunnel. The paper will review the main characteristics of supercolliders and of superconducting magnets for accelerators and will describe the main issue of the LHC dipoles. The construction of the accelerator is progressing well and installation has started.

1. Supercolliders : exceptional microscopes and more

To see the innermost components of the particles we need making use of particles more and more energetic, in order to break bonds that are stronger and stronger. This simple, although true, vision can be completed by a complementary point of view based on the resolving power. Like the resolving power of a microscope is limited by the wavelength of the light employed to illuminate the target, the resolving power of the accelerators is limited by the wave length of the De Broglie wave associated to the accelerated particle: $\lambda = h/p$. This explains why higher and higher energies or momenta are needed to discover finer and finer details. With the LHC, where protons are collided with a center-of-mass (c.o.m.) energy of 14 TeV, we'll be able to investigate details as fine as 10^{-19} m, in a controlled and reproducible way.

In addition to the vision of microscope we can perceive accelerators as peculiar "time backward" machines. Indeed with the LHC we'll reproduce conditions that were present only few instants after the big bang. A simple relation : $t_{\mu s} \cong 1/E^2_{GeV}$ tell us that with the LHC running at 7+7 TeV we'll be able to investigate phenomena that happened about 1 ps after the big bang. Like a telescope that more is able to see far away and more looks into the past, our machines more looks into the finer details and more "illuminate" our past.

Finally we can perceive our machine as a mass generator: indeed higher the energy and larger the mass of the particle that can be created or that can emerge from the smash of the proton smashed one against the others. Even if hadrons are composed particle and the 14 TeV must be subdivided – in average – in the various channels of the quark-quark collision, particle of more than 1 TeV of rest mass can be created, like if they were emerging from the remote time of $t = 10^{-12}$s after the beginning of our universe.

In case of heavy ions operation, the collision energy is subdivided into many more channels, and the LHC will be able to recreate quark-gluon soup, i.e. real collective status of matter, existing about 1 μs after the big bang.

2. The structure of supercolliders

2.1 Accelerator magnets

We have seen that the quest for energy is the primary goal for frontier accelerators. And while other accelerator parameters become more and more interesting, like luminosity, for a discovery machine there is **no substitute to energy**. For a circular accelerators, the beam energy is related to the field level in the bending magnets by the simple relation, valid in relativistic conditions: $E \cong 0.3BR$; *(TeV; T; km)*. So, accelerator physicists have looked in to superconductivity, as soon as it became a practical technology, in order to attain high magnetic fields. Thanks to the powerful 8.3 T dipoles (9 T at ultimate operation) that fill 2/3 of the 27 km underground tunnel the LHC beam will reach the unprecedented energy of 7 TeV (7.6 TeV at ultimate). But the field level, important as it is, is not all for our accelerators: beam optics requires also magnets that are able to focus the beam and to correct the chromatic errors and aberrations, similarly to what lens do for optical beam. Quadrupoles, sextupoles and higher order magnets do these corrections and constitute the necessary complement to the main dipoles keeping the beam in circular orbit. Furthermore all these magnets constitute a lattice that must be as accurate as possible: all main magnets of the same type must be equal in within a few 10^{-4}.

Finally we have to consider that since the magnets fill more than 80% of the tunnel, they bear the great part of the LHC cost. Then they must be designed to be as economic as possible, to stay inside the tight budget of today researches.

2.2 Detectors magnets

Before entering in the details of an accelerator it is worth to note that powerful magnets are needed also for the detectors that are placed at the collision point. Two main scopes justify the huge investment and the increase of complexity in the whole apparatus required by these magnets:

- Electrical charge identification
- Momentum spectrometry. From $p = qB\rho$ one can deduce that by crossing a magnetic field over a path length L each particle is bent of and angle $\phi = \frac{q}{p}BL$. Frequently BL is used for comparing the bending power of various magnets. However if momentum analysis is done by tracking inside the field volume on can show that the resolution in momentum depends on field and length as: $\frac{\Delta p}{p} \propto 1/BL^2$, i.e. the momentum resolution scales linearly with B and quadratically with the length. That is why for detector magnets a large volume pays off more that high field. Indeed in the LHC, the field of the two general-purpose detectors, ATLAS and CMS, is about half of the main dipoles but the volume is much larger. However, even detectors requires field in excess of the range of classical technique and most of it make use of large superconducting magnets.

The field in the detector magnets does not need to be as accurate as the one of colliders and usually a knowledge at 0.1-1% precision is enough: in many cases is simply computed and not even mapped in the volume but only checked in one or few points.

Detector magnets are sometimes placed inside the calorimetric part; therefore they need to be as much as possible thin, transparent to particles and radiation. Usually these are solenoids (like the ATLAS inner solenoid) and for them in these 30 years has been developed a very unique technique of conductor stabilization based on a matrix of a very pure (99.995%) aluminum. This technique has many advantages for large magnets and has been retained for almost all superconducting detector magnets, even for the ones that, being in the muon chamber like the ATLAS toroids and the CMS solenoid, do not need to be thin. In Fig.1 the Coils of the ATLAS barrel toroid, as assembled in the cavern, is shown.

Figure 1 The eight 25-m long, 5 m wide superconducting coils of the ATLAS barrel toroid just after assembly in the underground cavern

2.3 The main components of an accelerator

A collider is a very complex system and relies on the perfect functioning of many critical systems. However, conceptually, it needs four main systems, see Fig.2:

1) The vacuum tube: indeed without vacuum reaching values of 10^{-11} mbar the beam would be quickly lost. One has to go into space to have such a vacuum level in volumes of hundreds cubic meters like it happens in the LHC!
2) The accelerating system, which provide the energy to the charged particle beam. It is constituted of electromagnetic cavities working in the radiofrequency range, or, like it is envisaged for future colliders, even in the microwave range. The RF cavities are also manufactured with superconducting materials: in this case superconductivity is preferred not because it generates higher voltage (actually they provide a voltage which is usually less than copper cavity) but because they allow continuous wave operation, long bunches, and larger aperture, and all this means a better quality of the beam and higher bunch charge content. The beam

energy is then: $E = GL$, where GL is the integrated electric field gradient. For linear accelerator this is the basic formula for the beam energy.

3) The magnetic focusing system constituted by the quadrupoles magnets and the other higher order magnets. In the LHC there are correctors up to the dodecapole.
4) For circular accelerators, where the beam is bent to be re-circulated many times in the same RF cavities, the main component is the dipole magnet that provides the transverse field to guide the beam. As previously said, in presence of dipoles the collider energy depends on BR, which shows the necessity of pushing toward high fields. One should notice that pushing higher the energy means also having a more rigid beam bringing the necessity of stronger quadrupoles. Therefore, high field dipoles also demand superconducting quadrupoles and higher order correctors.

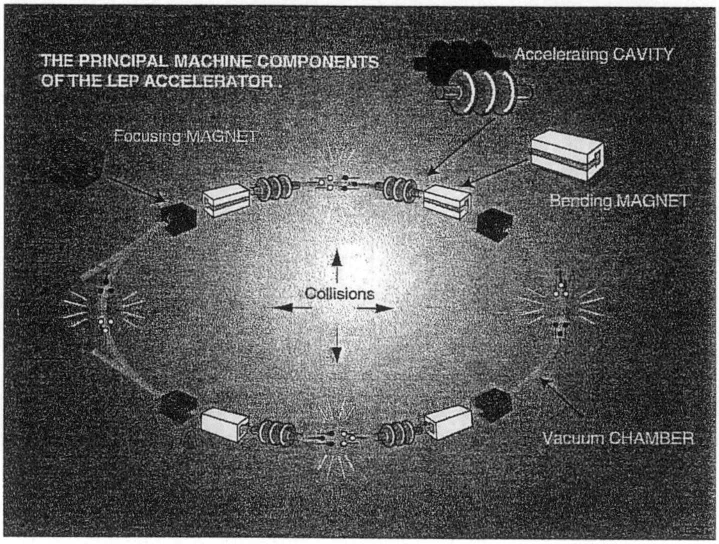

Figure 2 The main components of a circular collider.

3. Accelerator panorama and the LHC

Since the beginning of the thirties, when accelerators started to be built especially by E. O. Lawrence, which was awarded the Nobel prize in 1939 for the invention of the cyclotron, there has been the rush toward high energy. The famous Livingstone plot, see Fig.3, shows that when the energy is reduced to equivalent fixed target (to keep the compatibility with the first 40 years of accelerators) there is an exponential growth of energy vs. time, of almost 2 order of magnitude each decade. It is interesting to note that this growth has been sustained by jumps in technology: change of type of accelerators (from cyclotrons to synchrotrons and then to colliders) or change of technology of main components, namely the passage from normal to superconducting magnets. The increase of size cannot be the unique means of growth in energy.

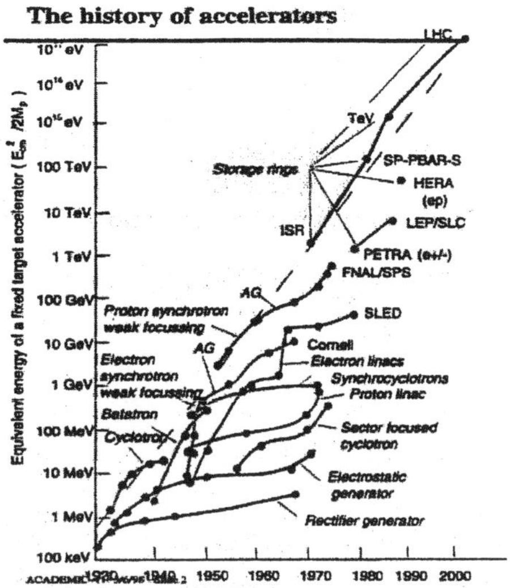

Figure 3 Livingstone plot, i.e. energy reached by various plot vs. time

4. The LHC Project

For the physics point of view the main parameters of the LHC project can be condensed in the following values at nominal (ultimate) operation: proton collision at 7+7 (7.54+7.54) TeV, with a peak luminosity of 1 (2.3) 10^{34} cm^2s^{-1} with an operation over about 6 months per year and a collision time of 2/3 of the total time, and a bunch-to-bunch spacing of 25 ns.

From the point of view of the novelties and of the challenges, in addition to the main one, that is the construction and operation of 22 km of high field superconducting magnets, many other very difficult points must be addressed in the frame of the LHC project:

- The use of the LEP tunnel, which is too small to accommodate well the LHC.
- A powerful cryogenic system capable to distribute superfluid helium almost isothermal all along the tunnel, maintaining at 1.9 K some 40,000 tonnes of magnets and associated systems.
- The underground caverns for the experiments. In particular two new huge caverns hosting the large multipurpose detectors, ATLAS and CMS have been built new, with large pits to allow equipment descent, (like the large magnet of Fig. 1).
- The handling of a beam with unprecedented energy and energy density. The energy of the beam can cause irreversible damage to many equipments, notably the magnets, if not fully controlled.
- The necessity to limit beam losses to avoid that the superconducting magnets are quenched too frequently with serious impact on reliability and then on integrated luminosity.

The machine is subdivided into 8 sectors, each one independently powered and serviced. Each sector is composed by a long arc, where lattice magnets are arranged, and by a much shorter straight section where the beam are prepared for collision. Actually four straight sections are allocate for experiments, the other four are dedicated to accelerator services, like beam dump, RF etc.

Superconductivity plays a key role in the LHC and in order to keep the magnets cold, cryogenics is the major infrastructure. We'll have about 140 kW of power installed at 4.5 K and about 20 kW of power at 1.9 K, with a consumption of about 40 MW of electric power from the grid. These numbers are impressive, and CERN is already today the largest cryogenic center of the world. However it is worth comparing what is the energy consumption in case of a normal conducting machine. To reach the same energy of 7 TeV per beam with magnets of 1.8 T, rather the 8.3 T of the LHC, we would have needed a 100 km tunnel and the power dissipated into the magnet coils would have been about 900 MW! This is the power produced by a nuclear plant, with obvious prohibitive capital and operation cost. Also, it would have had much larger impact on the environment, through the power generation and the power rejection to the ambient.
Indeed almost all leading edge accelerators of the last decades have been superconductive, both circular and linear (in this case superconducting RF cavities play a major role), as shown in Fig.4 on a special version of the Livingstone plot.

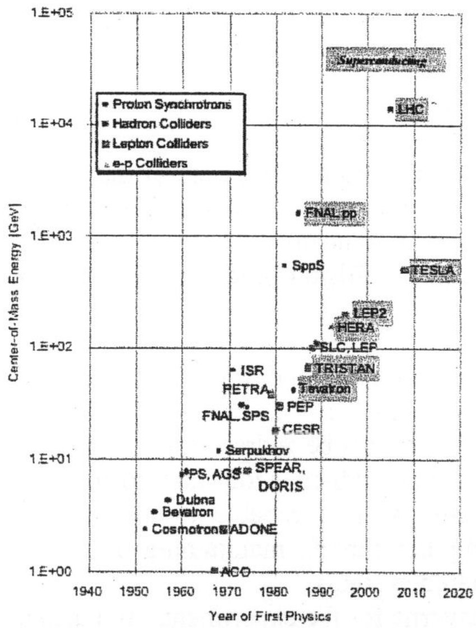

Figure 4 Extract of Livingstone plot with superconducting project in evidence

5. Superconductivity

H. Kammerlingh-Onnes in Leiden (NL) discovered superconductivity in 1911, however practical superconductors for applications in the magnet domain became available only at the end of the sixties. At the end of the seventies multi-filamentary wires became a real commodity and magnet technology passed from the pioneering stage to the industrial production. The materials that were made available in the seventies: the Nb-Ti alloy and the compound Nb_3Sn, are still by far the most used superconductors. The class of superconductors discovered in the eighties, ceramic compounds based on copper oxide, despite a high operating temperature allowing use of liquid nitrogen (at 77 K) have current performance still well below the classical superconductors. However, a considerable use of them in the LHC is planned for current leads from 60 to 13.000 A.

5.1 Critical current

A superconductive material is first characterized by its value of critical temperature, above which it is unconditionally in normal state. Below its critical temperature it is superconductor if the total field is not exceeding a value, that depends on temperature and is called critical field, $B_{c2}=B_{c2}(T)$. Once a superconducting sample is fed with current, the current flow resistanceless until it does reach a value, called critical current I_c, above which it gets back to normal state. For a fixed temperature I_c depends on the total field and of course is higher for weaker field and tends to zero when the field is approaching the critical field of the bath temperature. It is convenient, when com paring different material or conductor, to report the critical current density over the superconductive cross section, J_c, rather than I_c, to allow meaningful comparisons In Fig. 5 the critical current of various superconductors is plotted vs. field at fixed temperature. One can see that Nb-Ti is limited at field below 10 T: for very high field, Nb_3Sn is today the only viable material, although more and more test coil made with Bismuth-based material (in particular Bi-2212) show its possibility for field in excess of 18 T. What is limiting today these ceramic materials is: i) their limited mechanical strength; ii) the difficulty of their heat treatment conditions and; iii) their cost still too high.

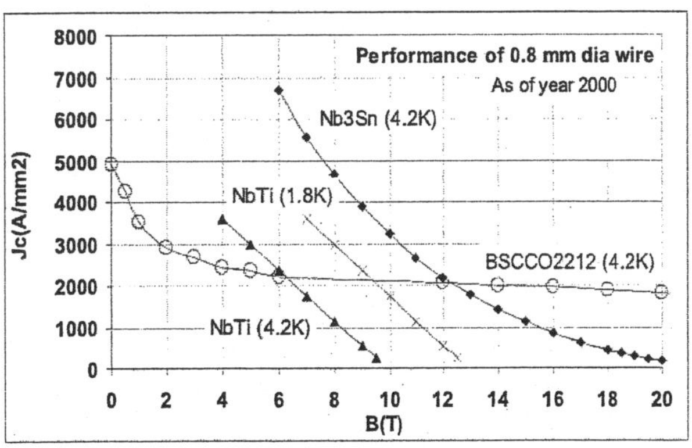

Figure 5 Critical current density vs. magnetic field of most performant superconductive materials. Nb-Ti and Nb3Sn are reliable products, available in long lengths, while Bi-2212 is still in R&D phase

In Fig. 6 the critical current density curve of the best Nb-Ti, both at 4.2 K (normal liquid helium) and 1.9 K (superfluid helium) temperatures, is reported in detail. One can note that by lowering the temperature the J_c is displaced toward high field of 3 tesla. LHC magnets are operated at 1.9 K, indeed, with a gain of about 1.5 T on the maximum field.

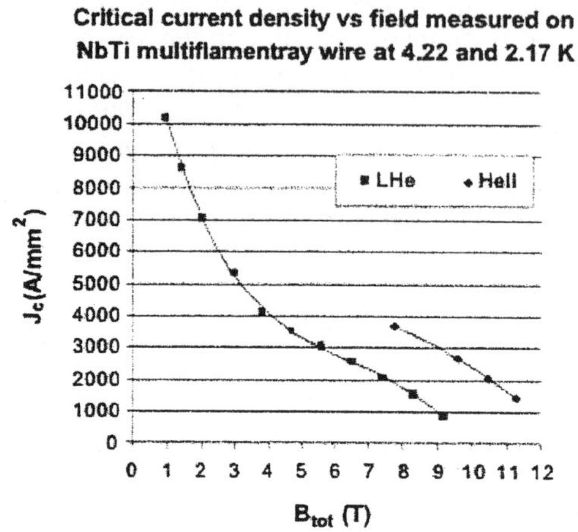

Figure 6 Critical current density of the best Nb-Ti commercial conductor in liquid helium (4.2 K) and at superfuid helium (2.17 K), measured at INFN-LASA (Milan) for the AMS magnet project.

5.2 Basic stability

Superconductors are not stable, if used alone. Small perturbations, of the order of 1μjoule or less, can push the material out of the superconducting zone into the normal zone where, due to the electrical resistivity, the superconducting state is irreversibly lost, see Fig.7. This event of sudden lost of superconductivity and return to normal conducting state is called "quench". Two are the measures taken to make stable a superconductor:

1) the superconducting element must be arranged in filaments of diameter smaller than 100 μm (or even than 50 μm, depending on the conditions). This is called adiabatic stability criterion.
2) The superconducting filament must be in intimate contact with a low resistivity material, to provide a low dissipation path to current that spill off the superconductor. This is called electrodynamic stability criterion.

Both conditions are met by a structure of conductor where the superconductive element is arranged in filaments immersed in a copper matrix, the so called multifilamentary wires reported in Fig. 8, where the most common configurations for superconducting wires used in HEP magnets, both accelerators and detectors, are shown. It worth underline that detector magnets requires conductors which are super-stabilized by a large matrix of pure aluminum surrounding the cable itself, composed by Nb-Ti/Cu wire.

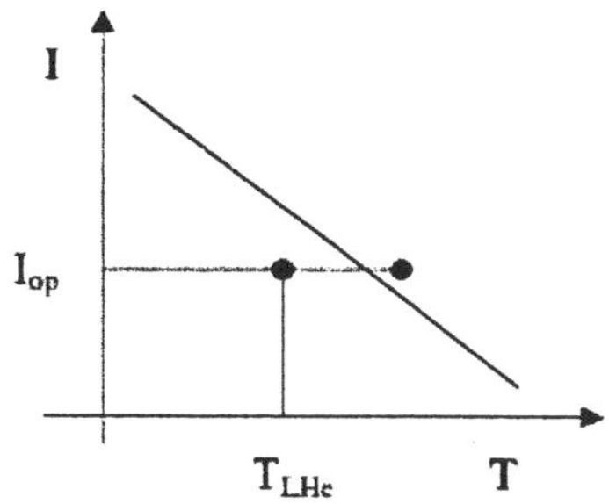

Figure 7 Stability scheme for a superconductor at constant current. The point at T_{LHe} and I_{op} is the operative point of the superconductor. Small energy perturbations can however heat the superconductor locally, bringing the operative point at constant current

Very different is the case of the accelerator, where the current density in the coils is the most important feature. So the basic element, the Nb-Ti/Cu wire is assembled in conductor, called Rutherford cable, without addition of other stabilizing material that then copper contained in the wire itself, see Fig. 8. Usually the ratio between copper and superconductor in the cross section is ranging from 1.5 to 2. The small amount of stabilizing material in accelerator magnets is partly compensated by the presence of liquid (or superfluid, like for the LHC) helium as direct coolant: latent heat of LHe and heat capacity of HEII are very high, 10-1000 time the value of specific heat of solid material at the low temperature, which give, in certain condition a fair stability margin.

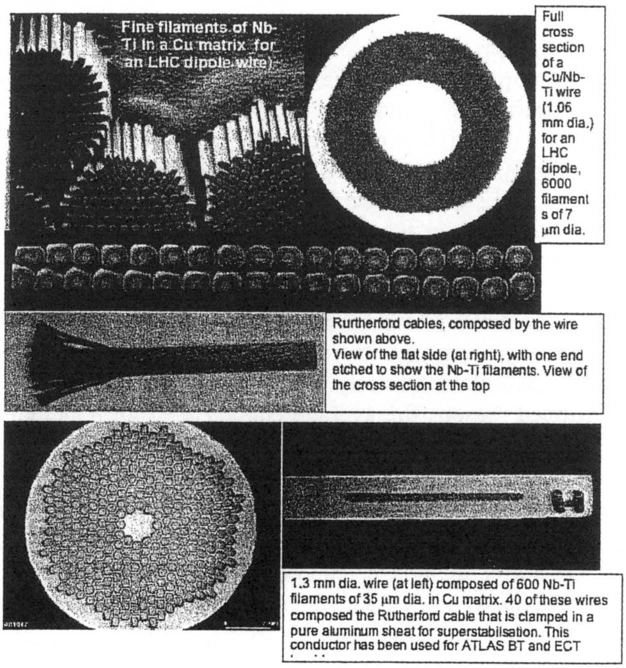

Figure 8 Picture of the superconducting wires and cables used for HEP.

Another important issue is the presence of persistent currents. A superconductor shows a hysteretic behavior, when cycled between low and high field. The width of the hysteresis loop is proportional to the current generating the magnetization: this current, induced by the field sweep, do not decay like the usual eddy current and generate field errors that, if not accurately controlled, may spoil the field quality of the magnets. Corrector magnets placed at the end of each dipoles are there to correct this and other field imperfections; however their range is limited to some 5 to 10 units of b_3 (a unit being 10^{-4} of the main field), therefore the width of the magnetization loop must be within strict limit, around 20-30 mT. This is achieved by two means:

1) A very fine subdivision of the superconducting cross sectional area: to achieve this in the LHC main magnets the strands used for the superconducting cables have 6000 or 10000 Nb-Ti filaments, each one of 6 or 7 μm diameter only. In Fig.8 one can see the much finer subdivision of filaments for an accelerator wire than for a detector wire,
2) An accurate control that these Nb-Ti fine filaments stay reasonably round without undergoing too much distortions in the extrusion and cold drawing processes that are necessary to bring the 200 mm diameter Nb-Ti ingots down to 6-7 μm final diameter in the composite Cu/Nb-Ti. In the LHC all wires are submitted to magnetization measurement and, in certain cases, magnetization exceeding the limits has been the main cause of material rejection.

5.3 Magnetization and interstrand resistance

Another important issue in the manufacturing of the superconductors for accelerator magnet is the necessity to control the average resistance among each single strand in a cable (a wire is called strands when part of a cable). Indeed when we raise the field in the magnets from minimum up to flat top value, i.e. when we really accelerate before passing in colliding mode, currents are induced in the numerous loop formed by each contact among different strands. If the resistance is too low, too much current is induced and this may spoil the field quality during ramp up and the heat generated in this resistive path can limit the stability of the superconductor. However if the contact resistance among the strands is too high the current transfer is inhibited. This is bad because a strand, in case of transition to normal state, must be able to transfer current to the adjacent strands: this transfer, if it happens on time scale of *ms*, helps to reduce the current level in the quenched strand and to recover to superconducting state. This phenomenon increases the stability limit of a strand in a cable with respect to the case of single strand, even if the effect is difficult to be quantified precisely.

For the above-mentioned reasons the average contact resistance among strands must be above 15 -30 μΩ (the upper limit being around 100-200 μΩ, and it is a target rather than a strict specification). For the LHC a new method has been set to reach these values: each wire is coated with a very fine, 0.3 ±0.1 μm thick, Sn-Ag alloy. Then, after cabling the whole cable length is treated in air, in homogeneous, ventilated ovens, at temperature around 200 °C for 4-8 hours, the precise conditions being determined by measurement of the actual coating thickness of that cable. The resulting oxidation gives reasonably predictable value of contact resistance and is very robust and stable all over the magnet fabrication cycle.

6. Magnet design

6.1 Magnetic design

It can be shown that intersecting ellipses carrying opposite uniform current density, like it is shown in Fig. 9 generate perfect dipole field in the central bore. The field is proportional to J, while the coil volume, and then the cost and the complexity, scale with the bore size. So it is clear that to have high field the primary factor is the increase of the current density as much as possible. For a quadrupole, a perfect field (i.e. a pure gradient) is generated by intersecting ellipses rotated by 90, as also shown in Fig. 9.

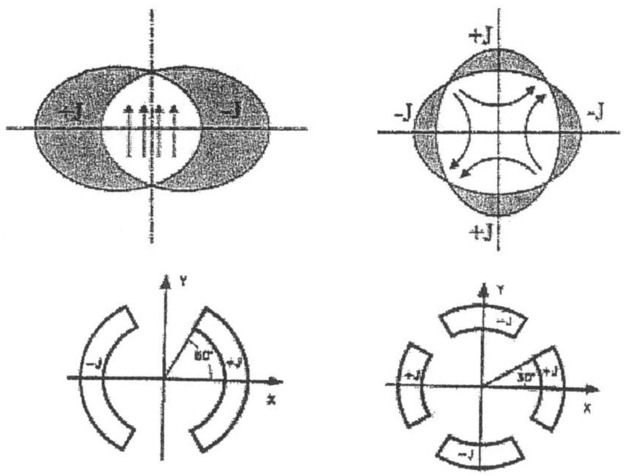

Figure 9 Intersecting ellipsis scheme (top) for dipole and quadrupole configuration and their practical implementation in actual magnets (below).

These type of configurations are called cosθ or cos2θ and, in practice, are approximate with one coil sector of a determined angular width (see Fig. 9 bottom) or by many coil sectors interleaved by copper spacers, as it can be see in the LHC dipole cross section shown in Fig. 10. The copper spacers (called also wedge for their shape) have great influence on the field quality and in the LHC magnet production have been used to control the harmonic content of the field. In Fig. 11 the evolution of the sextupole content in the main dipoles for about 80% of the production is reported with indications of the moment when intervention on the copper spacers was done in order to bring b_3 in the desired range. All design value of the components constituting the coils have to respected to within a few μm for repetitive components (cable, insulation) and within 20-30 μm for single component (wedge and collars). In total the accuracy of the coils must be within ±100 μm for average thickness and within 50 μm inside the same coil (waviness, see Fig.12).

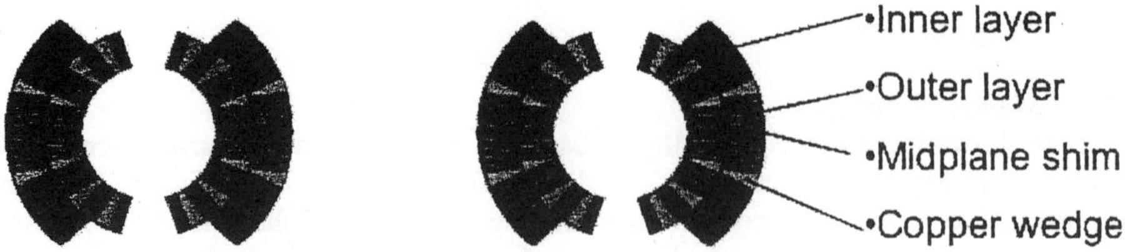

Figure 10 Cross section of the LHC dipole coils (straight section).

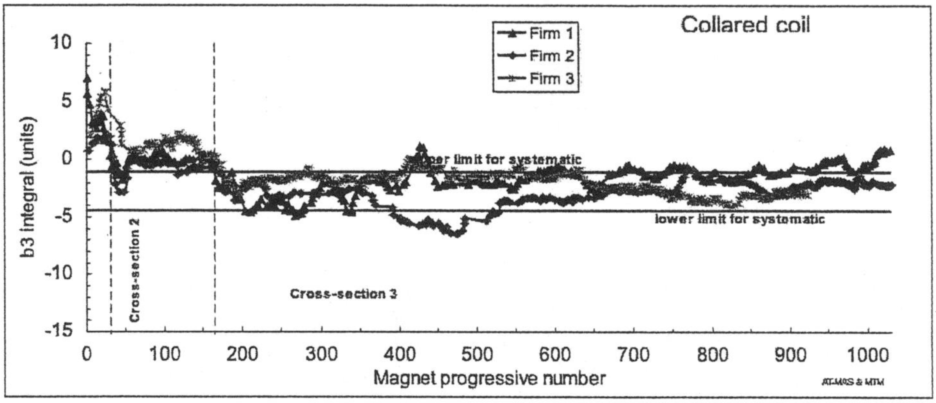

Figure 11 Trend of the sextupole components of the dipole field of the first 1000 magnets.

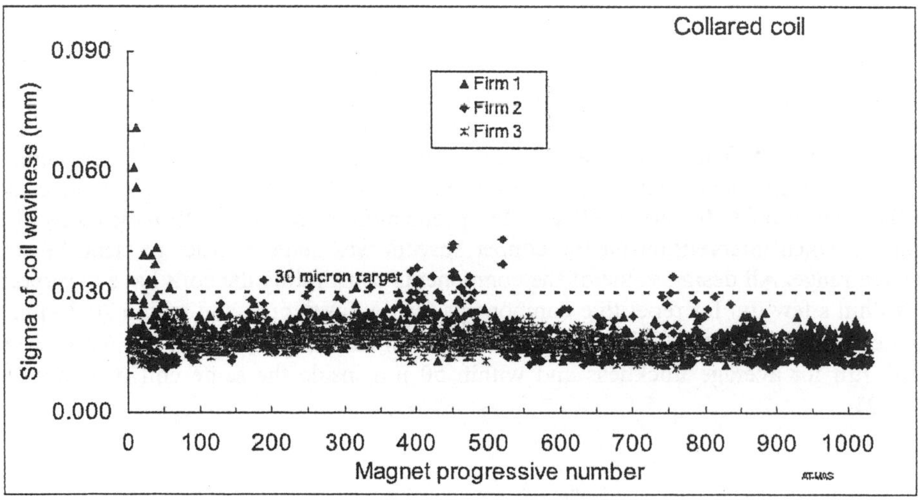

Figure 12 Trend of the coil waviness (peak-to-peak variation of thickness inside of a same coil) for all three production lines.

6.2 Mechanical design

The mechanical design of the dipole magnets is quite complex since tiny movements can release enough heat to start an irreversible transition. In presence of high field, 5 to 10 tesla and of very high current density, $J_{average} \cong 500$ A/mm^2, huge forces act on the coil package. The dipole configuration, as shown in the schematic of Fig. 13 is not self-sustaining (like it is in a solenoidal winding). Therefore a precise and strong system must be foreseen to contains the forces that tend to break apart the winding. In the LHC main dipoles, where the lateral force F indicated in Fig. 13 amounts to 400 tonnes per meter length, a system of collars made out of special austenitic steel (with a μ_r as low as 1.003 at 1.9 K) keep the coils in place, give them the right shape (field quality) and take care of almost all force containment. In addition an outer shell, which is the superfluid helium container, serves also as shrinking cylinder, transmitting to the collars, throughout the iron yoke, a further pre-stress that help to avoid coil movement at very high field (above 8 tesla) and contribute to the rigidity of the assembly. In Fig. 14 a cross section of the LHC main dipole is shown with indication of all components above mentioned. The cryostat looks very simple, since it is a mere temperature shield, to protect the precious 1.9 K superfluid helium against the "hot" ambient temperature. The main cryogen supply is provided by a dedicated cryogenic distribution line that run aside the magnets all along the tunnel.

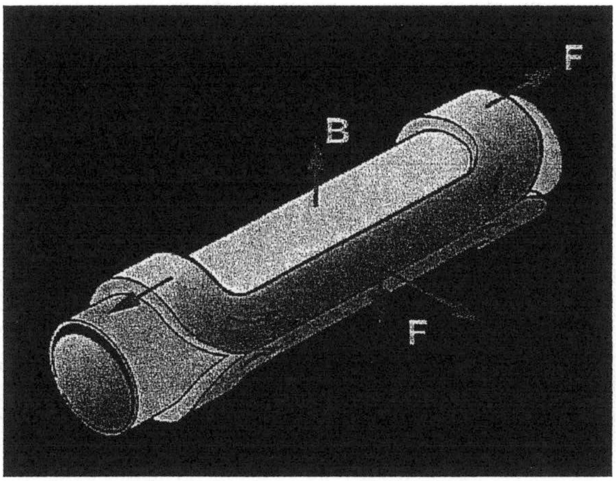

Figure 13 Sketch of dipole winding showing the direction of the total forces acting on the coils.

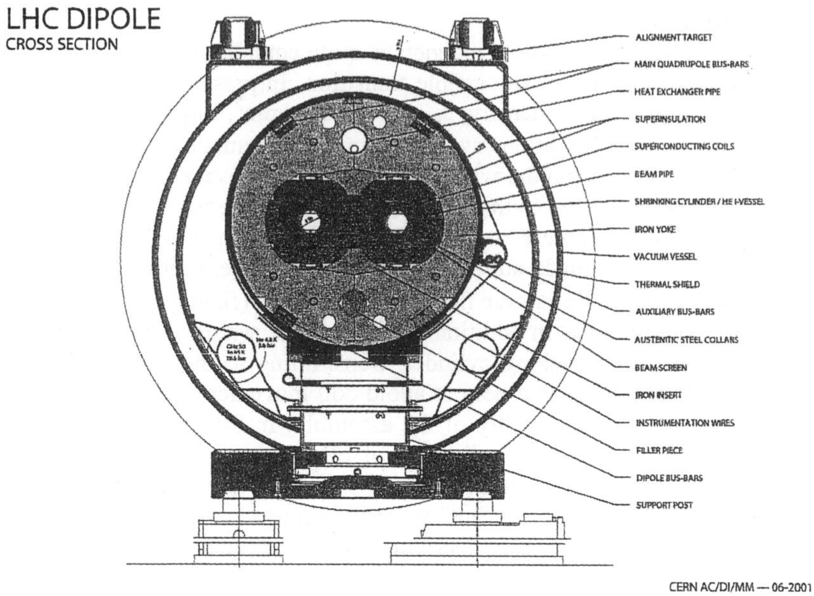

Figure 14 Complete LHC dipole cross section.

Figure 15 Snapshot at dipole production in industry: coil winding (top left), assembly of the coil in poles (top right), welding the magnet under press with a special designed robot (bottom left), cold mass finishing and storage before shipping to CERN (bottom right)

7. Magnet performance and schedule

7.1 Quench and training

The performance of a magnet is determined by two factors:

1) *Stability Margin by Design.* Superconductors, are shown in the previous Fig. 6 and 7, remains in the superconductive state only if they stay below a certain set of the parameters (T, B, J). In practice, the temperature is fixed by the coolant temperature, e.g. 1.9 K for the LHC magnets. At a fixed temperature the curve called critical current, I_c-B divides the region where the material is superconductive from the region where the material gets back to normal conducting state, i.e. one has to design its magnet to stay below the critical line of Fig. 16. Now we take the load line of the magnet in the peak field region, i.e. the I-B relation between the current flowing in the conductor and the maximum field generated in the coils by the current, and we draw it on the same graph of Fig. 16, we see that the magnet cannot go beyond the current and field given by the crossing of the two curve, otherwise it will quench, i.e., it will be driven normal. In reality we need to operate below this critical limit, sometime called short sample limit since the critical curve is usually deduced by measurements of short samples cut out from an end of a real conductor unit. Perturbations and various phenomena can inject energy into the coils, by friction (forces are very large), by dissipations of various types, by local heating due joints or current transfer among strands involving resistive paths. At low temperature the heat capacity is extremely low (Debye's Law) and energy release of μJ, or even nJ would suffice to drive the superconductor to normal. Stability, as reported in section 5.2 is obtained by immersing fine filaments of superconductor in a matrix of very good (pure) conductive metal, like copper. However this is not enough since the energy density, $\mathcal{E} = \mathcal{B}^2/2\mu_o$, is of the order of 10-100 J/cm^3, we need to have stability margin of the order of 1-10 mJ/cm^3, at least, to be stable against release of small fraction of the stored energy. The way to obtain this, without increasing in a prohibitive manner the low resistive stabilizer, is to operate at a certain distance from the limit point in the I_c – load line graph, see the point in Fig. 16 which is around 80-85% of critical limit (crossing between the two lines). It is may be worth noting that large detector magnets, like the ones for Atlas toroids and the CMS solenoid, need stability margin of the order of 1 J: this is achieved by increasing further the distance, i.e. operating at 60% of the critical limit and by adding a considerable amount of stabilizer, as shown in Fig.8.

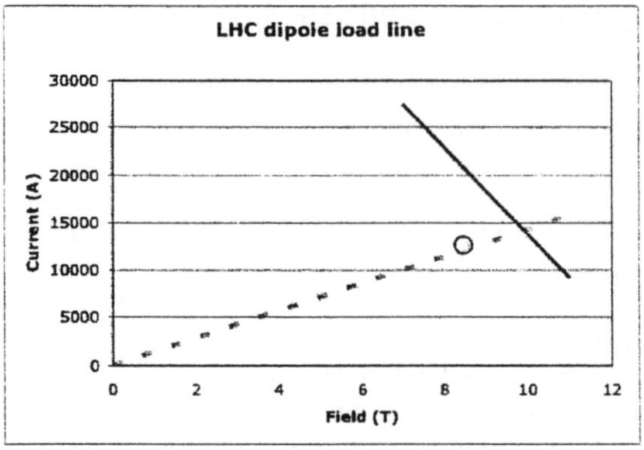

Figure 16 Critical current (continuous line) and load line (dashed line) of the LHC dipole. The circle is the operation point at nominal energy (7 TeV).

2) *Magnet Training (premature quench).* Superconducting magnets, designed to operate at a certain field, may (and frequently do) quench at current and field level well below the design value. In many cases more than one quench is needed, and each time a quench is triggered the magnet is violently warmed up, since a good part of the magnetic energy is dissipated as heat in the coils. Then the magnet is cooled again at cryogenic temperature and the current is ramped up again. This process is called training, since usually magnets get better each time, i.e. the energy release causing a quench happens at higher field. Usually magnets, once the design field is reached, stay stable at this field, until abnormal events, like heat deposition due to beam losses or debris coming from the collision points, are able to drive the coils normal. Sometimes may happen that the design value is not reached, in this case we talk of degradation. A quench, even if the magnet is well protected by a suitable system, is always a thermal and electric shock for the coils, so the ability to reach design value with little or no training is a figure of merit. In fig. 17 the number of training quench necessary to reach design value for the LHC main dipoles is reported. One can see that only one magnet is degraded and that the number of magnets needing more than 2 quenches to reach nominal field is 6% a very reasonable number that is getting better and better (it was 20% at the beginning of production).

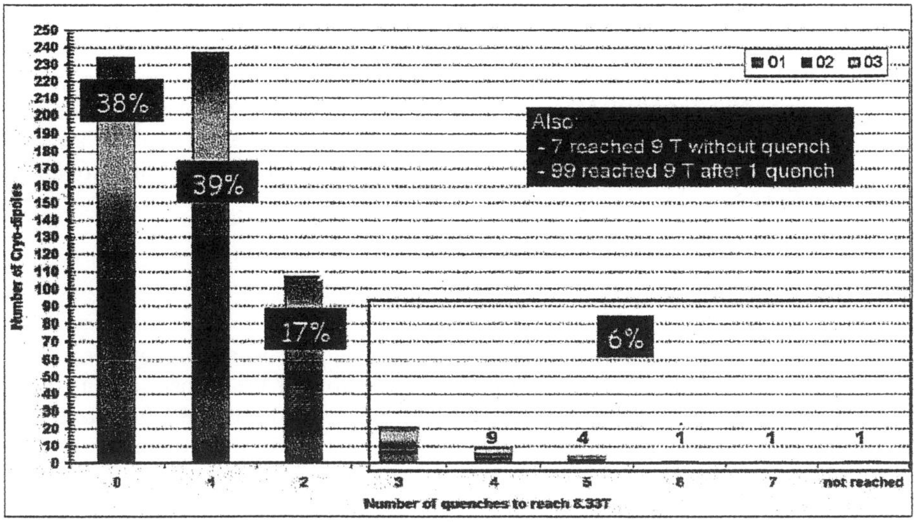

Figure 17 Hystogram of number of number of quench needed to reach nominal field for LHC dipole (virgin training curve).

7.2 Field Quality

Accelerator magnets are powered in large series: in an LHC sector 154 main dipoles are all powered by one power supply. Since the magnet array is a lattice for the beam, it must as perfect as possible and the field strength BL, of each dipole must be equal at the level of a few parts in 10^{-4}. Based on the experience of previous projects, the fact that for the LHC main dipoles three different firms or consortia, each one with many coil manufacturing lines, have been charged of the magnet construction, has been a source of concern, as far as equality of the BL all along production. Actually, thanks to good design, strict technical specifications of components (all procured by CERN) and of manufacturing process (which forced firms to design tooling accordingly) and thanks to a good system of surveying and checking production, the similarity in BL among the magnets is better than any initial expectation, as shown in Fig. 18.

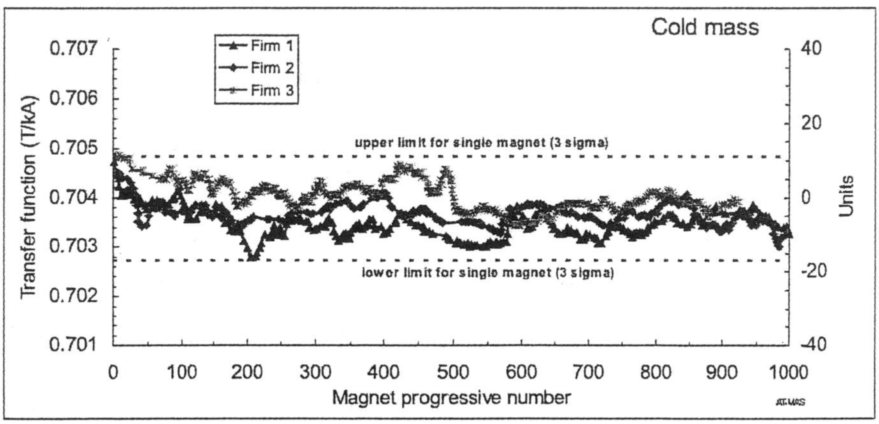

Figure 18 Bending strength (BL) for the LHC dipole (one curve per manufacturer).

In addition to the main component, each magnet is inevitably affected by higher order harmonics intrinsic in the design, due to the deviation from ideal *cosθ* configuration, that are called "allowed" harmonics. Moreover, tolerances of the components and on fabrication procedure, as well as errors and deviation from symmetry generate many other harmonics. In the section dedicated to magnet design the behavior of the sextupole component of the main dipoles (the first and most important allowed harmonic) was reported, see Fig.11.

7.3 Delivery

The production of LHC magnets is critical to the project, also because many operations at CERN are based on the magnet supply schedule: assembly of the cryostat on the magnet cold mass, cold testing, insertion of a beam screen in the cold bore tube (to intercept the synchrotron radiation emitted by circulating protons) and, finally, installation and interconnection among magnets.
The actual delivery of the magnet cold mass and its precursor, the collared coil, is shown in Fig. 19, where the end date is fixed at the end of October 2006, to allow the installation of the last magnet in the tunnel by end of February 2007. One should notice that although the main dipoles and quadrupoles fills some 22 km of the 27 km long LHC tunnel, many other super-magnets have been designed and build for the LHC: about 150 special insertion quadrupoles, 32 very high gradient quadrupoles for the collision regions (low-β triplets), a few special dipoles to separate beams and a zoo of about 8000 corrector single magnets. Production of all large magnets is proceeding extremely well and the low-β quadrupoles, produced by US and Japan in the frame of a world-wide collaboration, are all completed. Only a few type of corrector magnets have been delayed, either for technical reason or for lack of components, however all production here are now well on track and some completed since quite some time. In Fig. 20 a view of the LHC tunnel during installation of a magnet is proposed.

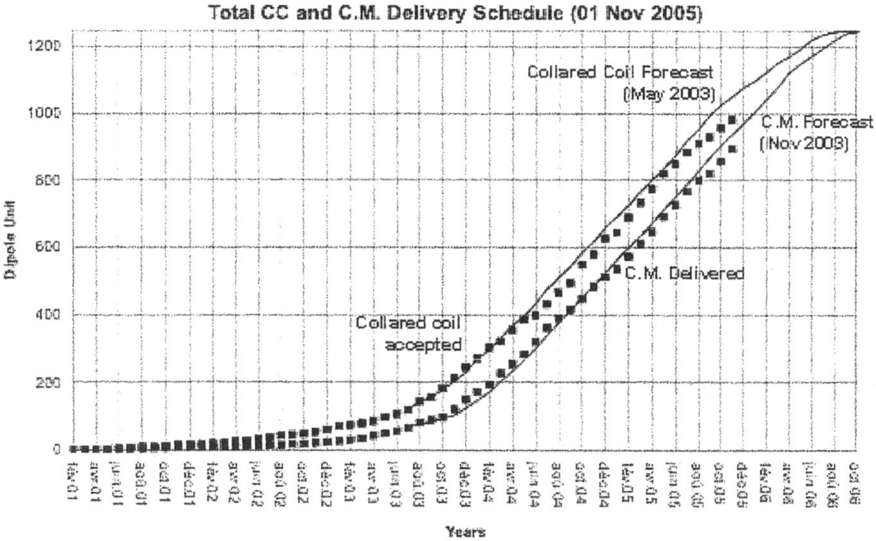

Figure 19 Deliver plan according to forecast of May 2003 (continuous lines) and actual delivery (marks) for the LHC main dipoles.

Figure 20 Installation of a dipole in the LHC main tunnel.

8. References

- LHC Design report, Vol. I The Main Ring, CERN-2004-003, 4 June 2004.
- Lucio ROSSI, "Superconducting magnets for accelerators and detectors", Cryogenics 43 (2003) 281-301, Elsevier Science Ltd.

CHAIRMAN: L. ROSSI

Scientific Secretaries: M. Bindi, R. Preghenella

DISCUSSION

- *Cifarelli:*

Can you tell us about the promising MgB_2 material for the construction of superconducting coils?

- *Rossi:*

We have basically three families of superconductors: The first family is composed by the old classical superconductor. Most of the practical superconductors has been worked out in the '60-'70. Among them we have the classical Nb-Ti and Nb_3Sn. Nb_3Sn is a material which is a little bit different then Nb-Ti, it is an intermetallic compound, not an alloy, which is much more difficult mechanically wise, but it is of the same family, based on Nb. Then you remember in '86 there was a kind of revolution, a new family of superconductors was discovered. Superconductivity was confined as a phenomena for liquid He for long time. In '86 there was an explosion, superconductivity went to the newspaper, got even the first page on the newspapers, because there was the discovery of a new class of superconductor, that were superconductors at liquid N. Liquid N is much "hotter" in temperature. Liquid He is 4 K, liquid N is 80 K, so is 20 times hotter. Room temperature is 300 K , so is only 3 times hotter than liquid N. So liquid N is much more easier, is hotter and the difference in cost between liquid He and liquid N is between champagne and mineral water: it is completely different to work with. So that's why there was a lot of work on this, there are therefore few applications but the properties they have so far make them not suitable for our applications (today. In a next future things may change!). I come finally to your question: three years ago another class of materials were discovered: MgB_2. This third class is so far the only one which has actually intermediate critical temperature (above 30 K). His critical temperature is less than ceramic compound, but is higher then classical superconductor. The class of superconductor is given by the relation between temperature and field. You see on the plot that there is a boundary line, actually is a surface if I would plot also the current density. You should not forget superconductivity is a real thermodynamic state, so like ideal gas you use temperature plus two other quantities to describe the state. The two other quantities, one is called intensive, usually is the pressure, and other is called extensive and this is the volume. This for ideal gas. You can do thermodynamic of everything until properties depends on temperature and if you can define an equation of state. For superconductivity the intensive quantity, so the role of pressure, is played by the magnetic field and the extensive quantity, the role of volume, is played by the current density. So we have a surface: if you stay below this surface you have superconductivity, if you stay above you are not superconducting, you are resistive. So you see this material, for example the YBCO or the BSCCO, this is based on Y and the other is based on Bi; these are the ceramics, basically sands! That's why there are nasty material, multi-components, five components, difficult to build, difficult to engineering, difficult to control in the right quantity each components. Mechanically they are not strong: but they have wonderful temperature properties so we can work with liquid N. But they are still very brittle, and very expensive, because they have to be stabilized, with Ag (and not with copper or aluminum like the classical superconductor) So you can imagine: even if the material itself we can be made at low cost, Ag we will be never made low cost. That's why these materials in this moment are still not usable really, not in accelerator applications.

But now we go to the real question.

So you see this material, MgB_2, this was discovered only 3-4 years ago. His critical temperature is only around 30 K, still is much higher than material based on Nb-Ti. With this material, we can dream for example, if we are able to increase the performance in J_c in this region, one day to use closed cycle with Ne. Why we want to use Ne? Ne is very rare, so is very expensive, but if you do in closed circuit, you can profit of the enormous heat capacity. Specific heat grows with the T^3, so between 30 K and 3 K, there is a factor 10 in temperature, there is a factor 1000 in specific heat, that means is much easier to be stable against perturbations, because our enemy is perturbations. Superconductivity is a wonderful state, but is very easy to loose it, to have a quench of the magnet. So this would be 1000 more stable; this material is also very attractive because the constituents are very simple, they are abundant on the earth, this is the cheapest superconducting material, very robust mechanically. But in this moment we are not able yet to transform its good quality of critical field into critical current, because we are not able to build the pinning. To have also current density we need to fix still the the flux lines, to pin down the flux lines; suppose that this is a slab of superconductor, we send current, you have the fluxoids (bundles of flux lines), the current goes through, you have field, you have current, so you have Lorentz force, electromagnetic force. The field lines are not bounded to a magnet, weighting tons, but really they are almost immaterial, that means also that the flux lines move. So to abolish dissipation I need to pin down this fluxoid, this bundle of flux. We are not able today to have this pinning high enough in this new material. It Is very difficult, actually nature has really gave us a grace making relatively easy technologically to have a very fine dense pinning in the classical superconductors, because we need a pinning at the level of 10 nm. We need to distribute this pinning centers at the level of 10 nm. This is reached with process which are difficult, but still are industrially possible. You can imagine, superconducting is not a rare phenomena, we have approximative 50000 superconductors, but of that 50000 only about 100 have reasonable critical temperature. Of that 100, only 10-20 have good properties in the plane of temperature and field. But of this 20-30 only for very few we are able to have a good engineering of the pinning. Superconductivity is very abundant as a phenomena, almost half of the material becomes superconductor, is enough to bring that down in temperature, but to make it stable you have to make this pinning, and this is something difficult.

- *Pinfold:*

What spin-off has originated from the development of superconducting magnets for moderns colliders?

- *Rossi:*

Any hospital of a certain rank has now an imaging resonance device. Actually should be called NMR, nuclear magnetic resonance device. Without the Tevatron it would not have been possible this development, because especially the Tevatron in the '80 really boosted the production of superconducting wires and cables that really make possible the technology breakthrough. When a technology was there, the technology proceeds by it's own. Second point: superconducting cyclotron are coming for medicine. In Europe in this moment two superconducting cyclotrons, have been installed, one in Switzerland and the other one in Germany, to start to do treatment of cancer with ions or with protons. There is also a project in Italy, already started with the synchrotron, but that is based on classical magnet, is called CNAU (old TERA project) and using ions. The cyclotron is using protons, but cyclotron can accelerate as well ions and this it is becoming reality. In the US is already very well reality, but is based on a classical cyclotron. Now finally superconductivity is entering also in this one. I am not saying that LHC as make it possible directly this, but are the researches that have been done ten years in Europe for the LHC that trained the companies and then the company has learn building magnet and other devices. This company ACCEL is preparing in Munich a center of one cyclotron and five lines for the treatment of the cancer. Thousand of people will be treated. Superconductivity make possible to have it compact, and at a reasonable cost: especially space compactness is very important for hospital that usually are in the town. You cannot build the CERN in a town!

- *Lindenbaum:*

I want to make the case, when i was in Washington and i tried to improve the budget, that all technology is spin-off from pure science. If you trace the tree back down to the roots you will find that all technologies as been spin-off from pure science.

- *Rossi:*

I fully agree. 3D calculations, for example, was basically invented in high energy physics, and especially for magnet. Has been, let say, brought by mechanical, structure calculation for nuclear power plant, and by calculation of magnet, electromagnetic field and so on. I fully agree, a lot coming from them, you must be able to sell a little bit better.

- *Lindenbaum:*

I was drafted when the budget was very bad. I was drafted to go to the ERDA which was the predecessor of the DOE to become scientific secretary, deputy of scientific affair, and to try to convince the congress men and senators that there is an economic advantage in supporting basic research. And i believe that the argument i gave, off coarse i did it in a much more sophisticated manner, i investigated the particular district of the congress men, took the districts of senators and pointed this specific example, is very important to his area that were really unknowingly historically that way, but in the future could that way. I think this is really the way that i tried to convince the powers that be in DOE and the ones in congress and the senate, that it is true that science is glorious, to scientists at least, but if you forget that, if you are just an hard nosed economist and you want to say "what he is doing for me economically?". "He is doing everything for you, economically". Off course, he has the bad part where technology is used for armored purposes, but that is not the responsibility of the scientists, that is been true since the beginning of man.

- *Munir:*

Is superconductivity any useful for linear colliders, because the mention of future collider CLIC was skipped in the mention of future colliders?

- *Rossi:*

As you know very well there are two types of linear colliders for the future. The near future is Tesla-Like, it is called now the ILC (International Linear Collider), is based upon superconducting cavities, this is another type of superconductivity. The cavities are made of pure Nb, inside there is an electromagnetic field. It is an electromagnetic wave that resonates, so we use, of course, the electric component of the electromagnetic wave to accelerate longitudinally the particle. Then there is another concept, that is worked out at CERN. This is the CLIC, which is based on cavities, there are not anymore cavities; there are really microwave. Here we are talking about 1.3 Ghz and in case of CLIC we are talking about 30 Ghz of the electromagnetic wave frequency. Completely different technology, this cavities becomes smaller and smaller, this is an advantage, but is also a big disadvantage for the alignment request. In this moment the community is all almost behind ILC project, except CERN. Who stayed outside LHC wants to have ILC. The point of this accelerator is that it stays below all the energies of the LHC, ten years later. Of course you do much more fine measurements, but is it worth to spend the same money for the LHC, even if more, to stay below all the energies of LHC? Personally I do not believe. For me it is difficult to tell our politicians to spend 5 billions of dollars, or euro, to stay around 500-1000 GeV. LHC is 14 TeV, you divide it by 10, because you have 3 quarks against 3 quarks, in average you have 1 TeV per elementary collision. In the LHC you have the probability that some collisions take more than the average energy, so you have really high energy events. CLIC is better because it can reach 5 times more energies, can reach 2-3 TeV, can go really behind the LHC. But CLIC is not possible before 2020-2025, in reality; that is why i think the hadron collider, we know how to build, we need to improve the magnetic field, but we know how to build, and I think that at least for CERN, it is possible to triple the energy of the LHC, and this will give an immense benefit. If we can have 40 TeV collider in CERN, at 10^{35} of the luminosity, and then thinking as a preparation for the next future that must be ELOISATRON or ELOISATRON like, that is kind of 300-500 TeV in the center of mass that really can bring us at the boundary of our knowledge.

Superconductivity may also in use for linear colliders for the final focusing magnets. You have to squeeze the beam, in the linear collider, at the level of 10 micron and for this one might need strong quadrupoles.

- *Faivre:*

ELOISATRON has been briefly mentioned, as a project of accelerator that would benefit from high magnetic fields, from superconductor technology. What would be it's characteristics? Meanwhile, it has been said that the synchrotron radiation would begin to be significant at LHC. Then, if ELOISATRON is meant to be “small”, thanks to the high magnetic fields available, would not it reach the synchrotron limit?

- *Rossi:*

You are fully right. Unfortunately synchrotron radiation scales with energy cube and with B linearly. Since the energy scales with B, practically we scale with B at the 4th power. This is the point we have to deal with and we are thinking how to solve this real problem of the hadron colliders. First of all we need to have the magnetic field at reasonable cost and then be able to deal with the synchrotron radiation. We have few ideas, if you have a cold temperature and if you intercept the synchrotron power at liquid He, is too bad.

- *Faivre:*

I was actually thinking about the effect that the energy you lose on the synchrotron radiation is equal at the energy you put in the accelerator.

- *Rossi:*

That is not a big problem, because anyway the quantity that you lose is not that much in such a way each time the beam recirculate, you have always the section of cavities that gives back the energy you lost. This is infinitely easier than for example in LEP. In LEP actually the field of the magnets was very low, because the limit to the energy was that the beam was losing so much irradiation, that was the maximum energy that we are able to give per turn to the beam to compensate the radiation losses. In case of proton is not a problem, the real problem is how to deal with the radiation, with the power that is emitted, this heat up our magnet and, as I said, specific heat is small and works against us.

- *Faivre:*

What would be the scheduled size for the ELOISATRON?

- *Rossi:*

ELOISATRON is a dream, that has already generated many accelerators. Is a dream launched by professor Zichichi 25 years ago, when I had your age. It is a dream to build up the ultimate frontier in high energy physics. To use the maximum magnetic field in the largest possible ring. And the maximum ring that you can have,. This is the longest ring we can have in Sicily, in a good stable region, so we can have a ring of 300 Km, 10 times the LHC. If we can put magnet that is two times the LHC magnet, and 10 times the ring, it means basically 20 times the energy. So we will reach effectively 300 TeV in the center of mass. ELOISATRON is a dream that is already generated a lot of discussion. Partially even the concept of CLIC and other linear colliders was generated here in many ELISATRON workshop. Has generated the idea to use high field magnet in hadron colliders, and practically one can see the LHC as a 10 or 20% test of the ELOISATRON. Only who dreams can go forward.

- *Zichichi:*

I can just tell you a telegraphic remark, because if the subnuclear physics community was a very organized community, we should have immediately started in 1979 this strategy for the following reasons. We do not really know what is the threshold for the Higgs and SUSY. Suppose that the threshold for Higgs and SUSY was above LHC, our community will be demolished by our enemies. If we would not started with 300 Km and then have 5 TeV, 10 TeV, ... , 300TeV, then this would be a continuity in the project, but you can not have a machine for each energy range. This is impossible. We must pray that the God is going to give us low energy SUSY, but if the SUSY is above the LHC potentiality, the subnuclear physics would be in big troubles. The concept of ELOISATRON was to avoid this possibility of suicide. But at present we have this enormous risk. What has LEP discovered? Nothing. Three families. I have been worked ten years on LEP. It could be that the threshold for new physics is above 50 TeV, which is fantastically interesting for the scientific point of view, but strategically speaking we subnuclear physicists had been not very smart because we should had agree all that the step to go was a very big ring, then start filling it, and then we go above in energy. You can not have a machine for each energy range. Let us hope that LHC will make fantastic discoveries, with new dimensions, this is our hope. Lucio Rossi says that you must dream, of course, we had a dream also. But the hope is that dream will become something.

The (Nearly) Perfect Fluid: A Primer on RHIC Physics

Peter Steinberg *

Abstract

This lecture is a brief introduction to RHIC physics and some of the striking phenomena that have been observed there. Attempts are made, where possible, to draw attention to similar behavior in heavy ion and elementary systems.

1 Introduction

Inside a nucleon, it is generally accepted that there are three "quarks" (uud) that conspire to carry its quantum numbers [1]. Moreover, the theory of Quantum Chromodynamics suggests that quarks are continually interacting via the exchange of gluons, the "colored" photons of the strong interaction. One well known consequence of the composite nature of the hadrons (baryons and mesons) is the huge number of bound states which arise from the various combinations of quarks, angular momentum, parity, etc. which are consistent with the basic constraints of QCD [2]. As soon as strongly-interacting states were revealed in experiments in the 1950's and 1960's, it was realized that the density of states, $\rho(M)$ was a strong function of mass. Rolf Hagedorn, based on his "bootstrap" principles, suggested that the mass spectrum may rise indefinitely, and with an exponential mass dependence, i.e. $\rho(M) \sim exp(m/T_0)$, with the parameter T_0 known as the "Hagedorn

*Current address: Physics Department, Brookhaven National Laboratory, Upton, NY 11973

Temperature" with a value estimated from data as $\sim 170MeV$ [3]. The striking conclusion from Hagedorn's work was that a system of hadrons in thermal equilibrium would show singular behavior at the Hagedorn temperature, since the partition function $Z = \int \rho(M)exp(-m/T)dm$ would diverge, rendering thermodynamics impossible. This suggested that new phenomena in the strong interaction of hadrons should occur at this temperature.

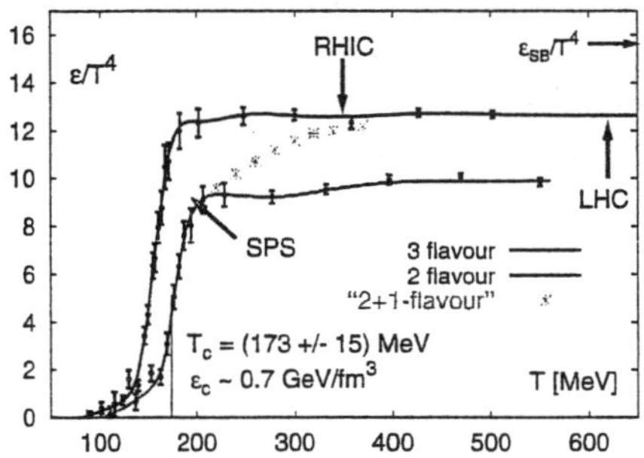

Figure 1: The scaled energy density (reflecting the number of thermodynamic degrees of freedom) as a function of temperature, from lattice calculations.

The arrival of QCD in the 1970's provided a new way to think about this problem [4]. However, QCD calculations at the Hagedorn temperature are notoriously difficult, since QCD is a non-Abelian gauge theory rendering low-energy perturbative techniques unreliable. Thus, the thermal properties of QCD have traditionally been explored using the techniques of "lattice QCD" (LQCD), where the quark and gluon fields are calculated numerically on a space-time lattice. These studies have found a jump in the number of thermodynamically active degrees of freedom by studying the energy density scaled by T^4 (the characteristic dependence of blackbody radiation). Interestingly, this jump takes place at a temperature of $T_c = 173 \pm 15$ MeV, corresponding to an energy density of $\epsilon_c \sim 700MeV/fm^3$ [5]. Thus, T_c is surprisingly close to the Hagedorn temperature, estimated in the 1960's, and ϵ_c is only slightly larger than mass density of the nucleon itself $\epsilon_p \sim 500MeV/fm^3$.

Above the critical temperature, one presumes that the degrees of freedom are no longer the bound hadrons in the theory, but something more to do with the quarks and gluons themselves. By itself, the lattice is silent about this question. Curiously, the "plateau" in ϵ/T^4 does not reach the Stefan-Boltzmann limit for a non-interacting gas of quarks and gluons, suggesting the presence of some residual correlations, but this will be discussed farther below. In the meantime, the state formed above T_c will be heuristically labeled the "Quark Gluon Plasma" (QGP), despite no proof that the degrees of freedom are in fact the objects described in perturbation theory [6].

To study the QGP, the RHIC facility has been built at Brookhaven National Laboratory (BNL) and has been running successfully every year since 2000 over a wide range of CM energies (19.6 - 200 GeV/NN collision) and colliding species (p+p, d+Au, Cu+Cu, Au+Au). The stages of the collision process are sketched out in Fig. 2. In a high-energy collision, the Lorentz contraction squeezes the nuclei along the beam direction, making the spherical nuclei into "pancakes". The initial collisions involve the hard scattering of quarks and gluons, as well as an overall deposition of energy into the collision region. The "middle" stage involves the evolution of the deposited energy, possibly through a strongly- interacting, hydrodynamic phase. The final stage is the freeze-out of hadrons from the system, which eventually impinge upon the detectors. The particles which register in the experiments originate in several of these stages, requiring us to get a handle on all of them in order to understand the relevant physics.

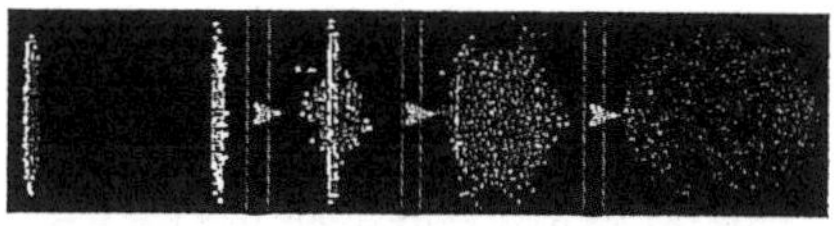

Figure 2: Schematic illustration of the stages of a RHIC collision.

At RHIC, there are four separate experiments, BRAHMS [7], PHENIX [8], PHOBOS [9], and STAR [10]. Two of these (PHENIX and STAR) are large multi-purpose spectrometers with a focus on particle production near mid-rapidity. The other two are narrow-band spectrometers with more of a focus on particles emitted at forward angles. This resulted in a complementary set of experiments able to elucidate many aspects of the rich physics available in heavy ion collisions.

2 Global Variables at RHIC

While nuclear collisions appear to be quite complicated many-body systems, much of the behavior can be systematized quite efficiently using very basic features of the initial collision. The Center-of-Mass energy per nucleon nucleon collision. This determines the energy available for particle production, the Lorentz contraction of the initial colliding nuclei, the nucleon-nucleon inelastic cross section, and the probabilities of "hard" processes occurring between the incoming partons. The impact parameter (b) determines the number of participating nucleons in the collision, the number of binary collisions per participant, and the overall shape of the reaction zone. While

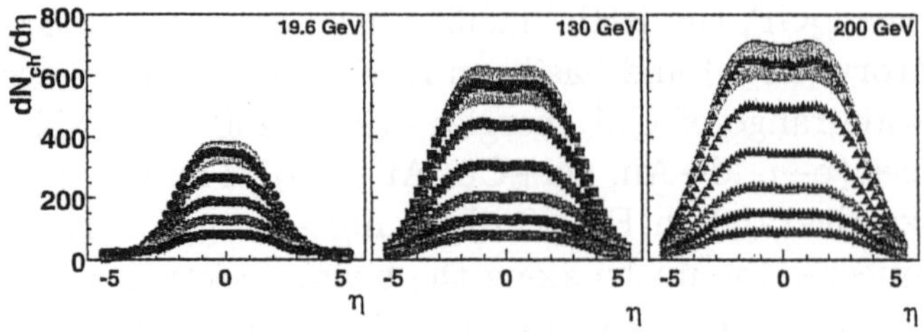

Figure 3: $dN/d\eta$ at three collision energies and over a range of centralities (impact parameter), from Ref. [11].

binary collisions are found to be important for hard processes (e.g. direct photons), it will be shown that participants are fundamental for soft particle production.

2.1 Particle Multiplicities

The simplest observable to connect to these initial-state features is the final-state inclusive multiplicity of charged particles, both integrated over the full solid angle as well as in angular bins, as shown in Fig. 3. If the transverse momentum available to emitted particles is limited, then most of the energy goes into longitudinal motion, which is best represented by the "rapidity" variable, $y = \tanh^{-1}\beta_z$. Furthermore, one can approximate the direct measurement of rapidity with "pseudorapidity", $\eta = -\log(\tan(\theta/2))$. The PHOBOS experiment has extensively mapped out distributions of $dN_{ch}/d\eta$ as a function of beam energy and collision geometry. Of course, many other observables have been measured in great detail in particular regions of phase space, but it will be argued that even very simple observables offer interesting, non-trivial information about the dynamical evolution of heavy ion collisions.

2.2 RHIC Hadrochemistry

To see how this can work, one can look in some detail at the properties of the final state, namely the particles that escape the collision zone and reach the detectors. The spectrum of these particles, as shown in raw form in Fig. 4 which presents the yield of particles near mid-rapidity as a function of transverse momentum, has a familiar blackbody shape, with a peak at around 200 MeV/c. Of course, this is not a blackbody spectrum holding among the photons emitted from the source, but of the hadrons themselves, which thus should be seen as the system's quanta of radiation.

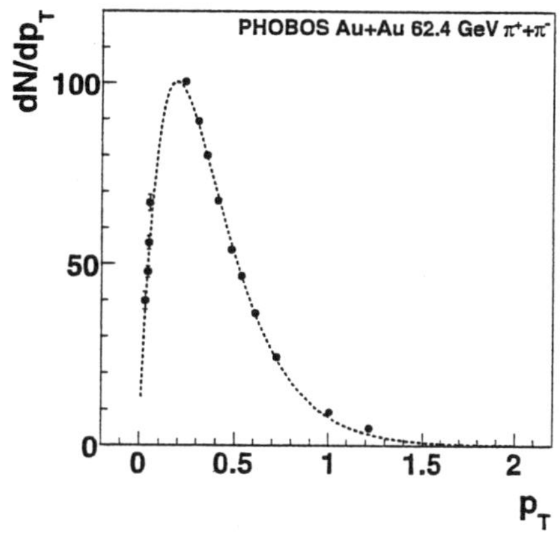

Figure 4: "Blackbody" spectrum of produced particles, from Ref.[12].

While this spectrum is dominated by final-state pions, we know from experimental data that many of the known hadron states are also being emitted. It thus becomes a natural question to ask if all of those particles are emitted from the same thermal source. The so-called "thermal models" [13] assume that all of the hadrons are emitted with a common temperature and baryochemical potential (representing the excess of baryons over anti-baryons in the initial colliding nuclei), and allows an integral over the blackbody spectrum for each particle species (fermion and boson) to calculate the total yield of each species, N_i: $N_i \propto g_i V \int \frac{d^3p}{2\pi^3} \frac{1}{e^{(\sqrt{p^2+m^2}-\mu_B)/T} \pm 1}$ From these, particle ratios can be constructed, such that the volume term disappears, leaving only the relative production of each particle type. Results of this are shown in the left panel of Fig. 5 from Ref.[10], which shows data on a large number of species from the light pions to the heavy Ω baryons. The deviation of $\overline{p}/p$ from unity (~ 0.7 at the top RHIC energy) suggests the presence of a substantial net baryon number near mid-rapidity at RHIC. Beyond that, non-identical particle yields allow a relatively precise estimate of T_{ch}, the temperature after which the hadrochemical abundances are fixed.

From various estimates made by several groups, this temperature tends to be around 170 MeV, strikingly close to the Hagedorn temperature and the quark-hadron transition temperature found on the lattice. Converting from MeV to Kelvin requires putting back the true value of k_B, which leads to the conclusion that the final state at RHIC exists in a thermalized state with a temperature of $2 \times 10^{12\circ}K$. But since this is the temperature as the system decouples, one can only presume that the earlier stages must have

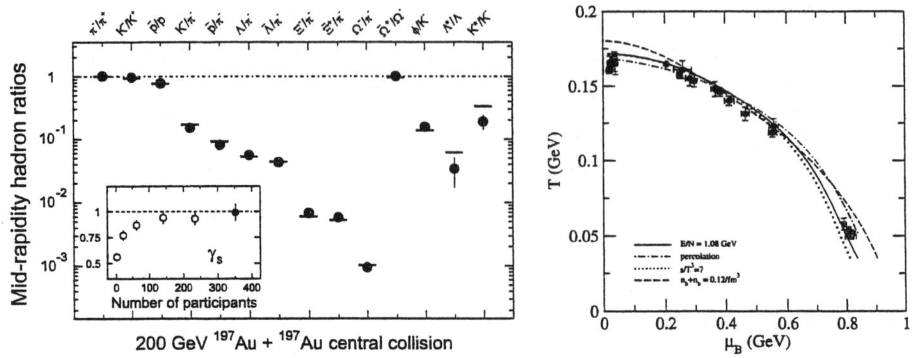

Figure 5: (left) Yields of hadron ratios fit to a thermal model, from Ref. [10]. (right) Thermal parameters at a variety of beam energies, from Ref. [14].

been even hotter, certainly beyond the temperature predicted for the QCD phase transition. This is another reason it is generally believed that the QGP *must* have been formed in RHIC collisions.

The baryochemical potential at RHIC is also found to be non-zero, $\mu_B \sim 30MeV$ but relatively small compared with the temperature. This is not the case at lower beam energies, as shown in the right panel of Fig. 5 from Ref.[14]. In general, as the beam energies decrease, the baryochemical potential rises and the temperature falls, both of which suggest that relatively more energy is directed towards maintaining the presence of heavy baryons (due to baryon number conservation) as opposed to producing lighter meson. And yet, it appears that despite this potentially complicated interplay between baryons and mesons, the relationship between T and μ_B is quite simple, as shown in Fig. The data can be described alternately as $\langle E\rangle/\langle N\langle$ (the average energy per particle) being about 1 GeV [15], or even a constant value of s/T^3, which is proportional to the number of active thermodynamic degrees of freedom per volume[16].

2.3 Total Entropy

While all discussions so far have been in terms of densities or particle ratios, these avoid the issue of the total entropy produced in the collision of two nuclei. Of course, the total entropy could be a complicated function of beam energy and collision geometry, considering the various factors that should

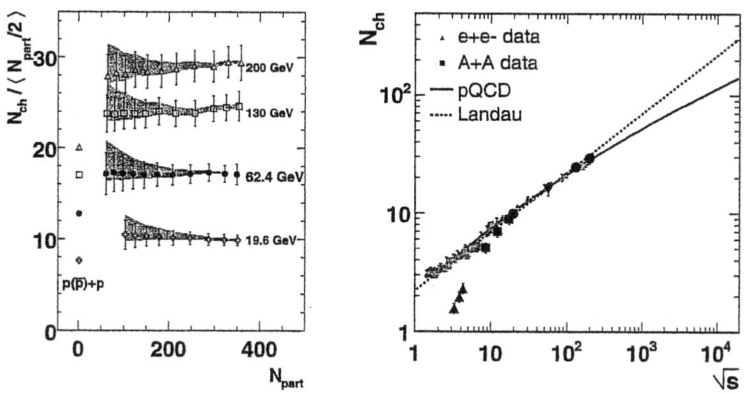

Figure 6: (left) Total multiplicity per participant pair as a function of N_{part} for Au+Au collisions at four beam energies, from Ref. [21], (right) Total multiplicity per participant pair vs. $\sqrt{s_{NN}}$ compared to e^+e^- data, adapted from Ref. citeBack:2006yw.

contribute: from the parton distributions, to the various reinteractions of partons and hadrons, to effects possible related to the final state freezeout stage.

That said, one could consider the opposite extreme of considering so many possible effects that they achieve at least an approximate local thermal equilibrium throughout the collision. Fermi [17] and Landau [18] followed this approach in the early 1950's, with no knowledge of the details of the strong interaction. They assumed full thermalization of all of the incoming energy within a volume determined by the overlap of the Lorentz-contracted nuclei ($V = V_0/\gamma$). At RHIC, this time is exceedingly small, less than 0.1 fm/c, with a resulting energy density that sounds staggering on its face: about 3 TeV per cubic fm. The reason this number gets so large is because energy density is $\epsilon = E/V$. The energy E increases linearly with $\sqrt{s}$, while the volume decreases as $1/\sqrt{s}$ due to the Lorentz contraction. This suggests that $\epsilon \propto s$, or the beam energy squared. To calculate the energy, one only needs to recall that blackbody radiation (with the simple EOS $p = \epsilon/3$) has the relation $s \propto \epsilon^{3/4}$ and $S = sV$, such that the total entropy scales as $s^{3/4}/\sqrt{s} = s^{1/4}$, or more simply the square root of the CM energy.

Of course, considering the factors mentioned before contributing to the total entropy, it is a non-trivial statement to suggest a linear relationship between that quantity and a bulk observable like the total inclusive charged-particle multiplicity. That said, it is typically assumed that the entropy is proportional to the total number of charged particles, and furthermore that the charged particle multiplicity is proportional to the total multiplicity, both in ways that are energy and geometry independent. This is a similar approach to what is called "Local Parton Hadron Duality" in e^+e^- physics [19].

Indirect evidence that these assumptions might make sense in nature come from the systematics of the total charged particle multiplicity measured in heavy ion collisions at RHIC by the PHOBOS experiment[20, 21]. These are essentially integrals of the distributions shown in Fig. 3 with some extrapolations into the regions beyond the detector acceptance. In Fig. it is shown that the multiplicity is linear with the number of participants. This is a natural outcome of the Landau/Fermi picture, since the volume should itself scale linearly with N_{part}, suggesting that the multiplicity scales linearly with the volume, like the entropy ($S = sV$). It also tells us that the scaled multiplicity $n_{ch} = N_{ch}/(\langle N_{part}/2\rangle)$ is a fundamental quantity for heavy ion collisions at all energies and collision geometries. In the right panel of Fig. 6, the total multiplicity per participant pair in central A+A collisions (Au+Au for RHIC and AGS data, and Pb+Pb for SPS data) are shown as a function of beam energy. Superimposed on these data are other experimental data from $p+p$, $\overline{p}+p$, and e^+e^- collisions, as well as a curve derived from the Landau/Fermi model: $N_{ch} = Ks^{1/4}$, with $K = 2.2$. One sees several interesting features:

- The e^+e^- data agrees with the Landau prediction fairly well over the full range of energies.
- The A+A data is similar to both e^+e^- data and the Landau prediction above 20 GeV.
- The $p + p$ and $\overline{p} + p$ data is similar to the e^+e^- data provided one corrects for the "leading particle effect" by rescaling the $\sqrt{s}$ by a factor of 2.

The apparent universality of the total multiplicity is a surprising outcome of these studies, and will be discussed further below.

The previous discussion has highlighted the important role of hadrochemistry in understanding RHIC collisions. The temperature of the system seems to reflect the blackbody temperature of the final state. The total entropy seems to be determined simply by the energy and geometry of the collision system. Finally, it can also be shown that the net baryon density is correlated with the total multiplicity in a systematic way, but this is too detailed for this lecture (those interested can consult Ref. [22]) And yet, all of this is a descriptive language, with no reference so far to any dynamical mechanism which could bring about this apparent equilibrium. More importantly, it is a *static* picture, with no discussion of space-time evolution. This latter issue will be addressed in the next section.

3 RHIC Hydrodynamics

The question this section will try and address is whether the system formed at RHIC is a "fireball", radiating isotropically into free space, or whether or not it is more "explosive" with true space-time dynamics preceding the freeze-out into hadrons. Hydrodynamics is generally thought of as a theory of liquid flow, but it is more generic than that. Any system for which one can define a local temperature and flow velocity is a *fluid*, amenable to a hydrodynamic description. The strongest assumption made in the hydrodynamic calculations discussed here is that the fluid is "ideal", i.e. zero viscosity and thermal conductivity. In other words, one assumes that the system is in perfect local equilibrium with itself and its neighbors. Viscous terms introduce non-equilibrium behavior, e.g. associated with resistance to flow, as with honey, or even water relative to superfluid ^3He.

Thus, in the context of high energy collisions, one defines "hydrodynamics" as indicating a three step process:

- One defines the "initial conditions", which is the spatial configuration of energy density thermalized in a particular volume.

- Under the presumption that adjacent phase space cells are in causal contact, the pressure gradients which develop are described by the divergenceless of the stress-energy tensor, with the system of equations closed by the Equation of State (EOS). In ideal (inviscid) hydrodynamics, the EOS provides the particular dynamical content of the system.

- As the system evolves,it expands and cools. The condition for freezeout is often stated in terms of a critical temperature or energy density

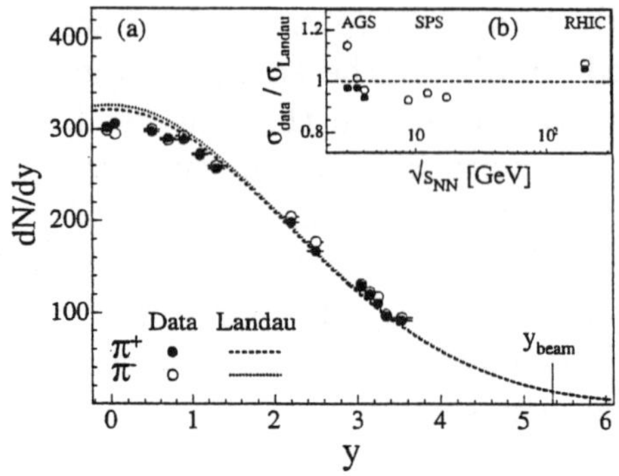

Figure 7: Rapidity distributions of charged pions in $\sqrt{s_{NN}} = 200$ GeV Au+Au collisions, from Ref. [28].

(which are related by $\epsilon \propto T^4$). This can correspond to the above-mentioned chemical freezeout temperature (T_{ch}) or sometimes even to a "thermal freezeout" temperature (T_{th}). At this point it is assumed that fluid cells decay isotropically into hadrons in their local rest frame.

Thus, the success of thermal models should not be taken to mean that there is one temperature that characterizes the entire system. Rather, the final temperature seems to be a *local* property of the fluid which terminates its hydrodynamic evolution.

The initial conditions in RHIC collisions are determined both by the Lorentz contraction and by the transverse overlap region given by the impact parameter of each particular collision. The enormous compression along the beam direction would naturally lead to much larger longitudinal pressure gradients than transverse, suggesting that the very earliest times are dominated by longitudinal flow. As time goes by, the finite transverse size will then lead to radial and elliptic flow on time scales of $O(R)$. Thus, we will discuss longitudinal flow first, since it provides the "initial conditions" for the later transverse flow.

In 1955, Landau and collaborators performed an important study of hydrodynamic evolution in proton and nuclear collisions, establishing much of the conceptual basis for the calculations made in the present day [23]. However, as QCD had not been developed in the 1950's, the equation of state was assumed to be of a simple blackbody form ($p = \epsilon/3$). While the cal-

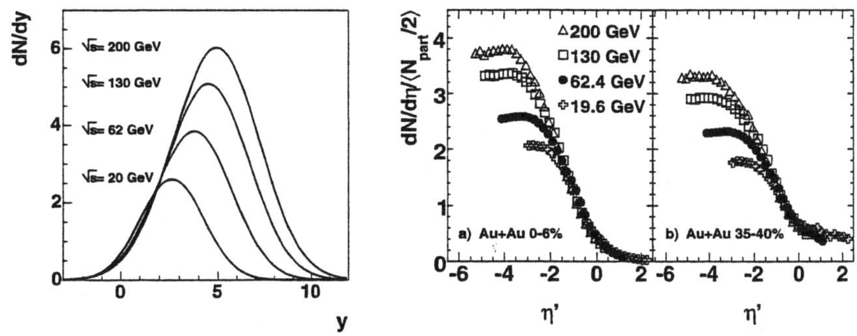

Figure 8: "Limiting fragmentation" in the context of Landau's hydrodynamic model (left) and in RHIC data (right).

culations were somewhat schematic, and based on some strong assumptions, the final result was quite simple. Hydrodynamic evolution of a lorentz contracted slab maps an initially-uniform distribution along the beam axis to a Gaussian in rapidity space, with variance $\sigma_y^2 = \log(\sqrt{s}/2M_P)$. Coupled with the previously-discussed energy dependence of the total multiplicity, Landau's hydrodynamics and Fermi/Landau's thermodynamics predict a compact form for the rapidity distribution of charged particles as a function of beam energy: $\frac{dN}{dy} = Ks^{1/4}\frac{1}{\sqrt{2\pi L}}exp(-\frac{y^2}{2L})$ where $L = \log(\sqrt{s}/2M_P)$[24, 25].

This rapidity distribution is quite nontrivial in that it does not show the "boost-invariance" predicted by Feynman[26]. In some sense, this is not surprising, as boost invariance suggests no causal connection between different regions of rapidity, a natural consequence of rapid longitudinal expansion. Landau's model, however, suggests that the system is fully coupled, albeit briefly, and then explodes collectively, leading to strong correlations between the forward and backward hemispheres. Given these stark differences, it is then interesting to realize that the general perception that boost invariance is seen in p+p collisions, even at ISR energies, should be treated a bit more critically than has been the case. Carruthers and collaborators found that even in the 1970's there was no clear evidence for boost invariance, and substantial evidence to the contrary[25].

This situation has been found to be equally interesting in heavy ion collisions. Rapidity distributions of pions have been measured from $\sqrt{s_{NN}} = 3$ GeV (AGS energies) all the way to $\sqrt{s_{NN}} = 200$ GeV (RHIC energies) [9, 28], the latter shown in Fig. 7. And surprisingly, the width of the distributions fits reasonably well to the function Landau predicted in 1955, with no free parameters, as shown in the inset to the figure.

The combination of the rapidity distribution with the evolution of the total multiplicity is particularly interesting in the rest frame of one of the nuclei. If the transformation $y' = y + y_{beam} = y + e^L$ is applied to Equation, it is seen that $dN/dy' \propto \frac{1}{\sqrt{L}} exp(-\frac{y'^2}{2L} - y')$ is primarily a function of y', with only a slow energy dependence near $y' = 0$, as shown in the left panel of Fig. 8 This is reminiscent of the phenomenon of "limiting fragmentation" seem in the pseudorapidity distributions of inclusive charged particles measured in proton-proton collisions, e.g. by UA5 in the early 1980's. They found that the yields were essentially invariant when viewed in the frame of one of the projectiles (proton or anti-proton). In fact, one finds a similar phenomenon in all multiparticle processes, whether in p+p, e^+e^- or A+A. Distributions for Au+Au collisions at a variety of energies are shown in the right panel of Fig. 8 [11].

Thus, it is possible that Landau's initial conditions are responsible for most of the basic features of the inclusive particle production data. While it would be interesting for these statements to hold exactly, they are probably only approximate. Several assumptions made here may not hold in detail:

- The entire volume may not be thermalized, but only some fraction. And yet for the multiplicity formulae to hold this would have to be violated in a way that does not depend strongly on beam energy.

- Lattice QCD calculations predict that the speed of sound is only $\sqrt{1/3}$ in particular high temperature regimes[29]. Thus, depending on the dynamical evolution of the system in both space and time, the equation of state may not be the ideal $p = \epsilon/3$ for the whole collision process.

- The assumption of zero viscosity may not hold true in general, and especially for high-p_T processes (heavy flavor and jets) [30]. It is a major goal of the RHIC II program to study these issues quantitatively, with real data.

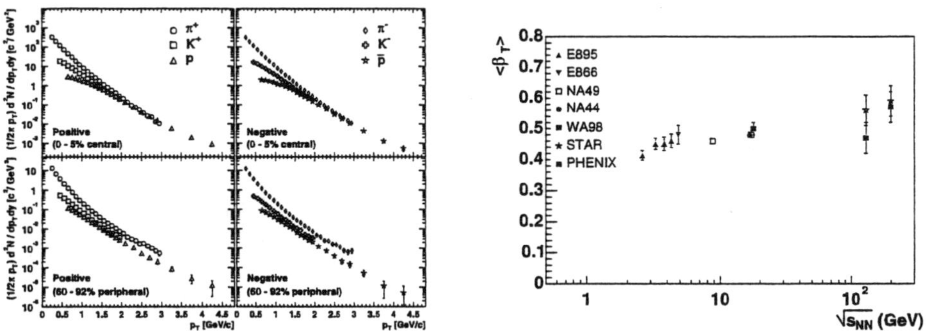

Figure 9: Particle spectra for different particle species as a function of impact parameter. Fits to spectra, extracting the mean transverse velocity, as a function of beam energy.

4 Transverse Expansion

So far, all of the distributions discussed have been dominated by longitudinal degrees of freedom. However, early thermalization naturally leads to *transverse* expansion controlled by the early-stage geometry and energy density. Intuitively, this is seen in the systematic "hardening" of the p_T spectra of various particles with increasing mass. That is, protons and antiprotons at low p_T have a much "flatter" spectrum relative to the pions measured in the same events, as seen in the left panel of Fig. 9. If one writes down the contributions of thermal and kinetic energy to the effective temperature $T_{eff} = T_0 + m\beta_T^2$, one can almost extract the average velocity of the flow by eye. However, it is typically estimated by hydrodynamic models, as well as models inspired by hydro (e.g. "blast-wave" models), to be about 60% of the speed of light, $\langle \beta_T \rangle \sim 0.6$, with a mild energy dependence as shown in Fig. 9.

The phenomenon known as "elliptic flow" directly relates the shape of the event-wise azimuthal distribution with the shape of the overlap region of the two nuclei as they collide. In a more peripheral collision, this shape is quite elliptical meaning that the pressure gradients are larger along one axis than the other. This leads to preferential emission of particles along the short axis and thus an overall modulation, an example of which is shown in

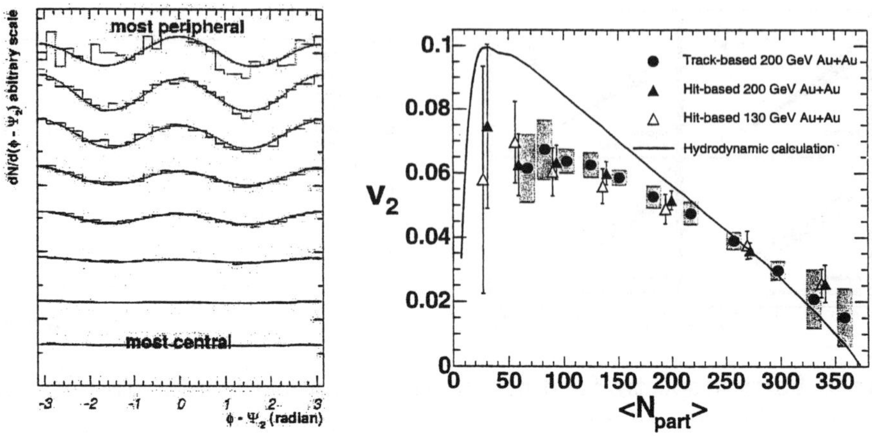

Figure 10: $dN/d\eta$ at three collision energies.

Fig. 10, which can be characterized by a Fourier expansion:

$$\frac{dN}{d\phi} = 1 + 2v_1 cos(\phi - \Psi_R) + 2v_2 cos(2[\phi - \Psi_R]) + \cdots$$

What has been striking about the RHIC data, shown in the right panel of Fig. 10, is the quantitative agreement reached between values of v_2 extracted in data as a function of the centrality of the collision, and hydrodynamic calculations requiring large energy densities (e.g. 30 GeV/fm^3) and short times (e.g. $\tau_0 = 0.6$ fm/c)[31]. Similar agreement with hydro is seen even in more differential quantities, such as $v_2(p_T)$ reflecting how strongly particles of different momenta are correlated with the overall reaction plane. At least up to 2 GeV, when one might expect viscous effects due to the presence of truly hard processes and jet fragmentation, hydro does a reasonable job of describing the data both qualitatively (e.g. the different particle masses) and quantitatively[32].

Nothing in the discussion so far makes any mention as to exactly what degrees of freedom are present in the fluid described by ideal hydrodynamics. Originally it was thought that perturbative quarks and gluons would be appropriate, but all attempts to use pQCD cross sections (which are small almost by definition, lest the perturbative framework be inapplicable) generate far less coupling between the constituents, making the system far more like a gas than a liquid[33]. Hadrons have much larger cross sections to interact among themselves, but it is not clear how to discuss their scattering processes in an environment with such high energy densities that there must be 10's of hadrons overlapping at any instant (at least in the early stages of the collision). It is thus a crucial question for the field to find observables which directly reflect the constituents of the stuff generally called the "strongly-coupled Quark Gluon Plasma" (sQGP) [38].

One of the more interesting concepts to emerge from the studies of elliptic flow of mass-identified particles is the role of "constituent quarks". If the sQGP is in fact made of composite objects, reflecting the quark degrees of freedom of the hadrons, then the flow of different final-state hadrons should reflect the recombination of those quarks into composite bound states [39]. This would give a v_2 and a p_T which both scale linearly with n_q, the number of quarks in each hadron (2 for mesons, 2 for baryons). Intriguingly, the plot of v_2/n_q vs. p_T/n_q is approximately the same for many measured species at RHIC energies. This is widely seen as evidence for constituent quark degrees of freedom, at least at the late stages of heavy ion collisions, but some ambiguities remain. For instance, models based on hadronic degrees of freedom can get similar systematic differences between mesons and baryons, which is attributed to the use of cross sections which are themselves linear with the number of constituent quarks (similar to the additive quark model) [40]. Thus, while constituent quarks may well prove to be important in understanding the dynamics of freezeout, it is not clear whether or not their status as bound or free objects can be determined based on existing data and theory.

5 The Limits of Hydrodynamics

From the RHIC data, one sees that hydrodynamics seems to have wide applicability. The initial collision parameters (nuclear size and beam energy) defines the initial state entropy and volume, as well as its shape. The longitudinal distributions are described approximately by Landau's initial conditions, implying a very short initial thermalization time ($\tau_0 << 1 fm/c$) and an enormous energy density. The transverse distributions develop on somewhat later time scales ($\tau_0 \sim 0.6 fm/c$) with high energy densities. At even later times, there even seems to be vestiges of non-hadronic degrees of freedom, in the constituent quark scaling. Given the usual assumptions that

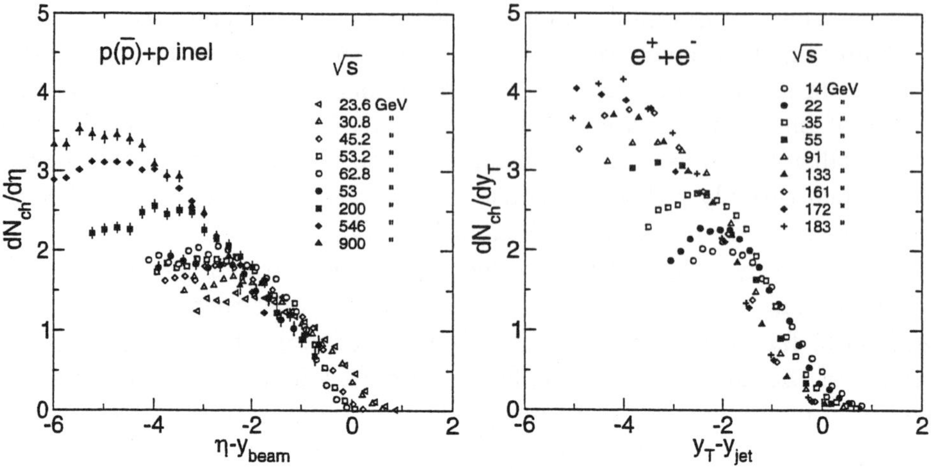

Figure 11: limfrag for $p + p$ and e^+e^-

strong interactions should be describable by pQCD cross sections, all of this would be "crazy" – if it didn't work to give quantitative descriptions of many features of the data.

Of course, all of this begs the question of the preconditions for hydrodynamic behavior. Is there a characteristic energy density or system size where the effects turn-off? Are there obvious differences in the data for elementary systems, such as e^+e^-, p+p and d+Au, which suggest a qualitative break with things seen in Au+Au? As seen previously, certainly the multiplicities per volume are quite similar (once one adjusts for the effective energy). The freezeout temperature extracted from hadronic yields is also essentially the same, at around $T \sim 160$ MeV. The same "limiting fragmentation" behavior, which was seen in Au+Au and connected (in one approach) with Landau's hydro model, is also seen, as shown in the left and right panels of Fig. 11.

Perhaps the coupling between the degrees of freedom liberated an all strongly-interacting multiparticle processes are so strong that thermalization occurs very rapidly, and so all of the above-mentioned systems proceed via a similar "perfect fluid" as seen at RHIC. The bulk features will be similar in these systems, but the details should differ, especially when the system size is already similar to the range of hadronic reactions (i.e. ~ 1 fm). This is not an original idea, as similar things have been suggested by several authors

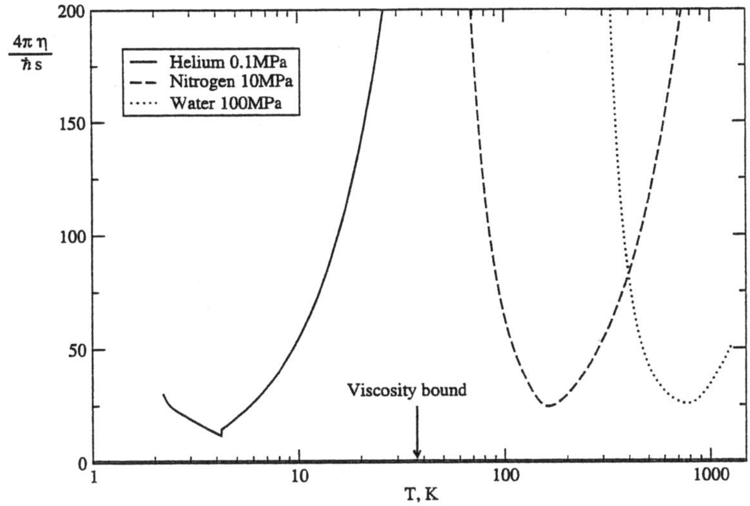

Figure 12: Viscosity

in the last 30 years [25], but it becomes more compelling with higher-quality data from RHIC. From this point of view, strong interaction phenomenology may not be so complicated despite the lack of a perturbative description. But how exactly does the system change from two colliding nuclei (or nucleons) to an exploding perfect fluid, and so fast?

Clearly the next stage of the RHIC program, both in experiment and theory, has to grapple with the problem of rapid thermalization and which degrees of freedom are able to achieve it. The concept of the "sQGP" at least focuses the discussion on strong coupling, but at present there is little well-motivated content to this picture. Some models suggest that colored bound states exist that have large cross sections but are not traditional hadrons[38]. These are intriguing proposals which have some success in modeling some aspects of lattice calculations, but strong counterarguments exist, based on the same lattice data [41]. In the end, the search for understanding the properties of the sQGP will push us to the frontiers of QCD itself.

One of these frontiers may already be accessible with string theory, via the AdS/CFT correspondence [42]. This idea suggests that strongly-coupled

Yang-Mills theories may be accessible through calculations done in a weakly-coupled regime of string theory. Such calculations find a mathematical connection between the quark-gluon liquid found at RHIC and 10+1D black holes. Careful study of the mathematics has yielded an interesting conjecture that the viscosity of the quantum field theory is bounded from below: $\eta/s > 1/4\pi$ [43]. All known liquids, even superfluid ^3He, are above this bound by factors of 10, as shown in Fig.12. RHIC collisions may well be one of the first man-made systems to saturate the viscosity bound, thus making one of the first experimental connections to string theory.

6 Acknowledgments

The author would like to thank the organizers for the invitation to speak about RHIC physics in the amazing environment of the Erice summer school. Thanks to Constantin Loizides for useful discussion on high-p_T physics. Thanks to Andrei Loginov and Annalisa Mastroserio for suffering through my rapid-fire slurs in transcribing my Q&A session. Finally, thanks to the editors of the Erice proceedings for their patience.

References

[1] M. Gell-Mann, Phys. Lett. **8**, 214 (1964).

[2] W. M. Yao *et al.* [Particle Data Group], J. Phys. G **33**, 1 (2006).

[3] R. Hagedorn, Nuovo Cim. Suppl. **3**, 147 (1965).

[4] D. J. Gross and F. Wilczek, Phys. Rev. D **8**, 3633 (1973).

[5] F. Karsch and E. Laermann, arXiv:hep-lat/0305025.

[6] J. L. Nagle, arXiv:nucl-th/0608070.

[7] I. Arsene *et al.*, Nucl. Phys. A **757**, 1 (2005).

[8] K. Adcox *et al.*, Nucl. Phys. A **757**, 184 (2005).

[9] B. B. Back *et al.*, Nucl. Phys. A **757**, 28 (2005).

[10] J. Adams *et al.*, Nucl. Phys. A **757**, 102 (2005)

[11] B. B. Back *et al.*, Phys. Rev. Lett. **91**, 052303 (2003).

[12] B. B. Back *et al.*, arXiv:nucl-ex/0610001.

[13] P. Braun-Munzinger, K. Redlich and J. Stachel, arXiv:nucl-th/0304013.

[14] J. Cleymans, H. Oeschler, K. Redlich and S. Wheaton, arXiv:hep-ph/0607164.

[15] J. Cleymans and K. Redlich, Phys. Rev. Lett. **81**, 5284 (1998).

[16] A. Tawfik, Europhys. Lett. **75**, 420 (2006).

[17] E. Fermi, Prog. Theor. Phys. **5**, 570 (1950).

[18] L. D. Landau, Izv. Akad. Nauk Ser. Fiz. **17**, 51 (1953).

[19] Y. L. Dokshitzer, V. A. Khoze and S. I. Troian, J. Phys. G **17**, 1585 (1991).

[20] B. B. Back *et al.*, Phys. Rev. C **74**, 021902 (2006).

[21] B. B. Back *et al.*, Phys. Rev. C **74**, 021901 (2006).

[22] J. Cleymans, M. Stankiewicz, P. Steinberg and S. Wheaton, arXiv:nucl-th/0506027.

[23] S. Z. Belenkij and L. D. Landau, Nuovo Cim. Suppl. **3S10**, 15 (1956) [Usp. Fiz. Nauk **56**, 309 (1955)].

[24] F. Cooper, G. Frye and E. Schonberg, Phys. Rev. Lett. **32**, 862 (1974).

[25] P. Carruthers, Annals N.Y.Acad.Sci. 229, 91 (1974).

[26] R. P. Feynman, Phys. Rev. Lett. **23**, 1415 (1969).

[27] J. D. Bjorken, Phys. Rev. D **27**, 140 (1983).

[28] I. G. Bearden *et al.*, Phys. Rev. Lett. **94**, 162301 (2005).

[29] G. Boyd, *et al*, Nucl. Phys. B **469**, 419 (1996).

[30] D. Teaney, Phys. Rev. C **68**, 034913 (2003).

[31] B. B. Back *et al.*, Phys. Rev. C **72**, 051901 (2005).

[32] P. F. Kolb and U. Heinz, arXiv:nucl-th/0305084.

[33] D. Molnar, arXiv:nucl-th/0406066.

[34] M. Bleicher and H. Stoecker, Phys. Lett. B **526**, 309 (2002).

[35] M. Basile *et al.*, Phys. Lett. B **92**, 367 (1980).

[36] G. J. Alner *et al.*, Z. Phys. C **33**, 1 (1986).

[37] J. Adams *et al.*, Phys. Rev. Lett. **93**, 012301 (2004).

[38] E. Shuryak, Nucl. Phys. A **774**, 387 (2006).

[39] R. J. Fries *et al*, Phys. Rev. Lett. **90**, 202303 (2003).

[40] Y. Lu *et al.*, J. Phys. G **32**, 1121 (2006).

[41] V. Koch, A. Majumder and J. Randrup, Phys. Rev. Lett. **95**, 182301 (2005).

[42] J. M. Maldacena, Adv. Theor. Math. Phys. **2**, 231 (1998).

[43] P. Kovtun, D. T. Son and A. O. Starinets, Phys. Rev. Lett. **94**, 111601 (2005).

CHAIRMAN: P. STEINBERG

Scientific Secretaries: A.Loginov, A.Mastroserio

DISCUSSION

- *Preghenella:*

You speak about the total charged particle multiplicity scaling with the center of mass energy. Do you think this could be a universal feature of the high-energy collision?

- *Steinberg:*

I tried to show that at high energy, at least where we have checked, it doesn't seem to depend on the system, which I still find to be a strange, but apparently true, fact (that is if I understood your question correctly). You can actually have DIS events also fit them reasonably well, if you look at multiplicity from ZEUS.

Whether or not this would break down is an interesting question. Some people argue that this agreement is actually a complete accident, that two trends having different directions in this plot happened to be crossing in this region. What is nice is that, unfortunately except for the e+e- part, the proton-proton and heavy ion data can and will definitely be checked at the LHC.

- *Preghenella:*

What's the possibility to predict the charge multiplicity at LHC energies based on this feature?

- *Steinberg:*

That's quite interesting. People have asked me that question and I used to think that I could answer it. Until when you really get down to it, you realize that you have a theoretical curve that both fits all the data, but which needs to be reliably extrapolated over a factor of 30. And just think about the factor of 30. When in the old days people said that it would be really extrapolate from 45 GeV to the Z mass with the pQCD formula, that's only a factor of 2.

You can see that almost any curve you can put on will fit, but that it's not that easy to constrain the extrapolation. I can tell you that two different predictions which I've done which is using this formula turns out to be about 160 from a hydro-inspired model, and 100 from the e+e- formula, which is a 60% difference. So different ways of extrapolating will give you reasonably good consistency with existing data but give very different numbers at the LHC. So it is not as easy as it sounds.

You also have to believe that the end of the day even QCD for e+e- just gives you an evolution, but the evolution variable is the energy, which must be divided it by a scale, called Λ_{QCD}. That's not a well-known number in general, and it's in principle not known theoretically, i.e. it's actually an empirically-determined number you have to plug into the formula. So, several things work against you, but in principle if you could control the systematics, then yes, you could use these ideas to try to constrain the physics at the LHC. And so yes, I guess, within

60% I can do it right now, but I don't think we can do it much better until the accelerator turns on.

- *Preghenella:*

OK, thank you.

- *Steinberg:*

It's always depressing: When I first started working with this, I actually thought I had a lot of control over the physics, because of pQCD, but it's not as straightforward as it looks.

- *Preghenella:*

Another question: may be also with the charge distribution in pseudorapidity distribution you show that also in this distribution there are universal features. What about the pseudorapidity distribution of the charged particles for different energies?

- *Steinberg:*

In these plots you see that it rises up the universal curve and then peels off. And so the question is what slope do you assign at the point of peeling-off in order to extrapolate it to higher energies. It will give you different slopes when you try to extrapolate it depending on the functional form you assume. And so you have to accept that this is presently an empirical observation and it's encouraging that it works, as well as it does and it seems to tell you a basic feature of particle production, but if you want to use this to extrapolate, then you have to really trust the slope you extract. Thus, I believe you can get a lower limit with this approach, but it is not clear how to get an upper limit.

- *Cifarelli:*

A comment: The extrapolation of the average charged particle multiplicity from RHIC to LHC energy is so far model dependent and affected by large uncertainty factors. May be the best solution would be to use the best fit of the e+e- vs. sqrt(s) data instead of models, because this is experimental. So you say let's be blind, and suppose nothing new happens, then the universality of interactions could tell you the true value. That's one possibility, because the extrapolations are model-dependent so far.

And the other thing is you have some blue points at low energy, and this morning you said let me explain to you, why they are below. They are simply not treated correctly.

- *Steinberg:*

To address the second point, it's actually possible to model this physics with concepts from chemistry. Remember that energy is when your sqrt(s) is about 20 GeV or below, and it gets very extreme if you go down to the sqrt(s)=3-4 at the AGS. Remember that the baryons you are colliding, which contribute to the total energy, have masses of about 1 GeV each, and these baryons have to be conserved in the final state. So at the AGS, when you make a collision, a lot of pions appear in association with a lot of baryons that are present in the final state, and one must account for that mass.

The way I tried to play this game was I wrote down the Gibbs potential, which is G = E + PV - TS (not the total entropy coming into that formula) and it's equal to the chemical potential times the number of baryons ($\mu_B N_B$). You can rearrange this such you get an enthalpy term, which you can treat as the baryon-free component of the total entropy. (I know this sounds a little bit strange, if you are not used to it!) But then you see that there is a μ_B term, which comes in with a negative sign. So as you increase μ_B it seems to suppress the total entropy by an amount which goes as μ_B/T, which is known quite well from thermal fits. So, you can actually try to fit the data by assuming that the suppression of the multiplicity scales as μ_B/T.

So this is almost like energy conservation at some level. You have the baryon mass that will take away some energy, which could have gone to pions.

You can implement this in two ways. This plot shows the charged multiplicity (divided by N_{part} for the A+A case) scaled down by the Landau formula ($2.2s^{1/4}$). In one way, you can correct the multiplicity using thermal parameters, which takes you from the open circles (the A+A data) to these closed circles. This is not a perfect description, but it definitely pushes up the A+A data to the overall level of the e+e- data. In the other way, you can also look at the entropy density as a function of μ_B in thermal models. They qualitatively do the same thing.

Of course, all of this is thermal phenomenology. And yet, everything we've seen from RHIC data says that, for very soft particles, thermal phenomenology seems to have a lot of power, because it really explains the various particle ratios. And so to some level, the argument I am making here (which is actually in a paper which I've submitted with some colleagues from South Africa) is that you can actually use this kind of phenomenology to understand why at low energies heavy-ions data have lower multiplicity than e+e- and pp.

- *Cifarelli:*

This is related to the baryon density, so again it is related to a kind of leading effect. Actually if you are not able to suppress this effect or to take it into account, then your definition of energy is not correct and therefore the points are out of scale. I mean, in proton-proton interactions, which is the simplest situation, if you don't do this, you don't sit on the same curve. So my impression is that it is more complicated of course because you are in heavy-ion collisions here, but nevertheless it is still related to the baryons, and so to the leading effects.

- *Steinberg:*

It is and it's not. The only way you get agreement with heavy ions and e+e- without additional scaling of the effective energy is by presuming that all of the energy is available, such that there is no leading effect in heavy-ion collisions.

If there is no leading effect, then all these baryons have been pushed in to the particle production region and the μ_B actually accounts for those baryons. At high energies ($\mathrm{sqrt(s)} >> m_p$) you are dominated by pions, and so few extra baryons don't have a dramatic effect on the overall particle production.

At low energy all of those stopped baryons, which deposit all the energy, have now essentially reduced the entropy density because they take the energy with their mass. The baryons are heavy and essentially that fact alone at low energies, when the baryon mass itself is comparable to the center of mass energy, will suppress the number of particles you are allowed to produce. So it's kind of trivial explanation, but it tells a consistent story, although then you have to believe that the leading effect is gone. Of course, there are different opinions about whether or not the leading effect persists in heavy ion collisions.

- *Zichichi:*

I don't know why you introduce chemistry. You should say "electromagnetism" in your previous graph. Chemistry is electromagnetism.

- *Steinberg:*

No, this is hadronic chemistry, and not real "chemistry".

- *Zichichi:*

Why do you call it chemistry?

- *Steinberg:*

As far as I know it's a correct word. It has to do with the partition of the energy among the various particle species. The baryon number is a conserved quantity present in the initial state and therefore the final state. And it affects the particle yields, so therefore can be considered as "hadro-chemistry".

- *Zichichi:*

Entropy is a sort of multiple interactions of the electromagnetic origin between particles.

- *Steinberg:*

In these collisions the strong interaction is at much shorter time scale than electromagnetic interactions, which are not active.

- *Zichichi:*

That's why I think you should not use this terminology.

- *Steinberg:*

This is why I use the term hadrochemistry. But you are right that it is not chemistry in a way we normally understand it.

- *Zichichi:*

You know, our grandfathers would be very happy and they indeed emphasized that chemistry was finally understood by physicists. Now, from your presentation it seems that we are lending something from chemistry. It is not correct: we are learning nothing from chemistry.

- *Steinberg:*

Again, if you are talking about real chemistry, you are right, we learn nothing. But you can think about techniques from chemistry which high energy physicists are not used to thinking about, and that's what I am trying to propose.

- *Zichichi:*

You should cancel chemistry.

- *Steinberg:*

OK, I will think about that.

- *Zichichi:*

This graph. It will be perfect if you give me the data and put the correct energy there. I think it should be pushed all on the right. All the points you have there, between 1 and 10 should be above hundred billion.

- *Steinberg:*

That's the thing, the original energy can be right but we must consider the fact that the baryons have a large mass.

- *Zichichi:*

The energy is not right.

- *Steinberg:*

To be honest, I can also correct for the mass of the protons themselves, but then I also should correct for the charge. At low energy if you correct for the mass of the protons and for the charge itself, this trend actually can remain the same. It is kind of amusing that those corrections cancel. So you are right, you could shift it, but this correction is not enough to move it around in the way I described previously.

- *Zichichi:*

Not because of the mass of the proton, because the effective energy.

- *Steinberg:*

The point I was trying to make is that I can arrive at an efficient explanation for the data, without an arbitrary change of the effective energy, which I can't explain.

- *Zichichi:*

We have shown without any doubt that e+e-, proton-proton, proton-antiproton, K-proton and so on, no matter what energy range they have, are absolute equivalent. There is no difference if you use the correct effective energy.

- *Steinberg:*

That's right.

- *Zichichi:*

These points must be wrong

- *Steinberg:*

Perhaps those systems don't have a net baryon density that affects the production of the other particles.

- *Zichichi:*

Something must be wrong there, it cannot be true!

- *Steinberg:*

What I am trying to propose is that there is a qualitative difference between heavy ions and p+p in this respect. In p+p there is a leading effect.

- *Zichichi:*

I'd like to convince you that you should forget about leading effect. The effective energy is the real world. You have a quark-quark interactions, in the subnuclear world, what counts is the effective energy.

- *Steinberg:*

I think I understand our difference of approach. I took the opposite conclusion that the leading effect controls the effective energy.

- *Zichichi:*

No, I am right, you are wrong. The leading effect is a consequence of the effective energy, which is the dominating quantity in QCD.

- *Steinberg:*

I interpret it that the incoming nucleons are stopped, and so deposit energy that leads to the effective energy as a consequence of that process, and I think that's the difference between us. I actually see that because in p+p collisions they collide once and, then, somehow there is a flat x distribution observed for the protons. In heavy ions there is no flat x distributions observed, and so it is very different from p+p distributions.

- *Zichichi:*

We've shown the universality features in subnuclear physics and you can not deny this. This is confusion.

- *Cifarelli:*

Could you show a pseudorapidity of this data? I think you have shown something this morning. Pseudorapidity distribution, this is instructive, I think, to remind us.

- *Steinberg:*

I have rapidity distributions at very low energies, and I only have pseudorapidity down to RHIC energy, down to 20 GeV. This is the rapidity density of pions, this is a top RHIC energy, these are for lower energies, down to AGS energies. If you want to see pseudorapidity at low energy, I don't have those available.

- *Zichichi:*

But these quantities of rapidity and pseudorapidity are tricks to avoid the understanding the basic problem. They mean nothing. The effective energy is the fundamental physics parameter.

- *Steinberg:*

What is amazing is that I think we agree on almost everything, but I think we have some subtle differences in our points of view. And I have exactly the same problems with other people I have argued with about this. I am interested to hear your point of view why it's fundamental, because I can't convince anybody that this physics issue is fundamental.

But again, if one sees the effective energy as a consequence of how the baryons stop, than it's a different story than yours. In that case, maybe my story becomes a bit more sensible.

- *Pinfold:*

The jet survival probability is a key probe of the Quark-Gluon Plasma. Given the complexity of the final state in the heavy ion interactions, how do you determine the jet survival probability?

- *Steinberg:*

It's an interesting question. At this stage of the game, nobody at RHIC reconstructs jets. What we rely on is the fact that we believe that perturbative QCD (pQCD) works. And we can calculate the spectrum of pions in p+p collisions from pQCD by relying on factorization formulas. So we can actually plug in structure functions, QCD cross sections and fragmentation functions. Now, from the fragmentation function, you know that, on average, some of the particles in the jet will be leading. When you go high momentum particles, those are typically biased towards being the leading particles from a high momentum jets. And so the idea is that when you look at some of these suppression variables you're essentially looking at the degradation of the leading particle, which then reflects an effective rescaling of the jet energy. If you believe QCD factorization should work in heavy-ion collisions, then each binary scattering should actually give you the chance of invoking a hard scattering, which might lead to jet production.

So if you look at the yields in heavy ion collisions, divided by the same yields from p+p, which is understandable perturbatively, and then you divide them by the number of binary collisions, this gives you the sense of the nuclear effect at high p_T. You wouldn't expect this to make sense for the soft particles, actually it scales like the number of participants, but if you look at high p_T, the data shows a large suppression. What is amazing is that when you look at the suppression factor, the things, which should factorize, like direct photons, do scale with the number of the binary collisions. I didn't think it would work as nicely as it does! But then if you look at the leading pion distributions, we see that the suppression factor is flat at very high p_T.

And so we are not reconstructing jet structure in any of these, but we are seeing what several perturbative calculations suggest, that the leading fragments of jets are suppressed by a simple constant value by the presence of a medium.

- *Zichichi:*

These values of p_T's, all these data, they do refer to the same primary energy or not?

- *Steinberg:*

This is totally inclusive, these are is inclusive spectra divided by each other, we can not isolate the primary energy of the jet. We don't know anything about the jet distribution, that's the trouble with all of this, because in heavy-ion collisions all the soft particles from the jet will be swamped out by the rest of the collision.

Two-particle correlations are also interesting. These are two particle measurements where you measure the trigger particle, which is high momentum, and you look at its correlation in azimuthal angle with an associated particle, which in this case is at a slightly lower momentum, but still above 2 GeV. What you get is a sort of correlation peak on the near side from the fact that high momentum particles like to be produced together. On the away-side in p+p you have a spread out peak, which is also seen in deuteron-gold collisions, when you essentially embed the collision in the environment of a single nucleus. But when you embed both of them in a nucleus-nucleus collision, then you actually get complete washing out of that correlation. This is considered to be the suppression of the "back-to-back" correlations.

And the really amazing part is that the reason why we think the jets have been suppressed, when you lower the cutoff, is because you are not just looking at semi-hard particles, above 2 GeV. When you lower the cutoff to a few MeV, you find a recoil from the near side jet, spread out in a huge region in an angular space. So in some sense you are always capturing the energy-momentum released by that jet, but it's completely spread out. This is a really hot topic these days, because it suggests that it is not simply just a recoil away from it, but actually things are happening to the energy as it is propagating through the nucleus (e.g. Mach cones).

- Faivre:

On the v^2/n versus p_T/n plot, is the polynomial fit motivated by a model or something?

- Steinberg:

It's motivated by having a line fitting the data. The last I remember, they just wanted something where they could compare the magnitudes of these various quantities relative to each other. Thus, there is no physical meaning of the curve itself. Nobody really knows how the primordial constituent quark dynamics work; even we have evidence that they are the important degrees of freedom.

- Csanad:

I have a comment about hydrodynamics. It describes the hadronic final state. It is a combination of initial state and equation of state (EoS). You have to fix EoS to access initial state. And penetrating probes (with photons) could help.

It's important to keep in mind that there are exact results with the same hadronic final state with different initial conditions and different equations of state, so hydro will never explore the initial state, unless we fix things somehow.

- Steinberg:

You can not play the movie backwards in a unique way, unless you have a prior knowledge of either of those two (EoS and initial conditions). I agree that hydro is really a consistency framework, not a really predictive framework. This is a very useful comment.

Highlights from CERN: The CLIC Project for a Future e^+e^- Linear Collider

Frank Tecker
CERN, Switzerland
E-mail: `Frank.Tecker@cern.ch`

Abstract

A high luminosity (10^{34}–10^{35} cm^2/s) linear electron-positron Collider (CLIC) with a nominal centre-of-mass energy of 3 TeV is under study in the framework of an international collaboration of laboratories and institutes, with the aim to provide the HEP community with a new facility for the post-LHC era.

After a brief introduction of the physics motivation, the CLIC scheme to extend Linear Colliders into the Multi-TeV colliding beam energy range will be described. In the following, the main challenges and the very promising achievements already obtained will be presented.

1 Introduction

In the history of particle accelerators, a fruitful interplay has always existed between hadron and lepton colliders. With constantly increasing energy, hadron machines have been typically at the energy frontier, while lepton machines have offered the possibility for precision physics experiments.

Hadron collisions are more difficult to analyze than lepton collisions due to the composite nature of the hadrons. Leptons are elementary particles with a well defined initial state in the collision. The different particles produced in the collision are generated democratically and the momentum conservation eases the decay product analysis.

Presently, a world-wide consensus exists that a linear e^+e^- collider (LC) with a centre-of-mass energy above 500 GeV is very interesting to complement LHC physics. Just as e^+e^- colliders provided an essential complement to hadron–hadron colliders in the 100 GeV energy range, establishing beyond doubt the validity of the Standard Model, so one expects that higher-energy e^+e^- colliders will be needed to help unravel the TeV physics. In the TeV scale, a lot of new territory is to be discovered beyond the Standard Model. The Higgs — or something to replace it — should be discovered by the Tevatron or LHC but an e^+e^- machine would explore its properties in detail. Many theories beyond the Standard Model, such as those postulating supersymmetry, extra dimensions or new strong interactions, predict the appearance of non-trivial new dynamics at the TeV scale. CLIC is the — presently unique — accelerator study that can reach the very high energy for probing the new physics up to the multi-TeV range.

For the study of this multi-TeV physics, a *CLIC Physics Study Group* [1] has been set up. Its detailed study of the physics potential of CLIC, including a close integration of experiments with the accelerator, can be found in [2]. In the following, this article will entirely concentrate on the description of the accelerator technology that will permit to access this energy range.

Two different linear collider projects are presently under study, the ILC (International Linear Collider) [3] and CLIC (Compact Linear Collider) [4]. The ILC is a large international project for a 500–1000 GeV centre-of-mass energy collider based on superconducting RF technology operating at an RF frequency of 1.3 GHz. It aims initially at an accelerating gradient of 31.5 MV/m and an energy of 500 GeV. With a later upgrade in gradient and extension of the linac, an energy reach of 1 TeV is foreseen. While the gradient for superconducting technology is fundamentally limited by the critical magnetic field in the accelerating structures, normal conducting technology can reach higher fields, higher gradients and thus a higher particle energy. CLIC is based on normal conducting accelerating structures. Yet many challenges have to be overcome to reach and operate stably at these very high fields.

2 The CLIC scheme and challenges

The CLIC design [5] is optimised for 3 TeV but can be built in stages. To limit its total length, CLIC uses normal conducting accelerating structures operating at a very high gradient (150 MV/m), and powered by high power (150 MW) 30 GHz RF pulses. Conventional RF sources do not cover this range and CLIC relies on a two-beam-acceleration concept [6]: A high current electron beam (drive beam) runs parallel to the main beams and is decelerated to produce the RF power. The 30 GHz RF power is extracted from the high current drive beam in special power extraction transfer structures (PETS) and transfered to the main beam (see Fig. 1). This leads to a very simple tunnel layout without any active RF components (i.e. klystrons). Both beams can be generated in a central injector complex and are transported along the linac. A sketch of the CLIC layout with the injector complex is shown in Fig. 2. An overview of important parameters of CLIC at 3 TeV is presented in Table 1.

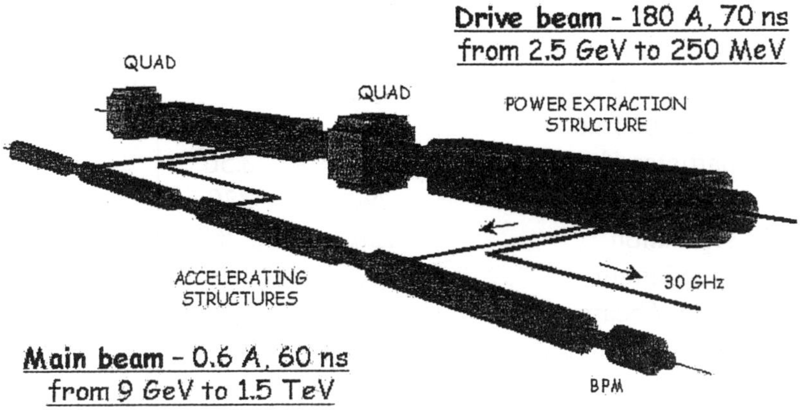

Figure 1: CLIC tunnel layout.

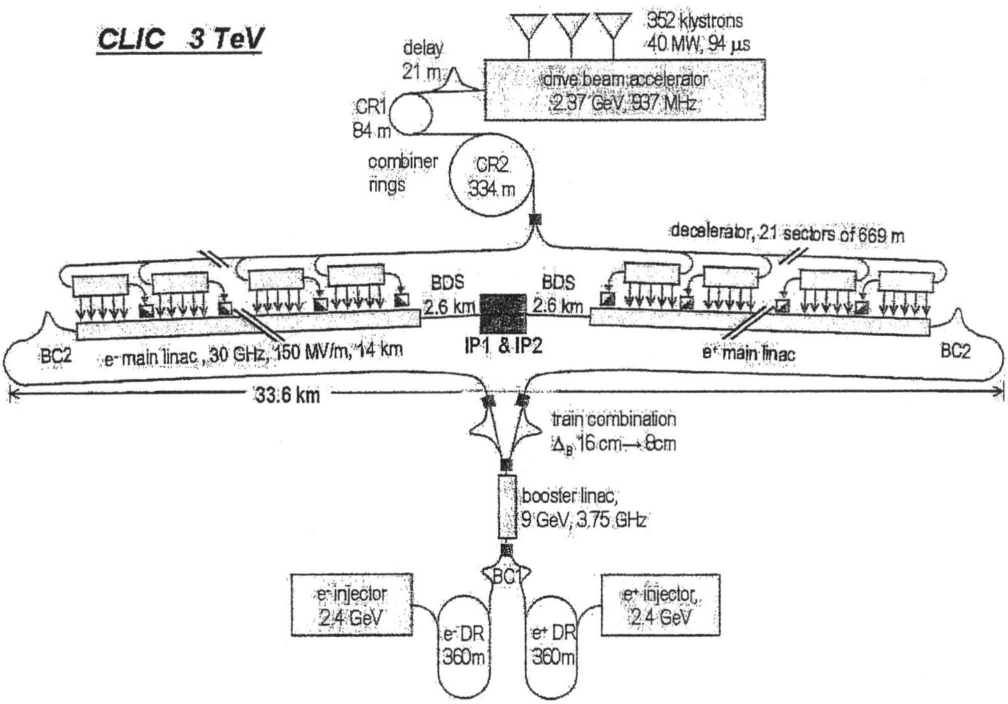

Figure 2: CLIC layout with the central main beam and drive beam production complexes.

Parameter	*Symbol*	**Value**	*Unit*
Overall Parameters			
Centre of mass energy	E_{CMS}	3000	GeV
Main Linac RF Frequency	f_{rf}	30	GHz
Luminosity	L	6.5	$10^{34}\text{cm}^{-2}\text{s}^{-1}$
Luminosity (in 1% of energy)	$\text{L}_{99\%}$	3.3	$10^{34}\text{cm}^{-2}\text{s}^{-1}$
Linac repetition rate	f_{rep}	150	Hz
No. of particles / bunch	N	2.56	10^9
No. of bunches / pulse	N_b	220	
Bunch separation	Δt_b	0.267	ns
Bunch train length	τ_{train}	58.4	ns
Beam power / beam	P_b	20.3	MW
Unloaded / loaded gradient	$G_{\text{unl/l}}$	172 / 150	MV/m
Overall two linac length	l_{linac}	28	km
Total beam delivery length	l_{BD}	2 x 2.6	km
Proposed site length	l_{tot}	33.2	km
Total site AC power	P_{tot}	418	MW
Wall plug to main beam power efficiency	η_{tot}	9.7	%

Table 1: CLIC parameters at 3 TeV centre-of-mass energy as of mid 2005.

2.1 Drive beam generation

The generation of high-intensity drive beam pulses with the right time structure is one of the main challenges in CLIC. A long pulse is accelerated in low frequency (937 MHz) normal-conducting accelerating cavities, operated in the "full beam loading" condition where the beam extracts more than 95% of the RF power. The electron bunches are then interleaved in stages by transverse RF deflectors in delay lines and rings to give the beam the desired bunch structure while increasing its intensity. A layout of the CLIC RF power source for the drive beam generation is shown in Fig. 3. By this bunch frequency multiplication, the bunch spacing is reduced from 64 cm to 2 cm, and the current increased accordingly from 5.7 A to 180 A.

The first stage of the interleaving is the so-called delay loop. Half of the bunches are deflected into the loop and combined with the later bunches in the initial bunch train. This doubles the bunch repetition frequency and the current. The principle is shown in Fig. 4. The initial bunch train has a time structure with sub-pulses that have a 180° RF phase difference with respect to each other. Every second sub-pulse is deflected into the delay loop by the RF deflector. Since the

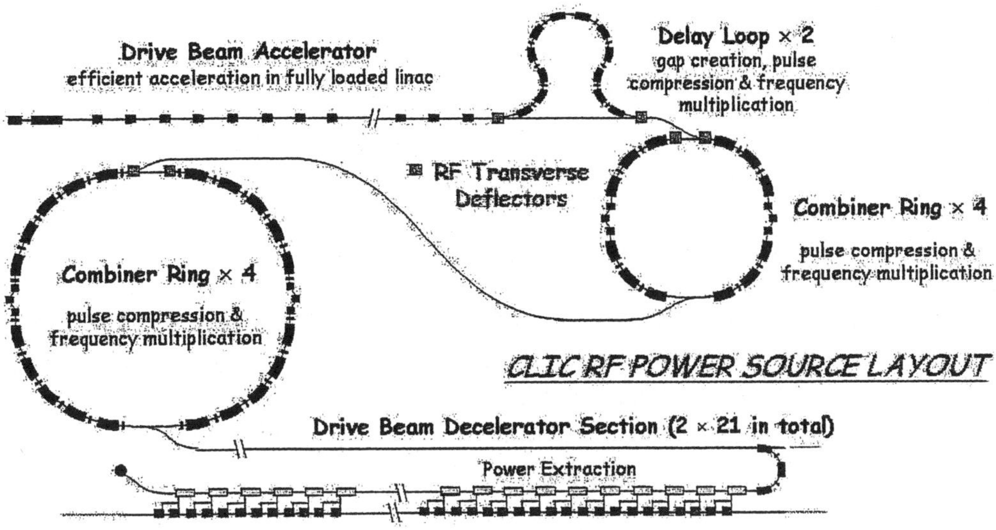

Figure 3: Schematic view of the CLIC drive beam generation.

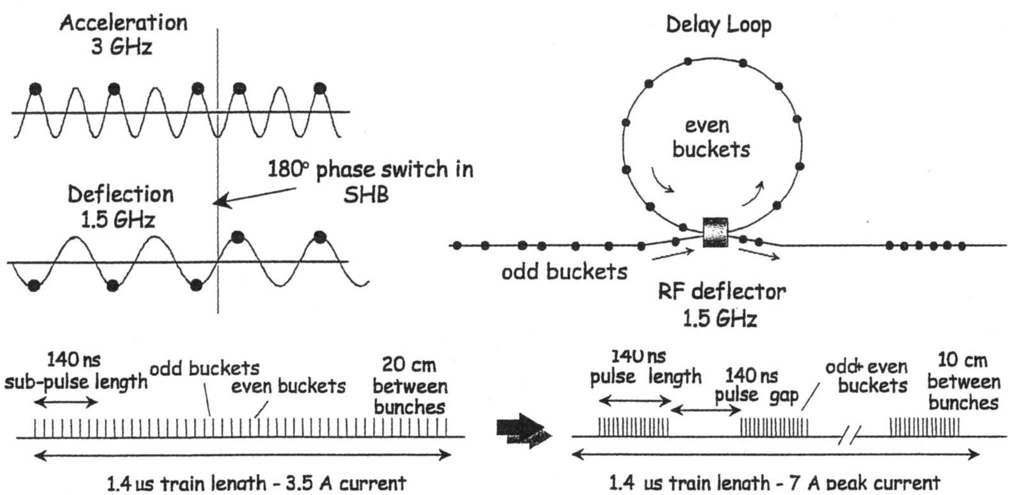

Figure 4: Principle of the bunch train combination by the delay loop.

length of the sub-pulses corresponds to the length of the delay loop, each sub-pulse is combined with the following, resulting in a time structure of pulses with twice the intensity and gaps of the same duration in between.

Higher combination factors up to 5 are reachable in a ring. Two RF deflectors form a time dependent closed orbit deformation and allow interleaving of the bunches as indicated in Fig. 5. The technical feasibility of increasing the drive beam current by this bunch interleaving was successfully demonstrated in 2001 and 2002 at low current (0.3 A) during the so-called preliminary phase of CTF3, a modification of an already existing storage ring [7] (see Figure 6). Up to five bunch trains were combined over five turns in a combiner ring to reach 1.5 A, without measurable losses. This combination will have to be repeated at a significantly higher current (nominal 35 A) in the present test facility CTF3.

The other important feature of the drive beam generation is the efficiency. The drive beam accelerator RF structures are designed for a very strong beam-loading. The RF power of the travelling wave is absorbed by the beam such that the field at the exit of the structure is almost zero for the nominal beam current (see Fig. 7). In this way, the RF power is completely transferred to the beam and – including ohmic losses – an overall RF-to-beam efficiency in excess of 95% can be obtained, larger than what is possible for superconducting structures, taking into account the cryogenic power. The price to pay for the high efficiency is an accelerating gradient that is about half of the unloaded gradient. This operation mode is working in routine operation of the CTF3 test facility (see below).

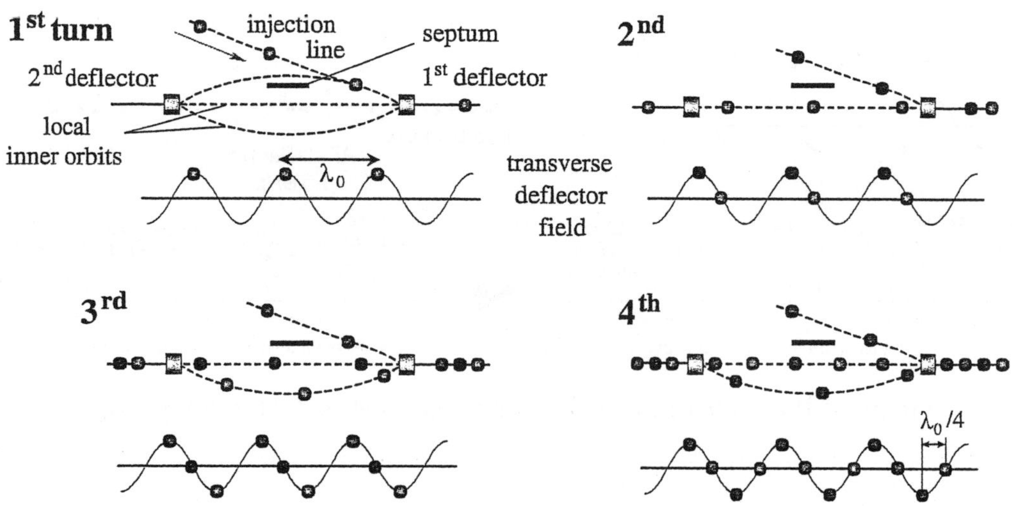

Figure 5: Principle of the bunch train combination in the combiner ring.

a)
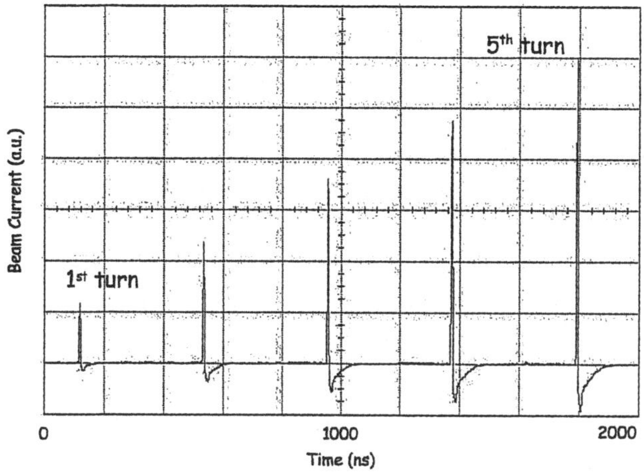

b)
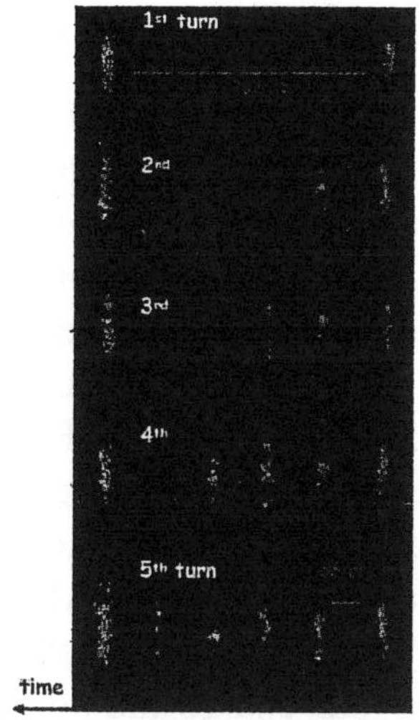

Figure 6: a) Oscilloscope trace from an intensity monitor in the CTF3 preliminary phase combiner ring, showing the circulating beam current for a beam combination test of factor five. The beam current increases over four turns from 0.3 A to 1.5 A. b) Streak Camera image of synchrotron radiation emitted in the ring, showing the interleaving of the bunches in the successive turns.

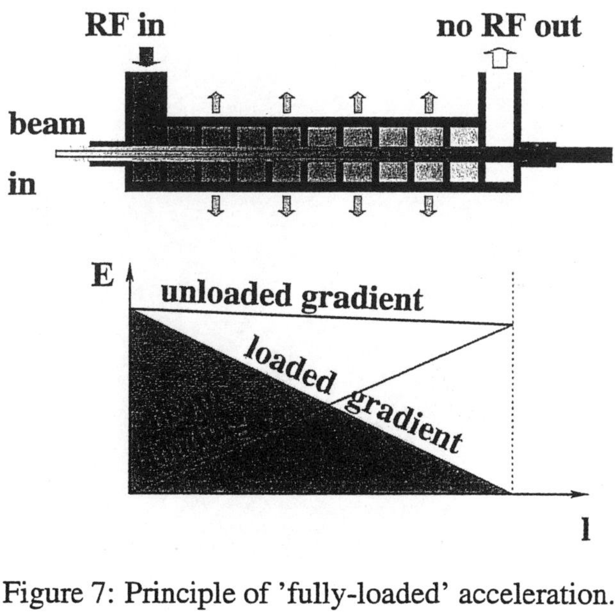

Figure 7: Principle of 'fully-loaded' acceleration.

2.2 Accelerating structures

Another challenge for CLIC are the accelerating structures capable of producing the required gradient, minimizing at the same time the wakefield effects that are a concern due to the small dimensions at 30 GHz. These structures are specially designed with a strong damping of higher order modes in order to efficiently damp the long-range wakefields so that later bunches are not perturbed.

Many experimental results have been obtained in the past years in previous test facilities (CTF, CTF 2), which have guided and confirmed the technical choices undertaken. The feasibility of two-beam acceleration was already demonstrated in the CLIC Test Facility CTF2 (see Fig. 8 and 9). There, an electron bunch was accelerated by 60 MeV using a string of 30 GHz structures powered by a drive beam. Such drive beam, produced directly from a photocathode gun, had however a total charge and pulse length much lower than the nominal CLIC values.

CTF2 has also been essential as a test bed for CLIC components. Its last two years of operation were dedicated to testing of accelerating structures after observing damage to their copper irises at high field. New structure designs, aimed

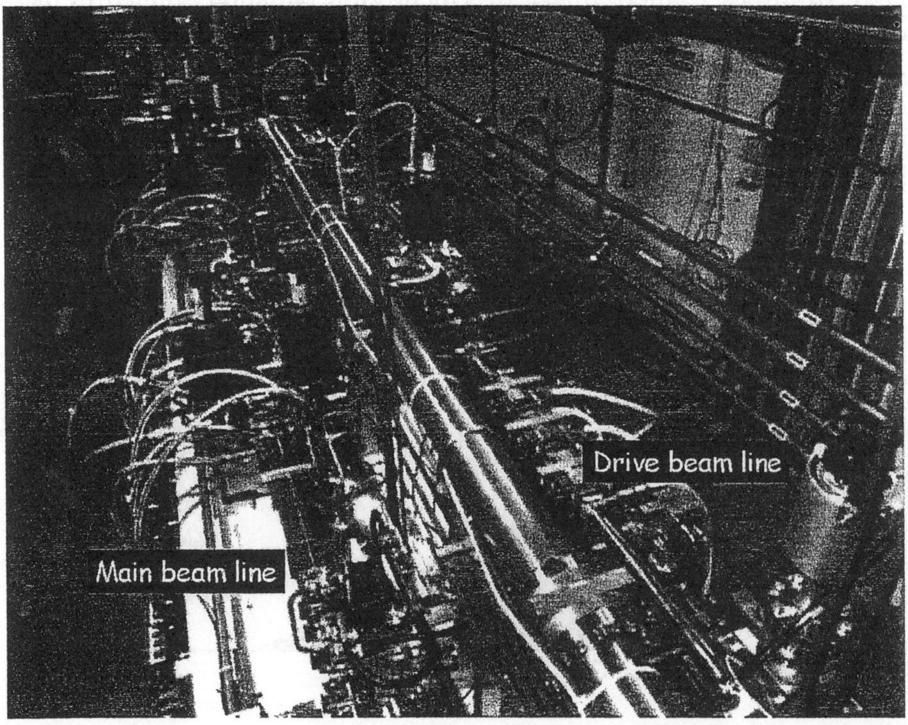

Figure 8: The CTF2 test facility. The high current drive beam is on the right and the 'main beam' on the left.

at reducing the peak surface field for a given accelerating gradient, were explored. A big improvement in performance was obtained by employing irises made of arc-resistant materials like tungsten or molybdenum (see Figure 10).

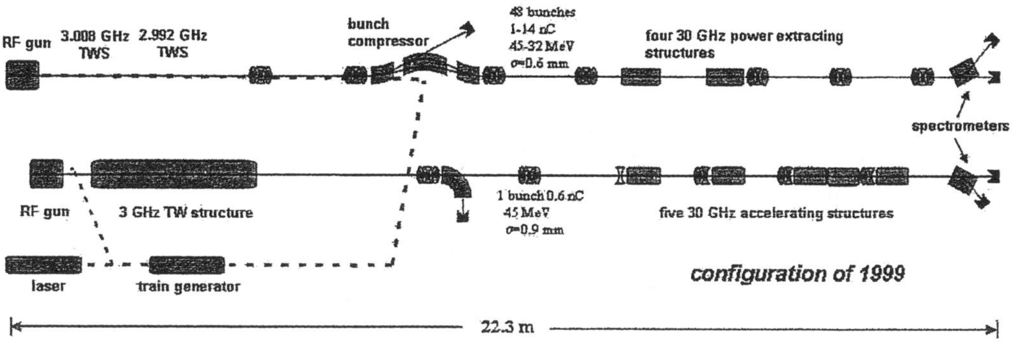

Figure 9: Schematic layout of CTF2. The high current drive beam is on the top and the 'main beam' on the bottom.

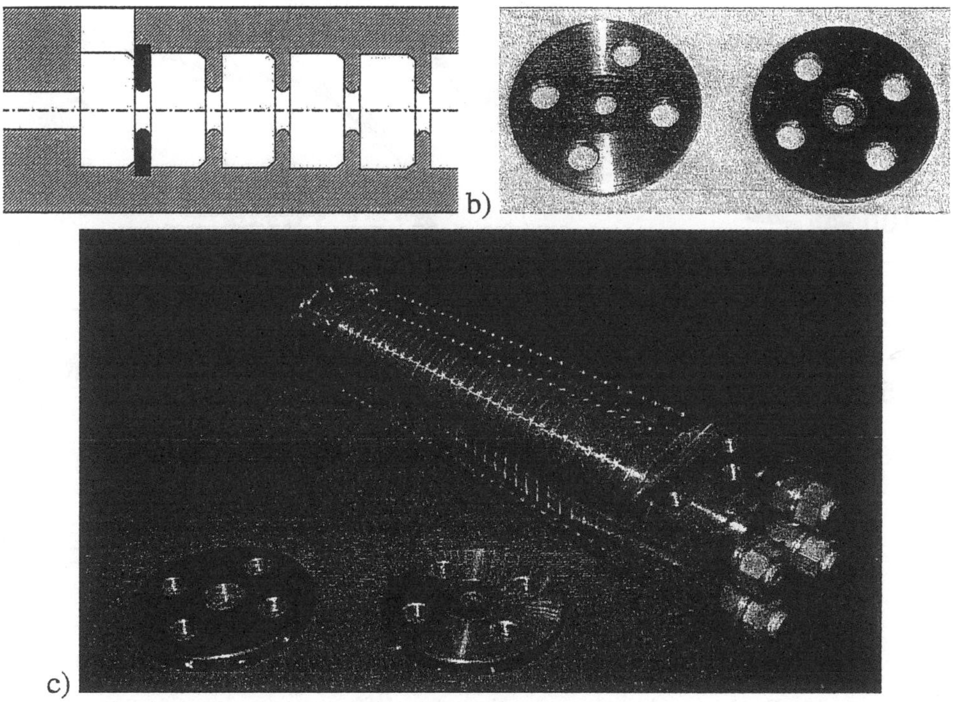

a) b) c)

Figure 10: a) Accelerating structure cross section showing the iris inlay. b) Disks with different iris material. c) Assembled structure.

Eventually, an accelerating gradient of 190 MV/m — well above the nominal CLIC design of 150 MV/m — was reached using molybdenum irises [10] (see Figure 11). The RF pulse length was however limited by the drive beam injector to 16 ns, well below the CLIC nominal value (70 ns). The CTF3 test facility is able to produce longer RF pulses to explore the behaviour at longer pulse length.

A new structure type was designed based on RF constraints and was extensively optimized in terms of luminosity per "wall plug" power. The new type is made of four quadrants that are clamped together instead of a large number of discs brazed, as for previous structures (see Fig. 12). Accordingly, it is much easier to assemble and it is cheaper in production.

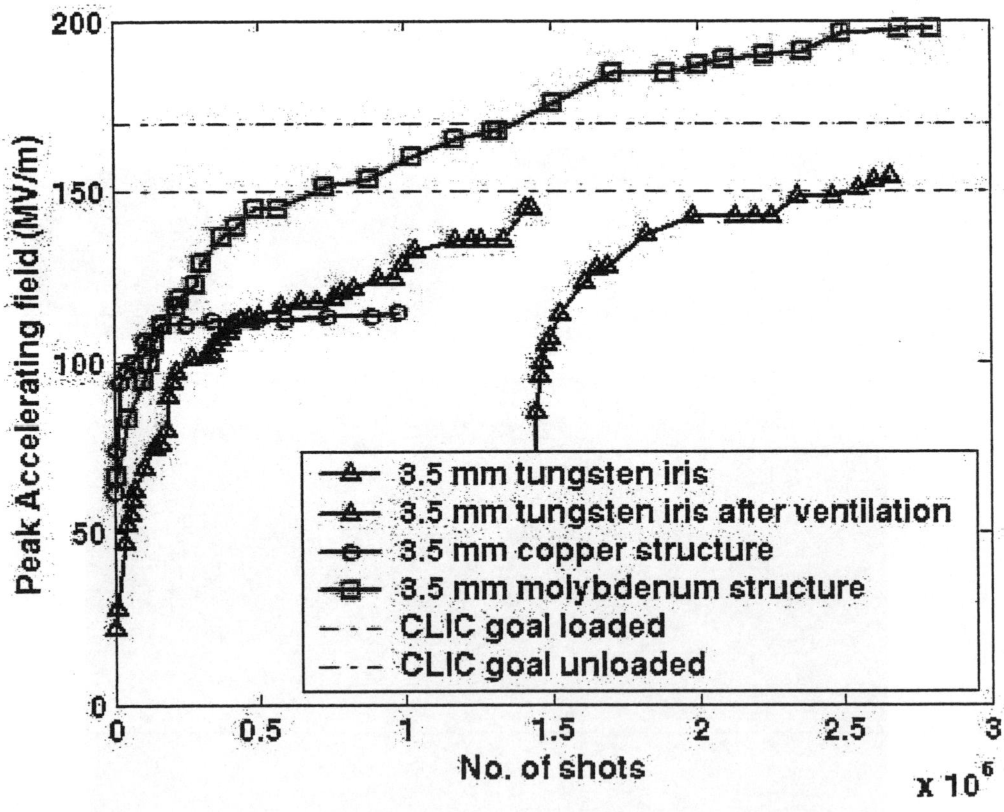

Figure 11: Peak accelerating gradient for different accelerating structures.

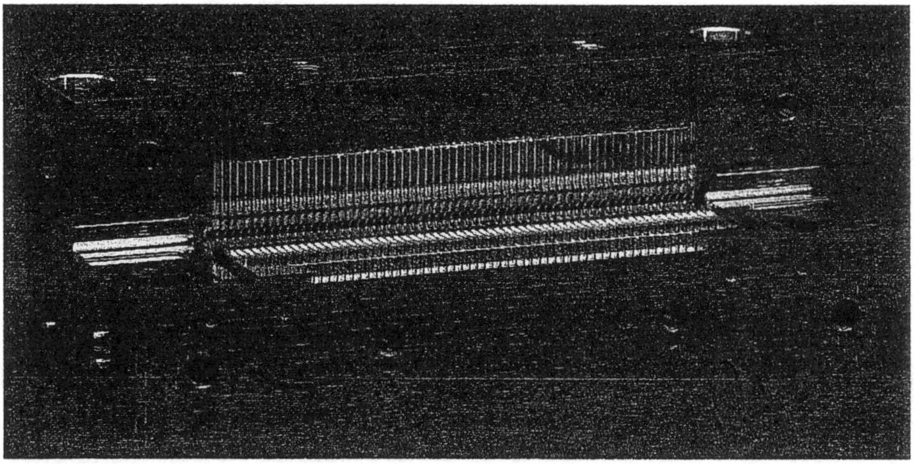

Figure 12: New type of accelerating structure built from four quadrants (of which three are shown here).

3 Power Extraction Structure (PETS)

The 30 GHz RF power required for the accelerating structures is extracted from the drive beam by the Power Extraction Structures (PETS). These structures have shallow sine-type corrugations (see Fig. 13) and the fields induced by the high current drive beam build up resonantly at 30 GHz as the beam travels along this structure. Each structure of 0.6 m active length will extract 640 MW RF power

a) 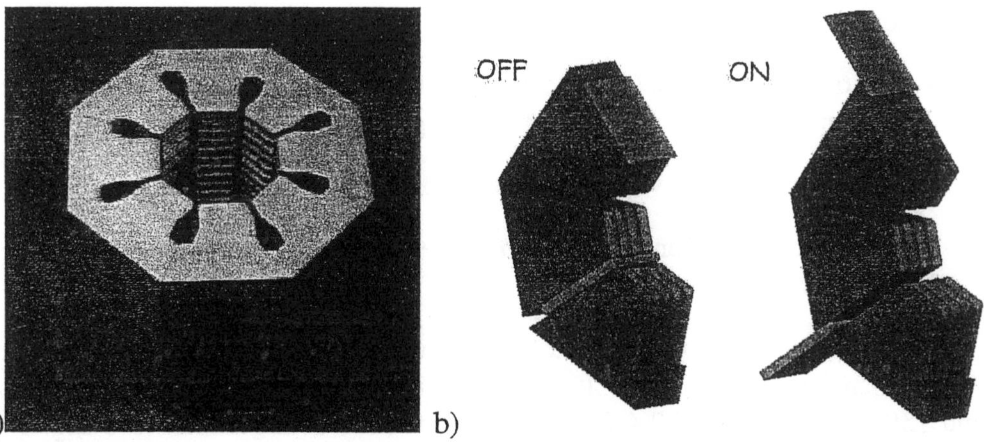 b)

Figure 13: a) PETS machining prototype b) PETS ON/OFF configuration with detuning wedges.

from the drive beam and power four accelerating structures for the main beam. The structures will have have a build-in mechanism to reduce or switch off the power generation, see Fig. 13 b): four wedges in the slots can be moved to detune the structure and stop RF production while the drive beam is still present. This allows to switch off a structure in case of breakdowns in a connected accelerating structure.

4 Test Facility CTF3

It is generally accepted that CLIC is the only viable technology for multi-TeV colliders. However, critical feasibility issues still need to be addressed. In the past years, the International Linear Collider Technical Review Committee [8] has ranked a number of crucial items to prove CLIC feasibility (the so-called R1 items), and to arrive at a conceptual design (the R2 items). The experimental program of the new CLIC Test Facility (CTF3) deals with all the CLIC technology-related R1 and R2 issues, while issues common to all linear collider are investigated in the framework of the EuroTeV collaboration. The goal is to demonstrate the CLIC feasibility before 2010. By then the first LHC results should be available and the energy reach required for new physics at a future linear collider better known.

CTF3 [9] is being built and commissioned in stages at CERN, by an international collaboration presently including Britain, Finland, France, Italy, Russia, Spain, Sweden, Turkey, and the USA. Recently, institutions from other countries have also shown interest in joining. CTF3 uses the buildings of the former LEP pre-injector complex (LPI) and makes maximum use of existing hardware. The main experimental program will focus on the R1 issues:

- Test of a prototype CLIC accelerating structure (including design features to damp higher order modes) at design gradient and pulse length.
- Validation of the drive beam generation scheme with a fully-loaded linac.
- Design and test of an adequately damped power-extraction structure, which can be switched on and off.

The CTF3 complex is designed to work at a lower beam current and a lower energy than foreseen for the CLIC drive beam (3.5 A instead of 5.7 A in the linac and 150 MeV instead of 2.4 GeV) [9]. In its final configuration (see Fig. 14), CTF3 will be composed of a 70 m long drive-beam linac followed by a 42 m delay loop and and a 84 m combiner ring, where the bunch train combinations will be carried out. The drive beam current after combination will reach 35 A,

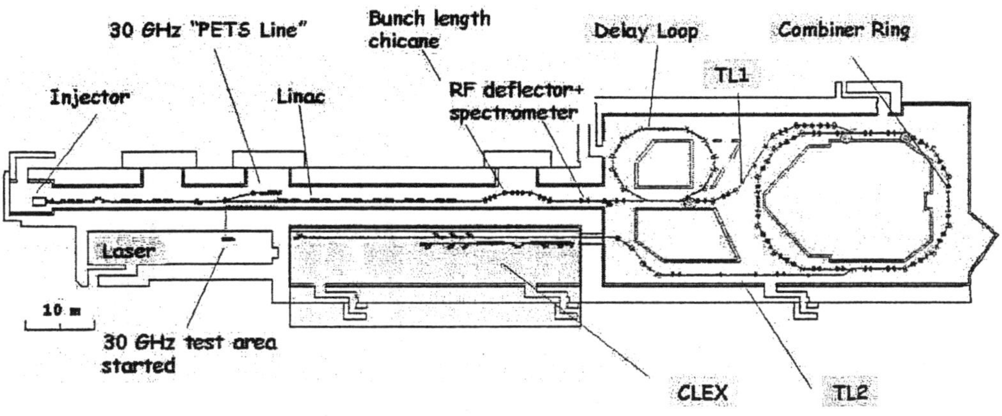

Figure 14: Layout of the CTF3 complex.

and will be transported to an experimental area (CLEX) to produce 30 GHz RF power in a high power test stand. A separate linac in the same area will provide a low charge 'main beam' for a representative CLIC two-beam module, and a test decelerator will be dedicated to drive beam stability studies.

The installation also includes already a high power test stand halfway along the linac ("PETS line") working at a lower beam current, needed to continue RF component testing. installation and commissioning.

In 2003 and 2004, the injector, the linac with its 30 GHz power production line and an end-of-linac chicane and instrumentation section have been installed and commissioned. In 2005, the installation of the delay loop was completed. The combiner ring and the experimental area will be put in operation in the following years.

In the linac, the design beam current and pulse length were reached rapidly. The beam is remarkably stable at all tested currents, successfully demonstrating the operation under full beam loading of the linac structures with their novel damping scheme [11] (see Fig. 15). The RF-to-beam efficiency was evaluated from input and output RF signals and confirmed by beam energy gain measurements. For a 4 A beam, an impressive efficiency of 94% was obtained, in agreement with simulations.

CTF3 has started its role as a 30 GHz power production facility with the first accelerating structure tests in 2005. These are ongoing at the time of the ISSP school 2005, and at present RF pulses up to 90 MW have been generated. Preliminary measurements shows that about 60 MW were delivered to a molybdenum iris test structure, corresponding to a gradient in the first cell above 150 MV/m. An ambitious test programme is foreseen to develop and test a prototype CLIC structure at design gradient and pulse length.

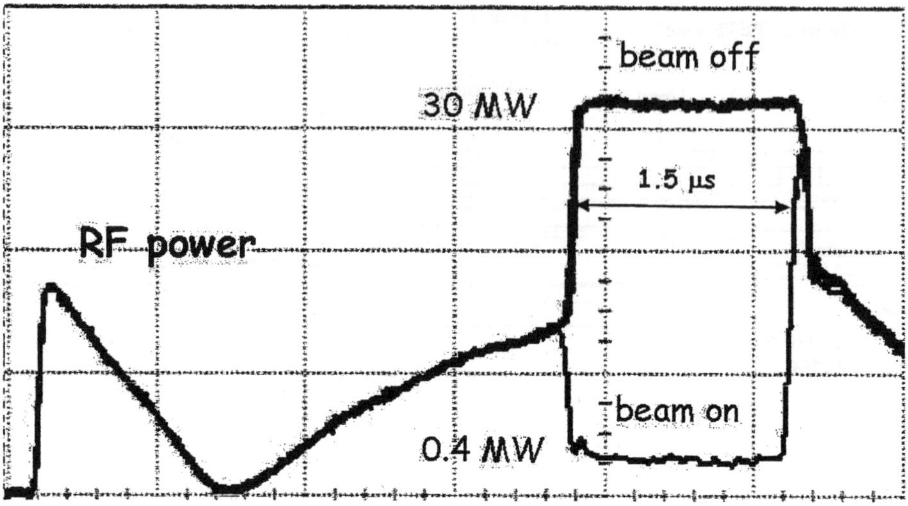

Figure 15: Demonstration of fully loaded operation. The traces show the RF power at the exit of an accelerating structure without and with beam. The signal almost goes to zero when the beam is present.

5 Conclusion

The aim of the CLIC R&D is to prove the feasibility of a future CLIC linear collider by 2010. At that time, the first physics results from the LHC should give a good indication of the required energy reach for a linear electron-positron collider. CLIC opens the door for pushing the physics frontier for a lepton collider into the very interesting multi-TeV region.

The normal conducting technology allows a high gradient and a reasonable size for a multi-TeV collider. The technology is not yet mature and still requires challenging R&D but many promising results have been obtained in past and present test facilities. The remaining key issues are well identified and are addressed in a multilateral collaboration of various institutes. The CTF3 test facility will demonstrate these remaining critical items.

References

[1] see http://cern.ch/CLICphysics

[2] M. Battaglia, A. de Roeck, J. Ellis, D. Schulte (editors), "Physics at the CLIC Multi-TeV Linear Collider : report of the CLIC Physics Working Group", CERN report 2004-005, hep-ph/0412251

[3] see http://www.linearcollider.org

[4] see http://cern.ch/clic-study

[5] CLIC Study Team, Ed. G. Guignard, *A 3 TeV e+e- linear collider based on CLIC technology*, CERN 2000-008 and I. Wilson, *The compact linear collider CLIC*, Physics Reports 404-404 (2004) 365-378.

[6] H. Braun et al., Ed. R. Corsini, *The CLIC RF power source*, CERN 99-06, 1999.

[7] R. Corsini, A. Ferrari, L. Rinolfi, P. Royer, F. Tecker, *Experimental results on electron beam combination and bunch frequency multiplication*, Phys. Rev. ST Accel. Beams 7, 040101 (2004).

[8] Editor G. Loew, *International linear collider technical review committee second report*, SLAC-R-606.

[9] G. Geschonke and A. Ghigo Eds, *CTF3 Design Report*, CERN/PS 2002-008 (RF).

[10] W. Wuensch, C. Achard, S. Döbert, H. H. Braun, I. Syratchev, M. Taborelli, I. Wilson, *A Demonstration of High Gradient Acceleration*, CERN-AB-2003-048-RF, Proc. PAC 2003.

[11] E. Jensen, *CTF3 Drive Beam Accelerating Structures*, Proc. LINAC 2002, 9-23 August 2002, Gyeongju (Kyongju), Korea, CERN/PS 2002-068 (RF) and CLIC Note 538.

CHAIRMAN: F. TECKER

Scientific Secretaries: B. O'Leary, J.-I. Noaki

DISCUSSION

- Preghenella:

You said that the CLIC collider will not use superconducting technology. This is also valid for the focusing magnets?

- Tecker:

There could be eventually superconducting technology for the final focusing quadrupoles where you need a very strong field. But for the majority of the quadrupoles, you are very well off with normal conducting quadrupoles. You do not have to go to superconducting technology there.

- Munir:

A 500 GeV center of mass energy is not a huge leap beyond the LEP energy and could be reached by the sub-process at the LHC although that is a hadron collider. So, what is the point of having an extra linear collider [like the ILC]?

- Tecker:

Good question. One has to say the ILC is a couple of years ahead of the CLIC technology in terms of feasibility. In the past, there have been several linear collider projects. For example, TESLA, which is based on superconducting technology was based at DESY, Hamburg. Then there was the NLC, the Next Linear Collider, based at SLAC, and the Japanese linear collider, based in Japan. When it was the time to evaluate these different projects, it was clear that the ILC, JLC and TESLA are more advanced than CLIC. We still have many feasibility issues which we have to demonstrate before we can really build CLIC one day. So, at that time the decision was taken for an international linear collider in the view that this machine can be built relatively soon. So right now, the ILC project is very much alive; right last week there was a big workshop in Snowmass, Colorado, where more than six hundred physicists were mainly discussing ILC physics and the machine. There is a big initiative, the global design effort behind it where at the beginning of this year, a director was elected. The ILC is a big project right now.

- Munir:

How much difference does it make if you have one collimated beam for the same length of the collider directed at a stationary point instead of accelerating two oppositely directed beams?

- Tecker:

As you can easily calculate from relativity, if you have just one beam and fixed target, then your energy scales only with the square root of the beam energy you have. The available energy is the square root of two times the mass times beam energy. If you have two beams, then you have exactly the sum of these two energies, if both beams are the same. So it is kinematically much more favorable to have two beams colliding than to have fixed target experiments.

- *Steinberg:*

What are the capabilities of CLIC to vary the beam energies basically from zero to the maximum as compared to the ILC? It is clearly an important feature in order to explore the landscape of the TeV-scale physics.

- *Tecker:*

You can vary the energy range by changing the accelerating gradient. In the ILC, this is based on cavities which are directly fed by klystrons. So you change the klystron power and have, more or less, full variation of the energy range straightforwardly. Of course, you have to rescale your magnets along the linac and rescale your final focus system.

In CLIC it is slightly more complicated, but not too much, because you can also change the gradient in the accelerating cavities by changing the current which you use to generate the fields. If you have a lower current there, you generate smaller fields for your main accelerating cavities and have a smaller gradient for lower energies. The only thing you have to keep in mind is that you want to generate the drive beam very efficiently. Then you have to lower the power of the drive beam klystrons so that you are still working in this fully beam-loaded regime. But it is possible.

- *Pinfold:*

What candidates sites are there for the CLIC machine?

- *Tecker:*

There have been some investigations about the ground at CERN where you could place it. You could place it along the Jura between the Jura mountains and Lake of Geneva. There has been a study done for CLIC.

- *Faivre:*

Is the probability to lose the beam for a linear collider high enough to worry about the possibility that the lost beam may fire the detector too often and damage some components? Or, is there anything to protect the experiment?

- *Tecker:*

There are some failure modes where you lose your beam. You have a huge system which is the order of a couple of kilometers long to collimate the beam. If you have some failures then most of failure modes are taken into account by this beam collimation system. This system would be designed in a couple of stages where the first blows up your beam and the next absorb the energy of the beams so that you do not damage your collimators. More details still have to be studied for CLIC.

- *Faivre:*

You mentioned in your talk that CLIC would use 415 MW of electrical power. That is between 1/2 and 1/3 of a nuclear power plant, or half a percent of the total electricity production of France. How much will the ILC need? How is this taken into account?

- *Tecker:*

You want to bring up the efficiency with which you collide the beams to a very high degree, because the final luminosity in your collision is determined by the beam power. You have to accelerate a number of particles to a certain energy. The beam power in CLIC is 40MW, which leads to the 400MW total power. Regardless the limit in power, you have to pay the power in order to get your luminosity. For the ILC, since it is at lower energy, it is somewhere between 100 and 200 MW.

- *Baldini Ferroli:*

In the luminosity formulae you only quoted the vertical emittance. Does it mean the beams are flat?

- *Tecker:*

Yes, the beams are flat in order to minimize the beamstrahlung parameter. The beamstrahlung parameter is inversely proportional to the [square of the] sum of the horizontal and vertical beam sizes and the luminosity is inversely proportional to the product of the horizontal and vertical sizes. By minimizing the product and keeping the sum relatively big with the flat beams, you can maximize the luminosity. In the case of CLIC, the horizontal beam size at the interaction point is of the order of 60nm while the vertical one is only below 1nm. So, you have flat beams. In the damping rings, it is much easier to generate the flat beams due to the synchrotron radiation, the majority of motion happening in the horizontal plane. You can achieve a much smaller beam size in the vertical plane in a damping ring. So, this fits very well with this design.

- *Hanagaki:*

50% of the luminosity is within 1% of the peak energy, which will be important for the energy scan. Is it better than the ILC if the energy is the same?

- *Tecker:*

I do not know the exact number for ILC right now, but I suspect that it might be about similar. At least ILC does not have a very sharp luminosity spectrum any more than LEP.

- *Munir:*

The beam that enters the main accelerating tube from the drive tub is of the order of 9GeV. Is that the best figure you can get or is that a calculated energy limit that you want the beam to have for some reason?

- *Tecker:*

Before going to the very small cavities which you have in your main beam which are of the the order of 2mm in diameter, you have to have a very small beam size. When beams are accelerated, all the transverse motion is damped down. So, the real emittance goes down with the energy. This is why you have to have a certain energy. If you choose a higher energy, you have to have much more of the ordinary linac to get to this energy. So this is why there is a certain energy range which is a good compromise for those two effects, and it determines the injection energy in your main linac.

- *Denisov:*

How will earth vibrations affect CLIC operation with beams of size 1nm? Is there any worry that every airplane in Geneva airport will kill the beam in the CLIC? Or, are there some active stabilization methods used?

- *Tecker:*

The CLIC tunnel would be far underneath the earth like the LHC tunnel. There is a lot of data about movement of the LEP tunnel and other tunnels at Fermilab and DESY to simulate the movements that you get for CLIC. With these data, you can simulate the dynamic behavior of the beam. In fact, you have to be fast enough with your feedback systems to correct all these movements in order to keep the beams in position in collision. This is a critical point, but it seems to be feasible with the present thinking.

- *O'Leary:*

Does CLIC have any capacity to produce asymmetric beams for similar purposes to Belle or BaBar?

- *Tecker:*

Not with the current design, where the same drive beam goes into both accelerators, but in principle two drive beam generators could produce different beams to create a beam asymmetry. It depends on how you generate the drive beam. If you just have one drive beam which you split into the two different linacs, then you are bound to having the same energy. This is the present design right now. But, if you want to keep the asymmetry option, you can change your drive beam generation scheme as well to have two separate drive beams which you would send one each into each side and, in principle, in the same way as you vary the energy, you can vary the current of the drive beam. So you could create an asymmetric collision.

- *Hanagaki:*

How big a fraction of the total amount of particles actually collides?

- *Tecker:*

This is actually a good question. I have never thought about how many of them really collide.

Gauge Charges from Supergravity

L. Andrianopoli

Centro E. Fermi, Rome, Italy;
PH-TH-Unit, CERN, CH-1211 Genève 23, Switzerland and
Dip. Fisica, Politecnico di Torino, C.so Duca degli Abruzzi, 24 I-10129 Torino, Italy and INFN, Sezione di Torino, Italy

E-mail: Laura.Andrianopoli@cern.ch

Abstract. Some recent results in the study of four dimensional supergravity flux compactifications are reviewed, discussing in particular the role of torsion on the compactification manifold in generating gauge charges for the effective four dimensional theories.

1. Introduction

Many efforts in theoretical physics have been devoted, in the last decades, to look for a quantum theory describing in a consistent and unified way gauge interactions with gravity. Such a theory, if any, should incorporate the Standard Model of elementary particles but is also expected to give answers to some problems left open after the huge theoretical and phenomenological successes of the Standard Model, like the understanding of the hierarchy problem, of confinement in QCD, of the tiny value of the cosmological constant, or of the black-hole entropy.

In this spirit, superstring theory appears nowadays as a very promising candidate for the quantum theory underlying Nature: it is indeed a consistent and finite theory for quantum gravity, giving unification of gauge interactions with gravity. It is described by a two dimensional conformal theory defined at the Planck scale, spanning a curved ten dimensional target space. When the spectrum is extended to include solitonic configurations (D-branes [1]) it may reveal one extra eleventh dimension (M-theory) [2, 3].

The massless excitations of superstrings are described, in a low energy limit, by effective supersymmetric field theories, named supergravity theories, which contain supersymmetric couplings of gravity with matter and gauge fields. Supergravity theories may be defined in any space-time dimensions up to eleven, however they get predictive power when interpreted as effective theories for superstrings or M-theory. If superstring theory has to provide a natural explanation of interactions in our real world, some trick must be present so that in particular the theory, at low energies, looks four dimensional. The simplest way to do so is to imagine that the vacuum configuration for space-time is not ten dimensional Minkowski space, but is instead of the form $\mathbb{R}^{(1,3)} \times M_6$, where M_6 is some six-dimensional compact manifold, whose size is so small that at the length-scales experimented in our low energy world it cannot be observed. Then, any tensor field of the ten dimensional theory looks, in the four dimensional compactified theory, as a collection of fields of various masses and spins [4] (the procedure of compactification from higher dimensions to four was first studied in five dimensions by Kaluza and Klein [5]). In particular, the ten dimensional fields give rise to a set of massless four dimensional scalars (*moduli*) describing metric and topological properties of the compactification manifold M_6. Any choice of the vacuum background of string theory selects one among a lot of possible inequivalent compactified space-times, and this implies giving a particular expectation value to the scalars. Note that these scalars are just the ones appearing in supergravity theories. Giving them some particular vacuum expectation values means, going down to very low energies, fixing the values of the physical observables in the Standard Model describing strong and electro-weak interactions.

We are going to focus on this intertwining role of supergravity between the stringy (Planck) scale and low energy physics. We will discuss in particular how gauge interactions arise in supergravity theories, with particular attention to the role of the torsion in the compactification manifold as gauge charge.

This contribution is organized as follows: In section 2 the role of scalar fields and the way gauge charges appear in supergravity are surveyed; in section 3 the example of no-scale models as gauged supergravity theories is discussed, with particular attention to the Scherk–Schwarz model, while section 4 focuses on the role of torsion on the compactification manifold in flux compactifications and on its relation to the Scherk–Schwarz model. Conclusions and outlook are left to the concluding section. The bibliography section is not exhaustive; somewhere we preferred to give reference to review papers where more extended references on the subject may be found. For a more complete list of references we refer to [6].

2. Scalars in supergravity and gauging

A peculiar feature of supergravity is that the spectrum includes fermionic and bosonic fields with spin ≤ 2, among which scalars. Supersymmetry strongly constrains the couplings, and in particular it implies that the scalars belong to non linear σ-models:

$$\mathcal{L}_{kin|scalars} = g_{\ell m}(\phi)\partial_\mu \phi^\ell \partial^\mu \phi^m. \tag{1}$$

The scalars play an important role in the field theory, as they generalize the Higgs field of the Standard Model. The low energy observables (gauge couplings, mass parameters, the cosmological constant, etc...) are given in terms of vacuum expectation values (v.e.v.) of scalars. The cosmological constant Λ, for example, is the v.e.v. of the scalar potential

$$\Lambda =< V(\phi) >, \tag{2}$$

while the gauge couplings and θ-term appear in the lagrangian as the v.e.v. of scalar dependent functions:

$$\mathcal{L}_{kin|vec} = \gamma_{\Lambda\Sigma}(\Phi)\mathcal{F}^{\Lambda} \wedge {}^*\mathcal{F}^{\Sigma} + \theta_{\Lambda\Sigma}(\Phi)\mathcal{F}^{\Lambda} \wedge \mathcal{F}^{\Sigma}. \tag{3}$$

However, the v.e.v. of the scalar fields fixing the value of the couplings in the effective supergravity theories are in fact determined by the choice of a vacuum in the corresponding underlying superstring theory. What happened, with the introduction of superstring theory as a unifying theory describing in particular particle physics, is that the Standard Model problem of having many free parameters to be fixed has turned into the problem of an enormous degeneracy of vacua for string theory. As we can see, in this way we do not gain any predictive power on the quantities characterizing our world; however, we obtain the important conceptual achievement of unifying gravity with the other interactions and to give a natural understanding for the origin of all the parameters involved (completely arbitrary in the Standard Model). Moreover, there is the hope that some non-perturbative stringy effect, inducing supersymmetry breaking, could also lift the vacuum degeneracy leaving fewer accessible vacua.

One is then driven to the problem of understanding how gauge charges and interactions arise in supergravity.

Supergravity theories are built from symmetry requirements: local invariance under the supersymmetry algebra. The bosonic sector includes gravity, abelian gauge fields and uncharged scalars; to get a theory invariant under the local supersymmetry algebra there is no need to introduce gauge charges. The gauge charges are introduced via a deformation of the theory named *the gauging procedure*, which corresponds to promote a global symmetry of the theory (an isometry of the scalar manifold) to gauge symmetry and to turn on interactions with the corresponding gauge fields of the theory (which may become non-abelian). Consistency of the deformation with supersymmetry implies the presence of a scalar potential, where the dynamical information on the model are codified.

However, if we want to study supergravity theories as effective theories of superstrings, it must be possible to interpret the gauge charges of a given supergravity model in terms of data corresponding to some superstring background. For example, *ungauged* supergravity models (with no gauge charges) appear as the low energy limit of perturbative Type II superstrings compactified on tori without any non-trivial flux for the tensor fields (p-forms) of the theory in the internal directions. On the other hand, starting from a ten dimensional theory, three different mechanisms have been identified to generate gauge charges in the four dimensional effective supergravity models [7, 8, 10, 6]:

- by turning on charges corresponding to non abelian gauge fields already in the ten dimensional theory (from gauge fields living on D-branes, for Type I or II theories, or from the heterotic sector, for the case of the heterotic string);
- by turning on non trivial p-form fluxes in the internal compact manifold;

- by turning on charges associated to the geometry of the compactification manifold (à la Scherk and Schwarz).

The emerging scenario is that the first class of charges should contain the gauge group of the Standard Model, while the other two classes of charges should be responsible for supersymmetry breaking, via the super-Higgs mechanism.

Note that the dualities among string theories still map one into the other the theories in the presence of charges, provided that also the charges are transformed accordingly.

3. An explicit example: the no-scale models

No-scale models are a particular class of supergravity theories emerged in the 80's for their phenomenological interest [11, 12]. They have seen a revived interest from theoretical physicists after that some recent progress in string theory (D-branes [1], Randall–Sundrum scenarios [13]) allowed to find them a nice interpretation in terms of the classes of ten dimensional charges introduced in the previous section.

No-scale supergravities are models with an automatically vanishing cosmological constant (at tree level), and where supersymmetry is spontaneously broken. The gravitino mass $m_{3/2}$ is a sliding scale at tree level, dynamically fixed by radiative corrections, with possibility of a hierarchical suppression with respect to the Planck scale. The characteristic features of the various no-scale models are codified in the choice of the gauge group and in the structure of the scalar potential. However, they all share the feature of having a semi-positive definite scalar potential with flat directions. From a purely supergravity point of view, no-scale models are typically related to gaugings of suitable non-semisimple global symmetry groups of the Lagrangian (*flat gaugings*) [14, 15]. A peculiar feature of this kind of models is that the spectrum appears separated in an *observable* sector, (where the standard Model is defined), and an *hidden* sector (which is decoupled for energy scales $E << M_P$) interacting with the observable sector only through gravity.

Phenomenologically viable no-scale models have been constructed in string compactifications [16] where the SM/GUT lives on space-time filling D-branes (observable sector) while the hidden sector is given by the bulk of the ten dimensional ambient space. A hierarchical supersymmetry breaking [17] with stabilization of (some of the) moduli may be achieved in the bulk (hidden sector) by turning on background fluxes in the compactification manifold. Interesting models of flux compactifications have been found mostly in Type IIB theory by switching on appropriate RR and NS three-form fluxes in the internal directions [18] (other examples were obtained in heterotic M-theory, due to a non-vanishing G-flux sourced by the boundaries [9], and on the Type IIA front [10, 19]).

Alternatively, no-scale models may be found as well from generalized dimensional reduction à la Scherk–Schwarz (SS) [20] of eleven dimensional supergravity (or any truncation thereof). This is a generalized type of dimensional reduction on tori, corresponding to a twist in the boundary conditions of the fields in the internal directions, with a phase dependent on a global symmetry of the theory (SS phase). The twist originates a no-scale model with non abelian couplings and a positive definite scalar potential, which, in general, yields spontaneous supersymmetry breaking. The charges obtained this way appear to be associated to the internal manifold geometry. However until recently they could not find a direct interpretation in the context of flux compactifications of superstring or M theory. We are going to see in the following section that such an interpretation is in fact possible in the presence of torsion in the internal manifold. Let us first briefly review what the SS mechanism is.

3.1. The Scherk-Schwarz mechanism

To describe how the mechanism works, let us consider the simple example of a free complex scalar field $\phi(x, y)$ on $\mathbb{R}^{1,3} \times S^1$, with lagrangian

$$\mathcal{L} = \frac{1}{2}\partial_{\hat{\mu}}\phi\partial^{\hat{\mu}}\phi^*, \tag{4}$$

where $\hat{\mu} = (\mu, 4)$; $\mu = 0, \ldots, 3$; $x^4 = y$. We want to interpret this model from a four dimensional point of view, supposing that the S^1 has a tiny radius. However, instead of the Kaluza–Klein (KK) ansatz, let us impose the following generalized ansatz of dimensional reduction:

$$\phi(x, y) = \mathrm{e}^{\mathrm{i}my} \sum_{n=-\infty}^{+\infty} \phi_n(x)\mathrm{e}^{\mathrm{i}ny/R}, \quad (y \sim y + 2\pi R). \tag{5}$$

We observe that this ansatz is not well defined, because $\phi(x, y)$ is multivalued on S^1. However, since the lagrangian (4) is invariant under global $U(1)$ transformations

$$\phi \longrightarrow \mathrm{e}^{\mathrm{i}\alpha}\phi \tag{6}$$

then the phase $\mathrm{e}^{\mathrm{i}2\pi mR}$ has no physical effects. The five dimensional kinetic lagrangian (4) generates, in the dimensional reduction to four dimensions, a "scalar potential" (a mass term, in fact, in this case) $V = m^2 > 0$. All the masses are shifted by m, and the zero-mode has a mass m.

The key ingredient, which allows to implement the SS mechanism and to introduce in the dimensionally reduced theory the "charge" m, is the global invariance of the lagrangian (4) under the symmetry group $U(1)$. The charge is then introduced as a y-dependent phase in the dimensional reduction ansatz (5). The same mechanism, applied to theories invariant under more general global symmetries, allows to turn on more general charges related to the internal geometry.

In particular, let us now study the case of generalized dimensional reduction of $4+n$ dimensional gravity on $\mathcal{M}_{4+n} = \mathbb{R}^{1,3} \times S^1 \times T^{n-1}$. The result will be compared, in the next section, with an ordinary KK dimensional reduction of $4 + n$ dimensional gravity on the same manifold $\mathcal{M}_{4+n}$ but in the presence of a torsion background.

Let us suppose to formally perform the reduction in two steps:

- first consider the KK compactification to *five dimensions* on T^{n-1}. One gets a five dimensional theory on $\mathbb{R}^{1,3} \times S^1$ with global invariance $SL(n-1)$, which is the isometry group of the torus T^{n-1}.
- then we may use this global invariance for building a SS phase-matrix for a *generalized* dimensional reduction *from five to four dimensions.*

From the $4+n$-dimensional vielbein one gets the following fields in the theory reduced to five dimensions: the five dimensional vielbein $V_{\hat{\mu}}{}^{\hat{a}}$, $n-1$ vectors $B^m_{\hat{\mu}}$ and $\frac{n(n-1)}{2}$ scalars ϕ^i, σ parametrizing the coset $\frac{SL(n-1)}{O(n-1)} \times O(1,1)$, whose coset-representatives are the internal components of the $4+n$ dimensional vielbein, $V_m{}^i(\phi^i, \sigma)$.

The SS ansatz for the dimensional reduction (restricted to the zero-modes) from five to four dimensions is (we take $x^4 \equiv y^1$ and no dependence of the fields on the extra internal coordinates $x^5 = y^2, \cdots, x^{n-1} = y^{n-4}$)

$$\begin{aligned} B^m_{\hat{\mu}}(x, y^1) &= U^m{}_n(y^1) B^n_{\hat{\mu}}(x) \\ V_m{}^i(x, y^1) &= (U^{-1})_m{}^n(y^1) V_n{}^i(x) \end{aligned} \tag{7}$$

where $U = \exp[My^1]$, $M \in SO(n-1)$. The theory reduced to four dimensions has non abelian gauge charges which come from the dimensional reduction of the Levi–Civita connection on $\mathcal{M}_{4+n}$, using the generalized ansatz (7). In particular one finds [20, 14] ($\underline{m} = (4, m)$; $\underline{i} = (4, i)$)

$$F^{\underline{m}}_{\mu\nu} = \partial_{[\mu} B^{\underline{m}}_{\nu]} + f^{\underline{m}}{}_{\underline{np}} B^{\underline{n}}_{[\mu} B^{\underline{p}}_{\nu]} \tag{8}$$

$$D_\mu V_{\underline{m}}{}^{\underline{i}} = \partial_\mu V_{\underline{m}}{}^{\underline{i}} + f^{\underline{n}}{}_{\underline{mp}} B^{\underline{p}}_\mu V_{\underline{n}}{}^{\underline{i}} \tag{9}$$

with structure constants given by the SS phases:

$$f^m{}_{4n} = M^m{}_n. \tag{10}$$

Let us make an observation which will be important in the next section. If the theory in $4+n$ dimensions contains also p-forms besides the metric, then the SS prescription requires that the scalars coming from the dimensional reduction of the p-forms have to be rotated as well with the SS phases.
For the case, *e.g.*, of a one-form A_m one has

$$A_m(x, y^1) = (U^{-1})_m{}^n(y^1) A_n(x), \tag{11}$$

so that

$$F_{4m}(x, y^1) = M_m{}^n A_n(x). \tag{12}$$

We may then conclude that the SS mechanism is surprisingly efficient in introducing geometrical charges in the theory. However it appears as a bit artificial prescription, which is not completely satisfying from the microscopical point of view.

We are going to see in the next section that the SS mechanism may be understood as an ordinary flux compactification if we assume the presence of a constant background torsion in the internal manifold.

4. The SS mechanism as a flux compactification with internal torsion

The purpose of the present section is to review how theories originating from SS dimensional reduction may alternatively be found as the result of a KK dimensional reduction in the presence of an internal torsion. In particular we shall consider a $D+n$ dimensional pure gravity theory compactified to D dimensions on a n–torus T^n showing that, by switching on an appropriate constant background torsion in the torus (a *torsion flux*), we may generate the same D-dimensional theory as the one described in section 3.1, originating from a SS reduction from $D+1$ dimensions with global symmetry generator chosen within the global symmetry algebra $\mathfrak{sl}(n-1,\mathbb{R})$. This is based on [6].

The relevance of torsion in flux compactifications was recently understood in the study of T-duality between Type IIB and Type IIA superstrings [21, 22]. Indeed it was found that the T-dual of the charge associated to the flux of $H^{(3)}_{NS}$ is not a p-form flux but internal torsion. Our geometrical interpretation allows to study also the no-scale SS models in the context of flux compactifications of superstring (or M) theory.

A delicate point in our analysis is the notion of gauge invariance in the presence of torsion. Indeed, we found that to have a precise identification of the SS models in terms of torsion a generalization of the notion of gauge transformation is needed, corresponding to the coupling of the torsion to the gauge fields.

4.1. Coupling torsion - gauge fields: the general idea

The torsion tensor may be defined as an antisymmetric contribution

$$T^P{}_{MN} \equiv \tilde{\Gamma}^P_{[MN]} = \frac{1}{2}(\tilde{\Gamma}^P_{MN} - \tilde{\Gamma}^P_{NM}) \tag{13}$$

to the affine connection $\tilde{\Gamma}$. In this case, the request of a metric connection ($\nabla_P\, g_{MN} = 0$) determines the connection coefficients as

$$\tilde{\Gamma}^P_{MN} = \Gamma^P_{MN} + K^P{}_{MN}, \tag{14}$$

where $\Gamma^P_{MN} = \Gamma^P_{NM}$ are the coefficients of the Levi-Civita connection while

$$K^P{}_{MN} = \frac{1}{2}(T^P{}_{MN} - T_M{}^P{}_N - T_N{}^P{}_M) \tag{15}$$

contains the torsion contribution. For further definitions and conventions, we refer to [6].

To discuss the origin of the coupling of the torsion to gauge fields let us consider, as an example, the case of a vector field A_M. The field strength feels the effect of the torsion background via the principle of general covariance:

$$F_{MN} \equiv \nabla_{[M} A_{N]} = \partial_{[M} A_{N]} + T^P{}_{MN} A_P. \tag{16}$$

When $T \neq 0$, F is not invariant under the usual gauge transformation

$$\delta A_M(X) = \partial_N \Lambda(X). \tag{17}$$

However, it is possible [23] to make the torsion compatible with the presence of gauge fields by introducing the generalized definition of gauge transformation

$$\delta_C A_M(X) = C_M^{\ N}(X)\partial_N \Lambda(X), \tag{18}$$

where the point-dependent matrix $C_M^{\ N}(X)$ has to be constrained by the request of (generalized) gauge invariance of the field strength

$$\delta_C F_{MN} = 0. \tag{19}$$

The general solution, found in [23] for a gauge field in d dimensions coupled to torsion in d dimensions is

$$C_M^{\ N} = \delta_M^N e^{\phi}, \qquad T^P_{\ MN} = \delta^P_{[M}\partial_{N]}\phi. \tag{20}$$

There is only one torsion degree of freedom allowed, the scalar field ϕ. Let us now see that this restriction may be largely relaxed if the idea of [23] is combined with the Kaluza–Klein ansatz in a dimensional reduction context, and that it is then possible to turn on an internal torsion with many independent components. The different components of the torsion generate geometrical charges that reproduce the SS phases while giving them a geometrical interpretation .

4.2. Coupling torsion / gauge fields and dimensional reduction

Let us consider a gauge field A_M in the $4+n$ dimensional manifold $\mathcal{M}_{4+n} = \mathbb{R}^{1,3} \times T^n$ $= \mathbb{R}^{1,3} \times S^1 \times T^{n-1}$, with coordinates $X^M = (x^\mu, y, y^m)$, $m = 1, \cdots n-1 \in T^{n-1}$ and $\mu = 0,1,2,3 \in \mathbb{R}^{1,3}$; y parametrizes S^1 (it is the same manifold as in section 3.1). We ask that the KK ansatz holds true:

$$A_M = A_M(x)\,; \quad \Lambda = \Lambda(x). \tag{21}$$

We suppose to have a constant torsion background on the torus T^n, and that the gauge field is coupled to the torsion as in (18). Then, $F = \nabla A$ is not gauge invariant, but it is invariant under the generalized gauge transformation à la HRRS :

$$\delta_C A_M = C_M^{\ N}\partial_N \Lambda = C_M^{\ \nu}\partial_\nu \Lambda. \tag{22}$$

If we further assume:

$$C_M^{\ N} = C_M^{\ N}(y), \tag{23}$$

then the generalized gauge invariance condition $\delta_C F = 0$ has now more general solutions (see [6] for details), allowing the following non-zero entries for the matrices C and T:

$$C_M^{\ N} : (C_\mu^{\ N}, C_M^{\ n})\,; \quad T^M_{\ NP} : (T^\mu_{\ \nu 4}, T^m_{\ NP}) \tag{24}$$

with the only constraint

$$C_\mu^{\ \nu}(y) = \delta_\mu^\nu e^{\phi(y)}\,; \quad T^\mu_{\ \nu 4} = \frac{1}{2}\delta^\mu_\nu \partial_4 \phi(y). \tag{25}$$

By choosing

$$T^m_{\ 4n} = M^m_{\ n} \in SO(n-1), \quad (\phi = 0), \tag{26}$$

we have a consistent theory, where

$$\delta_C A_\mu = \delta A_\mu = \partial_\mu \Lambda \,, \quad \delta_C A_4 = 0 \,, \quad \delta_C A_m = 0. \tag{27}$$

Moeover it gives

$$F_{4n} = M^m{}_n A_m \tag{28}$$

which precisely reproduces the effect of the geometrical SS phase for a vector, equation (12).

Let us now consider the KK dimensional reduction of the metric in the presence of the the torsion background (26), and let us compare the result with the one of Section 3.1. As in section 3.1, the theory reduced to four dimensions has non abelian gauge charges which come from the dimensional reduction of the affine connection on $\mathcal{M}_{4+n}$, and we find the same result as before

$$F^{\underline{m}}_{\mu\nu} \quad = \partial_{[\mu} B^{\underline{m}}_{\nu]} + f^{\underline{m}}_{\underline{np}} B^{\underline{n}}_{[\mu} B^{\underline{p}}_{\nu]} \tag{29}$$

$$D_\mu V_{\underline{m}}{}^{\underline{i}} = \partial_\mu V_{\underline{m}}{}^{\underline{i}} + f^{\underline{n}}_{\underline{mp}} B^{\underline{p}}_{\mu} V_{\underline{n}}{}^{\underline{i}} \tag{30}$$

However, the interpretation is now different. In this case the charges do not come from the dimensional reduction of the metric (which is in fact an ordinary KK reduction) but from the torsion contribution to the affine connection (14), which, differently from section 3.1, in this case is not any more Levi–Civita. The structure constants of the four dimensional gauge theory are now given by the components of the torsion

$$f^m_{\ 4n} = T^m_{\ 4n}. \tag{31}$$

The comparison between the two interpretations is summarized in the table below.

	S-S	Torsion
$f^m_{\ 4n} =$	$M^m_{\ n} = \text{SS phases}$	$T^m_{\ 4n} = \text{torsion}$
F_{4m} :	$A_m(x,y) = (e^{-My})^m{}_n A_n(x)$	$F_{4m} = T^m_{\ 4n} A_n$
$V_m{}^i$:	$(e^{-My})^n{}_m V_n{}^i(x)$	$V_n{}^i(x)$
$\Gamma^M_{\ NP}$:	$\Gamma^M_{\ [NP]} = 0$	$\Gamma^M_{\ [NP]} \neq 0$

5. Conclusions

We discussed how gauge charges and interactions do emerge in supergravity. We focused in particular on the role of the charges related to the geometry, and shown that the SS mechanism is in fact an ordinary compactification in the presence of an internal torsion background. The scenario that seems to emerge is that all the bulk charges, responsible for supersymmetry breaking, may be generated by compactifications in the presence of fluxes and internal torsion.

There are various directions for future investigations, which include a similar analysis of theories corresponding to compactifications on more general manifolds, the construction of a $D = 4$ gauged supergravity deriving from a general torsion background, and the effects of a dynamical torsion on this setting.

References

[1] J. Polchinski, **Phys. Rev. Lett. 75**, (1995) 4724.

[2] P. K. Townsend, **Phys. Lett. B 350** (1995) 184.

[3] For a review see, e.g.: M. Duff, **Int. J. Mod. Phys. A 11** (1996) 5623; J. Polchinski, **Rev. Mod. Phys. 68** (1996) 1245; J. Schwarz, **Nucl. Phys. Proc. Suppl. 55 B** (1997) 1; A. Sen, **Nucl. Phys. Proc. Suppl. 58** (1997) 5.

[4] E. Cremmer in "*Supergravity '81*", ed. by S. Ferrara and J. G. Taylor, Pag. 313; B. Julia in "*Superspace & Supergravity*", ed. by S. Hawking and M. Rocek, Cambridge (1981) pag. 331

[5] T. Kaluza, **Sitzungsber. Preuss. Akad. Wiss. Phys. Mat. K1** (1921) 966; O. Klein, **Z. Phys. 37** (1926) 895.

[6] L. Andrianopoli, M. A. Lledo and M. Trigiante, **JHEP 0505** (2005) 051.

[7] N. Kaloper and R. C. Myers, **JHEP 9905** (1999) 010.

[8] B. de Wit, H. Samtleben and M. Trigiante, **Nucl. Phys. B 655** (2003) 93; G. Dall'Agata and S. Ferrara, **Nucl. Phys. B 717** (2005) 223.

[9] G. Curio and A. Krause, **Nucl. Phys. B 602** (2001) 172; G. Curio and A. Krause, **Nucl. Phys. B 643** (2002) 131; M. Becker, G. Curio and A. Krause, **Nucl. Phys. B 693** (2004) 223.

[10] J. P. Derendinger, C. Kounnas, P. M. Petropoulos and F. Zwirner, **Nucl. Phys. B 715** (2005) 211.

[11] R. Barbieri, S. Ferrara and C. A. Savoy, **Phys. Lett. B119** (1982) 343. A. H. Chamseddine, R. Arnowitt and P. Nath, **Phys. Rev. Lett. 49** (1982) 970. L. J. Hall, J. Lykken and S. Weinberg, **Phys. Rev. D27** (1983) 2359.

[12] E. Cremmer, S. Ferrara, C. Kounnas and D.V. Nanopoulos, **Phys. Lett. B133** (1983) 61. J.R. Ellis, A.B. Lahanas, D.V. Nanopoulos and K. Tamvakis, **Phys. Lett. 134** (1984) 429. J. Ellis, C. Kounnas and D.V. Nanopoulos, **Nucl.Phys. B247** (1984) 373. For a review see A. B. Lahanas and D. V. Nanopoulos, **Phys. Rep. 145** (1987) 1.

[13] L. Randall and R. Sundrum, **Phys. Rev. Lett. f 83**, 3370 (1999); **Phys. Rev. Lett. 83**, 4690 (1999); **Nucl. Phys. B 557**, 79 (1999).

[14] L. Andrianopoli, R. D'Auria, S. Ferrara and M. A. Lledo, **JHEP 0207** (2002) 010.

[15] L. Andrianopoli, R. D'Auria, S. Ferrara, M. A. Lledo, **Nucl. Phys. B640** (2002) 63; L. Andrianopoli, R. D'Auria, S. Ferrara, M. A. Lledo, **Mod. Phys. Lett. A 18** (2003) 1001; L. Andrianopoli, S. Ferrara and M. A. Lledo, **JHEP 0406** (2004) 018.

[16] J. Polchinski and A. Strominger, **Phys. Lett. B 388**, 736 (1996); T. R. Taylor and C. Vafa, **Phys. Lett. B474**, 130 (2000). For an exausitive review on flux compactifications see A. R. Frey, arXiv:hep-th/0308156, and references therein.

[17] S. B. Giddings, S. Kachru and J. Polchinski, **Phys. Rev. D 66**, 106006 (2002); O. DeWolfe and S. B. Giddings, **Phys. Rev. D 67** (2003) 066008.

[18] R. D'Auria, S. Ferrara and S. Vaula, **New J. Phys. 4** (2002) 7; ; L. Andrianopoli, R. D'Auria, S. Ferrara, M. A. Lledo, **JHEP 0301** (2003) 045; L. Andrianopoli, R. D'Auria, S. Ferrara and M. A. Lledo, **JHEP 0303** (2003) 044 [arXiv:hep-th/0302174]; B. de Wit, H. Samtleben and M. Trigiante, **Phys. Lett. B 583** (2004) 338.

[19] C. Angelantonj, S. Ferrara and M. Trigiante, **Phys. Lett. B 582** (2004) 263.

[20] J. Scherk and J. H. Schwarz, **Nucl. Phys. B 153** (1979) 61.

[21] S. Gurrieri, J. Louis, A. Micu and D. Waldram, **Nucl. Phys. B 654** (2003) 61.

[22] S. Kachru, M. B. Schulz, P. K. Tripathy and S. P. Trivedi, **JHEP 0303**, 061 (2003)

[23] S. Hojman, M. Rosenbaum, M. P. Ryan and L. C. Shepley, **Phys. Rev. D 17** (1978) 3141.

Phase Structure and Instability Problem in Color Superconductivity

Kenji Fukushima

RIKEN BNL Research Center, Brookhaven National Laboratory, Upton, New York 11973, USA

Abstract. We address the phase structure of color superconducting quark matter at high quark density. Under the electric and color neutrality conditions there appear various phases as a result of the Fermi surface mismatch among different quark flavors induced by finite strange quark mass; the color-flavor locked (CFL) phase where quarks are all energy gapped, the u-quark superconducting (uSC) phase where u-quarks are paired with either d- or s-quarks, the d-quark superconducting (dSC) phase that is the d-quark analogue of the uSC phase, the two-flavor superconducting (2SC) phase where u- and d-quarks are paired, and the unpaired quark matter (UQM) that is normal quark matter without pairing. Besides these possibilities, when the Fermi surface mismatch is large enough to surpass the gap energy, the gapless superconducting phases are expected. We focus our discussion on the chromomagnetic instability problem related to the gapless CFL (gCFL) onset and explore the instability regions on the phase diagram as a function of the temperature and the quark chemical potential. We sketch how to reach stable physical states inside the instability regions.

1 Family of color superconducting phases

The phase structure of matter composed of quarks and gluons described by Quantum Chromodynamics (QCD) has been investigated for many years, and in the high temperature and low baryon (or quark if deconfined) density region which is accessible in heavy-ion collisions interesting discoveries have been reported both in theories and in experiments. In the high density and low temperature region, on the other hand, our knowledge is still poor as compared with the rich physics expected in this region. Heavy-ion collisions are not suitable for the purpose to probe dense and cold quark matter, and such a system could be realized, if any, only in the cores of compact stellar objects. The experimental data from the universe is, however, quite limited and there is no smoking-gun for color superconductivity so far. Nevertheless, the theoretical challenge to explore the QCD phase structure is of great interest on its own. Also it would be potentially important in studying the structure and evolution of neutron stars.

In order to look into a dense quark system, some of concepts known in condensed matter physics have been imported into QCD in hope of analogous phenomena taking place. In this sense the physics of dense quark matter is, so to speak, "condensed matter physics of QCD" as articulated clearly in the review [1]. Superconductivity is definitely one of them. In general the Cooper instability inevitably occurs wherever there are a sharp Fermi surface below which particles are degenerated and an attractive interaction between particles on the Fermi surface. Even QCD matter is not an exception and the condensation of quark Cooper pairs leads to color superconductivity.

A major difference between ordinary electric superconductivity in metals and color superconductivity in quark matter arises from the fact that quarks have three colors and three flavors in addition to spin, so that quark matter allows for many pairing patterns. The color and flavor degrees of freedom make the dense QCD phase structure so complicated that subtleties still remain veiled. In this article we shall argue what has been clarified by now and what should be solved in the future mainly following the discussions in my recent papers [2,3].

1.1 Pairing patterns

Many theoretical works have revealed that the predominant pairing pattern is anti-symmetric in spin (spin zero), anti-symmetric in color (color triplet), and anti-symmetric in flavor (flavor triplet). Moreover, the color-flavor locking

Gap Parameters	Phase
$\Delta_1 \neq 0,\ \Delta_2 \neq 0,\ \Delta_3 \neq 0$	CFL
$\Delta_1 = 0,\ \Delta_2 \neq 0,\ \Delta_3 \neq 0$	uSC
$\Delta_1 \neq 0,\ \Delta_2 = 0,\ \Delta_3 \neq 0$	dSC
$\Delta_1 = \Delta_2 = 0,\ \Delta_3 \neq 0$	2SC
$\Delta_1 = \Delta_2 = \Delta_3 = 0$	UQM

Table 1. A family of color superconducting phases; the color-flavor locked (CFL) phase, the u-quark superconducting (uSC) phase, the d-quark superconducting (dSC) phase, and the two-flavor superconducting (2SC) phase. UQM is unpaired quark matter. The sSC, 2SCsu, and 2SCds phases would not appear in the QCD phase diagram.

is known to be favored in energy, so that the color and flavor indices are locked together. Then there are three independent diquark condensates or gap parameters [4];

$$\langle \psi_i^a C\gamma_5 \psi_j^b \rangle \sim \Delta_1 \epsilon^{ab1}\epsilon_{ij1} + \Delta_2 \epsilon^{abs}\epsilon_{ij2} + \Delta_3 \epsilon^{ab3}\epsilon_{ij3}, \tag{1}$$

where (i,j) and (a,b) represent the flavor indices (u,d,s) and the color triplet indices (red,green,blue) respectively. The charge conjugation C and the Dirac matrix γ_5 are required to make (1) a Lorenz scalar. Of course you can consider other kind of pairing between quarks which are totally anti-symmetric under exchange, and even different types of condensates may coexist. Actually diquark condensates such as spin-zero pairing in the color symmetric (color sextet) channel and spin-one pairing between quarks of the same flavor have been analyzed quantitatively [5–8] and known to be much smaller than the predominant condensate. In this article we shall simply neglect them.

Under the pairing ansatz (1), Δ_1 is a gap parameter for the Cooper pairing between d and s flavors and green and blue colors. That is, Δ_1 is for bd-gs and gd-bs quarks and Δ_2 and Δ_3 are to be understood likewise;

Δ_1	*bd-gs*	*gd-bs*
Δ_2	*rs-bu*	*bs-ru*
Δ_3	*gu-rd*	*ru-gd*

Each gap parameter is either zero or finite and there are $2^3 = 8$ combinations accordingly. Only five of eight phase possibilities as listed in Table 1 are of our interest relevant to the QCD phase diagram. When three gap parameters are all nonzero, this state is called the color-flavor locked (CFL) phase. When only Δ_1 is zero, this is the u-quark superconducting (uSC) phase named after the fact that remaining Δ_2 and Δ_3 are gap parameters for pairing involving u-quarks. The d-quark superconducting (dSC) phase is understood in the same way. The two-flavor superconducting (2SC) phase has only Δ_3 which is nonvanishing. The question is; where and how they show up on the phase diagram. The next section is devoted to this issue.

It is worth mentioning that these phases can be characterized by global symmetry breaking patterns. In particular the second-order phase transitions between the CFL phase and the uSC or dSC phase belong to the universality class same as an O(2) vector model [9]. In QCD, neither the sSC (s-quark superconducting), 2SCsu (2SC of s- and u-quarks), nor 2SCds (2SC of d- and s-quarks) phase is realized actually because any pairing containing massive s-quarks is disfavored by the Fermi surface mismatch energy. (The solution branch of the gap equations belonging to the 2SCsu phase has been examined in Refs. [10,11] and confirmed to cost a larger energy.)

1.2 Gapless superconductors

The classification of superconducting phases under the pairing ansatz (1) is not complete until we will take account of the gapless superconducting states. They do not break any new symmetry and the phase transition to a gapless superconductor exists only at zero temperature [2]. In this sense the gapless CFL (gCFL) phase for instance which we will closely discuss later is not a totally new *phase* but can be regarded as a *variant* of the CFL phase augmented by the presence of gapless quarks. Before addressing the CFL problem, we shall see the simplest example that is actually enough for abstracting the essence.

Let us assume that there are two species of particles, 1 and 2, which have the Fermi momenta $\bar{\mu}-\delta\mu/2$ and $\bar{\mu}+\delta\mu/2$ respectively and they form a Cooper pair. The energy dispersion relations without pairing are shown by the solid and dotted lines in Fig. 1. In the presence of the 1-2 pair condensate Δ, the level repulsion by the energy Δ results in the dispersion relations which are smoothly connected between the hole state of 2 (or 1) and the particle state of 1 (or 2). In this simple model of superconductivity, therefore, the quasi-particle energy is expressed as

$$\epsilon^{\pm}(p) = \left|\sqrt{(p-\bar{\mu})^2+\Delta^2} \pm \tfrac{1}{2}\delta\mu\right|. \tag{2}$$

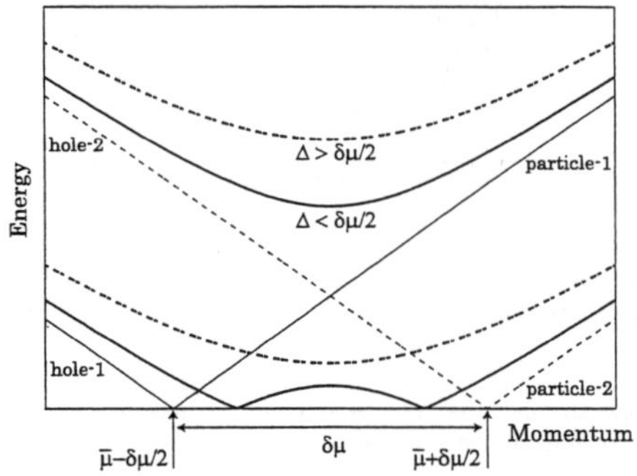

Fig. 1. The energy dispersion relations near the average Fermi surface $\bar{\mu}$ for 1 (solid lines) and 2 (dotted lines) species *without pairing* which have a Fermi surface mismatch $\delta\mu$. The dashed curves represent the energy dispersion relations *with pairing* between 1 and 2 species with the gap parameter $\Delta > \delta\mu/2$. When $\Delta < \delta\mu/2$ the dispersion relations cross zeros as shown by the solid curves and gapless quarks appear inside the blocking region.

As is obvious from Fig. 1 as well as from the expression (2), the dispersion relation comes to cross zeros when $|\frac{1}{2}\delta\mu| > \Delta$. The momentum region from one zero to another zero of the dispersion relation is called the blocking momentum region because the pairing within this region is hindered by degenerated particles. (In the case of Fig. 1 particle-2 is degenerated and particle-1 is absent in the blocking region.) Generally once a superconductor enters the gapless state, the gap parameter significantly decreases with increasing blocking momentum region.

In the language of physics, the condition for the gapless onset $|\frac{1}{2}\delta\mu| > \Delta$ means that a Cooper pair is not stable energetically. The pairing energy 2Δ is needed for breaking a Cooper pair into two particles and at the same time the mismatch energy $\delta\mu$ is released by doing so. If the mismatch is more expensive than the pairing, two particles would no longer form a Cooper pair. Roughly speaking, the realization of gapless superconductivity can be seen as a weak instability disrupting the Cooper pair only in a limited blocking region but not ruining the whole superconductivity.

We will end this subsection with one more remark. In this article (and in some literatures) $\delta\mu$ is frequently called the Fermi surface mismatch. It should be kept in mind that this mismatch is for the energy dispersion relations *without* pairing. In the presence of pairing, as observed in Fig. 1, the (approximate) Fermi surface is provided by the average $\bar{\mu}$. In other words two energy dispersions $\epsilon^{\pm}(p)$ in (2) have a common $\bar{\mu}$.

1.3 Strange quark mass effect and neutrality

We have so far illustrated what the gapless superconducting state is like. This gapless phase is often called the Sarma phase and was first demonstrated as a metastable state by Sarma [12]. The Sarma phase has recently attracted much interest since it was pointed out that it can be stable under the constraint to fix the particle number [13,14]. Also in the context of QCD the gapless 2SC (g2SC) phase first [15] and the gCFL phase next [16] have been shown to be stable under the electric and color neutrality conditions, that is; a potential *maximum* with respect to the gap parameters turns to be a *minimum* if additional gap parameter dependence through the neutrality conditions is properly taken into account. The gapless superconducting state stabilized in this way, however, turns out to be unstable after all, which is actually the main subject we shall argue spending the last half of this article.

Anyway, postponing the instability problem for a while, let us go on our discussion on the gCFL phase. Now we have to understand where $\delta\mu$ could come from. There are two distinct contributions from one origin; one is the direct M_s effect on the energy dispersion relations and the other is the induced M_s effect through the neutrality constraints.

The direct effect is simple. In the vicinity of the Fermi surface ($p \sim \mu$), the quark energy dispersion relation can be well approximated as

$$\sqrt{p^2 + M_s^2} - \mu \simeq p - \Bigl(\mu - \frac{M_s^2}{2\mu}\Bigr). \tag{3}$$

Thus the strange quark mass effect can be incorporated by a chemical potential shift proportional to M_s^2/μ. It is actually the case that the physics should change as a function of M_s^2/μ alone.

An explanation of the induced M_s effect needs some knowledge on the neutrality conditions. As far as a chunk of quark matter, like the cores of neutron stars which are of kilometer size, is concerned, neutrality with respect to charge associated with any gauge field must be required [17–19]. Besides, in the case of color charge in QCD, bulk quark matter must be a color singlet as a whole system. This is a more stringent condition than neutrality. The singletness is in principle to be implemented by an appropriate projection operator onto neutral states. There are many quarks in a macroscopic system and there are many thermodynamically equivalent states. If the number of singlet states grows in the large volume limit in the same way as the number of neutral states, the singletness condition can be simply ignored in thermodynamic properties. This is in fact the case in color superconducting quark matter [20]. Thus only the neutrality conditions for electric and color charges suffice for our purpose to explore the phase structure.

In the grand canonical ensemble, the number density of particles is specified by a chemical potential. We should introduce one chemical potential μ_e for the electromagnetic charge and eight chemical potentials μ_α for the color charges which are not gauge-invariant in non-abelian gauge theories. For the purpose of imposing neutrality it is enough to constrain only two eigenvalues of the color charge since zero charge remains to be zero charge if it is rotated by any gauge transformations. We thus only have to solve the color neutrality conditions with respect to two chemical potentials μ_3 and μ_8 [18].

Quark matter with three flavors would be automatically electric and color neutral if M_s is zero, meaning that μ_e, μ_3, and μ_8 should be of order M_s^2/μ just like the direct M_s effect.

In summary, the direct and induced M_s effects are concisely expressed in a form of the effective chemical potentials for respective quarks with color a and flavor i as

$$\mu_{ai} = \mu - \mu_e(Q)_{ii} + \mu_3(T_3)_{aa} + \mu_8\left(\tfrac{2}{3}T_8\right)_{aa} - \left(M^2\right)_{ii}/2\mu, \tag{4}$$

where $Q = \mathrm{diag}(\frac{2}{3}, -\frac{1}{3}, -\frac{1}{3})$ and $M = \mathrm{diag}(0, 0, M_s)$ in flavor (u,d,s) space and $T_3 = \mathrm{diag}(\frac{1}{2}, -\frac{1}{2}, 0)$ and $\frac{2}{\sqrt{3}}T_8 = \mathrm{diag}(\frac{1}{3}, \frac{1}{3}, -\frac{2}{3})$ in color (red,green,blue) space. From (4) the Fermi surface mismatch for all the pairings is easily inferred; for the bd-gs pairing for instance the mismatch is

$$\delta\mu_{bd-gs} = \mu_{bd} - \mu_{gs} = \tfrac{1}{2}\mu_3 - \mu_8 + \tfrac{M_s^2}{2\mu}. \tag{5}$$

In the CFL phase at zero temperature a model-independent argument [18] yields $\mu_e = \mu_3 = 0$ and $\mu_8 = -M_s^2/2\mu$, and together with the above expression, the gapless onset condition (where gapless quarks begin appearing) is

$$\delta\mu_{bd-gs} = M_s^2/\mu > 2\Delta_1. \tag{6}$$

Numerical calculations in a model study [2, 16] have confirmed this onset condition being a good estimate.

2 Phase diagram of dense quark matter

We will present the phase diagram first and then look closely at each phase in turn. Figure 2 is the phase diagram obtained in the Nambu–Jona-Lasinio (NJL) model. The model parameters are chosen as to yield $\Delta = 40$ MeV at $\mu = 500$ MeV and $M_s = T = 0$. The strange quark mass is fixed at $M_s = 150$ MeV which is the lowest estimate for the M_s effect in the intermediate density region. In Ref. [21] the authors solved the gap equations for the chiral condensates as well as the gap parameters and the chemical potentials. In the case of weak coupling, Δ is smaller and thus the critical value of the strange quark mass (or chemical potential) on the gCFL onset is smaller (or larger), and the gCFL phase is not substantially affected by the chiral dynamics then.

Even if the chiral condensate is not dealt with dynamically, a weird structure including a first-order phase boundary emerges on the phase diagram for large M_s^2/μ. As discussed in Ref. [2], however, the complicated structure seen at large M_s^2/μ is model and parameter dependent. Unfortunately there is no guiding principle to avoid unphysical artifacts in a certain parameter choice and one has to try and show all the cases for conclusive analyses [21]. In this article we dare not to touch that subtle issue and actually this is the reason why we present the phase diagram only for $\mu > 390$ MeV; a first-order phase boundary between the gCFL phase and the UQM is found at $\mu \simeq 388$ MeV.

The phase structure depicted in Fig. 2 is robust at least in a sense that the gross features would not be amended by assumptions we have made to draw this figure. Besides the chiral dynamics the important physics we have dropped is the K^0-condensation that could delay the gapless onset [22] and the 't Hooft interaction term that embodies the instanton induced $\mathrm{U_A}(1)$ breaking interaction in the NJL model Lagrangian. These two missing effects are actually closely linked. They could change the quantitative details of the phase diagram but no qualitative essence would be altered. We would not refer to the mixed phase possibility here, for it has already been settled [10].

We can see that in the lower density region near the gCFL phase the uSC region opens at finite temperature while the dSC phase is relevant at higher density. We can understand this behavior in a model-independent way and the existence of the "doubly critical point" is concluded [2]. In the subsequent subsections we shall discuss the important nature of the gCFL phase and how the uSC and dSC phases come out from the Fermi surface mismatch, and finally comment upon the doubly critical point.

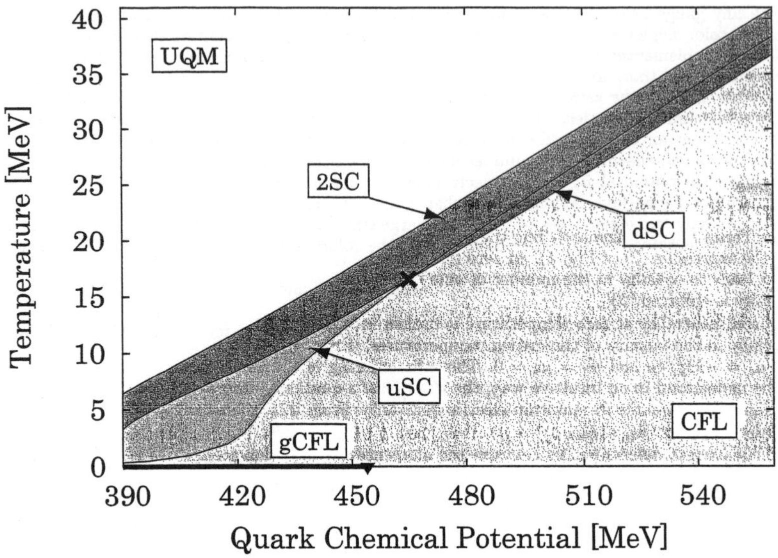

Fig. 2. Phase diagram of dense quark matter obtained in the NJL model with the parameters chosen to give $\Delta = 40$ MeV at $\mu = 500$ MeV and $T = M_s = 0$. The strange quark mass M_s is fixed at 150 MeV under the assumption that the first-order chiral phase transition occurred at some chemical potential below 390 MeV and $\langle \bar{s}s \rangle \propto M_s$ above that would not induce large corrections to M_s. The doubly critical point and the gCFL onset are marked by a cross and a triangle respectively.

2.1 gCFL phase

The gapless onset is located at $\mu \simeq 455$ MeV where $\Delta_1 = 24.9$ MeV and $\mu_8 = -24.7$ MeV and one can readily confirm that $\mu_8 \simeq -M_s^2/2\mu$ and $M_s^2/\mu \simeq 2\Delta_1$ as indicated by the condition (6). The gCFL phase has a definite meaning only at zero temperature because there is no clear distinction between gapless quarks and thermally excited quarks at finite temperature.

The onset condition (6) is only for *bd-gs* quarks with Δ_1 but gapless quarks can be present in another quark sector as well. In fact the energy dispersion relations of *rs-bu* quarks turn out to be identical to those of *bd-gs* quarks as long as the system remains to be in the CFL (not gCFL) phase. Thus *rs-bu* quarks become gapless at $M_s^2/\mu \simeq 2\Delta_2$ where $\Delta_2 = \Delta_1$ holds in the CFL phase. For larger M_s or smaller μ, the blocking momentum region in the *bd-gs* sector becomes wider, while the blocking region in the *rs-bu* sector is severely constrained by the neutrality conditions so that it must remain tiny (see Refs. [2,3,16] for details). As a result the *rs-bu* dispersion relations are kept to be almost quadratic in the entire gCFL region. Therefore only Δ_1 in the *bd-gs* quark sector significantly drops in the gCFL region due to the spreading blocking region with increasing M_s^2/μ.

Because this is an essential point in considering the instability problem later, we shall reiterate here;

bd-gs pairing with Δ_1 — gapless quarks appear at $M_s^2/\mu = 2\Delta_1$; the blocking region increases for larger M_s^2/μ.

rs-bu pairing with Δ_2 — gapless quarks appear at $M_s^2/\mu = 2\Delta_2$; the dispersion relations are kept to be almost quadratic in the entire gapless phase.

gu-rd pairing with Δ_3 — gapped quarks only.

ru-gd-bs pairing with all Δ's — gapped quarks only.

2.2 uSC phase

The uSC phase results from the presence of the gCFL phase at zero temperature. In the gCFL phase, as explained in the last section, Δ_1 is significantly reduced than Δ_2 and Δ_3. The pairing between *u-d* quarks is not reduced but Δ_3 is enhanced in the gCFL side as Δ_1 and Δ_2 decreases [2]. (Note that Δ_3 evaluated in the 2SC state is in general larger than Δ_3 in the CFL state for the same diquark interaction.) In this way the ordering $\Delta_1 < \Delta_2 < \Delta_3$ is realized.

It is a well-known fact that the critical temperature is of order of the gap parameter at zero temperature. We can anticipate that Δ_1 would melt first at finite temperature right above the gCFL region on the phase diagram. This means that the uSC phase where only Δ_1 is zero is expected when gCFL matter is heated. Figure 3 clearly shows that the phase structure is certainly as anticipated.

2.3 dSC phase

It is not the Fermi surface mismatch but the average Fermi momentum that is more relevant away from the gCFL region (see the expression (2) or Fig. 1). At zero temperature the average Fermi momenta are common in all the quark sectors, that leads to equality in the number of nine (three colors and three flavors) quarks, and thus the electric and color neutrality is enforced [23].

The enforced neutrality at zero temperature is broken at small temperatures of a few MeV [2]. At higher temperature, especially in the vicinity of the critical temperatures, it is a good approximation to estimate μ's in the normal phase, i.e., $\mu_e = -M_s^2/4\mu$ and $\mu_3 = \mu_8 = 0$. Then the ordering of the Fermi momenta, $\mu_s < \mu_u < \mu_d$ is concluded, which can be understood in an intuitive way; the number of s-quarks is suppressed by M_s and so d-quarks should be more abundant than s-quarks to maintain electric neutrality. From this, the average Fermi momenta should obey the following ordering; $\bar{\mu}_{su} < \bar{\mu}_{ds} < \bar{\mu}_{ud}$.

The gap equation to determine Δ's contains the momentum integration around the Fermi surface which effectively picks up the density of states at the Fermi momentum. The larger the density of states is, the greater the gap parameter becomes. In this way the ordering $\Delta_2 < \Delta_1 < \Delta_3$ is realized from the average Fermi momenta ordering, which follows the presence of the dSC phase accordingly, as first discussed in Ref. [24].

2.4 Doubly critical point

The phase boundary on which Δ_1 goes to zero crosses the phase boundary on which Δ_2 goes to zero at the "doubly critical point" where two phase transitions with respect to Δ_1 and Δ_2 take place simultaneously. Since the existence of the uSC and dSC phases are robust and model-independent, so is the doubly critical point.

We would shortly comment upon a puzzling question concerning the doubly critical point. The question is the following; what are the effects of gauge field fluctuations on the doubly critical point?

In weak coupling the gauge field fluctuations bring about an induced first-order phase transition and the critical temperature is shifted [25,26]. Between the CFL phase and the uSC or dSC phase, no manifest effects would be expected because eight gluons are all massive in both phases. Therefore we can conclude that the phase transitions from the CFL phase toward either the uSC or dSC phase is surely of second order belonging to the same universality class as an O(2) vector model.

The gauge field fluctuations play an important role, on the other hand, between the 2SC phase and the uSC or dSC phase and the phase transition is forced to be of first order.

The doubly critical point is the point at which two phase transitions meet and three gluons become massless around it since it faces the 2SC phase. This suggests that the phase transition must be of first order and the critical temperature should be shifted at the doubly critical point. But how is it possible if the phase boundaries facing the CFL phase are not shifted at all? Or, are they shifted near the doubly critical point by nearly massless gluons? To answer this question, further clarification on the treatment of the gauge field fluctuations is waited.

3 Chromomagnetic instability

The gapless superconductors could be stabilized by the neutrality conditions, at least, in the variational space spanned by gap parameters and chemical potentials, but after all, they have turned out to be unstable in a wider space including the gauge fields. The chromomagnetic instability signifies that the Meissner screening mass squared becomes negative, that is, the Meissner mass is imaginary. Before addressing the chromomagnetic instability problem first pointed out by Huang and Shovkovy [27] in the case of the g2SC phase and analyzed in Refs. [3,28,29] for the gCFL phase, we shall make a brief overview on the Debye and Meissner screening masses in the CFL phase, which would be useful to make a deeper insight into the instability problem. Our results are summarized in Fig. 3 and the goal of this section is to explain what is going on inside the instability regions presented by shadowed regions in this figure.

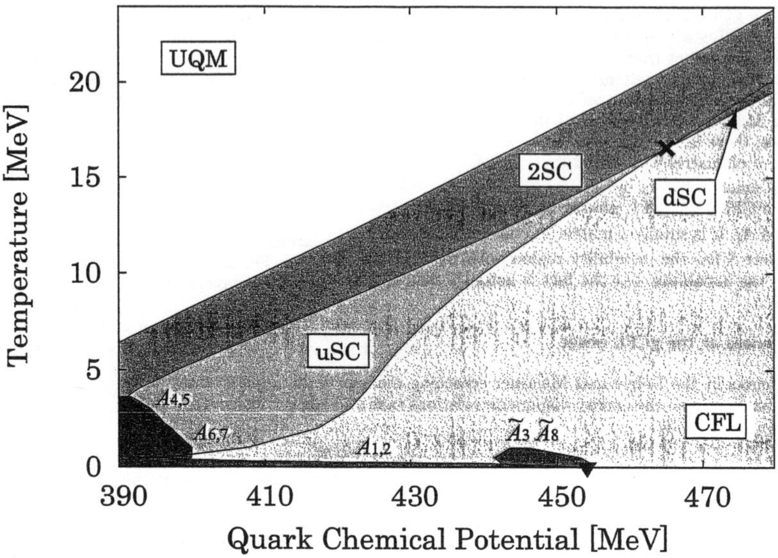

Fig. 3. A magnified drawing of the phase diagram around the gCFL and uSC phases with the instability regions overlaid.

3.1 Debye and Meissner screening masses

The Debye and Meissner screening masses are the screening masses for the longitudinal and transverse gauge fields respectively, which are defined by

$$m_{D,\alpha\beta}^2 = -\lim_{q\to 0} \Pi_{\alpha\beta}^{00}(\omega = 0, \boldsymbol{q}), \tag{7}$$

$$m_{M,\alpha\beta}^2 = \tfrac{1}{2}\lim_{q\to 0}(\delta_{ij} - \widehat{q}_i\widehat{q}_j)\Pi_{\alpha\beta}^{ij}(\omega = 0, \boldsymbol{q}), \tag{8}$$

where $\widehat{q}_i = q_i/|\boldsymbol{q}|$ and $\Pi_{\alpha\beta}^{\mu\nu}$ is the polarization tensor for the gauge fields A_α^μ with the color and Lorenz indices denoted by α and μ. At high enough density the polarization tensor is dominated by the quark-loop contributions alone and so we shall neglect the gluon and ghost loops that would not depend on μ at the one-loop level.

Let us take one example to elaborate the quark-loop contributions for each gluon. The A_1 and A_2 gluons with the color indices labeled according to the Gell-Mann matrices in color space couple red quarks to green quarks and vice versa, while the flavor is not changed at any gluon vertices. For example, a diagram graphically shown as

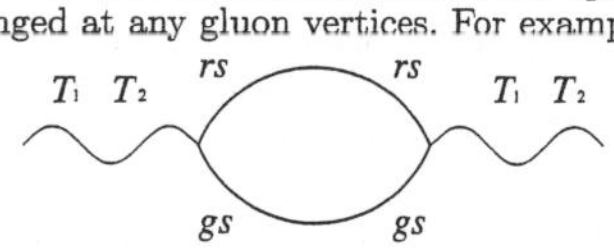

(Example 1)

can contribute to $\Pi_{11}^{\mu\nu}$ and $\Pi_{22}^{\mu\nu}$. Solid lines represent the quark propagator for rs and gs quarks in this specific case. From this kind of diagrammatic deliberation, it is easy to confirm that there is no mixing between gluons except A_3-A_8 mixing. Moreover, some algebra can readily tell us that the screening masses are degenerate for the gluon pairs (A_1,A_2), (A_4,A_5), and (A_6,A_7). In this article, like in Ref. [3], we shall denote $A_{1,2}$, $A_{4,5}$, and $A_{6,7}$ to mean either of the degenerated pairs.

3.2 Unstable channels and mixing with photon

If there is something funny in connection with gapless quarks in the quark-loop polarization like Example 1, it is expected to happen when both two quark propagators involve gapless energy dispersion relations. At the gapless onset, in fact, the quark-loop polarization consisting only of bd, gs, rs, and bu quarks is divergent, leading to the divergent contributions to the Debye and Meissner masses for $A_{1,2}$, A_3, and A_8 gluons. You can easily check that the polarization diagrams relevant to the $A_{4,5}$ and $A_{6,7}$ gluon channels must have at least one of gu, rd, ru, bd, and bs quarks which are gapped.

It should be noted here about mixing between the A_γ photon and the A_3 and A_8 gluons. In the symmetric CFL phase with $M_s = 0$, there is no mixing with respect to A_3, but in the present case with $M_s \neq 0$ and thus $\mu_e \neq 0$ in the gCFL phase, there is A_3-mixing as well due to the isospin symmetry breaking. The mass squared matrix for A_γ, A_3, and A_8 is a 3×3 matrix with nonvanishing off-diagonal components generating mixing. We will denote the eigenmodes of the mass squared matrix by $\widetilde{A}_\gamma$, $\widetilde{A}_3$, and $\widetilde{A}_8$. The rotated photon represented by the $\widetilde{A}_\gamma$ field is massless all the way because the CFL and gCFL phases preserve a rotated electromagnetic U(1) symmetry. Once mixing occurs among A_γ, A_3, and A_8, it is simply a matter of convention which eigenmode should be identified as $\widetilde{A}_3$ or $\widetilde{A}_8$. By this reason, though Figure 3 has the instability region with the label $\widetilde{A}_3$ and $\widetilde{A}_8$, it does not mean that both of two eigenmodes suffer from the instability, but the fact is actually that only one of them does.

3.3 Divergences at the gCFL onset

The divergences in the Debye and Meissner screening masses at the gapless onset derive from the density of states which is divergent when the energy dispersion relations take a quadratic form; $\epsilon(p) \sim (p-\bar{\mu})^2/2\Delta$. That is, the density of states $n(p)$ is given by

$$n(p) = 4\pi p^2 \left[\frac{d\epsilon(p)}{dp}\right]^{-1} \tag{9}$$

and obviously $n(p) \to \infty$ when the slope of $\epsilon(p)$ is zero at $p = \bar{\mu}$.

The next question is whether the divergence is positive or negative in the Debye and Meissner masses. A naive intuition would be that both are positive, for in the CFL phase it is well established that $m_M^2 = \frac{1}{3}m_D^2$ should hold [9, 30,31] and one might well consider that they are correlated in a similar way even in the presence of $M_s \neq 0$. This expectation is partly true but in an unexpected manner as we will shortly see below.

To find the relations between the Debye and Meissner masses, it is convenient to split the quark propagator into four distinct parts. If the quark mass effect is approximated as (3) near the Fermi surface, the energy projection operator divides the propagator into the *particle* part that is a function of $p - \bar{\mu}$ and the *antiparticle* part that is a function of $p + \bar{\mu}$. Moreover, in the Nambu-Gor'kov formalism, the quark propagator is a 2×2 matrix and its *diagonal* component is a normal propagation of particles, and its *off-diagonal* component is an abnormal propagation mediated by diquark condensates. There are thus four distinct combinations; diagonal–particle, diagonal–antiparticle, off-diagonal–particle, and off-diagonal–antiparticle.

The diagram shown in Example 1 is one example consisting of only the diagonal propagators. We can construct another example in the same gluon channel composed of only the off-diagonal propagators as follows;

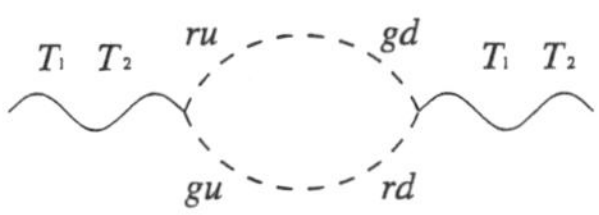

(Example 2)

where the upper propagator is the off-diagonal component proportional to Δ's connecting ru and gd quarks, and the lower one is the off-diagonal component proportional to Δ_3 connecting gu and rd quarks.

It is important to note that the possible singular behavior originating from gapless quarks can reside only in the particle parts because antiparticles would never be gapless. The Debye mass squared receives the contributions only from particles which can be separated into the diagonal part, $[m_D^2]_{\text{diag}}$, and the off-diagonal part, $[m_D^2]_{\text{off}}$. The Meissner mass, on the other hand, has all of the particle-particle, particle-antiparticle, and antiparticle-antiparticle contributions. In Ref. [3] interesting relations have been found;

$$\left[m_M^2\right]_{\text{diag(pp)}} = -\tfrac{1}{3}\left[m_D^2\right]_{\text{diag}}, \tag{10}$$

$$\left[m_M^2\right]_{\text{off(pp)}} = \tfrac{1}{3}\left[m_D^2\right]_{\text{off}}, \tag{11}$$

where $[m_M^2]_{\text{diag(pp)}}$ is a part of the Meissner mass squared coming from the diagrams with two diagonal particle propagators (like Example 1) and $[m_M^2]_{\text{off(pp)}}$ with two off-diagonal particle propagators (like Example 2). The relation (10) might be strange at a glance, for we already know that $m_M^2 = \frac{1}{3}m_D^2$ in the CFL phase at $M_s = 0$. Then, how can we retrieve the relation $m_M^2 = \frac{1}{3}m_D^2$? Actually, as long as the system stays in the CFL side, the diagonal particle-antiparticle contribution $[m_M^2]_{\text{diag(pa)}}$ is *twice larger* than $[m_M^2]_{\text{diag(pp)}}$ and changes the overall sign in the relation between the Debye and Meissner masses. The other contributions, $[m_M^2]_{\rm diag(aa)}$, $[m_M^2]_{\rm off(pa)}$, and $[m_M^2]_{\rm off(aa)}$ are all negligibly small.

It might be surprising that these relations (10) and (11) are satisfied as they are even in the gCFL phase! The diagonal and off-diagonal parts both can generate divergent contributions at the gCFL onset. As far as the QCD problem is concerned, the diagonal contributions are always larger than the off-diagonal ones, leading to the Debye mass squared diverging *positively* and Meissner mass squared diverging *negatively* at the gCFL onset. The negative sign in (10) cannot be compensated in the gCFL phase since, unlike in the CFL phase, $[m_M^2]_{\rm diag(pa)}$ is no longer a match for $[m_M^2]_{\rm diag(pp)}$.

To put it in another way, the key relation (10) can be stated as follows: In the diagonal part, we can say that the particle-particle loops tend to induce *paramagnetism*, while *diamagnetism* stems from the particle-antiparticle loops. Usually in the superconducting phase, the diamagnetic tendency is greater enough to exhibit the Meissner effect. In gapless superconductors, however, antiparticles are never gapless and only the particle-particle loops are abnormally enhanced due to the presence of gapless quarks. As a result of that, the diamagnetism gives way to the chromomagnetic instability.

3.4 Away from the gCFL onset

Figure 3 shows the instability regions for respective gluons. The difference between the $A_{1,2}$ behavior and the $\widetilde{A}_3$-$\widetilde{A}_8$ behavior can be understood from the difference between the *bd-gs* and *rs-bu* quark dispersion relations. It is only the diagram shown in Example 1 that causes singular behavior in $A_{1,2}$ around the gCFL onset. This diagram has two quark propagators; one of *gapless gs*-quarks and the other of *quadratic rs*-quarks. Since rs quarks are kept to be almost quadratic in the entire gCFL region, as we have explained before, the $A_{1,2}$-instability extends over the whole gCFL region at small temperatures. The instability would not persist into regions at higher temperature because the quadratic dispersion relations are easily affected by thermal excitations.

In contrast, the same species of quarks constitute the $\widetilde{A}_3$-$\widetilde{A}_8$ instability region. The instability caused by two quadratic quark propagations lies in the entire gCFL region like the $A_{1,2}$-instability but only at tiny temperatures of order eV. Instability boundaries at such low temperatures are not visible actually on the phase diagram. The instability caused by two gapless quark propagations is localized near the gCFL onset and spreads toward high temperatures than the $A_{1,2}$-instability because it has nothing to do with quadratic quarks.

Although we would not go further into details, the Meissner screening masses evaluated in the g2SC phase [27] indicate that the $\widetilde{A}_8$-instability occurs starting at the g2SC onset and there also arises the instability for $A_{4,5,6,7}$ gluons whose boundary is found not at the g2SC onset but inside the 2SC region. In our results the $A_{4,5}$ and $A_{6,7}$ instability regions are located at large Fermi surface mismatch and enter the uSC phase at higher temperature and then the 2SC phase farther. We conjecture that these instability regions for $A_{4,5}$ and $A_{6,7}$ would be linked to the instability found in the 2SC phase. The nature of the instability with respect to $A_{4,5}$ and $A_{6,7}$ is presumably different from the instability near the gCFL onset that we have seen in great details. In the aim of disclosing the QCD phase diagram in the intermediate density region, in particular, the $A_{4,5}$ and $A_{6,7}$ instability deserves further investigation.

4 Speculations

This final section is devoted to sketching some speculations on how to reach the stable states inside the instability regions on the phase diagram, which has been barely succeeded so far.

There are already some attempts to interpret and resolve the chromomagnetic instability problem [32–35]. The most important among them is, in my opinion, the observation pointed out by Giannakis and Ren in Ref. [32] that the instability with respect to gluons can be interpreted as the instability toward a plane-wave crystalline superconducting phase.

The most straightforward interpretation of the chromomagnetic instability is, of course, spontaneous generation of the expectation value for the transverse gluons. The gauge field itself is, however, not a physical quantity depending on the gauge choice. Actually, what Giannakis and Ren realized in Ref. [32] is that the spontaneously generated gauge fields can be absorbed in the phase of the gap parameters by means of the gauge transformation. It should be noted that whether the gauge invariance is maintained or not does not matter. The gauge transformation in this manipulation simply means the change of variables.

The crystalline phase had been already studied [36] before the chromomagnetic instability was discovered. It is called the "crystalline" phase because the gap parameter takes a form of

$$\Delta(x) = |\Delta| e^{i\boldsymbol{q}\cdot\boldsymbol{x}}, \tag{12}$$

which breaks the translational and rotational invariance. The essential point is that there is another quark basis where the phase factor of (12) vanishes at the price of the vector potential arising in the effective action. Therefore, taking care of a vector potential turns out to be equivalent with dealing with the gap parameter with a phase factor corresponding to the given vector potential.

Let us rephrase the above mentioned idea in a slightly different way. Supposing we performed the derivative expansion of the thermodynamic potential $\Omega_A[\Delta(x)]$ which is now *gauged* with gluons, then we have in general,

$$\Omega_A[\Delta(x)] \simeq \Omega_0[\Delta] \\ -\kappa^{ab}[\Delta]\mathrm{Tr}\big[(\partial_i+iA_i^*)\Delta^*(x)\big]^a\big[(\partial^i-iA^i)\Delta(x)\big]^b\,, \tag{13}$$

where a and b represent the color triplet indices and $A^i = A^i_\alpha T^\alpha$ with the color adjoint index α. The average over x is symbolically implied in Tr. Then, an alternative definition of the Meissner screening mass is immediately available from this thermodynamic potential as

$$m^2_{M,\alpha\beta} = \frac{1}{3}\sum_{i=1}^{3}\frac{\partial^2\Omega_A}{\partial A^i_\alpha \partial A^i_\beta}\Big|_{A=0} \sim 2\kappa^{ab}(T^\alpha)_{ca}(T^\beta)_{bd}\Delta^c\Delta^d\,, \tag{14}$$

where $\Delta(x)$ is assumed to be spatial constant. It might be instructive to see how this expression works actually. In the 2SC phase for example, $\Delta^a \propto \delta^{a3}$ and $(T^1)_{a3} = (T^2)_{a3} = (T^3)_{a3} = 0$, and hence one can instantly conclude that the A_1, A_2, and A_3 gluons are not Meissner screened. In the g2SC case the A_8 gluon is unstable, which can be stated by the condition $\kappa^{33} < 0$ because $m^2_{M,88} \sim \kappa^{33}(\Delta^3)^2$.

Now we shall assume the crystalline superconducting phase with (12). The curvature of the thermodynamic potential with respect to $\boldsymbol{q}$ is

$$\frac{1}{3}\sum_{i=1}^{3}\frac{\partial^2\Omega}{\partial q^i\partial q^i} \sim \kappa^{ab}\Delta^a\Delta^b. \tag{15}$$

In the 2SC phase the instability condition for $\boldsymbol{q}$ to grow is given by $\kappa^{33} < 0$ again. This argument works well only in the (g)2SC case in which the A_8-instability can be identified as the instability toward a crystalline superconducting phase equivalently. Essentially the above is what has been articulated in Refs. [32,33].

The generalization to the (g)CFL problem is possible with a simple extension of the crystalline ansatz (12). As suggested by Giannakis and Ren, we can consider a *colored* crystalline superconducting phase with the gap parameter taking a form of

$$\Delta(x) = |\Delta| e^{iT^\alpha \boldsymbol{q}^\alpha\cdot\boldsymbol{x}}, \tag{16}$$

then it is almost obvious that the curvature with respect to $\boldsymbol{q}^\alpha$ is identical with the Meissner mass squared. In this sense, we can identify the instability regions in Fig. 3 with the colored crystalline phase with nonvanishing $\boldsymbol{q}^\alpha$.

What we should do next is now apparent; $\boldsymbol{q}^\alpha$ are the new variational parameters to be determined so as to minimize the thermodynamic potential. This is a quite tough task, however. The number of the new variables is five corresponding to $A_{1,2}$, $A_{4,5}$, $A_{6,7}$, A_3, and A_8. To be worse, the rotational symmetry is broken by the direction of $\boldsymbol{q}^\alpha$, and the momentum angle integration cannot simplify in evaluating the thermodynamic potential. So, the numerical calculations are too time-consuming to be done. Moreover it is difficult to achieve an enough accuracy to get reliable outputs in the multi-dimensional numerical integration. Maybe we have to abandon this apparent strategy and instead need to invent a wiser simplification that would not lose the important physics.

One possibility would be to go to the higher orders in the derivative expansion (13). The calculation would be feasible because the expansion coefficients can be evaluated at vanishing $\boldsymbol{q}^\alpha$ and the rotational symmetry is not broken then. Although we certainly have a chance to find the energy minimum with some $\boldsymbol{q}^\alpha$, there is no guarantee that the next higher order terms can be adequate to stabilize the potential. In any case this kind of calculation has yet to be performed.

Finally, if I am excused to speak of my own opinion, all these theoretical efforts will be tested in the lattice QCD simulation someday when the sign problem at finite density will be solved and the lattice spacing can be small enough to describe high density matter. The phase diagram like Fig. 3 is to be confirmed then.

This article is based on the talk given for the new talent sessions at International School of Subnuclear Physics 43rd Course held at Erice in Italy from Aug. 29 to Sep. 7 in 2005. I thank all the organizers for stimulating lectures and discussions at school.

References

1. K. Rajagopal and F. Wilczek, arXiv:hep-ph/0011333.

2. K. Fukushima, C. Kouvaris and K. Rajagopal, Phys. Rev. D **71** (2005) 034002 [arXiv:hep-ph/0408322].
3. K. Fukushima, Phys. Rev. D **72** (2005) 074002 [arXiv:hep-ph/0506080].
4. M. G. Alford, K. Rajagopal and F. Wilczek, Nucl. Phys. B **537** (1999) 443 [arXiv:hep-ph/9804403].
5. M. G. Alford, J. Berges and K. Rajagopal, Nucl. Phys. B **558** (1999) 219 [arXiv:hep-ph/9903502].
6. S. B. Ruster, I. A. Shovkovy and D. H. Rischke, Nucl. Phys. A **743** (2004) 127 [arXiv:hep-ph/0405170].
7. M. Iwasaki and T. Iwado, Phys. Lett. B **350** (1995) 163.
8. For a review on spin-one color superconductivity, see, A. Schmitt, arXiv:nucl-th/0405076.
9. K. Fukushima and K. Iida, Phys. Rev. D **71** (2005) 074011 [arXiv:hep-ph/0501276].
10. M. Alford, C. Kouvaris and K. Rajagopal, arXiv:hep-ph/0407257.
11. H. Abuki and T. Kunihiro, arXiv:hep-ph/0509172.
12. G. Sarma, J. Phys. Chem. Solids **24** (1963) 1029.
13. W. V. Liu and F. Wilczek, Phys. Rev. Lett. **90** (2003) 047002 [arXiv:cond-mat/0208052].
14. E. Gubankova, W. V. Liu and F. Wilczek, Phys. Rev. Lett. **91** (2003) 032001 [arXiv:hep-ph/0304016].
15. I. Shovkovy and M. Huang, Phys. Lett. B **564** (2003) 205 [arXiv:hep-ph/0302142]; Nucl. Phys. A **729** (2003) 835 [arXiv:hep-ph/0307273].
16. M. Alford, C. Kouvaris and K. Rajagopal, Phys. Rev. Lett. **92** (2004) 222001 [arXiv:hep-ph/0311286]; Phys. Rev. D **71** (2005) 054009 [arXiv:hep-ph/0406137].
17. K. Iida and G. Baym, Phys. Rev. D **63** (2001) 074018 [Erratum-ibid. D **66** (2002) 059903] [arXiv:hep-ph/0011229].
18. M. Alford and K. Rajagopal, JHEP **0206** (2002) 031 [arXiv:hep-ph/0204001].
19. A. W. Steiner, S. Reddy and M. Prakash, Phys. Rev. D **66** (2002) 094007 [arXiv:hep-ph/0205201].
20. P. Amore, M. C. Birse, J. A. McGovern and N. R. Walet, Phys. Rev. D **65** (2002) 074005 [arXiv:hep-ph/0110267].
21. H. Abuki, M. Kitazawa and T. Kunihiro, Phys. Lett. B **615** (2005) 102 [arXiv:hep-ph/0412382].
22. A. Kryjevski and T. Schafer, Phys. Lett. B **606** (2005) 52 [arXiv:hep-ph/0407329]; A. Kryjevski and D. Yamada, Phys. Rev. D **71** (2005) 014011 [arXiv:hep-ph/0407350].
23. K. Rajagopal and F. Wilczek, Phys. Rev. Lett. **86** (2001) 3492 [arXiv:hep-ph/0012039].
24. K. Iida, T. Matsuura, M. Tachibana and T. Hatsuda, Phys. Rev. Lett. **93** (2004) 132001 [arXiv:hep-ph/0312363].
25. T. Matsuura, K. Iida, T. Hatsuda and G. Baym, Phys. Rev. D **69** (2004) 074012 [arXiv:hep-ph/0312042].
26. I. Giannakis, D. f. Hou, H. c. Ren and D. H. Rischke, Phys. Rev. Lett. **93** (2004) 232301 [arXiv:hep-ph/0406031].
27. M. Huang and I. A. Shovkovy, Phys. Rev. D **70** (2004) 051501 [arXiv:hep-ph/0407049]; Phys. Rev. D **70** (2004) 094030 [arXiv:hep-ph/0408268].
28. R. Casalbuoni, R. Gatto, M. Mannarelli, G. Nardulli and M. Ruggieri, Phys. Lett. B **605** (2005) 362 [Erratum-ibid. B **615** (2005) 297] [arXiv:hep-ph/0410401].
29. M. Alford and Q. h. Wang, J. Phys. G **31** (2005) 719 [arXiv:hep-ph/0501078].
30. D. T. Son and M. A. Stephanov, Phys. Rev. D **61** (2000) 074012 [arXiv:hep-ph/9910491]; Phys. Rev. D **62** (2000) 059902 [arXiv:hep-ph/0004095].
31. D. H. Rischke, Phys. Rev. D **62** (2000) 054017 [arXiv:nucl-th/0003063].
32. I. Giannakis and H. C. Ren, Phys. Lett. B **611** (2005) 137 [arXiv:hep-ph/0412015]; Nucl. Phys. B **723** (2005) 255 [arXiv:hep-th/0504053].
33. M. Huang, arXiv:hep-ph/0504235.
34. D. K. Hong, arXiv:hep-ph/0506097.
35. M. Alford and Q. h. Wang, arXiv:hep-ph/0507269.
36. M. G. Alford, J. A. Bowers and K. Rajagopal, Phys. Rev. D **63**, 074016 (2001) [arXiv:hep-ph/0008208]; J. .A. Bowers, J. Kundu, K. Rajagopal and E. Shuster, Phys. Rev. D **64**, 014024 (2001) [arXiv:hep-ph/0101067]; J. Kundu and K. Rajagopal, Phys. Rev. D **65**, 094022 (2002) [arXiv:hep-ph/0112206]; J. A. Bowers and K. Rajagopal, Phys. Rev. D **66**, 065002 (2002) [arXiv:hep-ph/0204079].

Spin-Bits and $\mathcal{N} = 4$ SYM

Alessio MARRANI

INFN - Laboratori Nazionali di Frascati (LNF),
Via E. Fermi 40, 00044 Frascati, Italy

marrani@lnf.infn.it

Abstract. We briefly review the spin-bit formalism, describing the non-planar dynamics of the $\mathcal{N} = 4, d = 4$ Super Yang-Mills $SU(N)$ gauge theory. After considering its foundations, we apply such a formalism to the $su(2)$ sector of purely scalar operators. In particular, we report an algorithmic formulation of a deplanarizing procedure for local operators in the planar gauge theory, used to obtain planarly-consistent, testable conjectures for the higher-loop $su(2)$ spin-bit Hamiltonians. Finally, we outlook some possible developments and applications.

1 Introduction

Large N physics [1] gained noticeable interest in the past few years (for a recent review see e.g. [2]) due to the AdS/CFT conjecture enlightenment [3,4] and, more recently, to the consideration of various limits of this correspondence ([5]-[13]). Initially formulated in the $N \to \infty$ limit, the conjecture in its strong form extends to finite N. It relates the strongly coupled regime of $\mathcal{N} = 4$ SYM to the weakly coupled string theory and viceversa. This property, which makes out of this correspondence a very strong and efficient predictive tool, appeared to be an obstacle in proving the duality in itself.

Berenstein, Maldacena and Nastase studied in [5,6] the correspondence in the neighborood of null geodesics of $AdS_5 \times S^5$, where the geometry appears to resemble that of a gravitational wave [14]-[16]. On the CFT side this corresponds to focusing on SYM operators with a large $\mathcal{R}$-charge. The possibility to find a solution of string theory [17,18] in such a background allows for a quantitative comparison with predictions coming from perturbative SYM computations [19](see [20] for recent reviews on the BMN correspondence and references). This led to an intensive study of the anomalous dimensions of local gauge-invariant (g.i.) composite operators in $\mathcal{N} = 4$, 4-dimensional ($d = 4$) Super Yang–Mills (SYM) theory [21]. A real breakthrough was the discovery of the integrability of the Hamiltonians governing anomalous dimensions in the planar limit $N \to \infty$ [22,23]. Then, these results were extended to 2 and higher loops [24,25]. Indeed, the dynamics in the sector of single-trace bosonic operators of SYM can be mapped into that of the Heisenberg SO(6) spin one model, so that the matrix of planar one-loop anomalous dimensions is identified with the spin chain Hamiltonian [22]. The Bethe Ansatz techniques used for diagonalizing the Hamiltonian become then a powerful tool in determining anomalous dimensions in the gauge theory. As it is now clear, there is a one-to-one correspondence between single-trace operators in SYM theory and spin states in spin-chain models. The improved understanding of SYM overshadowed, to some extent, the study of nonplanar contributions. The latter has to correspond through AdS/CFT to considering the string production on the AdS side. String bits [26] were proposed as a model which mimics this feature out of (but not very far from) the BMN limit. Although the string bit model yields a good tool for the computation of the relevant bosonic quantities, it is affected by serious consistency problems related to the fermionic doubling [13,27].

On the SYM side the exact one-loop dilation operator was derived in [21,22,28]. When non-planarity is taken into account, single and multi-trace operators get mixed. This could still not be tested in the string dual picture. Waiting for a better understanding of string physics on AdS space, one could hope to learn about string interactions there by exploiting the dual gauge theory picture. This was the main motivation of the work carried out by the LNF research group over the past two years. We studied the corresponding spin system which mixes the integrable spin approach and the string bits one. Such a theory can be called a *spin bit* model. Since it allows for dynamical splitting and joining of chains and its variable content is given by spins, the spin bit model differs from the spin chain and the string bit ones, although it can be considered as a mixture of them. In particular, there is no fermion doubling, and supersymmetry in the spin bit model is consistently implemented.

At $N \to \infty$ the spin Hamiltonian is a local and integrable operator. The Hamiltonian and the total spin generator represent the first two charges, in the tower of commuting ones, predicted by integrability [28]. Higher charges are given in terms of higher powers of next-to-nearest spin generators summed up over the chain. Corrections in $\frac{1}{N}$ spoil locality and integrability. The Hamiltonian and its higher spin analogs, which can be interpreted as broken symmetries of the would-be integrable system, can still be defined in terms of powers of spin generators. However, now there is no more restriction to next-to-nearest interactions and the corresponding charges are no longer commuting among themselves. The role of these broken charges in the theory near the "integrable" point $N \to \infty$ remains to be understood.

If the non-planar contributions are considered, the single-trace sector is not conserved anymore, and one ends up with trace splitting and joining in the operator mixing [29]. Even in this case one can still consider a one-to-one map, the so-called spin-bit map, between local g.i. operators and a spin system [30,31]. In this case one has to introduce a set of new degrees of freedom, beyond the spin states, which describes the linking structure of the sites in the spin-chain. This can be encoded in a new field, taking values in the spin-bits permutation group and introducing a new gauge degree of freedom [31].

In this paper we use conventions and notations of [30–32].

2 The $\mathcal{N} = 4, d = 4$ SYM $SU(N)$ gauge theory

In the following we will consider a particular quantum field theory, namely the $\mathcal{N}$=4,d = 4 SYM $SU(N)$ gauge theory. Such a theory has the noteworthy property to be conformally-invariant, due to the vanishing of its beta function (see e.g. [33]). It has the following field content:

$$F_{\mu\nu} \text{ gauge field strength, } \mu,\nu = 0,1,2,3;$$

$$\phi^i \text{ real scalars, } i = 1,...,6 \text{ (vector repr. of } SO(6));$$

$$\lambda^{\mathcal{A}}_{\alpha}, \overline{\lambda}_{\mathcal{A}\dot{\alpha}} \text{ gauginos, } \mathcal{A} = 1,...,\mathcal{N} = 4, \quad \alpha,\dot{\alpha} = 1,2. \tag{2.1}$$

All the fields take value in the adjoint representation (repr.) of the gauge group $SU(N)$, i.e. for example $\phi^i = \phi^i_a T^a$, where T^a $(a = 1,...,N^2-1)$ are the generators of $SU(N)$ in the adjoint. The scalars also span the vector repr. of $SO(6)$, which is the maximal compact bosonic subgroup of the whole $\mathcal{N} = 4$ supergroup $SU(2,2 \mid 4)$; moreover, the underlying algebra $so(6) \sim su(4)$ is the automorphism, or $\mathcal{R}$-symmetry, algebra of the whole $\mathcal{N} = 4, d = 4$ superconformal algebra (SCA) $psu(2,2 \mid 4)$. In the following we will use a compact notation for the $SU(N)$-gauge covariant derivatives of the fields, namely ($s \in N \cup \{0\}$)

$$\nabla^s\phi \equiv \nabla_{\mu_1\mu_2...\mu_s}\phi^i,$$

$$\nabla^s\lambda \equiv \nabla_{\mu_1\mu_2...\mu_s}\lambda^{\mathcal{A}}_{\alpha}, \tag{2.2}$$

and so on. All the elementary fields, as well their derivatives, can be obtained by acting with generators of the $\mathcal{N} = 4$ SCA on the "primary" fields ϕ^i.

By adopting a convenient "philological" nomenclature, we may say that the $\mathcal{N} = 4, d = 4$ SYM "alphabet" is composed by the set of "letters"

$$W_A \equiv \{\nabla^s F, \nabla^s\phi, \nabla^s\lambda, \nabla^s\overline{\lambda}\}. \tag{2.3}$$

The components of W_A transform in the so-called "singleton" (infinite-dimensional) repr. V_F of the $\mathcal{N} = 4$ SCA. Out of the "letters" W_A one can build $SU(N)$ g.i. "words", i.e. single-trace composite operators given by traces (in the adjoint of $SU(N)$) of a sequence of "letters" W_A's. Examples are given by

$$\begin{aligned} \mathcal{O}^{i_1 i_2 ... i_n} &\equiv Tr\left(\phi^{i_1}\phi^{i_2}...\phi^{i_n}\right); \\ \mathcal{O}^{i_1 i_2 ... i_{m+1}}_{\alpha\dot{\alpha}} &\equiv Tr(\phi^{i_1}\phi^{i_2}...\phi^{i_m}\nabla_{\mu_1...\mu_n}\phi^{i_{m+1}} \\ &\qquad \nabla^{\mu_1...\mu_{n-2}}\lambda^{\mathcal{A}}_{\alpha}\nabla^{\mu_{n-1}\mu_n}\overline{\lambda}_{\mathcal{A}\dot{\alpha}}) \end{aligned} \tag{2.4}$$

with

$$\nabla^{\mu} = \eta^{\mu\nu}\nabla_{\nu}, \tag{2.5}$$

where $\eta^{\mu\nu}$ is the 4-dim. Minkowski metric. Moreover, out of "words" one can produce "sentences", which are sequences of "words", i.e. products of single-trace composite operators, given by products of traces (in the adjoint of $SU(N)$) of sequences of "letters". Some examples are

$$\begin{aligned} \mathcal{O}^{i_1 i_2 ... i_{n_1} j_1 j_2 ... j_{n_2}} &\equiv Tr\left(\phi^{i_1}\phi^{i_2}...\phi^{i_{n_1}}\right) Tr\left(\phi^{j_1}\phi^{j_2}...\phi^{j_{n_2}}\right); \\ \mathcal{O}^{i_1...i_{n_2}\nu}_{\dot{\alpha}\beta} &\equiv Tr\left(\nabla_{\mu_1...\mu_{n_1}}\phi^{i_1}...\phi^{i_{n_2}}\right) \cdot \\ &\qquad \cdot Tr\left(\nabla^{\mu_1...\mu_{n_1-1}}\overline{\lambda}_{\mathcal{A}\dot{\alpha}}\right) Tr\left(\lambda^{\mathcal{A}}_{\dot{\beta}} F^{\mu_{n_1}\nu}\right). \end{aligned} \tag{2.6}$$

The length of a "word" or "sentence" in $\mathcal{N} = 4$ SYM is defined as the number of "letters" W_A's composing the considered trace structure.

Summarizing, the above introduced "words" and "sentences" correspond to (possibly multitrace) $SU(N)$ g.i. polynomial composite operators in $\mathcal{N} = 4, d = 4$ SYM. As we will see further below, the spin-bit map gives an one-to-one spin description of such (multi)trace structures, and thus allows one to perturbatively calculate the non-planar $(N < \infty)$ anomalous dimensions in (certain sectors of) the considered conformally-invariant gauge theory.

3 The spin-bit model

A generic M-trace g.i. polynomial composite operator of length L will have the general form

$$\begin{aligned} \mathcal{O} \equiv\ & Tr\left(W_{A_1}...W_{A_{L_1}}\right) Tr\left(W_{A_{L_1+1}}...W_{A_{L_1+L_2}}\right)... \\ & ...Tr\left(W_{A_{L-L_M+1}}...W_{A_L}\right). \end{aligned} \tag{3.1}$$

Let us now consider an element of the permutation group S_L (of rank $L!$) of L elements, namely

$$\gamma \equiv (\gamma_1\gamma_2...\gamma_L) : \begin{pmatrix} a_1 & a_2 & ... & a_L \\ a_{\gamma_1} & a_{\gamma_2} & ... & a_{\gamma_L} \end{pmatrix}, \qquad (3.2)$$

or equivalently

$$S_L \ni \gamma = (L_1)(L_2)...(L_M) : \sum_{r=1}^{M} L_r = L, \qquad (3.3)$$

where $S_{L_r} \ni (L_r)$ is a cyclic permutation of L_r elements ($r = 1, ..., M$). Actually, Eq. (3.3) has a deeper meaning, because in general S_L is split in equivalence classes (labelled by $L_1, L_2...L_k$ such that $\sum_{r=1}^{k} L_r = L$) of permutations consisting of cycles of respective lengths. By reducing to (products of the) minimal, non-trivial permutational "bricks", that is to (products of the) pair-site permutations σ_{kl}, $(k,l) \in \{1, ..., L\}^2$ (which simply exchange the k-th and l-th elements), the decomposition expressed by Eq. (3.3) leads to (planar and) non-planar (pair-site) permutational identities, extensively treated in [34].

Thus, by suitably choosing γ, the operator $\mathcal{O}$ may be rewritten as

$$\mathcal{O} = (W_{A_1})^{a_1 a_{\gamma_1}} (W_{A_2})^{a_2 a_{\gamma_2}} ... (W_{A_L})^{a_L a_{\gamma_L}}, \qquad (3.4)$$

where the matrix structure (with values in the adjoint repr. of $SU(N)$) of the "letters" W_A's is manifest (by convention the first upper index is a row index, whereas the second is a column one).

As it may be easily seen, the fundamental features we have to take into account are the length L of the operator, the number M of traces, and the "linking" configuration expressed by γ. By thinking each "letter" as sitting on a distinct spin-chain site, we may therefore write the following equivalence relation (which we will extensively comment in the following)

$$\mathcal{O} \equiv |A_1, ..., A_L; \gamma\rangle, \qquad (3.5)$$

where A_k is the direction in V_F determined by the "letter" W_{A_k}, i.e. the direction of the spin state at the k-th spin-chain site. The r.h.s. of Eq. (3.5) represents a state in the Hilbert space of a spin-chain model, but with an explicit extra degree of freedom (represented by γ), properly describing the structure of the interconnections among spin-chain sites: such an "improved" spin-chain model will be called "spin-bit" model. Due to the separating action of the semicolon in Eq. (3.5), the spin-bit Hilbert space $\mathcal{H}_{sb}$ naturally gets divided in an usual spin-part $\mathcal{H}_{sc}$ (the usual spin-chain Hilbert space) and in a so-called "linking" part. As we will see later, this latter will allow one to correctly take into account also the non-planar contributions to anomalous dimensions. Eq. (3.5) defines the so-called spin-bit map in $\mathcal{N} = 4, d = 4$ SYM, whose isomorphicity we are going to discuss.

3.1 Spin part of $\mathcal{H}_{sb}$: the spin-chain picture

As previously mentioned, the spin part of the "improved" spin-chain state, i.e. of the spin-bit state, is given by

$$\mathcal{H}_{sc} \ni |A_1, ..., A_L\rangle = |S_1, ..., S_L\rangle = |S_1\rangle \otimes ... \otimes |S_L\rangle. \qquad (3.1.1)$$

As it is well known, the spin-chain Hilbert space $\mathcal{H}_{sc}$ is the tensor product of the one-spin (or, equivalently, one-site) Hilbert spaces $\mathcal{H}_k$'s. Consequently, as shown by Eq. (3.1.1), a generic spin-chain state is given by the tensor product of the one-spin states at each spin-chain site. $\mathcal{H}_k$ is given by the representation space of the considered symmetry (which in the case at hand will be described by a subalgebra of $psu(2,2 \mid 4)$) at the k-th site. Indeed, $A_k \equiv S_k$ is the direction in $\mathcal{H}_k$ determined by the symmetry of the $\mathcal{N} = 4, d = 4$ SYM "letter" W_{A_k} in V_F

$$\mathcal{H}_{sc} = \mathcal{H}_1 \otimes \mathcal{H}_2 \otimes ... \otimes \mathcal{H}_L = (V_F)_1 \otimes (V_F)_2 \otimes ... \otimes (V_F)_L, \qquad (3.1.2)$$

where $(V_F)_k$ stands for the "singleton" representation space of the (relevant subalgebra of the) SCA $psu(2,2 \mid 4)$ at the k-th site.

In the following treatment we will assume that the same representation of the same symmetry algebra will hold at each spin-chain site. With such an assumption we get

$$\mathcal{H}_{sc} = (V_F)^L, \qquad (3.1.3)$$

where V_F is the "singleton" repr. space of the (relevant subalgebra of the) SCA $psu(2,2 \mid 4)$, which will determine the symmetry of the spin-chain model.

In $\mathcal{N} = 4, d = 4$ SYM theory, the most general symmetry is given by the whole SCA $psu(2,2 \mid 4)$, implying that the dimension of the one-spin Hilbert space $\mathcal{H}_k$ is the same for every spin-chain site, and it is infinite, because $\dim(V_F) = \infty$

$$\dim(\mathcal{H}_{sc}) = \dim\left((V_F)^L\right) = \infty^L. \qquad (3.1.4)$$

The situation changes if compact symmetries (having finite-dimensional unitary representations) are considered (an example is the $su(2)$ sector which will be extensively treated later). Also the possibilities to have non-ultralocalizations and/or different representations of symmetries from site to site along the chain could be taken into account, but here we will not deal with such cases (however, see e.g. [35]).

Of course, $\mathcal{H}_k$ can be endowed with a consistent scalar product $\langle \cdot \mid \cdot \rangle$. Consequently, by choosing an orthonormal basis $\{e_\alpha\}$ (with α ranging in a numerable set)

$$\langle e_\alpha \mid e_\beta \rangle = \delta_{\alpha\beta}, \qquad (3.1.5)$$

we get

$$S_k = S_k^\alpha e_\alpha, \qquad (3.1.6)$$

denoting the decomposition of the spin vector S_k along the orthonormal basis $\{e_\alpha\}$ of the "singleton" repr. space V_F of the $\mathcal{N} = 4$ SCA $psu(2,2 \mid 4)$ at the spin-chain site labelled by k. Therefore, we may rewrite Eq. (3.1.1) as follows:

$$|A_1,...,A_L\rangle = S_1^{\alpha_1}...S_L^{\alpha_L}\,|e_{\alpha_1}\rangle \otimes ... \otimes |e_{\alpha_L}\rangle = \\ = \prod_{k=1}^{L} \otimes S_k^{\alpha_k}\,|e_{\alpha_k}\rangle \equiv |S\rangle \in \mathcal{H}_{sc}. \quad (3.1.7)$$

3.2 The "linking" part and the permutational redundance in $\mathcal{H}_{sb}$

On the other hand, by considering only the "linking" part of the spin-bit state, we obtain

$$|\gamma\rangle \in \zeta_L, \quad (3.2.1)$$

where ζ_L is nothing but the representation space of the permutation group S_L. It may be considered a metrizable space, too, and consequently it may be endowed with a consistent scalar product $\langle\gamma \mid \gamma'\rangle = \delta_{\gamma\gamma'}$.

Thus, it would seem reasonable to define the whole Hilbert space of the spin-bit model $\mathcal{H}_{sb}$ as the tensor product of the spin part $\mathcal{H}_{sc}$ (given by Eq. (3.1.7)) and of the "linking" part ζ_L (given by Eq. (3.2.1))

$$|A_1,...,A_L;\gamma\rangle \equiv |A_1,...,A_L\rangle \otimes |\gamma\rangle \in \mathcal{H}_{sc} \otimes \zeta_L. \quad (3.2.2)$$

But, by simply doing a tensor product, we would then over-estimate the spin-bit Hilbert space $\mathcal{H}_{sb}$. Indeed, an extra symmetry exists, given by the action of S_L on the direct tensor product $\mathcal{H}_{sc}\otimes\zeta_L$ and determining the following equivalence relation of "permutational conjugation"[1]

$$|A_1,...,A_L;\gamma\rangle \sim |A_{\sigma_1},...,A_{\sigma_L};\sigma\cdot\gamma\cdot\sigma^{-1}\rangle, \\ \forall(\gamma,\sigma) \in (S_L)^2; \quad (3.2.3)$$

in particular, for $\sigma = \gamma$ we obtain the property of cyclicity of the trace:

$$|A_1,...,A_L;\gamma\rangle \sim |A_{\gamma_1},...,A_{\gamma_L};\gamma\rangle, \quad \forall\gamma \in S_L. \quad (3.2.4)$$

Otherwise speaking, we may define the representation of the action of the permutational symmetry group S_L on the factorized Hilbert space $\mathcal{H}_{sc} \otimes \zeta_L$ with the operator

$$\widehat{\Sigma}_\sigma\left(|A_1,...,A_L\rangle \otimes |\gamma\rangle\right) \equiv |A_{\sigma_1},...,A_{\sigma_L}\rangle \otimes |\sigma\cdot\gamma\cdot\sigma^{-1}\rangle. \quad (3.2.5)$$

Actually, the action of $\widehat{\Sigma}_\sigma$ corresponds to nothing but an S_L-covariant relabelling of the spin-chain site indices. It is a symmetry of the spin-bit model, in the sense that it can be easily checked that the r.h.s.'s of Eqs. (3.2.2) and (3.2.5) describe the same (multi)trace polynomial composite g.i. operator of length L.

It should also be noticed that, due its very definition, the operator $\widehat{\Sigma}_\sigma$ may be naturally decomposed as the direct product of two independent operators, acting on distinct spaces ($\forall\sigma \in S_L$)

$$\widehat{\Sigma}_\sigma = U_\sigma \otimes \widetilde{\Sigma}_\sigma, \quad (3.2.6)$$

with the definitions

$$U_\sigma\,|A_1,...,A_L\rangle \equiv \\ \equiv \left(P_{1,\sigma_1} \otimes P_{2,\sigma_2} \otimes ... \otimes P_{L,\sigma_L}\right)|A_1,...,A_L\rangle = \quad (3.2.7) \\ = |A_{\sigma_1},...,A_{\sigma_L}\rangle,$$

where $P_{k,l} \equiv P_{kl}$ is the pair-site index permutation operator, acting in the spin-chain Hilbert space $\mathcal{H}_{sc}$ as follows (upperscripts denote the site positions):

$$P_{kl}\left|A_1,...,\overset{k}{A_k},...,\overset{l}{A_l},...,A_L\right\rangle \equiv \\ \equiv \left|A_1,...,\overset{l}{A_k},...,\overset{k}{A_l},...,A_L\right\rangle = \quad (3.2.8) \\ = \left|A_1,...,\overset{k}{A_l},...,\overset{l}{A_k},...,A_L\right\rangle,$$

and

$$\widetilde{\Sigma}_\sigma\,|\gamma\rangle \equiv |\sigma\cdot\gamma\cdot\sigma^{-1}\rangle. \quad (3.2.9)$$

Thus, in order to make the spin-bit map a one-to one (i.e. isomorphic) map, we have to quotient by this extra symmetry S_L, obtaining the following rigorous definition of (state in the) spin-bit Hilbert space:

$$\mathcal{H}_{sb} \equiv \left\{(V_F)^L \otimes \zeta_L\right\}/S_L \ni |A_1,...,A_L;\gamma\rangle \equiv \\ \equiv \left\{|A_1,...,A_L\rangle \otimes |\gamma\rangle\right\}/S_L \equiv |A_1,...,A_L\rangle \otimes_{S_L} |\gamma\rangle, \quad (3.2.10)$$

where $\otimes_{S_L}$ stands for the direct tensor product, modulo the action of S_L represented by $\widehat{\Sigma}_\sigma$.

Therefore, given an arbitrary factorized basis element $|A_1,...,A_L\rangle \otimes \otimes |\gamma\rangle$, one can find the corresponding element of the quotient space $\left\{(V_F)^L \otimes \zeta_L\right\}/S_L$, i.e. the corresponding state in the spin-bit Hilbert space $\mathcal{H}_{sb}$, by "averaging" with respect to the action of S_L

$$|A_1,...,A_L\rangle \otimes_{S_L} |\gamma\rangle \equiv |A_1,...,A_L;\gamma\rangle \equiv \\ \equiv \tfrac{1}{|S_L|}\textstyle\sum_{\sigma\in S_L} \widehat{\Sigma}_\sigma\left(|A_1,...,A_L\rangle \otimes |\gamma\rangle\right) = \\ = \tfrac{1}{|S_L|}\textstyle\sum_{\sigma\in S_L}\left(|A_{\sigma_1},...,A_{\sigma_L}\rangle \otimes |\sigma\cdot\gamma\cdot\sigma^{-1}\rangle\right) \equiv \\ \equiv \widehat{\Pi}\left(|A_1,...,A_L\rangle \otimes |\gamma\rangle\right), \quad (3.2.11)$$

where $\widehat{\Pi}$ is the cyclic symmetry operator, defined as

$$\widehat{\Pi} \equiv \frac{1}{|S_L|}\sum_{\sigma\in S_L} \widehat{\Sigma}_\sigma = \frac{1}{|S_L|}\sum_{\sigma\in S_L}\left(U_\sigma \otimes \widetilde{\Sigma}_\sigma\right), \quad (3.2.12)$$

[1] Here and below products in S_L are understood as

$$\gamma\cdot\sigma \equiv \gamma\sigma = \gamma\cdot(\sigma_1,...,\sigma_L) = (\sigma_{\gamma_1},...,\sigma_{\gamma_L}).$$

and $|S_L| = L!$ is the rank of S_L.

From Eqs. (3.2.7), (3.2.9) and (3.2.12), it is not hard to check that $\widehat{\Pi}$ is actually a projective operator

$$\left(\widehat{\Pi}\right)^2 = \widehat{\Pi}, \tag{3.2.13}$$

and that it commutes with permutationally-invariant operators. Therefore the state $|A_1, ..., A_L; \gamma\rangle$, defined by Eq. (3.2.10), is S_L-invariant, as it has to be in order to correctly estimate the spin-bit Hilbert space $\mathcal{H}_{sb}$, and therefore to make the spin-bit map an isomorphic one.

At 1 loop in SYM perturbation theory, it can be also explicitly shown that the "extra" symmetry S_L of $\mathcal{H}_{sb}$ is nothing but a "gauge" symmetry, in the sense that the spin-bit model may be seen as arising from the corresponding spin-chain model by "gauging" with respect to the permutational symmetry S_L [31], where, as previously mentioned, L is the length of the considered operator, i.e. the total number of spin-chain sites, and also the total length of the spin-chain (if unit distance between neighboring sites is assumed).

3.3 Canonical reduction of S_L

By recalling Eq. (3.3), a generic element $\gamma \in S_L$ may be decomposed (uniquely, up to some possible pair-site permutational identities [34]) as follows:

$$S_L \ni \gamma = (L_1)(L_2)...(L_M) : \sum_{r=1}^{M} L_r = L,\ M \leqslant L, \tag{3.3.1}$$

where (L_r) is a cyclic permutation of L_r elements ($r = 1, ..., M$). Due to the "extra" symmetry S_L determining the equivalence relation (3.2.3), by choosing $\sigma \in S_L$ (in a suitable way, depending on the starting element $\gamma \in S_L$) it is always possible to reduce γ to its canonical form, i.e. to the form where each spin-chain site index is sent to the immediate next one modulo cyclicity

$$\begin{aligned} \gamma &= (L_1)(L_2)...(L_M) \\ &\downarrow \\ \overset{(\sim)}{\gamma} &= \sigma \cdot \gamma \cdot \sigma^{-1} = \sigma (L_1)(L_2)...(L_M)\sigma^{-1} : \\ k_r &\longmapsto [k_r + 1] \equiv k_r + 1,\ mod.\ L_r, \end{aligned} \tag{3.3.2}$$

where k_r is a spin-chain site index running inside the r-th trace.

Even though in what follows we will not restrict ourselves to consider (only) the canonical form of the permutations, one should bear in mind that the S_L-covariant relabelling of site indices corresponding to the permutational conjugation given by Eq. (3.2.3) determines an equivalence relation which makes the switching to canonical permutational forms not implying any loss of generality.

4 The dilatation operator in $\mathcal{N} = 4, d = 4$ SYM

The anomalous dimensions of g.i. operators in the conformally-invariant $\mathcal{N} = 4, d = 4$ SYM gauge theory are given by the action of the dilatation operator Δ.

In perturbation theory, it may be written as

$$\Delta(g_{YM}) = \sum_{n=0}^{\infty} H_{2n}\lambda^n, \tag{4.1}$$

where g_{YM} is the Yang-Mills coupling constant, and

$$\lambda = \lambda(g_{YM}) \equiv \frac{g_{YM}^2 N}{16\pi} \tag{4.2}$$

is the 't Hooft coupling. H_{2n} is the n-loop effective vertex, determined by an explicit evaluation of the divergencies of the n-loop, 2-point function Feynman amplitudes $\langle \mathcal{O}(0)\mathcal{O}(x)\rangle$ in $\mathcal{N} = 4, d = 4$ SYM.

The first few effective vertices read [21]

$$n = 0 \text{ (tree level): } H_0 = \Delta_{0A} Tr\left(W_A \check{W}^A\right); \tag{4.3}$$

$n = 1$ (1-loop level):

$$H_2 = -\tfrac{2}{N}\textstyle\sum_{j=0}^{\infty} h(j)\,(P_j)_{CD}^{AB} : Tr\left[W_A, \check{W}^C\right]\left[W_B, \check{W}^D\right] :, \tag{4.4}$$

where

$$\left(\check{W}^A\right)_{ab} \equiv \frac{\partial}{\partial (W_A)^{ba}} \tag{4.5}$$

is the "letter" operatorial derivative in $\mathcal{N} = 4, d = 4$ SYM, such that

$$\left(\check{W}^A\right)_{ab}(W_A)^{bc} = \delta_a^c, \tag{4.6}$$

and :: denotes the "normal-ordering" of the operators inside, namely the fact that the derivatives $\left(\check{W}^A\right)_{ab}$ never act on the "letters" from the same group inside the colons. Moreover, Δ_{0A} stands for the classical (bare) dimension of the "letter" W_A. For the elementary fields previously mentioned, it is $\Delta_0 = 1$ for each scalar field ϕ^i and each ($SU(N)$-covariant) derivative, $\Delta_0 = \frac{3}{2}$ for the gauginos and $\Delta_0 = 2$ for the gauge field strength.

$(P_j)_{CD}^{AB}$ is the (rank 4) $psu(2,2 \mid 4)$ projector to the irreducible module V_j in the expansion of the tensor product of two ∞-dim. "singleton" V_F representations of the $\mathcal{N} = 4, d = 4$ SCA $psu(2,2 \mid 4)$

$$V_F \otimes V_F = \sum_{j=0}^{\infty} V_j. \tag{4.7}$$

In general, the first modules V_0, V_1 and V_2 contain the symmetric, antisymmetric and trace components in the tensor product of two SYM scalars and their superpartners. Higher modules V_j, $j \geqslant 3$, contain spin $(j-2)$ currents and their superpartners. Finally, $h(j)$ is the j-th harmonic number, defined as

$$h(j) \equiv \sum_{s=1}^{j} \frac{1}{s}, \; h(0) \equiv 0. \tag{4.8}$$

5 The spin-bit Hamiltonian at 1 loop

In general, by applying the isomorphic spin-bit map to the dilatation operator of $\mathcal{N}=4, d=4$ SYM $SU(N)$ gauge theory, we obtain an operator acting on the spin-bit Hilbert space $\mathcal{H}_{sb}$. Such an operator may be identified with the spin-bit Hamiltonian; as we will see, it yields a "deplanarized" form of the related spin-chain Hamiltonian, in the sense that it perfectly reproduces the known results from the theory of spin-chains in the planar limit $N \to \infty$.

Since on the SYM side the dilatation operator is perturbatively known, we will correspondingly obtain a perturbatively expanded expression of the spin-bit Hamiltonian.

At tree level, we trivially get (see Eqs. (4.1) and (4.3))

$$\Delta_{n=0} = H_0 = \Delta_{0A} Tr \left(W_A \check{W}^A \right); \tag{5.1}$$

by applying the spin-bit map, i.e. by applying H_0 on a generic spin-bit state, we get that the tree-level spin-bit Hamiltonian $H_{0,sb}$ is simply proportional to the identity, namely:

$$H_{0,sb} = \Delta_0 1, \tag{5.2}$$

where, as previously mentioned, Δ_0 is the total classical dimension of the $SU(N)$-g.i. $\mathcal{N}=4$ SYM (composite) operator uniquely associated to the considered spin-bit state.

Let us now consider $n=1$, i.e. the 1-loop contribution to Δ; from Eqs. (4.1), (4.2) and (4.4) we obtain

$$\begin{aligned} &\Delta_{n=1} = \\ &= -\tfrac{g_{YM}^2}{8\pi^2} \textstyle\sum_{j=0}^{\infty} h(j) \left(P_j\right)_{CD}^{AB} \\ &: Tr \left[W_A, \check{W}^C \right] \left[W_B, \check{W}^D \right] :; \end{aligned} \tag{5.3}$$

in order to obtain the 1-loop spin-bit Hamiltonian, we simply have to apply the 1-loop $\mathcal{N}=4$ SYM effective vertex H_2 (given by Eq. (4.4)) to a generic spin-bit state

$$H_2 \left| A_1, ..., A_L; \gamma \right\rangle . \tag{5.4}$$

In such a way we will map (by means of the spin-bit isomorphic correspondence) H_2 to the 1-loop spin-bit Hamiltonian $H_{2,sb}$. Clearly, since H_2 is a second-order differential operator (it contains two operatorial derivatives), the Leibnitz rule will decompose the result in a sum over all possible couples of spin-chain sites:

$$\begin{aligned} &H_2 \left| A_1, ..., A_L; \gamma \right\rangle = \\ &= -\tfrac{2}{N} \textstyle\sum_{j=0}^{\infty} h(j) \left(P_j\right)_{CD}^{AB} \\ &: Tr \left[W_A, \check{W}^C \right] \left[W_B, \check{W}^D \right] : \left| A_1, ..., A_L; \gamma \right\rangle = \\ &= \textstyle\sum_{k,l=1}^{L} H_{2,kl} \left| A_1, ..., \overset{k}{A_k}, ..., \overset{l}{A_l}, ..., A_L; \gamma \right\rangle , \end{aligned} \tag{5.5}$$

where $H_{2,kl}$ is nothing but the restriction of the 1-loop effective vertex to the couple of sites $(k,l) \in \{1,...,L\}^2$, and it will be later identified, by the spin-bit map, with the two-site 1-loop spin-bit Hamiltonian.

A number of technical, permutational results are used in the explicit calculations; they respectively read:

$$\text{Fission formula}: \quad Tr \left(A \check{W}^C B W_D \right) = \delta_D^C Tr \left(A \right) Tr (B); \tag{5.6}$$

$$\text{Fusion formula}: \quad Tr \left(A \check{W}^C \right) Tr \left(W_D B \right) = \delta_D^C Tr \left(AB \right), \tag{5.7}$$

where A and B are supposed not to depend on W's. An useful property (holding true for any permutation γ and for any pair-site permutation σ_{kl} in S_L) is

$$\sigma_{kl} \gamma = \gamma \sigma_{\gamma_k \gamma_l}; \tag{5.8}$$

by using it, Eq. (3.2.3) yields

$$\begin{aligned} &\left| A_1, ..., \overset{k}{B}, ..., \overset{l}{A}, ..., A_L; \gamma \sigma_{kl} \right\rangle = \\ &= \left| A_1, ..., \overset{k}{A}, ..., \overset{l}{B}, ..., A_L; \gamma \sigma_{\gamma_k \gamma_l} \right\rangle \end{aligned} \tag{5.9}$$

or, in terms of operators

$$P_{kl} \Sigma_{kl} = \Sigma_{\gamma_k \gamma_l}, \tag{5.10}$$

where P_{kl} is the (k,l)-site permutation operator acting on $\mathcal{H}_{sc}$ defined by Eq. (3.2.8), and $\Sigma_{k,l} \equiv \Sigma_{kl}$ is the (1-loop) chain "splitting and joining" (or "twist") operator, acting on ζ_L, and defined as

$$\Sigma_{kl} \left| \gamma \right\rangle \equiv \begin{cases} \left| \gamma \sigma_{kl} \right\rangle , & k \neq l \\ N \left| \gamma \right\rangle , & k = l \end{cases}, \tag{5.11}$$

or equivalently

$$\Sigma_{kl} = N \delta_{kl} + (1 - \delta_{kl}) \overline{\Sigma}_{kl}, \text{ with } \overline{\Sigma}_{kl} \left| \gamma \right\rangle = \left| \gamma \sigma_{kl} \right\rangle . \tag{5.12}$$

The factor N in the case $k=l$ appears because splitting and joining a trace/chain at the same site leads to a new chain of length zero, whose corresponding trace is $Tr\,(1) = N$, because 1 stands for the identity in the adjoint repr. of the gauge group $SU(N)$.

The final result is the 1-loop spin-bit Hamiltonian

$$H_{2,sb} = \frac{1}{2N} \sum_{\substack{k,l=1 \\ (k \neq l)}}^{L} H_{kl} \left(\Sigma_{\gamma_k l} + \Sigma_{k\gamma_l} - \Sigma_{kl} - \Sigma_{\gamma_k \gamma_l} \right) \tag{5.13}$$

or, using the canonical form of the permutation $\gamma \in S_L$,

$$H_{2,sb} = = \tfrac{1}{2N} \textstyle\sum_{\substack{k,l=1 \\ (k \neq l)}}^{L} H_{kl} \left(\Sigma_{[k+1],l} + \Sigma_{k,[l+1]} - \Sigma_{k,l} - \Sigma_{[k+1],[l+1]} \right), \tag{5.14}$$

where $H_{k,l} \equiv H_{kl} \equiv H_{kl,sb}$ is the two-site Hamiltonian, acting on $\mathcal{H}_{sc}$, and defined as follows ($k \neq l$):

$$H_{kl} \left| A_1, ..., A_L \right\rangle \equiv \equiv 4 \textstyle\sum_{j=0}^{\infty} h(j) \left(P_j \right)^{AB}_{A_k A_l} \left| A_1, ..., \overset{k}{A}, ..., \overset{l}{B}, ..., A_L \right\rangle. \tag{5.15}$$

Notice that P_{kl} and H_{kl} act on $\mathcal{H}_{sc}$, whereas Σ_{kl} acts on ζ_L, and therefore

$$\left[P_{kl}, \Sigma_{mn} \right] = 0 = \left[H_{kl}, \Sigma_{mn} \right], \quad \forall (k,l,m,n) \in \{1, ..., L\}^4. \tag{5.16}$$

By comparing Eq. (5.5) with Eqs. (5.13) and (5.14), and by disregarding the degenerate case of coinciding sites $k = l$ (this can be shown not implying any loss of generality), we may conclude that

$$\begin{aligned} &H_2 \left| A_1, ..., \overset{k}{A_k}, ..., \overset{l}{A_l}, ..., A_L; \gamma \right\rangle = \\ &= \textstyle\sum_{\substack{k,l=1 \\ (k \neq l)}}^{L} H_{2,kl} \left| A_1, ..., \overset{k}{A_k}, ..., \overset{l}{A_l}, ..., A_L; \gamma \right\rangle = \\ &= \tfrac{1}{2N} \textstyle\sum_{\substack{k,l=1 \\ (k \neq l)}}^{L} H_{kl} \left(\Sigma_{\gamma_k l} + \Sigma_{k\gamma_l} - \Sigma_{kl} - \Sigma_{\gamma_k \gamma_l} \right) \\ &\left| A_1, ..., \overset{k}{A_k}, ..., \overset{l}{A_l}, ..., A_L; \gamma \right\rangle \Leftrightarrow \\ &\Leftrightarrow H_{2,kl}(\gamma) \equiv H_{2,kl} = \\ &= \tfrac{1}{2N} H_{kl} \left(\Sigma_{\gamma_k l} + \Sigma_{k\gamma_l} - \Sigma_{kl} - \Sigma_{\gamma_k \gamma_l} \right), \\ &\forall (k,l) \in \{1, ..., L\}^2, k \neq l, \end{aligned} \tag{5.17}$$

where, as previously mentioned, $H_{2,kl}$ is the (γ-dependent) restriction of the 1-loop spin-bit Hamiltonian to the couple of sites $(k,l) \in \{1, ..., L\}^2$, $k \neq l$.

5.1 The planar limit

From Eq. (5.12), the planar limit $N \to \infty$ affects just the (1-loop) "twist" operator in the following way:

$$\lim_{N \to \infty} \frac{1}{N} \Sigma_{kl} = \delta_{kl}. \tag{5.1.1}$$

Therefore the planar contributions to the 1-loop spin-bit Hamiltonian come from terms involving Σ_{kk}, i.e. from the cases $l = \gamma_k$ and $k = \gamma_l$ (because $k \neq l$); the final result is

$$\lim_{N \to \infty} H_{2,sb} = \sum_{k=1}^{L} H_{k\gamma_k} = \sum_{k=1}^{L} H_{k,[k+1]} = H_{2,sc}. \tag{5.1.2}$$

Otherwise speaking, by construction the planar limit of the 1-loop spin-bit Hamiltonian coincides with the 1-loop spin-chain Hamiltonian $H_{2,sc}$.

6 Spin-bits in the $su(2)$ sector of $\mathcal{N} = 4$ SYM

We will now consider the "minimal" sector of $\mathcal{N} = 4$, $d = 4$ SYM $SU(N)$ gauge theory, made by "purely-scalar" g.i. (polynomial) operators, i.e. by operators generated only by two holomorphic combinations of the real SYM scalars, which may be defined as

$$\begin{cases} \Phi \equiv \phi^5 + i\phi^6, \\ Z \equiv \phi^1 + i\phi^2. \end{cases} \tag{6.1}$$

Thus, Φ and Z will be the only SYM "letters" used to compose "words" and "sentences" in such an operator sector, which may be shown to be closed under the operator mixing due to perturbative renormalization of the theory. Φ and Z will transform in the 2-dim. $s = 1/2$ fundamental repr. of $su(2)$, and therefore the whole sector will be $su(2)$-symmetric. Notice that $su(2)$ is the smallest non-trivial bosonic compact subalgebra of the whole $\mathcal{N} = 4$ SCA $psu(2,2 \mid 4)$, and the following chain of inclusions holds:

$$su(2) \subset so(6) \sim su(4) \subset so(4,2) \oplus su(4) \subset psu(2,2 \mid 4). \tag{6.2}$$

A generic M ($\leqslant L$)-trace g.i. operator in such a $su(2)$ closed subsector reads

$$\mathcal{O} \equiv \underbrace{Tr\,(\Phi Z \Phi\Phi ...)}_{L_1 \text{ "letters"}} \underbrace{Tr\,(\Phi\Phi\Phi Z ...)}_{L_2 \text{ "letters"}} ... \underbrace{Tr\,(\Phi Z Z \Phi Z ...)}_{L_M \text{ "letters"}}, \tag{6.3}$$

with $\sum_{r=1}^{M} L_r = L$.

As previously shown, we may use the isomorphic spin-bit map to equivalently represent $\mathcal{O}$ as

$$\begin{aligned} \mathcal{O} &= \left(\phi^{i_1}\right)^{a_1 a_{\gamma_1}} \left(\phi^{i_2}\right)^{a_2 a_{\gamma_2}} ... \left(\phi^{i_L}\right)^{a_L a_{\gamma_L}} \equiv \\ &\equiv \left| S; \gamma \right\rangle \equiv \left| S \right\rangle \otimes_{S_L} \left| \gamma \right\rangle, \end{aligned} \tag{6.4}$$

where now the range of all "i-indices" of the scalars is $\left\{ \widehat{1}, \widehat{2} \right\}$, with $\phi^{\widehat{1}} \equiv \Phi$ and $\phi^{\widehat{2}} \equiv Z$ by convention. In the case of Eq. (6.3) we have

$$S_L \ni \gamma = (L_1)(L_2) ... (L_M), \tag{6.5}$$

with (L_r) denoting a cyclic permutation of L_r elements, $r = 1, ..., M \leqslant L$. The correspondence operated by the spin-bit map is completed by associating to each spin-chain site the spin value $|-1/2\rangle$ if we find Φ there, and $|1/2\rangle$ if we find Z.

By specializing the general expression (5.15) of the two-site Hamiltonian to the case of 2-dim. $s = 1/2$ (representation of) $su(2)$ symmetry, the final result is simply [21]

$$H_{kl,su(2)} = 2\left(1 - P_{kl}\right), \tag{6.6}$$

and therefore, by substituting it in the general formulae (5.13) and (5.14), we obtain the 1-loop $su(2)$ spin-bit Hamiltonian $H_{2,sb,su(2)} \equiv H_{2,su(2)}$

$$\begin{aligned} H_{2,su(2)} &= \frac{1}{2N}\sum_{k,l=1}^{L} H_{kl,su(2)}\left(\Sigma_{\gamma_k l} + \Sigma_{k\gamma_l} - \Sigma_{kl} - \Sigma_{\gamma_k\gamma_l}\right) = \\ &= \frac{2}{N}\sum_{k,l=1}^{L}\left(1 - P_{kl}\right)\Sigma_{k\gamma_l}. \end{aligned} \tag{6.7}$$

In the planar limit, by recalling Eq. (5.1.1) we get

$$\begin{aligned} &\lim_{N\to\infty} H_{2,su(2)} = 2\textstyle\sum_{k,l=1}^{L}\left(1 - P_{kl}\right)\lim_{N\to\infty}\frac{1}{N}\Sigma_{k\gamma_l} = \\ &= 2\textstyle\sum_{k,l=1}^{L}\left(1 - P_{kl}\right) \\ &\lim_{N\to\infty}\tfrac{1}{N}\left[N\delta_{k\gamma_l} + \left(1 - \delta_{k\gamma_l}\right)\overline{\Sigma}_{k\gamma_l}\right] = \\ &= 2\textstyle\sum_{k=1}^{L}\left(1 - P_{k\gamma_k}\right) = \sum_{k=1}^{L} H_{k\gamma_k,su(2)} = \\ &= H_{2,sc,su(2)}. \end{aligned} \tag{6.8}$$

In other words, the planar limit of the 1-loop $su(2)$ spin-bit Hamiltonian coincides with the integrable $XXX_{s=1/2}$ Heisenberg $su(2)$ spin-chain Hamiltonian $H_{2,sc,su(2)}$ [36].

The expression (6.7) of $H_{2,su(2)}$ might also be obtained by starting from the known combinatorial formula of the 1-loop $\mathcal{N} = 4, d = 4$ SYM effective vertex on the $su(2)$-symmetric closed subsector of "purely scalar" composite (polynomial) g.i. operators, reading [28]

$$H_2 = -\frac{4}{N} : Tr\left([\Phi, Z]\left[\check{\Phi}, \check{Z}\right]\right) :, \tag{6.9}$$

where

$$\begin{cases} \left(\check{\Phi}\right)^{ab} \equiv \frac{\partial}{\partial\Phi^{ba}}, \\ \left(\check{Z}\right)^{ab} \equiv \frac{\partial}{\partial Z^{ba}}. \end{cases} \tag{6.10}$$

7 $su(2)$ spin-bits at 2 loops

The $su(2)$ sector of $\mathcal{N} = 4, d = 4$ SYM is the only one, as far as we know, for which a detailed treatment of non-planar ($N < \infty$) 2-loop anomalous dimensions has been given. This is due to the noteworthy fact that a combinatorial formula of the 2-loop $\mathcal{N} = 4, d = 4$ SYM effective vertex on the $su(2)$-symmetric closed subsector of "purely scalar" composite (polynomial) g.i. operators is known [28]:

$$H_4 = \frac{2}{N^2}\begin{bmatrix} : Tr\left([Z, \Phi]\left[\check{Z}, \left[Z, \left[\check{Z}, \check{\Phi}\right]\right]\right]\right) : + \\ + : Tr\left([Z, \Phi]\left[\check{\Phi}, \left[\Phi, \left[\check{Z}, \check{\Phi}\right]\right]\right]\right) : + \\ +2N : Tr\left([\Phi, Z]\left[\check{\Phi}, \check{Z}\right]\right) : \end{bmatrix}. \tag{7.1}$$

By applying such a combinatorial formula on a generic $su(2)$-symmetric $\mathcal{N} = 4$ SYM operator/spin-bit state $|S;\gamma\rangle$, we obtain, after some permutational algebra and technical tricks, the following expression for the 2-loop $su(2)$ spin-bit Hamiltonian $H_{4,sb,su(2)} \equiv H_{4,su(2)}$:

$$H_{4,su(2)} = \frac{2}{N^2}\sum_{k,l,m=1}^{L}\left(2P_{lm} + 2P_{kl} - P_{km} - 3\right)\Sigma_{klm}(\gamma), \tag{7.2}$$

where $\Sigma_{klm}(\gamma)$ is the ($su(2)$) 2-loop "twist" operator, acting on ζ_L, and defined as

$$\Sigma_{klm}(\gamma) \equiv \Sigma_{k\gamma_l}\Sigma_{l\gamma_m}. \tag{7.3}$$

7.1 The planar limit

From Eq. (5.12), the planar limit $N \to \infty$ affects just $\Sigma_{klm}(\gamma)$ in the following way:

$$\lim_{N\to\infty}\frac{1}{N^2}\Sigma_{klm}(\gamma) = \delta_{k\gamma_l}\delta_{l\gamma_m}. \tag{7.1.1}$$

Therefore, we obtain

$$\begin{aligned} &\lim_{N\to\infty} H_{4,su(2)} = 2\textstyle\sum_{k=1}^{L}\left(4P_{k\gamma_k} - P_{k\gamma_k^2} - 3\right) = \\ &= 2\textstyle\sum_{k=1}^{L}\left(4P_{k,[k+1]} - P_{k,[k+2]} - 3\right) = H_{4,sc,su(2)}. \end{aligned} \tag{7.1.2}$$

Otherwise speaking, by construction the planar limit of the 2-loop $su(2)$ spin-bit Hamiltonian coincides with the integrable $su(2)$ spin-chain Hamiltonian $H_{4,sc,su(2)}$, which in turn corresponds to an integrable (higher-order) deformation of the previously mentioned $XXX_{s=1/2}$ Heisenberg $su(2)$ spin-chain Hamiltonian $H_{2,sc,su(2)}$ [25,28,37].

8 $su(2)$ spin-bits beyond 2 loops

8.1 The "deplanarizing operator lifts" (d.o.l.) method

$su(2)$-symmetric spin-chain Hamiltonians are known explicitly up to (and including) 5 loops [25,28]. Unfortunately, combinatorial formulae for the SYM effective vertices in the $su(2)$ sector are not known beyond 2 loops, and therefore higher-loop $su(2)$ spin-bit Hamiltonians are not directly obtainable as in the cases of 1 and 2 loops. Thus, other approaches have to be pursued in order to derive them. After the failure of the elegant and geometrically meaningful "spin-edge differences" Ansätze [34], the only planarly-consistent, fully testable set of conjectures for the higher-loop $su(2)$ spin-bit Hamiltonians are those obtained by applying the recently proposed [34, 38] "deplanarizing operator lifts" ("d.o.l.") method, eventually with the additional hypothesis of "symmetrization of deplanarizing operator splittings" (hp. "s.d.o.s.").

In the following we will present such a deplanarizing approach in a sketchy, algorithmic way, addressing the interested reader to the original literature for further elucidations.

The d.o.l. algorithm may be realized through the following steps:

1) we have to start from the known planar results for the Hamiltonian. Since the non-planar 1- and 2-loop orders are already known and have been previously treated, we have to consider the 3-, 4- and 5-loop expressions of the planar $su(2)$ spin-chain Hamiltonian [25, 28] (*input* of the deplanarizing algorithm);

2) then, we have to perform the (non-reductive) conventional site-index identifications

$$l = k + 1, \ m = k + 2, \ ...; \tag{8.1.1}$$

3) therefore, we have to consider all possible products of operators P's that, in the planar limit, would give the considered planar permutational term; this will determine some proper "deplanarizing operator lifts". All such non-planar permutational terms will come with free (real) coefficients, constrained by two requests:

3.i) their algebraic sum must give the right numerical known coefficient of the considered planar permutational term;

3.ii) they must make the spin part of the non-planar Hamiltonian completely symmetric under the particular, inverting exchange of site indices

$$(k, l, ..., r, s) \leftrightarrow (s, r, ..., l, k)\,, \tag{8.1.2}$$

as requested by the site index structure determined by the "linking" part.

4) Indeed, for what concerns the "linking" part of the Hamiltonian, i.e. the "twisting" operators Σ's, we may generalize the "linking" part of Eqs. (6.7) and (7.2) by introducing the ($S_L \ni \gamma$-dependent) "splitting and joining chain operator of order n" as

$$\Sigma_{k_1 k_2 ... k_{n+1}}(\gamma) \equiv \Sigma_{k_1 \gamma_{k_2}} \Sigma_{k_2 \gamma_{k_3}} ... \Sigma_{k_{n-1} \gamma_{k_n}} \Sigma_{k_n \gamma_{k_{n+1}}}\,. \tag{8.1.3}$$

5) Generally, at higher-loop orders, some (real) parameters still remain undetermined at this step. In order to obtain a completely determined expression of the higher-loop $su(2)$ spin-bit Hamiltonian as output of the proposed deplanarization procedure, we may proceed as follows.

As a reasonable conjecture, we may formulate an additional assumption, that we are going to call hypothesis of "symmetrization of deplanarizing operator splittings" (hp. "s.d.o.s."). This conjecture has to be applied *after* the symmetrization of the non-planar terms with respect to the peculiar renaming of spin-chain site indices given by (8.1.2); it amounts to say that each of the sets of non-planar terms arising from a considered planar term in the deplanarization procedure will *equally* contribute in the planar limit $N \to \infty$.

For example, if, after the symmetrization with respect to (8.1.2), a planar term $\Im$ is deplanarized by the 3-fold splitting

$$\Im \to a_1 \Im_{M_1} + a_2 \Im_{M_2} + a_3 \Im_{M_3}\,, \tag{8.1.4}$$

where $\Im_{M_1}$, $\Im_{M_2}$ and $\Im_{M_3}$ are respectively sets consisting of M_1, M_2 and M_3 non-planar terms made by permutation operators, then we will assume that

$$M_1 a_1 = M_2 a_2 = M_3 a_3. \tag{8.1.5}$$

Hence the contribution of $\Im_{M_1}$, $\Im_{M_2}$ and $\Im_{M_3}$ to the planar limit $\Im$ is the same, and therefore the operator splitting given by (8.1.4) may be considered symmetric.

As it will be seen explicitly further below at 3-loops, this additional hypothesis will allow to fix *all* the free real parameters, otherwise necessarily introduced by the deplanarizing operator lifts acting at the considered higher-loop order. Notice that the constraints 3.i and 3.ii are implied by the hp. s.d.o.s. .

Thus, a complete Ansatz for the $su(2)$ spin-bit Hamiltonian at the considered loop order is obtained[2] (*output* of the deplanarizing algorithm).

8.2 Application at 3 loops

Let us consider an explicit example of application of the described method, in order to build a consistent Ansatz for the expression of the 3-loop $su(2)$ spin-bit Hamiltonian. Let us follow the previously mentioned steps:

1) we start from the known expression of $H_{6,sc,su(2)}$, namely the 3-loop, integrable, perturbative deformation of the $XXX_{s=1/2}$ Heisenberg $su(2)$ spin-chain Hamiltonian $H_{2,sc,su(2)}$ [25, 28, 37], given by (*input* of the deplanarizing

[2] It should be noticed that here we assume that (eventually rather structurally complicated) non-planar permutational terms, such that their planar limit is zero, do *not* exist; indeed, for the time being, their existence may not be guessed by an inferring approach starting from the planar level, such as the one adopted in this paper.

algorithm for $n =$ 3-loop order)

$$H_{6,sc,su(2)} = = 4\sum_{k=1}^{L} \begin{bmatrix} 15 - 26P_{k,k+1}+ \\ +6\left(P_{k,k+1}P_{k+1,k+2} + P_{k+1,k+2}P_{k,k+1}\right)+ \\ +P_{k,k+1}P_{k+2,k+3}+ \\ -(P_{k,k+1}P_{k+1,k+2}P_{k+2,k+3}+ \\ +P_{k+2,k+3}P_{k+1,k+2}P_{k,k+1}) \end{bmatrix}; \tag{8.2.1}$$

2) we conventionally identify (without loss of generality) the spin-chain site indices in the following way:

$$l \equiv k+1, \quad m \equiv k+2, \quad n \equiv k+3; \tag{8.2.2}$$

3) therefore, we have to find *all* possible non-planar permutational terms giving rise, in the planar limit $N \to \infty$, to each of the permutational terms of $H_{6,sc,su(2)}$ given by Eq. (8.2.1).

3.*i*) We have that:

3.*i.a*) the planar term $P_{k,k+1}$ receives three contributions from the non-planar level, respectively from $P_{k,k+1} = P_{kl}$, $P_{k+1,k+2} = P_{lm}$ and $P_{k+2,k+3} = P_{mn}$, whence the proper "deplanarizing operator lift" of $P_{k\gamma_k}$ reads ($\xi_1, \xi_2 \in R$)

$$-26P_{k,k+1} \to \xi_1 P_{kl} + \xi_2 P_{lm} - (26 + \xi_1 + \xi_2)P_{mn}; \tag{8.2.3}$$

3.*i.b*) the planar-level product $P_{k,k+1}P_{k+1,k+2}$ instead receives contribution just from two non-planar terms, i.e. $P_{k,k+1}P_{k+1,k+2} = P_{kl}P_{lm}$ and $P_{k+1,k+2}P_{k+2,k+3} = P_{lm}P_{mn}$, whence the proper "deplanarizing operator lift" of the term $P_{k,k+1}P_{k+1,k+2}$ reads ($\xi_3 \in R$)

$$6P_{k,k+1}P_{k+1,k+2} \to \xi_3 P_{kl}P_{lm} + (6 - \xi_3)\, P_{lm}P_{mn}; \tag{8.2.4}$$

3.*i.c*) analogously, for the other terms of $H_{6,sc,su(2)}$ we obtain the following proper "deplanarizing operator lifts" ($\xi_4 \in R$):

$$\begin{aligned} &6P_{k+1,k+2}P_{k,k+1} \to \xi_4 P_{lm}P_{kl} + (6 - \xi_4)\, P_{mn}P_{lm}, \\ &P_{k,k+1}P_{k+2,k+3} \to P_{kl}P_{mn}, \\ &P_{k,k+1}P_{k+1,k+2}P_{k+2,k+3} \to P_{kl}P_{lm}P_{mn}, \\ &P_{k+2,k+3}P_{k+1,k+2}P_{k,k+1} \to P_{mn}P_{lm}P_{kl}; \end{aligned} \tag{8.2.5}$$

3.*ii*) thence, we impose the symmetry of the spin part under the site index exchange

$$(k,l,m,n) \leftrightarrow (n,m,l,k); \tag{8.2.6}$$

the imposition of such a condition on the spin part decreases the number of free (real) parameters from four to two, thence renamed η_1 and η_2;

4) finally, we put

$$\frac{1}{N^3}\Sigma_{klmn}(\gamma) = \frac{1}{N^3}\Sigma_{k\gamma_l}\Sigma_{l\gamma_m}\Sigma_{m\gamma_n} \tag{8.2.7}$$

as the linking variable part.

Thus, we may finally write the most general expression of the 3-loop $su(2)$ spin-bit Hamiltonian ($\eta_1, \eta_2 \in R$):

$$H_{6,su(2)}(\eta_1,\eta_2) = \frac{4}{N^3}\sum_{k,l,m,n=1}^{L} \tag{8.2.8}$$

$$\begin{bmatrix} 15 + \eta_1\left(P_{kl} + P_{mn}\right) - 2\left(\eta_1 + 13\right)P_{lm}+ \\ +\eta_2\left(P_{kl}P_{lm} + P_{mn}P_{lm}\right)+ \\ +\left(6 - \eta_2\right)\left(P_{lm}P_{kl} + P_{lm}P_{mn}\right)+ \\ +P_{kl}P_{mn} - P_{kl}P_{lm}P_{mn} - P_{mn}P_{lm}P_{kl} \end{bmatrix} \Sigma_{k\gamma_l}\Sigma_{l\gamma_m}\Sigma_{m\gamma_n}.$$

Formulating the hp. s.d.o.s. we get $(\eta_1,\eta_2) = \left(-\frac{13}{2}, 3\right)$, and therefore we obtain a *completely fixed* expression for the 3-loop $su(2)$ spin-bit Hamiltonian (*output* of the deplanarizing algorithm for $n =$ 3-loop order):

$$H_{6,su(2)} = \frac{4}{N^3}\sum_{k,l,m,n=1}^{L} \tag{8.2.9}$$

$$\begin{bmatrix} 15 - \frac{13}{2}\left(P_{kl} + P_{lm} + P_{mn}\right)+ \\ +3\left(P_{kl}P_{lm} + P_{mn}P_{lm}+\right. \\ \left.+P_{lm}P_{kl} + P_{lm}P_{mn}\right)+ \\ +P_{kl}P_{mn} - P_{kl}P_{lm}P_{mn} - P_{mn}P_{lm}P_{kl} \end{bmatrix} \Sigma_{k\gamma_l}\Sigma_{l\gamma_m}\Sigma_{m\gamma_n}.$$

Analogous, more and more involved, expressions for the 4- and 5-loops $su(2)$ spin-bit Hamiltonian (with or without the additional hp. s.d.o.s.) have been obtained by applying the d.o.l. method [34,38].

9 Outlook and further developments

The d.o.l. method [38] is fully compatible with (independently obtained) known results at the 1- and 2-loop, non-planar level [30–32], and it allows one to obtain explicit formulae for the 3-, 4- and 5-loop $su(2)$ spin-bit Hamiltonians. By construction, such expressions are planarly-consistent, i.e. they have the correct planar limit, matching the known results reported in the literature (see e.g. [25,28,37]).

It is also worth noticing that, by construction, all the higher-loop $su(2)$ spin-bit Hamiltonians (for 3-loops, see Eqs. (8.2.8) and (8.2.9)) show an explicit full factorization in the spin and chain-splitting parts; as already pointed out in [32], such a property is expected to hold at every loop order, since the Hilbert space of the spin-bit model $\mathcal{H}_{sb}$ is given by the direct product (modulo the action of the permutation group S_L) of the spin-chain Hilbert space $\mathcal{H}_{sc}$ and of the linking space ζ_L.

Attention must also be paid to the fact that, while (both at non-planar and planar level) the 1- and 2-loop formulae for the $su(2)$ Hamiltonians are *linear* in the site permutation operators P's, the 3-, 4- and 5-loop level expressions, both at non-planar and planar level, show a non-linearity (and non-linearizability) in P's. For example, the non-linearizability of the 3-loop $su(2)$ spin-chain Hamiltonian (8.2.1) caused the failure of the elegant and geometrically meaningful "spin edge-differences" [34] approach to higher-loop Ansätze.

Thus, the non-linearity (and non-linearizability) in site permutation operators seems to be a crucial and fundamental feature, starting to hold at the 3-loop order, of the spin part of the Hamiltonian of the $su(2)$ spin-bit model, underlying the non-planar dynamics of the $su(2)$ sector of the $\mathcal{N}=4, d=4$ SYM theory. Reasonably, one would expect that such a breakdown of "permutational linearizability" at 3 loops (for the first evidences from 3-loop calculations, see e.g. [25,28]; for further subsequent developments see e.g. [2,8,39]) gives rise, by means of the AdS/CFT correspondence [3,4], to some "new" features in the dynamics of the (closed) superstrings in the bulk of $AdS_5 \times S^5$. Actually, in the AdS/CFT correspondence framework, there is a problem of discrepancy between the calculations made with fast spinning, semiclassical strings (i.e. in the so-called Frolov-Tseytlin limit) and the calculations made in the thermodynamical limit of long spin-chains with a large number of excitations (i.e. the so-called Berenstein-Maldacena-Nastase limit) (see e.g. [37] and Refs. therein). Such a disagreement starts to hold at 3 loops, and it is one of the most intriguing "mysteries" of the AdS/CFT conjecture. Recently, an explanation for such 3- and higher-loop disagreement has been proposed: it should be related to an "order-of-limit" non-commutation problem in the perturbative expansions and thermodynamical asymptotical regimes or, equivalently, to the presence of operational "wrapping" interactions (see e.g. [37,40]). Conjecturally, we may here put forward the suggestion that the breakdown of "permutational linearizability" of $su(2)$ spin-chain/spin-bit Hamiltonians, which starts to hold at 3 loops, could be related to such a "3-loop discrepancy mystery" in AdS/CFT, and possibly it could be extended also to larger symmetries inside $psu(2,2|4)$.

Also, the application of the d.o.l. method, originally introduced for the Hamiltonian, to the higher-order charges of the $su(2)$ spin-chain model [38] raises some interesting and intriguing questions, such as:

1) the d.o.l. appears to be consistent only for odd higher-order charges, sharing the symmetry properties of the Hamiltonian. Thus, the extension of the deplanarization procedure to even higher-order charges, and in general to antisymmetric operators, should be needed, in order to have a complete deplanarizing algorithm for the local operators in $\mathcal{N}=4, d=4$ SYM theory;

2) we know that the spectrum of the (perturbatively expanded) Hamiltonian of the spin-chain/spin-bit model is related to the spectrum of the (perturbatively expanded) anomalous dimensions and mixing of local operators on the $\mathcal{N}=4, d=4$ $SU(N)$ SYM gauge theory side of AdS/CFT. Then a natural question [28] to ask is: *do the spectra of the higher-order planar charges (and of their deplanarized counterparts) have a physical meaning in $\mathcal{N}=4$ SYM ?*

The deep question is evidently: *why does exact integrability seem to hold for every loop-order in the planar $\mathcal{N}=4, d=4$ SYM* [24,37,41], *and why and how is it lost at the non-planar level?*

By the way, even if the exact (classical) integrability is lost when deplanarizing, i.e. when passing from the (all-loop) $su(2)$ Heisenberg spin-chain to the (all-loop) $su(2)$ spin-bit model, nevertheless we may put forward the following intriguing suggestion: could the $su(2)$ spin-bit model still be an integrable model, but in a sort of generalized, broader sense, e.g. in the sense of *quasi-integrability* and *quasi-exact solvability* (see e.g. [42]) or *quantum-integrability* (see e.g. [43])?

If yes, then in general the deplanarization procedure here presented might algebraically correspond to some kind of *"deformation"* of the (dynamical) symmetries of the system being considered, and thence of the structure and properties of its (eventually conserved) charges.

In our opinion, this is an interesting problem, strictly related to the consistent definition of the higher-order charges of the spin-bits as the deplanarization of the infinitely many conserved charges of the spin-chains, and it is currently under study.

Moreover, we notice that it would be interesting, following recent research directions, to extend the considered deplanarizing method to other operatorial sectors of the $\mathcal{N}=4, d=4$ SYM theory [44]. Indeed, sectors with non-compact symmetries have been shown to be relevant, in order to describe the renormalization in the large N_c (non-)SUSY QCD, also in relation with the attempts to construct a string description of QCD (see e.g. [45] and Refs. therein).

Finally, some possible additional directions for further research are briefly summarized as follows:

- All the spin-chain/spin-bit models treated so far are characterized by periodic boundary conditions, corresponding to closed chains. The possibility to modify the boundary conditions yields open spin-chain models, corresponding to a dynamical discretization of the open strings [46]. The issue of deplanarizability, and the related problem of integrability, of such models remains to be discussed.

- In general, the spin-bits may be considered as a dynamical polymer model with decaying and fusing chains; potential applications to relativity theory [47], field theory [48] and biophysics could be addressed.

- Interesting analogies could be explored with the spin-network approach to discrete quantum gravity [49].

- The above mentioned non-planar permutational identities could be linked to the random graph theory on a lattice [34].

- The spin representation of the permutation operators in the $su(2)$ sector leads to some "generalized" Fierz identities for Pauli σ matrices [34], deserving a more detailed analysis.

- An alternative approach to the calculation of anomalous dimensions in $\mathcal{N}=4$ SYM is based on matrix models [12,50], which can be also formulated in terms of a non-commutative field theory on a torus, and whose equivalence with spin-chain/spin-bit models in the $N \to \infty$ (planar) and $N \to 0^+$ limits is still under study. An interesting direction of research to be pursued would be the formulation of such matrix models on other "fuzzy" manifolds, such as the "fuzzy sphere" [51], and the study of the relation of such non-toroidal "fuzzy" matrix models with the initial non-commutative torus representation.

Acknowledgements

Most of the results here presented have been obtained during the last years by the LNF team, directed by S. Bellucci, and composed, beside the author, by P.-Y. Casteill, F. Morales and C. Sochichiu. I would like to gratefully acknowledge stimulating discussions with all of them.

References

1. G. 't Hooft, "A planar diagram theory for strong interactions", *Nucl. Phys.* **B72** (1974) 461.
2. A. A. Tseytlin, "Semiclassical strings and AdS/CFT", contribution to the Proceedings of Cargese Summer School, June 7-19, 2004, hep-th/0409296.
3. J. M. Maldacena, "The large N limit of superconformal field theories and supergravity", *Adv. Theor. Math. Phys.* **2** (1998) 231, hep-th/9711200.
4. S. S. Gubser, I. R. Klebanov, and A. M. Polyakov, "Gauge theory correlators from non-critical string theory", *Phys. Lett.* **B428** (1998) 105, hep-th/9802109.
5. D. Berenstein, J. M. Maldacena, and H. Nastase, "Strings in flat space and pp waves from $\mathcal{N}=4$ super Yang Mills", *JHEP* **04** (2002) 013, hep-th/0202021.
6. D. Berenstein and H. Nastase, "On lightcone string field theory from super Yang-Mills and holography", hep-th/0205048.
7. D. Berenstein, E. Gava, J. M. Maldacena, K. S. Narain, and H. Nastase, "Open strings on plane waves and their Yang-Mills duals", hep-th/0203249.
8. A. A. Tseytlin, "On semiclassical approximation and spinning string vertex operators in $AdS_5 \times S^5$", *Nucl. Phys.* **B664** (2003) 247, hep-th/0304139.
9. S. Frolov and A. A. Tseytlin, "Multi-spin string solutions in $AdS_5 \times S^5$", *Nucl. Phys.* **B668** (2003) 77, hep-th/0304255.
10. S. Frolov and A. A. Tseytlin, "Rotating string solutions: AdS/CFT duality in non- supersymmetric sectors", *Phys. Lett.* **B570** (2003) 96, hep-th/0306143.
11. N. Beisert, S. Frolov, M. Staudacher, and A. A. Tseytlin, "Precision spectroscopy of AdS/CFT", *JHEP* **10** (2003) 037, hep-th/0308117.
12. S. Bellucci and C. Sochichiu, "On matrix models for anomalous dimensions of super Yang-Mills theory", *Nucl. Phys.* **B726** (2005) 233-251, hep-th/0410010.
13. S. Bellucci and C. Sochichiu, "On the dynamics of BMN operators of finite size and the model of string bits", Contribution to the BW2003 Workshop, 29 August - 02 September, 2003 Vrnjacka Banja, Serbia, hep-th/0404143 • S. Bellucci and C. Sochichiu, "Can string bits be supersymmetric?", *Phys. Lett.* **B571** (2003) 92, hep-th/0307253 • S. Bellucci and C. Sochichiu, "Fermion Doubling and Berenstein–Maldacena–Nastase Correspondence", *Phys. Lett.* **B564** (2003) 115, hep-th/0302104.
14. M. Blau, J. Figueroa-O'Farrill, C. Hull, and G. Papadopoulos, "A new maximally supersymmetric background of IIB superstring theory", *JHEP* **01** (2002) 047, hep-th/0110242.
15. M. Blau, J. Figueroa-O'Farrill, C. Hull, and G. Papadopoulos, "Penrose limits and maximal supersymmetry", *Class. Quant. Grav.* **19** (2002) L87, hep-th/0201081.
16. M. Blau, J. Figueroa-O'Farrill, and G. Papadopoulos, "Penrose limits, supergravity and brane dynamics", *Class. Quant. Grav.* **19** (2002) 4753, hep-th/0202111.
17. R. R. Metsaev, "Type IIB Green-Schwarz superstring in plane wave Ramond–Ramond background", *Nucl. Phys.* **B625** (2002) 70, hep-th/0112044.
18. R. R. Metsaev and A. A. Tseytlin, "Exactly solvable model of superstring in plane wave Ramond–Ramond background", *Phys. Rev.* **D65** (2002) 126004, hep-th/0202109.
19. C. Kristjansen, J. Plefka, G. W. Semenoff, and M. Staudacher, "A new double-scaling limit of $N=4$ super Yang–Mills theory and pp-wave strings", *Nucl. Phys.* **B643** (2002) 3, hep-th/0205033 • D. J. Gross, A. Mikhailov, and R. Roiban, "Operators with large R charge in $N=4$ Yang–Mills theory", *Annals Phys.* **301** (2002) 31, hep-th/0205066 • N. R. Constable *et. al.*, "pp-wave string interactions from perturbative Yang–Mills theory", *JHEP* **07** (2002) 017, hep-th/0205089 • N. Beisert, C. Kristjansen, J. Plefka, G. W. Semenoff, and M. Staudacher, "BMN correlators and operator mixing in $N=4$ super Yang–Mills theory", *Nucl. Phys.* **B650** (2003) 125, hep-th/0208178 • N. R. Constable, D. Z. Freedman, M. Headrick, and S. Minwalla, "Operator mixing and the BMN correspondence", *JHEP* **10** (2002) 068, hep-th/0209002 • M. Spradlin and A. Volovich, "Note on plane wave quantum mechanics", *Phys. Lett.* **B565** (2003) 253, hep-th/0303220.
20. M. Spradlin and A. Volovich, "Light-cone string field theory in a plane wave", Lectures at the ICTP Spring School on Superstring Theory and Related Topics, Trieste, 31 March–8 April 2003, hep-th/0310033 • R. Russo and A. Tanzini, "The duality between IIB string theory on pp-wave and $N=4$ SYM: A status report", *Class. Quant. Grav.* **21** (2004) S1265, hep-th/0401155.

21. N. Beisert, "The complete one-loop dilatation operator of $\mathcal{N} = 4$ super Yang-Mills theory", *Nucl. Phys.* **B676** (2004) 3, hep-th/0307015.
22. J. A. Minahan and K. Zarembo, "The Bethe–Ansatz for $\mathcal{N} = 4$ super Yang–Mills", *JHEP* **03** (2003) 013, hep-th/0212208.
23. N. Beisert, "The dilatation operator of $\mathcal{N}=4$ super Yang–Mills theory and integrability", *Phys. Rept.* **405** (2005) 1, hep-th/0407277.
24. D. Serban and M. Staudacher, "Planar $\mathcal{N} = 4$ gauge theory and the Inozemtsev long range spin chain", *JHEP* **0406** (2004) 001, hep-th/0401057.
25. N. Beisert, "Higher loops, integrability and the near BMN limit", *JHEP* **0309** (2003) 062, hep-th/0308074.
26. H. Verlinde, "Bits, matrices and 1/N", *JHEP* **0312** (2003) 052, hep-th/0206059 • J.-G. Zhou, "pp-wave string interactions from string bit model", *Phys. Rev.* **D67** (2003) 026010, hep-th/0208232 • D. Vaman and H. Verlinde, "Bit strings from N = 4 gauge theory", *JHEP* **0311** (2003) 041, hep-th/0209215.
27. U. Danielsson, F. Kristiansson, M. Lubcke, and K. Zarembo, "String bits without doubling", *JHEP* **0310** (2003) 026, hep-th/0306147.
28. N. Beisert, C. Kristjansen, and M. Staudacher, "The dilatation operator of $\mathcal{N}=4$ super Yang–Mills theory", *Nucl. Phys.* **B664** (2003) 131, hep-th/0303060.
29. N. Beisert, C. Kristjansen, J. Plefka, and M. Staudacher, "BMN gauge theory as a quantum mechanical system", *Phys. Lett.* **B558** (2003) 229, hep-th/0212269.
30. S. Bellucci, P. Y. Casteill, J. F. Morales, and C. Sochichiu, "Spin bit models from non-planar $\mathcal{N} = 4$ SYM", *Nucl. Phys.* **B699** (2004) 151, hep-th/0404066.
31. S. Bellucci, P. Y. Casteill, J. F. Morales, and C. Sochichiu, "Chaining spins from (super)Yang–Mills", contribution to the XI International Conference on Symmetry Methods in Physics (SYMPHYS-11) (2004), hep-th/0408102.
32. S. Bellucci, P. Y. Casteill, A. Marrani and C. Sochichiu, "Spin bits at two loops", *Phys. Lett.* **B607** (2005) 180, hep-th/0411261.
33. V. E. R. Lemes, M. S. Sarandy, S. P. Sorella, O. S. Ventura and L. C. Q. Vilar, "An algebraic criterion for the ultraviolet finiteness of quantum field theories", *J. Phys. A: Math. Gen.* **34** (2001) 9485, hep-th/0103110.
34. S. Bellucci and A. Marrani, "Non-Planar Spin Bits beyond two loops", hep-th/0505106.
35. A. Kundu, "New nonultralocal quantum integrable models through gauge transformation", hep-th/0207036 • A. Kundu, "Ultralocal solutions for quantum integrable nonultralocal models", *Phys. Lett.* **B550** (2002) 128, hep-th/0208147.
36. L. D. Faddeev, "How algebraic Bethe Ansatz works for integrable models", in *"Les Houches 1995, Relativistic gravitation and gravitational radiation"*, 149 (1971) hep-th/9605187.
37. N. Beisert, V. Dippel and M. Staudacher, "A Novel Long-Range Spin Chain and Planar $\mathcal{N}=4$ super Yang-Mills", *JHEP* **0407** (2004) 075, hep-th/0405001.
38. S. Bellucci and A. Marrani, "Deplanarization methods for Hamiltonian and higher-order charges in spin-bit models", in preparation.
39. E. D'Hoker, P. Heslop, P. Howe and A. V. Ryzhov, "Systematics of quarter BPS operators in $\mathcal{N}=4$ SYM", *JHEP* **0304** (2003) 038, hep-th/0301104 • A. V. Ryzhov and A. A. Tseytlin, "Towards the exact dilatation operator of $\mathcal{N}=4$ super Yang–Mills theory", *Nucl. Phys.* **B698** (2004) 132, hep-th/0404215 • M. Kruczenski, A. V. Ryzhov and A. A. Tseytlin, "Large spin limit of $AdS_5 \times S^5$ string theory and low energy expansion of ferromagnetic spin-chains", *Nucl. Phys.* **B692** (2004) 3, hep-th/0403120.
40. N. Beisert, "Higher-Loop Integrability in $\mathcal{N}=4$ Gauge Theory", *Comptes Rendus Physique* **5** (2004) 1039, hep-th/0409147.
41. N. Beisert and M. Staudacher, "Long Range $psu(2,2|4)$ Bethe Ansätze for Gauge Theory and Strings", hep-th/0504190.
42. G. Falqui, C.-M. Viallet, "Singularity, complexity, and quasi-integrability of rational mappings", *Commun. Math. Phys.* **154** (1993), 111, hep-th/9212105 • S. Klishevich, "Quasi-exact solvability and intertwining relations", hep-th/0410064 • D. Gomez-Ullate, N. Kamran and R. Milson, "Quasi-exact solvability and the direct approach to invariant subspaces", *J. Phys.* **A38** (2005), 2005, nlin.SI/0401030.
43. V. A. Kazakov and K. Zarembo, "Classical/quantum integrability in non-compact sector of AdS/CFT", *JHEP* **0410** (2004) 060, hep-th/0410105 • V. A. Kazakov, A. Marshakov, J. A. Minahan and K. Zarembo, "Classical/quantum integrability in AdS/CFT", *JHEP* **0405** (2004) 024, hep-th/0402207 • O. Babelon and M. Talon, "Riemann surfaces, separation of variables and classical and quantum integrability", *Phys. Lett.* **A312** (2003), 71, hep-th/0209071.
44. S. Bellucci, P. Y. Casteill, J. F. Morales, and C. Sochichiu, "$SL(2)$ spin chain and spinning strings on $AdS_5 \times S^5$", *Nucl. Phys.* **B707** (2005) 303, hep-th/0409086 • S. Bellucci, P. Y. Casteill, and J. F. Morales, "Superstring sigma models from spin chains: the $SU(1,1|1)$ case", *Nucl. Phys.* **B** (in press), hep-th/0503159.
45. G. Ferretti, R. Heise and K. Zarembo, "New integrable structures in Large-N QCD", *Phys. Rev.* **D70** (2004) 074024, hep-th/0404187 • N. Beisert, G. Ferretti, R. Heise and K. Zarembo, "One Loop QCD Spin Chain and its Spectrum", *Nucl. Phys.* **B717** (2005) 137, hep-th/0412029.
46. B. Chen, X.-J. Wang and Y.-S. Wu, "Integrable Open Spin Chain in Super Yang-Mills and the Plane-wave/SYM duality", *JHEP* **0402** (2004) 029, hep-th/0401016 • B. Chen, X.-J. Wang and Y.-S. Wu, "Open Spin Chain and Open Spinning String", *Phys. Lett.* **B591** (2004), 170, hep-th/0403004.
47. A. Ashtekar, "Polymer geometry at Planck scale and quantum Einstein equations", *Int. J. Mod. Phys.* **D5**, 629 (1996), hep-th/9601054 • S. Kalyana Rama, "Size of balck holes through polymer scaling", *Phys. Lett.* **B424** (1998), 39, hep-th/9710035 • R. R. Khuri, "Black holes and strings: the polymer link", *Mod. Phys. Lett.* **A13** : 1407 (1998), gr-qc/9803095.
48. O. Bergman and C. B. Thorn, "The size of a polymer of string bits: a numerical investigation", *Nucl. Phys.* **B502** (1997) 309, hep-th/9702068 • A. Ashtekar, J. Lewandowski and H. Sahlmann, "Polymer and Fock representations for a scalar field", *Class. Quant. Grav.* **20**: L11-1 (2003), gr-qc/0211012 • F. Ferrari and I. Lazzizzera, "Polymer topology and Chern-Simons field theory", *Nucl. Phys.* **B559** (1999) 673.

49. R. Penrose, "Angular momentum: an approach to combinatorial space-time", in *"Quantum Theory and Beyond"*, Cambridge University Press, 1971 • C. Rovelli and L. Smolin, "Spin networks and quantum gravity", *Phys. Rev.* **D52** (1995), 5743, gr-qc/9505006.
50. C. Sochichiu, "Continuum limit(s) of Berenstein-Maldacena-Nastase matrix theory: Where is the (non-abelian) gauge group?", *Phys. Lett.* **B574** (2003) 105, hep-th/0206239.
51. P. Valtancoli, "Stability of the fuzzy sphere solution from matrix model", *Int. J. Mod. Phys.* **A18** (2003) 967, hep-th/0206075.

A Maximal CP Violating Phase for a Maximal Atmospheric Mixing

Isabella Masina

Centro Studi e Ricerche "E. Fermi", Via Panisperna 89/A, Rome, Italy and INFN, Sezione di Roma, P.le A. Moro 2, Rome, Italy

Abstract

I present an interesting mechanism to predict a maximal atmospheric angle, which is based on a maximal CP violating phase difference between second and third lepton families in the flavour symmetry basis. In this framework, a discussion of the general formulas for θ_{12}, $|U_{e3}|$, δ and their possible correlations in some limiting cases is also provided.

1 Introduction

The atmospheric mixing angle has experimentally turn out to be quite large [1] and, according to recent analysis [2], its 1σ range is $\theta_{23} = 38^\circ - 47^\circ$. From the theoretical point of view this invites to speculate on the possibility of maximal flavour violation for second and third lepton families. In flavour model building, achieving a maximal θ_{23} is far from trivial, taking also into account that $|U_{e3}|$ is small and the solar mixing angle large but not maximal: $\theta_{13} \leq 7^\circ$ and $\theta_{12} = 33^\circ - 36^\circ$ at 1σ [2]. The aim of this talk is to display which mechanisms naturally generate a maximal atmospheric mixing[1].

If an underlying flavour symmetry exists, it selects a *privileged* flavour basis for fermion mass matrices. The lepton mixing matrix results from combining the unitary matrices which - in that basis - diagonalise left-handed charged lepton and neutrino mass matrices respectively, $U_{MNS} = U_e^\dagger U_\nu$. Obtaining a maximal atmospheric angle because of a conspiracy between many large mixings present in U_e and U_ν appears to be quite a fortuitous explanation, especially in the case of effective neutrino masses as in the seesaw mechanism[2]. A better starting point for a flavor model is to predict one between θ_{23}^e and θ_{23}^ν to be maximal, so that the goal is reached when the other parameters present in $U_e^\dagger U_\nu$ marginally affect this maximal 23-angle.

[1]To be quantitative (and rather subjective), we are going to ask $|1 - \tan\theta_{23}| \lesssim 5\%$, corresponding to an uncertainty of about 2° in θ_{23}.

[2]Exceptions are the A_4 models [3], where tunings are eventually displaced in the neutrino spectrum.

Models so far proposed [4] along these lines adopted the strategy of having, in the flavour symmetry basis, a maximal θ_{23}^{ν} together with a negligibly small θ_{23}^{e}, or viceversa. Since the upper bound on θ_{13} naturally suggests θ_{12}^{e} and θ_{13}^{e} to be small and since the latter affect θ_{23} at second order, this is in principle a simple and effective framework to end up with a maximal atmospheric angle. However, the drawback in model building is the difficulty in managing such a huge hierarchy among θ_{23}^{e} and θ_{23}^{ν}.

In this talk I focus on an alternative mechanism to achieve a maximal atmospheric angle [5], which is based on the presence of a maximal CP violating phase difference between second and third lepton families and which does not require any particular hierarchy among θ_{23}^{e} and θ_{23}^{ν}, provided that one of them is maximal. If θ_{12}^{e} and θ_{13}^{e} are small as suggested by the bound on θ_{13}, then $\theta_{23} = \pi/4$ is robustly predicted. This mechanism is based on maximal CP violation in the sense that, denoting by $g/\sqrt{2}\ (e^{-iw_2}\bar{\mu}_L\ \gamma^\lambda \nu_\mu + \bar{\tau}_L\ \gamma^\lambda \nu_\tau)\ W_\lambda^- +$ h.c. the weak charged currents of the second and third lepton families in the flavour symmetry basis (before doing the rotations in the $\mu - \tau$ and $\nu_\mu - \nu_\tau$ planes to go in the mass eigenstate basis), it requires the phase difference w_2 to be $\pm\pi/2$. For three families and Majorana neutrinos, U_{MNS} contains three CP violating phases, which turn out to be complicated functions of w_2 and of other phases potentially present. We will focus in particular on the connection between δ and w_2 and on the expectations for $|U_{e3}|$ and θ_{12}. This mechanism also has remarkable analogies with the quark sector.

2 On the Origin of the Atmospheric Mixing

In the basis of the unknown flavour symmetry the leptonic sector is described by

$$\mathcal{L} = -\frac{1}{2}\,\nu^T m_\nu^{eff} \nu \; - \; \bar{e}_R^T m_e e_L \; + \; \frac{g}{\sqrt{2}}\,\bar{e}_L^T \gamma^\lambda \nu\, W_\lambda^- \; + \text{ h.c.} \tag{1}$$

$$m_\nu^{eff} = U_\nu^* \hat{m}_\nu U_\nu^\dagger \;, \qquad m_e = U_e^R \hat{m}_e U_e^{L\dagger}$$

where a hat is placed over a diagonal matrix with real positive eigenvalues whose order is established conventionally by requiring $|m_2^2 - m_3^2| \geq m_2^2 - m_1^2 \geq 0$, and the U's are unitary matrices. The MNS mixing matrix is $U_{MNS} = U_e^{L\dagger} U_\nu$. We find it convenient to write unitary matrices in terms of a matrix in the standard CKM parameterization [6] multiplied at right and left by diagonal matrices with five independent phases, defined for definiteness according to

$$U_\ell = e^{i\alpha_\ell}\, \mathcal{W}_\ell\, U_\ell^{(s)}\, \mathcal{V}_\ell \qquad \ell = e, \nu, \mathrm{MNS} \tag{2}$$

where, omitting the ℓ index, $\mathcal{W} = \mathrm{diag}(e^{i(w_1+w_2)}, e^{iw_2}, 1)$, $\mathcal{V} = \mathrm{diag}(e^{i(v_1+v_2)}, e^{iv_2}, 1)$, $U^{(s)} = R(\theta_{23})\Gamma_\delta R(\theta_{13})\Gamma_\delta^\dagger R(\theta_{12})$, $\Gamma_\delta = \mathrm{diag}(1, 1, e^{i\delta})$, angles belong to the first quadrant and phases to $[0, 2\pi[$.

Upon phase redefinitions for ν and e_L fields and a unitary transformation for e_R fields, one can go into the basis where the Lagrangian (1) reads

$$\mathcal{L} = -\frac{1}{2}\,\nu^T (U_\nu^{(s)*} \hat{m}_\nu \mathcal{V}_\nu{}^{*2} U_\nu^{(s)\dagger})\nu - \bar{e}_R^T(\hat{m}_e U_e^{L(s)\dagger})e_L + \frac{g}{\sqrt{2}}\,\bar{e}_L^T \gamma^\lambda \mathcal{W}^* \nu\, W_\lambda^- + \text{ h.c.} \;, \tag{3}$$

where $\mathcal{W}^* = \mathcal{W}_e^* \mathcal{W}_\nu = \text{diag}(e^{-i(w_1+w_2)}, e^{-iw_2}, 1)$. The phases w_2 and w_1 can be chosen in $]-\pi, \pi]$ and represent the phase difference between the second and third generations of leptons, first and second respectively, in the basis (3), namely before shuffling the flavours by means of $U_\nu^{(s)}$ and $U_e^{L(s)}$ to go in the mass basis. They are a source for CP violation and, in spite of the particular convention adopted here to define angles and phases[3], it has to be recognized that they are univocally determined by the flavour symmetry and cannot be removed. Clearly w_2 and w_1 are not directly measurable, but nevertheless play a quite profound role for the MNS mixing matrix:

$$\begin{aligned} U_{MNS} &= U_e^{L(s)\dagger} \mathcal{W}^* U_\nu^{(s)} \mathcal{V}_\nu \\ &= \underbrace{R^T(\theta_{12}^e)\Gamma_{\delta^e} R^T(\theta_{13}^e)\Gamma_{\delta^e}^\dagger}_{S_e} \; \underbrace{R^T(\theta_{23}^e)\mathcal{W}^* R(\theta_{23}^\nu)}_{L} \; \underbrace{\Gamma_{\delta^\nu} R(\theta_{13}^\nu)\Gamma_{\delta^\nu}^\dagger R(\theta_{12}^\nu)\mathcal{V}_\nu}_{S_\nu} \quad . \end{aligned} \tag{4}$$

None of the 12 parameters dictated by the flavour symmetry and appearing in the r.h.s of eq. (4) is actually measurable, because only 9 combinations of them are independent - see eq. (2) -, among which 3 can be absorbed, α_{MNS} and $\mathcal{W}_{MNS}$. Note that CP violation through δ can be generated in the limit where only w_1 and/or w_2 are present, as well as in the limit where there are just δ^e and/or δ^ν ($\mathcal{V}_\nu$ solely contributes to Majorana CP violating phases). Different flavour symmetries can thus predict the same leptonic (and even hadronic) physics and there is no way to discriminate between them unless adopting theoretical criteria like, e.g., absence of tunings and stability of the results. Adhering to such criteria, we now ask which sets of flavour symmetry parameters more robustly predict the atmospheric angle to be maximal.

It is natural to expect the leading role to be played by the core of the MNS mixing matrix, denoted by L in eq. (4):

$$L = \begin{pmatrix} e^{-i(w_1+w_2)} & 0 & 0 \\ 0 & e^{-iw_2} c_e c_\nu + s_e s_\nu & e^{-iw_2} c_e s_\nu - s_e c_\nu \\ 0 & e^{-iw_2} s_e c_\nu - c_e s_\nu & e^{-iw_2} s_e s_\nu + c_e c_\nu \end{pmatrix} \tag{5}$$

where from now on $s_{e,\nu} = \sin\theta_{23}^{e,\nu}$, $c_{e,\nu} = \cos\theta_{23}^{e,\nu}$ for short. If also S_e in eq. (4) had large mixings, bringing it at the right of L would in general induce large contributions to all three MNS mixings. Both the experimentally small θ_{13} and a potentially maximal θ_{23} would then result from a subtle conspiracy between the many angles and phases involved. At the price of some tuning in the neutrino spectrum, this may happen in the case of tribimixing models [3], which are often based on an A_4 flavour symmetry. On the contrary, the bound on θ_{13} is naturally fulfilled if each mixing in S_e and $R(\theta_{13}^\nu)$ does. Denoting them with $\varphi = \sin\theta_{12}^e$, $\psi = \sin\theta_{13}^e$ and $\xi = \sin\theta_{13}^\nu$, it turns out that they induce second order corrections in the 23 block of L, hence smaller than $O(5\%)$ given the bound on θ_{13}.

Adopting the latter point of view, the atmospheric mixing angle can be identified with the one of L up to corrections smaller than about 2°. It turns out to depend on w_2 and, symmetrically, on θ_{23}^e and θ_{23}^ν:

$$\tan^2\theta_{23} = \frac{c_e^2 s_\nu^2 + s_e^2 c_\nu^2 - 2\cos w_2 \; c_e s_e c_\nu s_\nu}{s_e^2 s_\nu^2 + c_e^2 c_\nu^2 + 2\cos w_2 \; c_e s_e c_\nu s_\nu} \quad . \tag{6}$$

The crucial role played by the phase w_2 is manifest: only in the case $w_2 = {}^0_\pi$ one has the simple relation $\theta_{23} = |\theta_{23}^\nu \mp \theta_{23}^e|$. A maximal atmospheric angle requires the following relation

[3]Needless to say, the standard parameterization for unitary matrices is as noble as other possible ones.

among three parameters to be fulfilled

$$\cos w_2 = -\frac{1}{\tan(2\theta^{\nu}_{23})\tan(2\theta^{e}_{23})} \quad . \tag{7}$$

Notice that maximality is generically lost by slightly varying one of the parameters involved. However, one realises from the above expressions that for some exceptional values of two parameters the atmospheric angle turns out to be maximal independently from the value assumed by the third parameter:

i) for $\theta^{e(\nu)}_{23} = \pi/4$ and $w_2 = \pm\pi/2$, independently of $\theta^{\nu(e)}_{23}$;

h) for $\theta^{e(\nu)}_{23} = \pi/4$ and $\theta^{\nu(e)}_{23} = 0$ or $\pi/2$, independently of w_2.

All this is graphically seen in fig. 1 which, for different values of $\theta^{e(\nu)}_{23}$, shows the region of the plane $\{w_2, \theta^{\nu(e)}_{23}\}$ allowed at $1, 2, 3\sigma$ by the experimental data on the atmospheric angle.

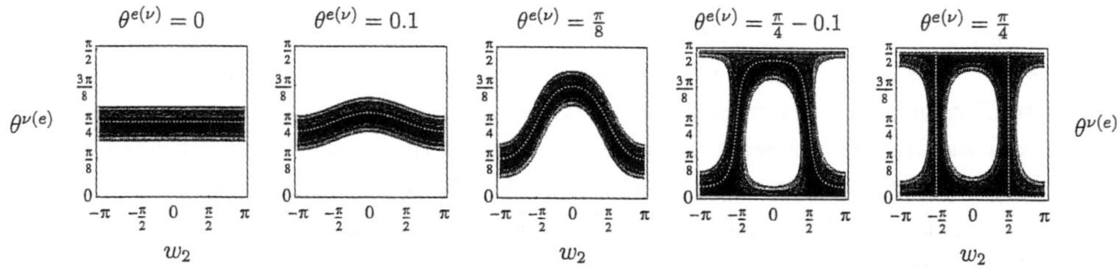

Figure 1: Region of the plane $\{w_2, \theta^{\nu(e)}_{23}\}$ allowed at $1, 2, 3\sigma$ by the experimental data on the atmospheric angle [2] for $\theta^{e(\nu)}_{23} = \{0, 0.1, \pi/8, \pi/4 - 0.1, \pi/4\}$. The dotted curve is the surface where the atmospheric angle is maximal. The plots also correspond to $\theta^{e(\nu)}_{23} = \pi/2 - \{0, 0.1, \pi/8, \pi/4 - 0.1, \pi/4\}$ upon the substitution $\theta^{e(\nu)} \to \pi/2 - \theta^{e(\nu)}$.

The possibility h) has been widely exploited in flavour model building. The difficulty of this approach is not to predict a maximal $\theta^{e(\nu)}_{23}$, but rather to manage in having a sufficiently small $\theta^{\nu(e)}_{23}$, say less than 2°. For instance, a negligible θ^{ν}_{23} is somewhat unnatural in the case of hierarchical neutrinos because the ratio between the corresponding eigenvalues is not so small: $m_2/m_3 \sim 1/6$. Notice also that in seesaw models θ^{ν}_{23} is an effective angle which depends on both the Dirac and Majorana Yukawa couplings.

The possibility i) has not (to our knowledge) been singled out so far[4]. It has the advantage that it does not require a huge hierarchy between θ^{e}_{23} and θ^{ν}_{23}. Indeed, in the case that $g/\sqrt{2}\,(\bar{\tau}_L\gamma^{\lambda}\nu_{\tau} \pm i\,\bar{\mu}_L\gamma^{\lambda}\nu_{\mu})\,W^{-}_{\lambda} +$ h.c. are the charged currect interactions of the second and third lepton families in the basis (3) - namely before applying $R(\theta^{e}_{23})$ and $R(\theta^{\nu}_{23})$ to go in the mass eigenstate basis - the maximal phase difference i shields a maximal $\theta^{e(\nu)}_{23}$ from any interference due to $\theta^{\nu(e)}_{23}$. This remarkable fact can be visually seen in fig. 2, where we plot $\tan\theta_{23}$ as a function of w_2 for different values of θ^{e}_{23} and θ^{ν}_{23}.

[4]We thank however the authors of ref. [7] for pointing out some interesting analogies with one of their quaternion family symmetry models.

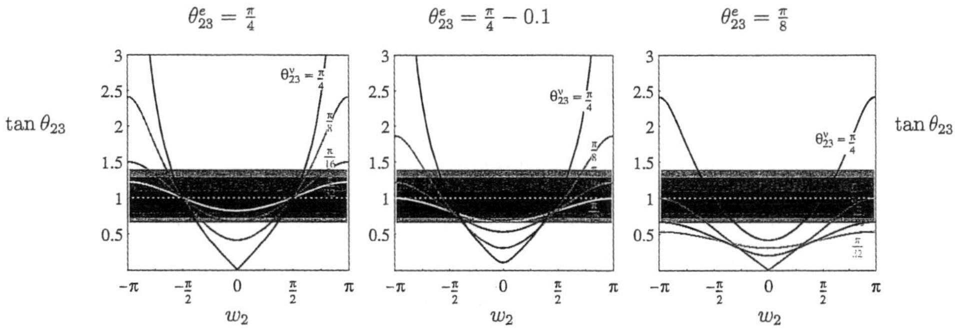

Figure 2: Values of $\tan\theta_{23}$ as a function of w_2 for $\theta^e_{23} = \pi/4, \pi/4 - 0.1, \pi/8$. Colored curves correspond, as marked, to $\theta^\nu_{23} = \pi/32, \pi/16, \pi/8, \pi/4$. The experimental range at $1, 2, 3\sigma$ [2] is also shown. The same plot holds for $e \leftrightarrow \nu$.

It is worth to stress that a maximal CP violating phase difference in between fermion families is likely to be at work in the quark sector too, in particular for the the Cabibbo angle [8] - by the way, again the largest angle of the mixing matrix. Neglecting the presumably very small 13 and 23 mixings, the charged currect interaction of the lighter families in the basis analog of (3) is $g/\sqrt{2}\ (\bar{c}_L\ \gamma^\lambda s_L + e^{-iw^q_1}\ \bar{u}_L\ \gamma^\lambda d_L)\ W^+_\lambda +$ h.c.. Expanding at first order in $s_u = \sin\theta^u_{12}$ and $s_d = \sin\theta^d_{12}$, the analog of eq. (6) reads

$$\tan^2\theta_C = s^2_u + s^2_d - 2\cos w^q_1\ s_u s_d + O(s^4) \quad . \tag{8}$$

In the plausible case that $s_d = \sqrt{m_d/m_s}$, $s_u = \sqrt{m_u/m_c}$, one has

$$|V_{us}| = |\sqrt{\frac{m_d}{m_s}} - e^{-iw^q_1}\sqrt{\frac{m_u}{m_c}}| \tag{9}$$

and it is well known that experimental data strongly indicate $w^q_1 = \pm\pi/2$. Models of this sort have been studied where also $\alpha \approx w^q_1$ [9], so that the unitarity triangle turns out to be rectangular.

3 Some general relations and limiting cases

Here we discuss the general formulas for $|U_{e3}|$, θ_{12} and the MNS phase δ under the assumptions that the bound on θ_{13} is naturally fulfilled because so do $\varphi = \sin\theta^e_{12}$, $\psi = \sin\theta^e_{13}$ and $\xi = \sin\theta^\nu_{13}$. This allows S_e to commute with L and the dependence on the mechanism responsible for a maximal atmospheric angle is encoded in $w_2, \theta^e_{23}, \theta^\nu_{23}$. Since $L_{22} = e^{-iw_2}L^*_{33}$, $L_{32} = -e^{-iw_2}L^*_{23}$ and a maximal atmospheric angle implies $|L_{ij}| = 1/\sqrt{2}$ for $i, j = 2, 3$, this dependence is equivalently expressed in terms of w_2, λ_{23} =Arg(L_{23}), λ_{33} =Arg(L_{33}). We collect in table 1 the values of $\lambda_{23}, \lambda_{33}$ for the cases h) and i) discussed previously.

	$\theta^e_{23}=\pi/4$ $\theta^\nu_{23}=0\ (\pi/2)$	$\theta^\nu_{23}=\pi/4$ $\theta^e_{23}=0\ (\pi/2)$	$\theta^e_{23}=\pi/4$ $w_2=\pm\pi/2$	$\theta^\nu_{23}=\pi/4$ $w_2=\pm\pi/2$
λ_{23}	$\pi\ (-w_2)$	$-w_2\ (\pi)$	$\pm\theta^\nu_{23}+\pi$	$\mp(\theta^e_{23}+\pi/2)$
λ_{33}	$0\ (-w_2)$	$0\ (-w_2)$	$\mp\theta^\nu_{23}$	$\mp\theta^e_{23}$

Table 1

Introducing the quantities

$$v_\varphi=\frac{\varphi}{\sqrt{2}}\,e^{i(w_1-\lambda_{33})}\ ,\quad v_\psi=\frac{\psi}{\sqrt{2}}\,e^{i(w_1-\lambda_{23}-\delta^e)}\ ,\quad v_\xi=\xi\,e^{-i(w_2+\lambda_{23}+\lambda_{33}+\delta^\nu)} \tag{10}$$

one has $\theta_{23}=\pi/4+O(v^2)$ with $v^2\sim 5\%$, together with the general formulas

$$\begin{aligned}\theta_{12}&=\theta^\nu_{12}-\mathrm{Re}(v_\varphi-v_\psi)+O(v^2)\\ U_{e3}&=v_\varphi+v_\psi-v_\xi+O(v^3)\\ \delta&=\pi-\mathrm{Arg}U_{e3}+O(v\sin(\mathrm{Arg}U_{e3}))\ .\end{aligned} \tag{11}$$

The above expressions allow to complete the phenomenological study of our framework and generalise previous studies that assumed bimixing [10] or tri-bimixing [11] for U_ν. As already stressed, a measure of these independent observable quantities cannot reveal from which mechanism they come from, nor whether the v's interfere in originating them. In order to find some potential correlations, additional hypothesis have to be introduced.

In the limit $\xi\gg\varphi,\psi$ the above expressions simplify to

$$|U_{e3}|\approx\xi\ ,\qquad \delta\approx w_2+\lambda_{23}+\lambda_{33}+\delta^\nu\ ,\qquad \theta_{12}\approx\theta^\nu_{12}\ . \tag{12}$$

Notice that there are no correlations between $|U_{e3}|$, θ_{12} and δ. The latter does not depend on w_1 and is rather related to the mechanism at work for the atmospheric angle, even though it cannot reveal which one is actually at work. As can be seen from table 1, in the case i) with $\theta^e_{23}=\pi/4$, $\delta=w_2+\pi+\delta^\nu$.

Interesting correlations emerge for $\psi\gg\varphi,\xi$ in which case

$$|U_{e3}|\approx\frac{\psi}{\sqrt{2}}\ ,\qquad \delta\approx\pi+\lambda_{23}+\delta^e-w_1\ ,\qquad \theta_{12}\approx\theta^\nu_{12}-|U_{e3}|\cos\delta\ , \tag{13}$$

and for $\varphi\gg\psi,\xi$ in which case

$$|U_{e3}|\approx\frac{\varphi}{\sqrt{2}}\ ,\qquad \delta\approx\pi+\lambda_{33}-w_1\ ,\qquad \theta_{12}\approx\theta^\nu_{12}+|U_{e3}|\cos\delta\ . \tag{14}$$

Notice that these situations are phenomenologically equivalent provided $\delta\leftrightarrow\delta+\pi$. Both δ and θ_{12} depend on the mechanism at work for the atmospheric angle. For instance, for $\varphi\gg\psi,\xi$ and with $\theta^{e(\nu)}_{23}=\pi/4$, $w_2=\pm\pi/2$, it turns out that δ depends on w_1 and the 23-angle whose magnitude is irrelevant for the atmospheric angle: $\delta\approx\pi-w_1\mp\theta^{\nu(e)}_{23}$.

In the following we focus on the possibility that $\sin\theta^e_{12}=\varphi$ dominates. This is a particularly interesting scenario because naturally compatible with a grandunified picture. The correlations are shown in fig. 3 by plotting, for different values of θ^ν_{12}, the region of the $\{\delta, |U_{e3}|\}$ plane allowed by the present range of θ_{12} at $1, 2, 3\sigma$. The case of a maximal θ^ν_{12} is particularly interesting from the theory point of view. As shown by the plot, present data [2] suggest $\delta \approx \pi$ and $|U_{e3}| \approx 0.2$, dangerously close to its 3σ bound and interestingly close to the Cabibbo angle θ_C. Notice that the so-called "quark lepton complementarity" proposal $\theta_{12} = \pi/4 - \theta_C$ [12] corresponds precisely to $\delta = \pi$ and $|U_{e3}| = \theta_C$, i.e. exact CP and $\varphi = \sqrt{2}\theta_C$. Anyway, also $\varphi = \theta_C$, i.e. U_{e3} close to its 1σ bound, falls inside the 2σ window for θ_{12} provided $\delta = \pi$. Remarkably enough, it turns out that a maximal δ strongly favours $\tan\theta^\nu_{12} \approx 1/\sqrt{2}$, with a mild dependence on $|U_{e3}|$. The possibility $\varphi = \theta_C/3$, particularly relevant for grandunified models, is thus well compatible with tribimixing and maximal CP violation.

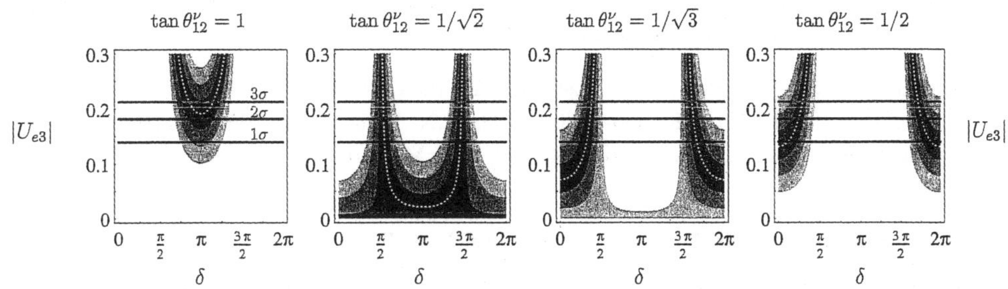

Figure 3: Region of the $\{\delta, |U_{e3}|\}$ plane allowed by the present range of θ_{12} at $1, 2, 3\sigma$ [2] for different values of $\tan\theta^\nu_{12}$. The dotted line corresponds to the best fit of θ_{12}. Also shown are the $1, 2, 3\sigma$ bounds on $|U_{e3}|$.

4 Conclusions and Outlook

We pointed out that if CP and flavour are maximally violated by second and third lepton families in the flavour symmetry basis, a maximal atmospheric angle is automatically generated when the bound on $|U_{e3}|$ is fulfilled in a natural way. This mechanism has two advantages with respect to the one usually exploited: it is very suggestive of the quark sector and it does not require one between θ^ν_{23} and θ^e_{23} to vanish, which could be difficult to achieve especially for seesaw models. Under the assumption that the bound on $|U_{e3}|$ is naturally fulfilled, we also discussed the general formulas for $|U_{e3}|$, θ_{12} and the CP violating phase δ, clarifying in particular its relation with the phases among lepton families in the flavour symmetry basis.

We think that such a mechanism deserves more studies, both from the point of view of grandunified theories and flavour symmetries.

Acknowledgements

I thank G. Altarelli and C.A. Savoy for enlightening discussions.

References

[1] Super-Kamiokande Collaboration: Y. Fukuda *et al.*, Phys. Rev. Lett. **81** (1998) 1562; Y. Ashie *et al.*, Phys. Rev. Lett. **93** (2004) 101801; hep-ex/0501064. K2K collaboration: M.H. Ahn *et al.*, Phys. Rev. Lett. **90** (2003) 041801; E. Aliu *et al.*, Phys. Rev. Lett. **94** (2005) 081802.

[2] G.L. Fogli, E. Lisi, A. Marrone and A. Palazzo, hep-ph/0506083. See also: M.C. Gonzalez-Garcia, M. Maltoni and A. Yu. Smirnov, Phys.Rev. **D 70** (2004) 093005, hep-ph/0408170.

[3] An incomplete list: P.F. Harrison, D.H. Perkins and W.G. Scott, Phys. Lett. **B 530** (02) 167, hep-ph/0202074; Z.Z. Xing, Phys. Lett. **B 533** (02) 85, hep-ph/0204049; K.S. Babu, E. Ma and J.F.W. Valle, Phys. Lett. **B 552** (03) 207, hep-ph/0206292; X.G. He and A. Zee, Phys. Lett. **B 560** (03) 87, hep-ph/0301092; G. Altarelli and F. Feruglio, hep-ph/0504165; K.S. Babu and Xiao-Gang He, hep-ph/0507217.

[4] K.S. Babu and S. M. Barr, Phys. Lett. **B 525** (2002) 289, hep-ph/0111215; W. Grimus *et al*, JHEP **0408** (04) 078, hep-ph/0408123; R.N. Mohapatra, JHEP **0410** (2004) 027, hep-ph/0408187; S.F. King, hep-ph/0506297; I. de Medeiros Varzielas and G.G. Ross, hep-ph/0507176.

[5] I. Masina, hep-ph/0508031, to appear on Phys. Lett. B.

[6] The Review of Particle Physics, S. Eidelman *et al*, Phys. Lett. **B 592** (2004) 1.

[7] M. Frigerio, S. Kaneko, E. Ma and M. Tanimoto, Phys. Rev. **D 71** (2005) 011901, hep-ph/0409187.

[8] See for instance: H. Fritzsch and Z.Z. Xing, Phys. Lett. **B 413** (1997) 396, hep-ph/9707215; R.G. Roberts, A. Romanino, G.G. Ross and L. Velasco-Sevilla, Nucl. Phys. **B 615** (2001) 358, hep-ph/0104088; G.C. Branco, M.N. Rebelo and J.I Silva-Marcos, Phys. Lett. **B 597** (2004) 155, hep-ph/0403016; H. Fritzsch, J.Korean Phys. Soc. **45** (2004) S297, hep-ph/0407069.

[9] R. Barbieri, L.J. Hall and A. Romanino, Phys. Lett. **B 401** (1997) 47, hep-ph/9702315; H. Fritzsch, hep-ph/9706243.

[10] P.H. Frampton, S.T. Petcov and W. Rodejohann, Nucl. Phys. **B 687** (2004) 31, hep-ph/0401206; G. Altarelli, F. Feruglio and I. Masina, Nucl. Phys. **B 689** (2004) 157, hep-ph/0402155; A. Romanino, Phys. Rev. **D 70** (2004) 013003, hep-ph/0402258; S.T. Petcov and W. Rodejohann, Phys. Rev. **D 71** (2005) 073002, hep-ph/0409135.

[11] F. Plentinger and W. Rodejohann, hep-ph/0507143.

[12] M. Raidal, Phys. Rev. Lett. **93** (2004) 161801, hep-ph/0404046; H. Minakata and A.Yu. Smirnov, Phys. Rev. **D 70** (2004) 073009, hep-ph/0405088; S. Antusch, S.F. King and R.N. Mohapatra, Phys. Lett. **B 618** (2005) 150, hep-ph/0504007; C. Jarlskog, hep-ph/0507212.

Higgs Phenomenology in the Next-to-Minimal Supersymmetric Standard Model

Shoaib Munir

High Energy Physics Group, School of Physics & Astronomy,
University of Southampton, Southampton, SO17 1BJ, UK

July 9, 2007

Abstract

The NMSSM contains a Higgs singlet in addition to the two Higgs doublets typical of the MSSM, thus resulting in a total of seven physical Higgs mass states. The phenomenology of the NMSSM Higgs sector varies considerably from that of the MSSM and there are good prospects of finding in regions of the NMSSM parameter space Higgs signals that cannot be reproduced in the MSSM. We examined the two-photon decay mode of a Higgs boson and found that up to three Higgs bosons could be observable over sizable regions of NMSSM parameter space, through the conventional detection channels used at the LHC.

preprint SHEP-05-36

1 Introduction

The Minimal Supersymmetric Standard Model (MSSM) [1]–[4] effectively solves the Hierarchy and Naturalness problems of the Standard Model (SM). In the MSSM, by postulating a new symmetry, called Supersymmetry (SUSY), between bosons and fermions, one introduces new contributions to the otherwise quadratically divergent quantum corrections to the Higgs mass. SUSY ensures that the contributions of the new particles exactly cancel the quadratic divergences of their SM partners, removing the sensitivity of the Higgs mass to the cut-off scale in a natural way.

The introduction of SUSY significantly enhances the particle spectrum of the SM. In the MSSM, two Higgs doublets are needed in order to separately provide masses for the up-type and down-type fermions. As a result, in addition to having a counterpart for every SM particle, the MSSM ends up with five physical Higgs bosons after electroweak (EW) symmetry breaking (and corresponding sparticles). These include: a charged pair $H^{\pm}$; two neutral CP-even scalars H and h (where, by convention, $m_H > m_h$); and a neutral CP-odd scalar A [5]–[6]. From the expression for m_h in the MSSM, at tree-level one obtains an upper bound on the lightest Higgs mass as

$$m_h \;\leq\; min(m_A, m_Z)\cdot|\cos 2\beta| \;\leq\; m_Z,$$

which, with inclusive two-loop corrections, is raised to $m_h \leq 130$ GeV [7]. From a phenomenological perspective, the search for a Higgs boson in the MSSM at future colliders has been summed up into a 'No-Lose Theorem' for MSSM Higgs bosons, which states that at least one Higgs boson should be observed via the usual production and decay channels at the Large Hadron Collider (LHC) throughout the entire MSSM parameter space [8]–[10].
However, there is still a problem of phenomenological significance with the MSSM. The most general MSSM scalar potential can be written as [11]

$$\begin{aligned} V_{Higgs} &= (|\mu|^2 + m_{H_u}^2)|H_u|^2 + (|\mu|^2 + m_{H_d}^2)|H_d|^2 \\ &\quad -B\mu(H_uH_d + h.c.) + \frac{g_2^2+g_1^2}{8}(|H_u|^2 - |H_d|^2)^2 \\ &\quad +\frac{1}{2}g_2^2|H_d^\dagger H_u|^2. \end{aligned}$$

The term '$B\mu H_uH_d$', mixing the two Higgs doublets, emerges as a manifestation of soft SUSY breaking in the MSSM. But since it is present in the superpotential before the EW symmetry is broken, μ has the dimensions of mass, so its natural value would be either 0 or the Planck mass M_P. But $\mu = 0$ would mean no mixing is generated between Higgs doublets at any scale and the minimal of the Higgs potential would occur for $\langle H_d \rangle = 0$. This would, in turn, result in massless down-type fermions and leptons after SU(2) symmetry breaking. Conversely, $\mu \cong M_P$ would reintroduce the fine-tuning problem since the Higgs scalars would acquire a huge contribution $\sim \mu^2$ to their squared masses. Therefore, the values of this (theoretically arbitrary) parameter μ are phenomenologically constrained to be close to $M_{\rm SUSY}$ or M_W [12].

The most elegant solution to the 'μ problem' is to introduce a new singlet Higgs field S, and replace $B\mu H_uH_d$ by an interaction $\lambda S(H_uH_d)$. When the extra scalar field S acquires a vacuum expectation value (VEV), an effective $B\mu$ term, naturally of the EW scale, is generated automatically as $B\mu = \lambda\langle S \rangle$. This idea has been implemented in the Next-to-Minimal Supersymmetric Standard Model (NMSSM) [13]–[15], described by the superpotential

$$\begin{aligned} W &= \hat{u}^c h_u \hat{Q}\hat{H}_U - \hat{d}^c h_d \hat{Q}\hat{H}_d - \hat{e}^c h_e \hat{L}\hat{H}_d \\ &\quad +\lambda\hat{S}(\hat{H}_u\hat{H}_d) + \frac{1}{3}\kappa\hat{S}^3, \end{aligned}$$

where the last term is required to explicitly break the dangerous Peccei-Quinn U(1) symmetry [16].

2 Higgs Phenomenology in the NMSSM

As a result of the introduction of an extra complex singlet scalar field, which only couples to the two MSSM type Higgs doublets, the Higgs sector of the NMSSM comprises of a total of seven mass eigenstates; a charged pair $h^{\pm}$, three CP-even Higgses $h_{1,2,3}$ ($m_{h_1} < m_{h_2} < m_{h_3}$), and two CP-odd Higgses $a_{1,2}$ ($m_{a_1} < m_{a_2}$) [17]. Consequently, Higgs phenomenology in the NMSSM is plausibly different from that of the MSSM and is extremely rich. For example, the mass expressions for the CP-even Higgses in the NMSSM can be translated into a slightly modified upper bound on the lightest Higgs mass, m_{h_1}, as [18]

$$m_{h_1}^2 \leq min\{m_Z^2, \frac{1}{2}\kappa v_s(4\kappa v_s + \sqrt{2}A_\kappa)\}(\simeq 140\text{GeV}).$$

Besides, the higher order corrections are similar to those in the MSSM. Thus, the upper bound is different from one model to the other.

Quite some work has been dedicated to probing the NMSSM Higgs sector over recent years. In an attempt to extend the No-Lose Theorem to the NMSSM [19], it was realized that at least one Higgs boson should be observable at the LHC over the parameter space that does not allow Higgs to Higgs decay. For the parameter range allowing the Higgs to Higgs decay through one of the modes

$$\begin{array}{c} i)\ h \to \acute{h}\acute{h}, \quad ii)\ h \to aa, \quad iii)\ h \to h^{\pm}h^{\mp}, \\ iv)\ h \to aZ, \quad v)\ h \to h^{\pm}W^{\pm}, \quad vi)\ \acute{a} \to ha, \\ vii)\ a \to hZ, \qquad viii)\ a \to h^{\pm}W^{\mp}, \end{array}$$

(where, by convention, $h(\acute{h})$ denotes one of the three CP-even and $a(\acute{a})$ stands for either of the two CP-odd Higgses, respectively; $m_{\acute{h}} > m_h$ and $m_{\acute{a}} > m_a$) the light Higgs boson, in particular a SM-like one, may not have a signal of statistical significance.

3 Parameter Space Scan

The choice of parameter space largely depends on the version of NMSSM under consideration, or more specifically, on the implication of unification of parameters at some very high scale. This, in turn, leans on the mechanism adopted for the breaking of SUSY, since it still remains undetected in nature. The advantage of assuming unification of masses and couplings in a SUSY model is twofold. Firstly, it enormously reduces the labour from a phenomenological point of view by minimizing the number of parameters required to extrapolate physical information from the model. Secondly, and theoretically more crucially, it caters to the fundamental objective of constructing a Grand Unification Theory (GUT), which was, to a great extent, responsible for devising SUSY in the first place.

A particular case of a low energy NMSSM with gravity mediated SUSY breaking was studied in [20]. As usual, in such a model, SUSY is assumed to be broken in some hidden sector and then mediated to the physical sector through gravitational interactions. Such a model implies unification of the couplings and soft masses at the GUT scale, which are then run down, using renormalisation group equations, to the weak scale or to some other scale at which the theory is being tested. This results in only a handful of parameters to deal with. In that paper, it was shown that with these settings, at least 100 events for 2 light CP-even and one light CP-odd Higgses might be observable at the Tevatron Run-II, for an expected luminosity of 15 fb^{-1}, while all the 5 neutral NMSSM Higgses may be produced at the LHC after an expected luminosity of 300 fb^{-1} is achieved.

For a more general study of the NMSSM Higgs sector with a wider range of parameters, we used here the publicly available fortran code NMHDECAY developed by C. Hugonie et al. [21]. This program computes the masses, couplings and decay widths of all the Higgs bosons of the NMSSM in terms of its parameters at the EW scale. For our purpose, instead of postulating unification, and without taking into account the SUSY breaking mechanism, we fixed the soft SUSY breaking terms to a very high value, so that they have little or no contribution to the outputs of the scans. Consequently, we are left with only six free parameters, almost as much as in the MSSM, despite the fact that there are at least four extra parameters in a general NMSSM scalar potential as compared to the MSSM. Our parameter space includes the Yukawa couplings λ and κ, the soft trilinear terms A_λ and A_κ, plus tanβ and $\mu_{\text{eff}} = \lambda\langle S\rangle$. The computation of the spectrum includes leading two loop terms, EW corrections and propagator corrections. The decay widths, however, do not include three body decays. The program also implicitly takes into account theoretical as well as experimental constraints from negative Higgs searches at LEP-II, along with the unconventional channels relevant for the NMSSM.

The code NMHDECAY scans over a user defined region of the NMSSM parameter space, which in our case is described as follows:

$$\lambda : 0.0001 \text{ - } 0.75, \quad \kappa : -0.65 \text{ - } +0.65, \quad \tan\beta : 1.6 \text{ - } 54,$$
$$\mu,\ A_\lambda,\ A_\kappa : -1000 \text{ - } +1000 \text{ GeV}.$$

Remaining soft terms which are fixed in the scan include:

- $m_{Q_3} = m_{U_3} = m_{D_3} = m_{L_3} = m_{E_3} = 2$ TeV
- $A_{U_3} = A_{D_3} = A_{E_3} = 1.5$ TeV
- $m_Q = m_U = m_D = m_L = m_E = 2$ TeV
- $M1 = M2 = M3 = 3$ TeV

This particular range of parameters was borrowed from a recent paper in which C. Hugonie et al. singled out the difficult scenarios for NMSSM Higgs discovery at the LHC [22] (using the mentioned code). In such a paper they also concluded, as a follow up of the No-Lose Theorem, that the region where no Higgs is observable due to its decays into the non-conventional channels mentioned earlier, comprises of only about 1% of the entire parameter space of the NMSSM. Our idea is to explore the same range of parameter space and look for the 'best case scenarios' where the discovery of one or more Higgses at the LHC is possible above and beyond what is predictable in the MSSM.

We used the NMHDECAY program to scan over millions of randomly selected points in the specified parameter space. The output, as stated earlier, contains masses, branching ratios and couplings of the NMSSM Higgses, for all the points which are not forbidden by the various constraints. The points which violate the constraints are directed to separate files, along with the warnings. The data with the 'good points' was then used to determine the cross-sections of the Higgses for different production channels (which include, gluon-gluon fusion, associated production with W, Z boson or heavy quarks, and vector boson fusion) using locally developed codes. The allowed decay modes are:

$$h, a \to gg, \quad h, a \to \mu\mu, \quad h, a \to \tau\tau, \quad h, a \to bb, \quad h, a \to tt,$$
$$h, a \to ss, \quad h, a \to cc, \quad h \to WW, \quad h \to ZZ,$$
$$h, a \to \gamma\gamma, \quad h, a \to Z\gamma, \quad h, a \to Higgses, \quad h, a \to sparticles.$$

(Notice that for the pseudoscalar Higgses, the decay into vector boson pairs is not allowed due to CP-conservation.)

4 Scan Results and Discussion

As an initial step towards the analysis of the data, we plotted the 'total cross-section times branching ratio' (BR) into $\gamma\gamma$ pair against each of the six parameters of the NMSSM, for each Higgs boson. A brief look at some of these plots readily revealed subtle differences between the MSSM and NMSSM Higgs sectors although, evidently, the Higgs BRs follow basically the same trend in both models. The dominant decay channel being the $h \to bb$ for the lightest (SM-like) Higgs, while for the heavier Higgs, this turns out to be the decay into lighter Higgses, depending on the kinematics.

We are here focusing on the $\gamma\gamma$ decay mode since it is the most promising channel for the discovery of a (neutral) Higgs boson at the LHC in the moderate Higgs mass range (say, below 140 GeV). In fact, since the tail of the $\gamma\gamma$ background rapidly falls with increasing invariant mass of the di-photon pair, signal peaks for heavier Higgses could also be visible in addition to the lightest one, although the cross-section for these processes is relatively very low [7, 23]. Based on the ATLAS analysis of Ref. [24], we argue that cross-section times branching ratio rates of 10 fb or so may yield visible signal events, the more so the heavier the decaying Higgs state (also because the photon detection efficiency grows with the Higgs mass). Tab. 1 recaps the potential observability of one or more NMSSM Higgs states in the di-photon mode at the LHC, under the above assumptions. It is obvious from the table that one light CP-even Higgs is observable almost throughout the NMSSM parameter space. However, there is also a fair number of points where two Higgses are visible simultaneously (h_1 and h_2 or – more rarely – h_1 and a_1), while production and decay of three Higgses, although possible, occurs at only an insignificant number of points in the parameter space. Furthermore, the percentage of points for which only the second lightest Higgs state is visible is also non-negligible. These last two conditions are clearly specific to the NMSSM, as they are never realized in the MSSM.

We have then plotted the cross-sections times branching ratios for the three observable Higgses against the various NMSSM parameters. These plots, shown in figs. 1–7, reveal that the distribution of visible points in the parameter space is quite homogeneous and not very fine-tuned at all. Moreover, by studying the cross-section times branching ratio of the Higgses where two or more of them are observable against their respective mass differences, one sees that the latter are higher than the typical mass resolution, so that the two decaying objects should indeed appear as separate resonances. (We have also verified, though not shown here either, that their decay widths are small compared to the detector resolution in $m_{\gamma\gamma}$.) Eventually, by doing a proper signal-to-background ratio analysis, we believe we should be able to single out a reasonable number of points in the NMSSM parameter space where the aforementioned signals, unavailable in the MSSM, can be established with a 5σ probability [25].

References

[1] P. Fayet, Nucl. Phys. B **90**, (1975) 104; Phys. Lett. B **64**, (1976) 159; Phys. Lett. B **69**, (1977) 489 and Phys. Lett. B **84**, (1979) 416.

[2] K. Inoue, A. Komatsu and S. Takeshita, Prog. Theor. Phys **68**, (1982) 927; (E) *ibid.* **70**, (1983) 330.

[3] H.E. Haber and G.L. Kane, Phys. Rep. **117**, (1985) 75.

[4] For reviews on the MSSM, see: P. Fayet and S. Farrera, Phys. Rep. **32**, (1977) 249; H. P. Nilles, Phys. Rep. **110**, (1984) 1; R. Barbieri, Riv. Nuov. Cim. **11**, (1988) 1; R. Arnowitt and P. Nath, Report CTP-TAMU-52-93; J. Bagger, Lectures at TASI-95, hep-ph/9604232.

[5] J. F. Gunion, H. E. Haber, G. Kane and S. Dawson, *The Higgs Hunter's Guide* (Westview Press,Oxford 1990).

[6] S. P. Martin, hep-ph/9709356.

[7] A. Djouadi, hep-ph/0503173.

[8] J. Dai, J.F. Gunion, R. Vega, Phys. Lett. B **315**, (1993) 355 .

[9] J. Dai, J.F. Gunion, R. Vega, Phys. Lett. B **345**, (1995) 29.

[10] J.R. Espinosa, J.F. Gunion, Phys. Rev. Lett. **82**, (1999) 1084 .

[11] J.F. Gunion and H.E. Haber, Nucl. Phys. B **272**, (1986) 1; (E) hep-ph/9301205.

[12] J.E. Kim and H.P. Nilles, Phys. Lett. B **138**, (1984) 150.

[13] M. Dine, W. Fischler and M. Serdnicki, Phys. Lett. B **104**, (1981) 199; H.P. Nilles, M. Serdnicki and D. Wyler, Phys. Lett. B **120**, (1983) 346; J.M. Frere, D.R. Jones and S. Raby, Nucl. Phys. B **222**, (1983) 11; A.I. Veselov, M.I. Vysotsky and K.A. Ter-Martirosian, Sov. Phys. JETP **63**, (1986) 489 [Zh. Eksp. Teor. Fiz. **90**, (1986) 838].

[14] J. P. Derendinger and C.A. Savoy, Nucl. Phys. B **237**, (1984) 307.

[15] U. Ellwanger, M. Rausch de Traubenberg and C.A. Savoy, Phys. Lett. B **315**, (1993) 331.

[16] R.D. Peccei and H.R. Quinn, Phys. Rev. Lett. **38**, (1977) 1440; Phys. Rev. D **16**, (1977) 1792.

[17] A. Dedes, C. Hugonie, S. Moretti and K. Tamvakis, Phys. Rev. D **63**, (2001) 055009.

[18] D.J. Miller, R. Nevzorov and P.M. Zerwas, hep-ph/0304049.

[19] U. Ellwanger, J.F. Gunion, C. Hugonie and S. Moretti, hep-ph/0305109.

[20] C. Hugonie and S. Moretti, hep-ph/0110241.

[21] U. Ellwanger, J.F. Gunion and C. Hugonie, hep-ph/0406215.

[22] U. Ellwanger, J.F. Gunion and C. Hugonie, hep-ph/0503203.

[23] V. Büscher and K. Jacobs, hep-ph/0504099.

[24] ATLAS Collaboration, Technical Proposal CERN/LHCC/94-43, LHCC/P2, 15 December 1994.

[25] S. Moretti and S. Munir, in preparation.

Higgs Flavor	Points Visible		Percentage
h_1	Total:	1346137	99.7655
	Alone:	1345398	99.7108
	With h_2:	567	0.042
	With a_1:	177	0.0131
	With h_2 and a_1:	5	0.0004
h_2	Total:	1305	0.0971
	Alone:	734	0.0543
	With h_1:	567	0.042
	With a_1:	9	0.0007
	With a_1 and a_1:	5	0.0004
h_3	Total:	0	0
a_1	Total:	187	0.0139
	Alone:	6	0.0004
	With h_1:	177	0.0131
	With h_2:	9	0.0007
	With h_1 and h_2:	5	0.0004
a_2	Total:	0	0

Table 1: Higgs events expected to be visible at the LHC through the $\gamma\gamma$ decay mode.

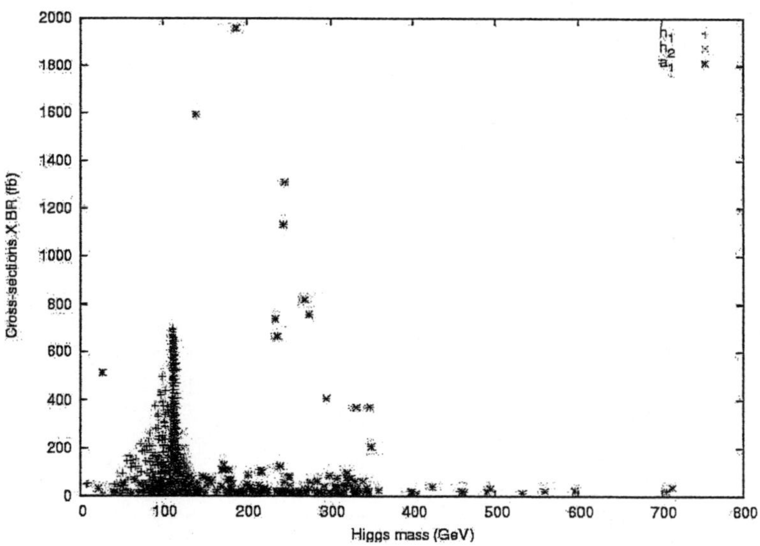

Figure 1: Higgs mass vs. cross-section times branching ratio.

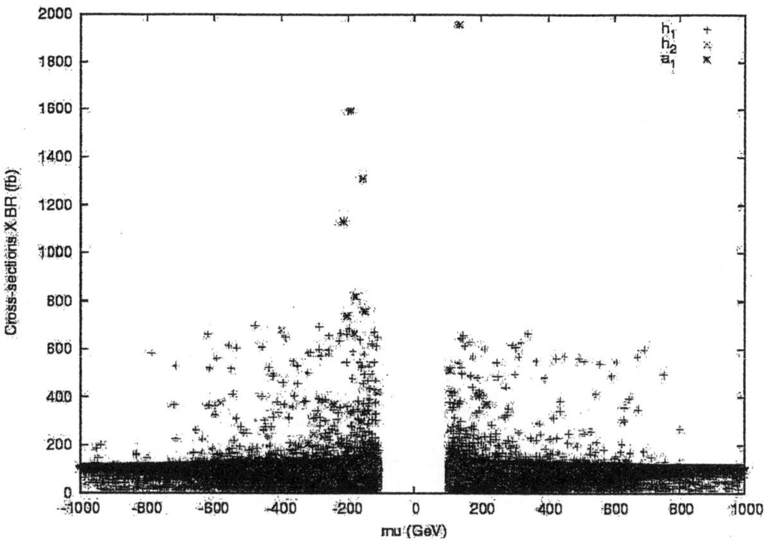

Figure 2: $\mu_{\rm eff}$ vs. cross-sections times branching ratio.

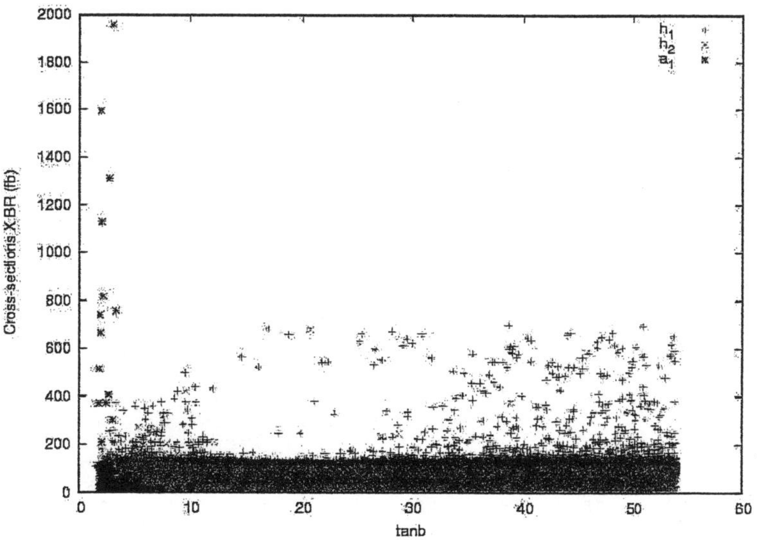

Figure 3: $\tan\beta$ vs. cross-section times branching ratio.

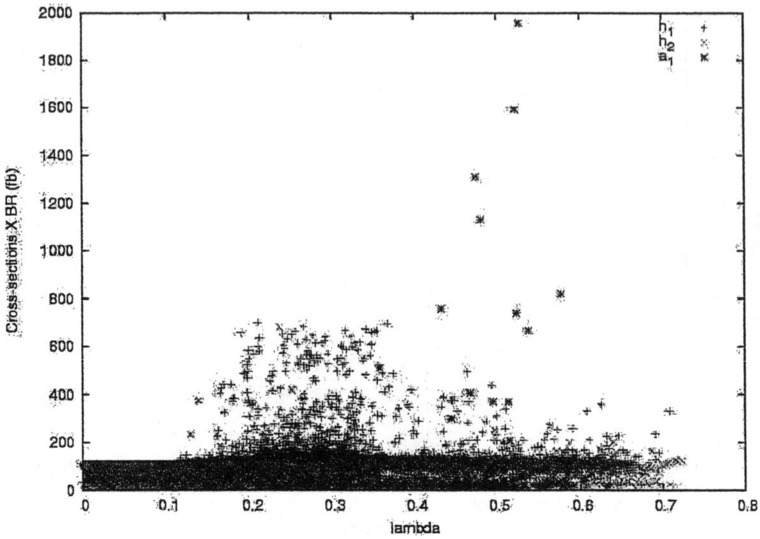

Figure 4: λ vs. cross-section times branching ratio.

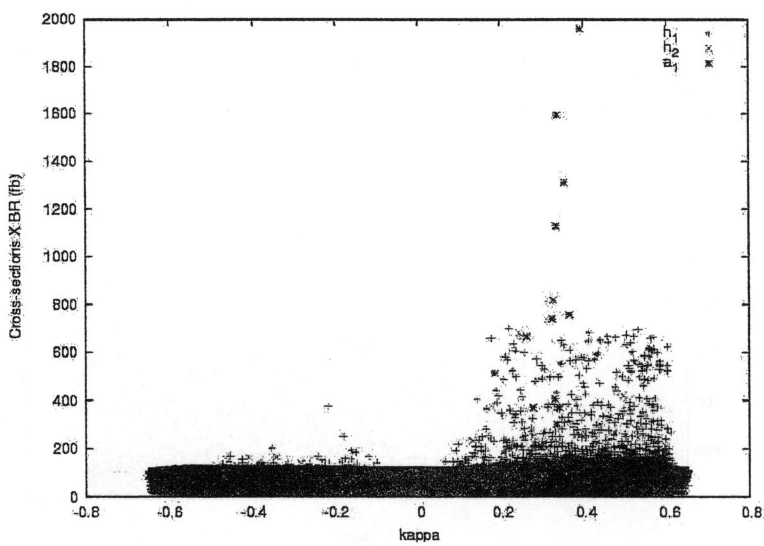

Figure 5: κ vs. cross-section times branching ratio.

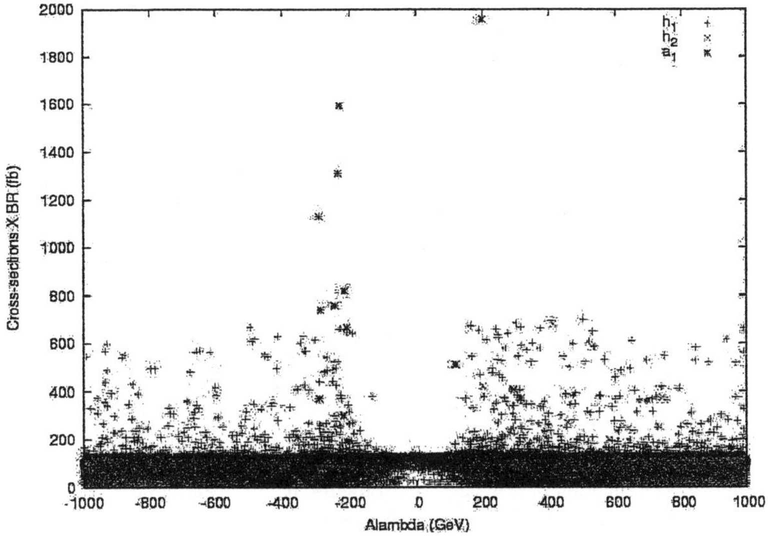

Figure 6: A_λ vs. cross-section times branching ratio.

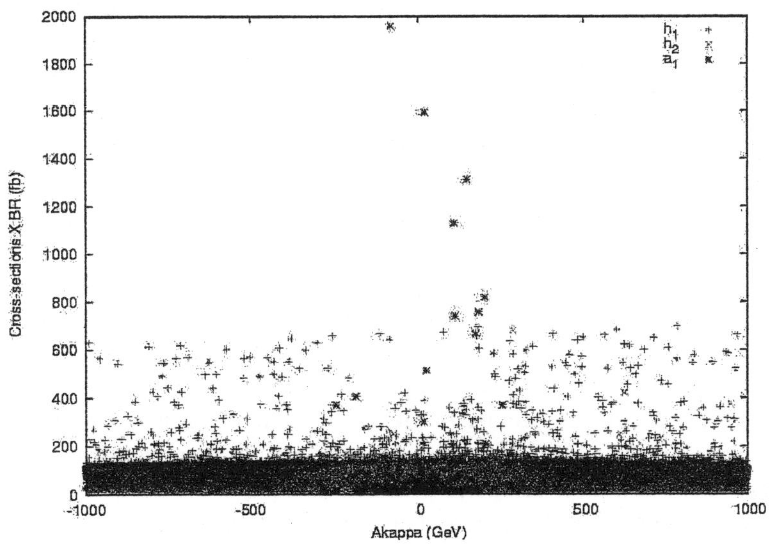

Figure 7: A_κ vs. cross-section times branching ratio.

Hadronic Multiplicity in Ion Collisions at High Energies

Marzia Nardi

Enrico Fermi Centre,
Compendio Viminale, 00184 Roma (Italy)

Abstract. We give a theoretical interpretation of experimental data on hadronic multiplicity, as a function of centrality and rapidity, in $Au - Au$ and $d - Au$ collisions at RHIC, in the framework of the Parton Saturation Model.

1. High energy strong interactions

Hadronic scattering at high energies allow us to explore the region of small Bjorken's x, where the coupling constant α_s of the strong interactions is small but the density of partons (prevalently gluons) is so large that the conventional perturbative approach is not applicable because of strong non-linear effects. In fact, the one-gluon exchange approximation, the basis of normal perturbative QCD (pQCD) calculations, is valid only when the gluons interact independently from each other, and this can not be true if their density is too high.

A more precise way of stating the problem is to note that the partons taking part in a high energy scattering have energies between E_{hadron} (the hadron energy) and a minimum value of the order of m_π (pion mass). The Bjorken's x, i.e. the fraction of longitudinal momentum carried by the parton, is $x = E_{parton}/E_{hadron}$ and it ranges from m_π/E_{hadron} to 1. It is evident that a higher incident energy E_{hadron} provides experimental access to regions of smaller x.

On the other hand, the number of gluons per unit rapidity in a proton, $dN/dy = xG(x, Q^2)$, as measured at HERA [1] for various x and Q^2, rises rapidly at small x. This rise becomes more and more dramatic with increasing Q^2.

It is evident that at very high Q^2 and at very small x the normal evolution equations, based on pQCD, are no longer valid: the parton distribution functions can not grow unlimitedly, because this would correspond to an uncontrolled growing of the cross-section (calculated as a convolution of the parton distribution functions) which would eventually violate the unitarity bound. One concludes, therefore, that the number of gluons $xG(x, Q^2)$ (the contribution from quarks and antiquarks in these small-x regions is irrelevant) must saturate. We then see how by a simple, logical argument, the idea of *parton saturation* arises naturally in high energy strong interactions. We will give here a brief sketch of the theory underlying the parton saturation model, namely the Color

Glass Condensate. A more satisfactory presentation can be found in [2] and references therein.

2. The Color Glass Condensate

The matter controlling high energy strong interactions is universal, independent of the hadron (including heavy nuclei) under consideration. It exists over sizes larger than $l \sim 1$ fm, $\tau \sim 10^{-23}$ sec.

Its name, Color Glass Condensate (CGC), originates from:

- *Color* : the gluons are colored.
- *Glass* : the gluons at small x are emitted by other partons at larger x. In the infinite momentum frame, these larger momentum partons travel very fast and their natural time scales are Lorentz dilated. As a consequence, the low x gluons evolve very slowly compared to their own natural time scales. Thus they form a "glass".
- *Condensate* : the parton saturation occurs when the two competing processes, gluon emission and gluon recombination, compensate each other. The emission rate is proportional to the gluon density ρ, while the recombination rate depends on the interactions between gluons, namely it is proportional to ρ^2 times the coupling constant α_s. Therefore we have saturation when the condition $\rho \sim \alpha_s \rho^2$ is fulfilled, giving $\rho \propto 1/\alpha_s$, characteristic of Bose condensation phenomena.

It follows that there is a critical momentum scale Q_s which controls the occupation number through $1/\alpha_s(Q_s)$: this is the *saturation scale*. The transverse phase space density $n \equiv (1/\pi R^2)(dN/dyd^2p_T)$ is constant, i.e. "saturated", for $p_T < Q_s$(R is the transverse size of the hadron or the nucleus under consideration; n is the two-dimensional transverse density, the density in the longitudinal direction will be assumed uniform : this is a common assumption in heavy ion collisions phenomenology); for $p_T > Q_s$ the gluon density is smaller than the critical one, the gluons therefore interact independently and the perturbative result $n \propto \alpha_s(p_T)/(R^2 p_T^2)$ is recovered.

The CGC picture is based on a classical effective theory, originally proposed by McLerran and Venugopalan [3] to describe the gluon distribution in large nuclei: the valence quarks of the hadrons (fast partons) are treated as a source for a classical color field representing the small-x (soft) gluons. The classical approximation is appropriate since the saturated gluons have large occupation numbers. The theory implies a non linear renormalization group equation [4]. The CGC approach is justified in the limit of $Q_s \gg \Lambda_{QCD}$: as we will see in the following, at RHIC energies this condition is only marginally satisfied but LHC will provide a better opportunity to test this theory.

In the recent years, a significant experimental evidence has been found in support of the hypothesis of the CGC [2], mainly from HERA and RHIC. In particular the RHIC data concern the multiplicity of produced hadrons and will be the focus of the rest of this paper.

3. QCD in the high density regime and heavy ion collisions

We will introduce here the basic ideas of parton saturation in an intuitive way. Consider the wave function of a nucleus boosted to a large momentum. The nucleus is Lorentz–contracted, and partons "live" on a thin sheet in the transverse plane. Each parton occupies the transverse area π/Q^2 determined, by uncertainty principle, by its transverse momentum Q, and can be probed with the cross section $\sigma \sim \alpha_s(Q^2)/Q^2$. On the other hand, the entire transverse area of the nucleus is $S_A \sim \pi R_A^2$. Therefore, if the number of partons exceeds

$$N_A \sim \frac{S_A}{\sigma} \sim \frac{\pi}{\alpha_s(Q^2)}\, Q^2 R_A^2, \tag{1}$$

they will begin to overlap in the transverse plane and start interacting with each other, which prevents further growth of parton densities. This happens when the transverse momenta of the partons are on the order of

$$Q_s^2 \sim \alpha_s(Q_s^2)\, \frac{N_A}{\pi R_A^2} \sim A^{1/3}, \tag{2}$$

which is called the "saturation scale". In the saturation regime, as is apparent from equations. (1) and (2), the multiplicity of the produced partons should be proportional to

$$N_s \equiv xG_A(x, Q_s^2) \sim \frac{\pi}{\alpha_s(Q_s^2)}\, Q_s^2 R_A^2 \sim A. \tag{3}$$

($xG_A(x, Q_s^2)$ is the gluon distribution function for a nucleus of atomic number A). From equation (1) it is also evident that the saturation scale Q_s depends on the nucleus: the heavier the nucleus, the larger the saturation scale. This is not in contrast with the fact that the CGC is an universal form of matter: the matter is the same, for all hadrons and nuclei, but the momentum scale at which it starts to reveal itself changes with the hadron and with the incident energy, as we will see in the following. This is also the reason why the experiments with high energy heavy ions are so interesting: they allow a larger Q_s for the starting of saturation phenomena, they are more "efficient" because in order to see the same effect in hadron collisions, one would need much higher energies.

In the weak coupling regime, the density of partons ($\rho \propto 1/\alpha_s$) becomes very large. In the first approximation, the multiplicity in this high density regime scales with the number of participants. There is, however, an important logarithmic correction to this from the evolution of parton structure functions with the saturation scale Q_s^2, which we discuss below.

While this "derivation" has been very simplistic, the formulas (2),(3) can be reproduced by more sophisticated methods [5, 6], which also allow to reconstruct the coefficient of proportionality in (2):

$$Q_s^2 = \frac{8\pi^2 N_c}{N_c^2 - 1}\, \alpha_s(Q_s^2)\, xG(x, Q_s^2)\, \frac{n_{part}(b)}{2}, \tag{4}$$

where $N_c = 3$ is the number of colors and n_{part} is the density of participants in the transverse plane. We divide n_{part} by 2 to get the density of those nucleons in a single nucleus which will participate in the collision at a given impact parameter. equation (4) implies that Q_s^2 depends also on the impact parameter b, i.e. on the centrality of the collision, through the density n_{part}, as we will see in the next section.

4. Hadronic multiplicity

We will see now how to apply the saturation idea presented in section 3 to calculate the multiplicity production in heavy ion collisions.

First of all, we have to give a numerical estimate of the saturation scale: we refer, for this purpose, to the first RHIC data (central $Au - Au$ collision at $\sqrt{s} = 130$ GeV and pseudorapidity $|\eta| < 1$ [7]).

Equation (4) can be solved by iterations; a self–consistent solution can be found at $Q_s^2 \simeq 2$ GeV^2 if we use $xG(x, Q_s^2) \simeq 2$ [8] at $x \simeq 2Q_s/\sqrt{s} \simeq 0.02$, with $\alpha_s(Q_s^2) \simeq 0.6$. We use an explicit expression [6] for the number of produced partons

$$\left.\frac{d^2N}{d^2bd\eta}\right|_{|\eta|<1} = c\,\frac{N_c^2-1}{4\pi^2 N_c}\,\frac{1}{\alpha_s}\,Q_s^2, \tag{5}$$

where c is the "parton liberation" coefficient accounting for the transformation of virtual partons in the initial state to the on–shell partons in the final state. Integration over the transverse coordinate and the use of (4) yield simply

$$\left.\frac{dN}{d\eta}\right|_{|\eta|<1} = c\ N_{part}\ xG(x, Q_s^2). \tag{6}$$

If we assume that $dN/d\eta \simeq 3/2\ dn_{exp}/d\eta$, take $xG(x, Q_s^2) \simeq 2$ at $Q_s^2 \simeq 2$ GeV^2 and $x \simeq 2Q_s/\sqrt{s} \simeq 0.02$ [8], and use $N_{part} \simeq 339$ for the 6% centrality cut, the experimental number [7] $dn/d\eta = 555 \pm 12(stat) \pm 35(syst)$ translates into the following value of the "parton liberation" coefficient:

$$c = 1.23 \pm 0.20. \tag{7}$$

This number appears to be close to unity, as expected by Mueller [6], which implies a very direct correspondence between the number of the partons in the initial and final states. Moreover, this may imply that the number of particles is conserved through the parton–to–hadron transformation – a miraculous fact first noted in the context of the "local parton–hadron duality" hypothesis [9].

The value (7) can be compared to the lattice calculation [10] by Krasnitz and Venugopalan, which yields $c = 1.29 \pm 0.09$. An analytical calculation for c has been presented by Kovchegov [11], with the result $c = 2\ln 2 \simeq 1.39$.

To compute the centrality dependence, we still need to know the evolution of the gluon structure function with the density of partons, which is proportional to the mean density of participants in the transverse plane. We will assume that this evolution is governed by the DGLAP equation [12], and take

$$xG(x, Q_s^2) \sim \ln\left(\frac{Q_s^2}{\Lambda_{QCD}^2}\right). \tag{8}$$

The dependence (8) emerges when the radiation of gluons is treated classically, and so is consistent with equation (4). The equations (4) and (6) can now be used to evaluate the centrality dependence; with the parameters described above we get

$$\frac{2}{N_{part}} \left.\frac{dn}{d\eta}\right|_{|\eta|<1} \simeq 0.82 \ \ln\left(\frac{Q_s^2}{\Lambda_{QCD}^2}\right), \tag{9}$$

where we take $\Lambda_{QCD} \simeq 200$ MeV and use the values of Q_s obtained from equation (4).

From equation (9) it is possible, now, to calculate the centrality dependence of the hadron multiplicity at midrapidity (or, more precisely, for $|\eta| < 1$) and $\sqrt{s} = 130$ GeV. Let us see how to extend it at different energies and also how to take into account the rapidity dependence [13, 14, 15, 16].

The energy dependence of the saturation scale is assumed to be the same observed at HERA [17]; at $y = 0$:

$$Q_s^2(x) = Q_{s0}^2 \left(\frac{x}{x_0}\right)^{-\lambda} = Q_{s0}^2 \left(\frac{\sqrt{s}}{\sqrt{s_0}}\right)^{\frac{\lambda}{1+\lambda/2}} \tag{10}$$

(the r.h.s. is obtained with $x = Q_s/\sqrt{s}$) with $\lambda = 0.288$. equation 10 has been derived analytically in [18], and the estimate given for λ is in excellent agreement with the experimental fit.

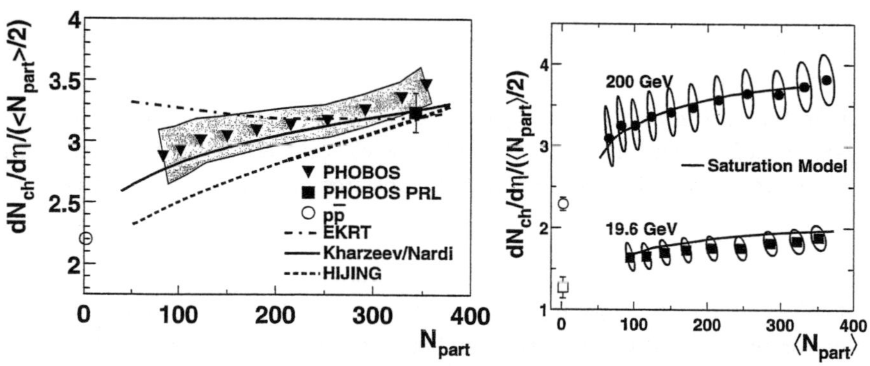

Figure 1. The centrality dependence (measured in terms of the number of participants) of the hadronic multiplicity at RHIC energies in the interval of pseudorapidity $|\eta| < 1$.(from [19, 20])

Figure 1 shows the experimental data on the hadronic multiplicity measured by PHOBOS in $Au - Au$ collisions at different incident energies as a function of the number of participants (related to the impact parameter b). The left plot shows the data at $\sqrt{s} = 130$ GeV [19] compared to the theoretical predictions of a few models. The saturation model (labeled "Kharzeev/Nardi") is indicated with the solid line. The normalization factor in equation (9) was chosen to fit the preliminary data published in [7] (solid square); the N_{part} dependence was then a genuine prediction of the model [21] and was later confirmed by the experimental data. The right plot shows the analogous results at $\sqrt{s} = 200$ and 19.6 GeV. The same excellent agreement can be seen in the comparison with PHENIX data.

To describe the rapidity distribution, we have to note first that the Bjorken variables x_1 and x_2 for the interacting partons in the two colliding nuclei are related to y by :

$$x_1 = \frac{Q}{\sqrt{s}} \exp^{-y} \qquad \text{and} \qquad x_2 = \frac{Q}{\sqrt{s}} \exp^{y} \ . \tag{11}$$

This implies that two different saturation scales must be considered: $Q_s(x_1)$ and $Q_s(x_2)$, with

$$Q_s^2(x_{1,2}) = Q_{s0}^2 \left(\frac{x_{1,2}}{x_0}\right)^{-\lambda} = Q_{s0}^2 \left(\frac{\sqrt{s}}{\sqrt{s_0}}\right)^{\frac{\lambda}{1+\lambda/2}} \exp\left\{\frac{\mp\lambda\, y}{1+\lambda/2}\right\} . \tag{12}$$

Different values of y correspond to different regions of x for the two nuclei: it is even possible that one of the two nuclei is in the saturation region and the other is not (this happens when y is close to $\pm\, y_{beam}$).

Finally, we must introduce the unintegrated gluon distribution function $\varphi(x, k_t)$, related to $xG(x, Q^2)$ by :

$$xG(x, Q^2) = \int^{Q^2} dk_t^2\, \varphi(x, k_t) \tag{13}$$

and giving the probability of having a gluon, inside a nucleon, with longitudinal momentum fraction x and transverse momentum k_t. The lower limit Λ_{QCD} in the integral in equation (13) (and in the following equations) is omitted for simplicity. The corresponding function for a nucleus will be denoted by $\varphi_A(x, k_t)$.

According to the above considerations, we assume that $\varphi(x, k_t)$ saturates at small p_T; for simplicity, we will assume the following (crude!) analytical expression for φ :

$$\varphi_A(x, k_t^2) = \begin{cases} \kappa'\kappa \dfrac{S_A}{\alpha_s} (1-x)^4 & \text{for} \quad k_t < Q_s(x) \ . \\[2ex] \kappa \dfrac{\alpha_s}{\pi} \dfrac{1}{k_t^2} (1-x)^4 & \text{for} \quad k_t > Q_s(x) \end{cases} \tag{14}$$

In the region $k_t > Q_s$ there is no saturation, the unintegrated gluon distribution function shows normal bremsstrahlung radiation. However, in the saturation region, $k_t < Q_s$, instead of growing, as pQCD would predict, φ_A *saturates* (the integral of (14), in this region, gives equation (3)). The factor S_A is the overlap area in the transverse plane of the colliding nuclei. The constant κ is simply an overall normalization factor: it is included in the constant c of equation (6) and it is, therefore, irrelevant (κ' guarantees the continuity of φ at $k_t = Q_s$).

The gluon distribution function $xG_A(x, p_t^2)$ is given by :

$$xG_A(x, p_t^2) = \begin{cases} \dfrac{\kappa'\kappa}{\alpha_s} S_A\, p_t^2\, (1-x)^4 & \text{for } p_t < Q_s(x) \\[2ex] \dfrac{\kappa}{\alpha_s} S_A\, Q_s^2(x)\, (1-x)^4 & \text{for } p_t > Q_s(x) \ , \end{cases} \tag{15}$$

where a term proportional to α_s has been neglected (its contribution is numerically small as compared to the one coming from the region below Q_s).

With these ingredients it is easy to write the formula for the inclusive production [22, 13]:

$$E\frac{d\sigma}{d^3p} = \frac{4\pi N_c}{N_c^2-1}\frac{1}{p_t^2}\int^{p_t} dk_t^2 \alpha_s\, \varphi_{A_1}(x_1,k_t^2)\, \varphi_{A_2}(x_2,(\mathbf{p}-\mathbf{k})_t^2), \tag{16}$$

We can compute the multiplicity distribution by integrating Eq. (16) over p_t, namely,

$$\frac{dN}{dy} = \frac{1}{S}\int d^2p_t E\frac{d\sigma}{d^3p}; \tag{17}$$

S is either the inelastic cross section for the minimum bias multiplicity, or a fraction of it corresponding to a specific centrality cut.

Equation (17) (with (16)) can be integrated numerically to get the results on hadronic multiplicity that can be compared to experimental data; the results are shown in the following plots. However we want to present here a simplified version of this formula to extract some physical idea. First, we note that in equation (16) the largest contribution to the integral comes from the two regions $k_t \ll p_t$ and $|\mathbf{k}_t - \mathbf{p}_t| \ll p_t$. In the first case one can write $\varphi_{A_2}(x_2,(\mathbf{p}-\mathbf{k})_t^2) \sim \varphi_{A_2}(x_2,p_t^2)$, in the second case, by changing the integration variable $\mathbf{k}_t \to \mathbf{k}_t - \mathbf{p}_t$ one has $\varphi_{A_1}(x_1,k_t^2) \to \varphi_{A_1}(x_1,(\mathbf{p}-\mathbf{k})_t^2) \sim \varphi_{A_1}(x_1,p_t^2)$. Therefore equation (17) becomes:

$$\begin{aligned}
\frac{dN}{dy} &= \frac{1}{S}\frac{4\pi N_c\alpha_s}{N_c^2-1}\int \frac{d^2p_t}{p_t^2}\Big(\varphi_{A_1}(x_1,p_t^2)\int^{p_t} dk_t^2\, \varphi_{A_2}(x_2,k_t^2) \\
&\qquad + \varphi_{A_2}(x_2,p_t^2)\int^{p_t} dk_t^2 \varphi_{A_1}(x_1,k_t^2)\Big) = \\
&= \frac{1}{S}\frac{4\pi N_c\alpha_s}{N_c^2-1}\int \frac{d^2p_t}{p_t^2}\Big(\varphi_{A_1}(x_1,p_t^2)\, x_2G_{A_2}(x_2,p_t^2) \\
&\qquad + \varphi_{A_2}(x_2,p_t^2) x_1G_{A_1}(x_1,p_t^2)\Big) = \\
&= \frac{1}{S}\frac{4\pi N_c\alpha_s}{N_c^2-1}\int \frac{d^2p_t}{p_t^2}\frac{d^2}{dp_t^2}\Big(x_1G_{A_1}(x_1,p_t^2) x_2G_{A_2}(x_2,p_t^2)\Big) \\
&= \frac{1}{S}\frac{4\pi N_c\alpha_s}{N_c^2-1}\int \frac{d^2p_t}{p_t^4} x_1G_{A_1}(x_1,p_t^2)\, x_2G_{A_2}(x_2,p_t^2).
\end{aligned}$$

With the substitution of (15) one observes that the hadronic multiplicity for any y is always proportional to $SQ_{s,min}^2$ ($Q_{s,min}$ is the minimum between $Q_s(x_1)$ and $Q_s(x_2)$ according to (12)), i.e. it is proportional to the number of participating nucleons of the *saturated* nucleus [15, 16]. The property of the scaling according to the number of participants is peculiar of this model and it is confirmed by experiment.

Usually the experimental results are given in terms of the pseudorapidity rather than y: the conversion between the two variables is straightforward, see for instance [13].

In the formulas of this section the impact-parameter dependence has been omitted: one should however remember that the saturation scales always depends on b according to (4), therefore dN/dy also depends on b. The comparison with experimental data requires also the restriction to limited intervals of centrality.

Figure 2 shows the results obtained in this model compared to the experimental data of $Au - Au$ collisions at different energies (by PHOBOS and BRAHMS Collab.). Again it should be stressed that the overall normalization is guaranteed by fitting the first PHOBOS data [7] at midrapidity for the 6% most central collisions. The shapes of the curves are the *predictions* of the model.

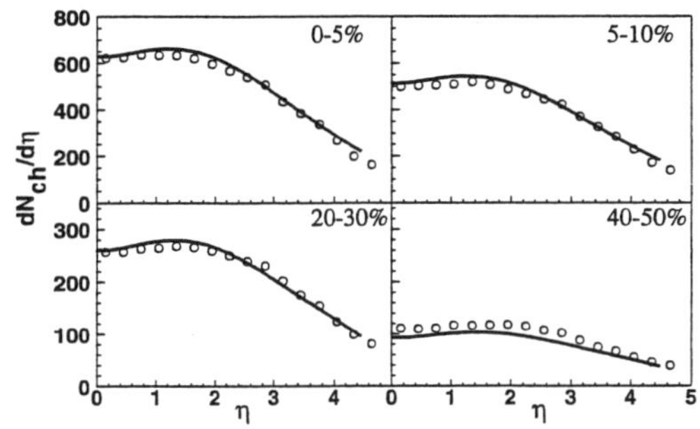

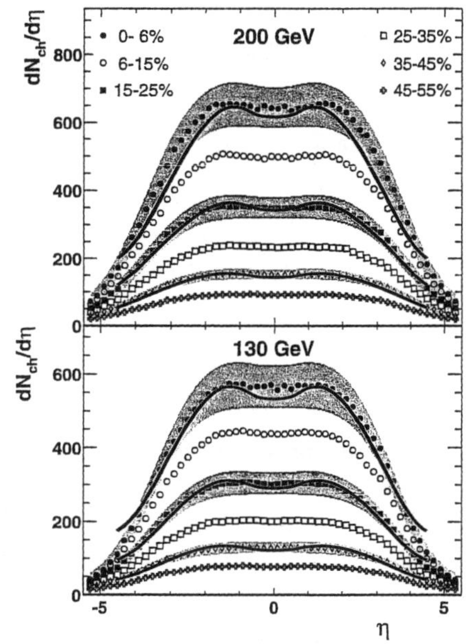

Figure 2. Rapidity distribution of charged particle in $Au - Au$ collisions at RHIC for various centrality intervals. Left: BRAHMS data at $\sqrt{s} = 200$ GeV [24]. Right: PHOBOS data at $\sqrt{s_{NN}} = 130$ and 200 GeV [23]. The solid lines are the predictions of the saturation model.

We can conclude that the present model reproduces very well all the data for the highest energies in the pseudorapidity interval $|\eta| < 4$. At the lowest energy, the data for the less central collisions are underestimated. This is not a surprise, since when the saturation scale becomes too small (this is the case in the peripheral collisions, but also in the region near $\pm y_{beam}$ at higher energies) the saturation effects are no longer dominants in the production of particles.

The case of $d - Au$ collisions requires some care, since the deuteron can not be treated as a heavy nucleus. The Glauber approximation used here to compute the number of participating nucleons is, therefore, not accurate. In this case, just to test our saturation model we "borrowed" the number of participants calculated by the experimental collaborations with Monte Carlo methods. The results are then in good agreement with the experimental data as, for instance, one can see in figure 3, for $\eta > -2$. The region $\eta < -2$ is the fragmentation region of the Au nucleus, where the present approximation is not justified. In the right plot, the dashed line is obtained assuming $dN/d\eta = N_{part}^{Au} \times dN_{pp}/d\eta$ in the Au fragmentation region (see [15]).

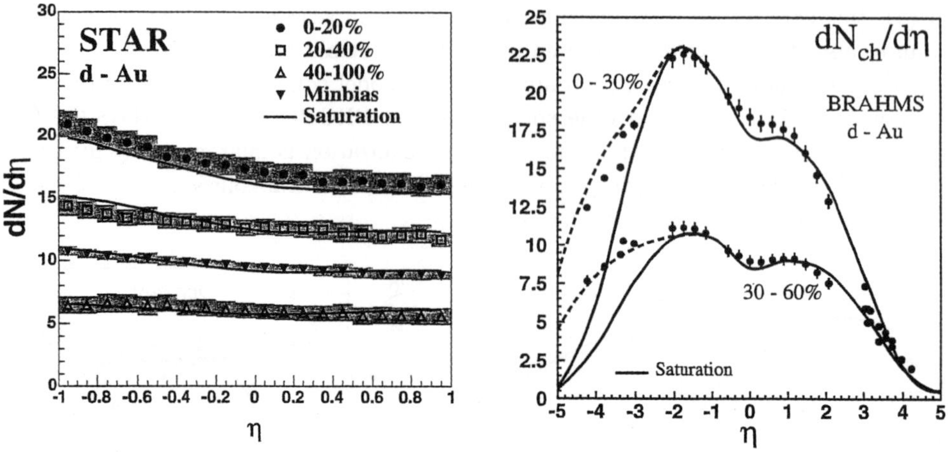

Figure 3. Rapidity distribution of charged particle in $d - Au$ collisions at $\sqrt{s}$ = 200 GeV for various centrality intervals, compared to experimental data from STAR (left) and BRAHMS (right) Collaborations ([25, 26]).

5. Conclusions

The results presented here and in many other recent works (see [2]) support the idea that the dense matter produced in the early stage of heavy ion collisions at RHIC can be described as a Color Glass Condensate. The CGC is interesting by itself, because it represents the universal form of high energy QCD wavefunctions at small x, not only of hadrons but heavy nuclei as well. It allows to study the QCD in the regime of high density and small coupling constant. It is important also for its relation with the study of deconfined matter which is the main goal of heavy ion collisions at high energy: it provides a theoretical description, based on QCD, of the initial state of the nucleus-nucleus system, from which the subsequent quark-gluon-plasma evolves.

The CGC hypothesis can be tested in a wide range of experimental environments not restricted to heavy ion collisions; further data, at RHIC and at LHC but also in $e - p$ interactions (HERA, eRHIC) are still needed both to corroborate this conclusion and to study with higher accuracy the properties of the CGC.

References

[1] J. Breitweg *et al.* [ZEUS Collaboration], Eur. Phys. J. C **6** (1999) 603 [arXiv:hep-ex/9808020].

[2] L. McLerran, arXiv:hep-ph/0311028. M. Gyulassy and L. McLerran, Nucl. Phys. A **750** (2005) 30 [arXiv:nucl-th/0405013]. L. McLerran, arXiv:hep-ph/0402137.

[3] L. D. McLerran and R. Venugopalan, Phys. Rev. D **49** (1994) 2233 [arXiv:hep-ph/9309289]. L. D. McLerran and R. Venugopalan, Phys. Rev. D **50** (1994) 2225 [arXiv:hep-ph/9402335].

[4] J. Jalilian-Marian, A. Kovner, L. D. McLerran and H. Weigert, Phys. Rev. D **55** (1997) 5414 [arXiv:hep-ph/9606337]. J. Jalilian-Marian, A. Kovner, A. Leonidov and H. Weigert, Nucl.

Phys. B **504** (1997) 415 [arXiv:hep-ph/9701284]. J. Jalilian-Marian, A. Kovner, A. Leonidov and H. Weigert, Phys. Rev. D **59** (1999) 014014 [arXiv:hep-ph/9706377].

[5] J. P. Blaizot and A. H. Mueller, Nucl. Phys. B **289** (1987) 847. A. H. Mueller and J. w. Qiu, Nucl. Phys. B **268** (1986) 427. L. D. McLerran and R. Venugopalan, Phys. Rev. D **59** (1999) 094002 [arXiv:hep-ph/9809427]. Y. V. Kovchegov, Phys. Rev. D **54** (1996) 5463 [arXiv:hep-ph/9605446]. Y. V. Kovchegov and D. H. Rischke, Phys. Rev. C **56** (1997) 1084 [arXiv:hep-ph/9704201]. Y. V. Kovchegov, Phys. Rev. D **60** (1999) 034008 [arXiv:hep-ph/9901281]. M. Gyulassy and L. D. McLerran, Phys. Rev. C **56** (1997) 2219 [arXiv:nucl-th/9704034]. E. Iancu, A. Leonidov and L. D. McLerran, Nucl. Phys. A **692** (2001) 583 [arXiv:hep-ph/0011241].

[6] A. H. Mueller, Nucl. Phys. B **572** (2000) 227 [arXiv:hep-ph/9906322].

[7] B. B. Back *et al.* [PHOBOS Collaboration], Phys. Rev. Lett. **85**, 3100 (2000) [arXiv:hep-ex/0007036].

[8] A. D. Martin, R. G. Roberts, W. J. Stirling and R. S. Thorne, Eur. Phys. J. C **18**, 117 (2000) [arXiv:hep-ph/0007099].

[9] Y. L. Dokshitzer, V. A. Khoze, A. H. Mueller and S. I. Troian,

[10] A. Krasnitz and R. Venugopalan, Phys. Rev. Lett. **86**, 1717 (2001) [arXiv:hep-ph/0007108].

[11] Y. V. Kovchegov, [arXiv:hep-ph/0011252].

[12] Y. L. Dokshitzer, Sov. Phys. JETP **46**, 641 (1977) V. N. Gribov and L. N. Lipatov, Phys. Lett. B **37**, 78 (1971). G. Altarelli and G. Parisi, Nucl. Phys. B **126**, 298 (1977).

[13] D. Kharzeev and E. Levin, Phys. Lett. B **523**, 79 (2001) [arXiv:nucl-th/0108006].

[14] D. Kharzeev, E. Levin and M. Nardi, Phys. Rev. C **71**, 054903 (2005) [arXiv:hep-ph/0111315].

[15] D. Kharzeev, E. Levin and M. Nardi, Nucl. Phys. A **730**, 448 (2004) [Erratum-ibid. A **743**, 329 (2004)] [arXiv:hep-ph/0212316].

[16] D. Kharzeev, E. Levin and M. Nardi, Nucl. Phys. A **747**, 609 (2005) [arXiv:hep-ph/0408050].

[17] K. Golec-Biernat and M. Wusthoff, Phys. Rev. D **59**, 014017 (1999) [arXiv:hep-ph/9807513]. K. Golec-Biernat and M. Wusthoff, Phys. Rev. D **60**, 114023 (1999) [arXiv:hep-ph/9903358]. A. M. Stasto, K. Golec-Biernat and J. Kwiecinski, Phys. Rev. Lett. **86**, 596 (2001) [arXiv:hep-ph/0007192].

[18] D. N. Triantafyllopoulos, Nucl. Phys. B **648**, 293 (2003) [arXiv:hep-ph/0209121]. A. H. Mueller and D. N. Triantafyllopoulos, Nucl. Phys. B **640**, 331 (2002) [arXiv:hep-ph/0205167].

[19] B. B. Back *et al.* [PHOBOS Collaboration], Nucl. Phys. A **698**, 555 (2002).

[20] B. B. Back *et al.* [PHOBOS Collaboration], Phys. Rev. C **70**, 021902 (2004) [arXiv:nucl-ex/0405027].

[21] D. Kharzeev and M. Nardi, Phys. Lett. B **507** (2001) 121 [arXiv:nucl-th/0012025].

[22] L. V. Gribov, E. M. Levin and M. G. Ryskin, Phys. Rept. **100**, 1 (1983).

[23] B. B. Back *et al.*, Phys. Rev. Lett. **91**, 052303 (2003) [arXiv:nucl-ex/0210015].

[24] I. G. Bearden *et al.* [BRAHMS Collaboration], Phys. Rev. Lett. **88**, 202301 (2002) [arXiv:nucl-ex/0112001].

[25] J. Adams *et al.* [STAR Collaboration], Phys. Rev. C **70**, 064907 (2004) [arXiv:nucl-ex/0408016].

[26] I. Arsene *et al.* [BRAHMS Collaboration], Phys. Rev. Lett. **94**, 032301 (2005) [arXiv:nucl-ex/0401025].

Nucleon Form Factors and Dispersion Relations

S. Pacetti

Laboratori Nazionali di Frascati dell'INFN, Frascati, Italy

Abstract

We define and use two dispersive techniques, based on data, analyticity and dispersion relations to handle the nucleon form factors.

In the first case, we use the available information on the ratio between the electric and magnetic proton form factors, coming from recent space-like and time-like data and theory, as inputs in a dispersive approach that gives as outcome a complex description of this ratio, valid in the whole q^2 plane.

In the second case, by inverting a dispersion relation for the magnetic nucleon form factors and by considering the data and the pQCD asymptotic behaviour, we reconstruct these form factors in the unphysical region. We found resonances and phases in agreement with the expectations.

1 Introduction

Nucleon electromagnetic form factors (ff) play a fundamental role in our understanding of the hadronic dynamics. Hence, for over 40 years, many attempts [1] have been made through dispersion relations (DR) either to unravel these structure functions in the unphysical region or to obtain information on the asymptotic behaviour.

We consider two applications where we use DR's to handle nucleon ff's.

In the first case, the dispersive technique is used in "direct way" in order to give a complete description of the ratio $R(q^2) = \mu_p G_E^p(q^2)/G_M^p(q^2)$. By means of the DR for the imaginary part, theoretical and experimental constraints have been imposed on an analytic parameterization of $R(q^2)$, in such a way we obtain a function which describes the ratio in the whole q^2-complex plane.

In the second application we are focused on the magnetic proton and neutron ff's G_M^p and G_M^n. In this case a special kind of DR, i.e. the DR for the logarithm, has been inverted in order to define an integral equation, which has, as inputs, the space-like data, the time-like data with the asymptotic behaviour provided by the pQCD [2] and, as outcome, the modulus of the ff's in the unphysical region.

2 Dispersive description of $G_E^p(q^2)/G_M^p(q^2)$

2.1 Motivations

Measurements of the ratio $G_E^p(q^2)/G_M^p(q^2)$ performed in 1999 and 2000 at Jlab [3] and MIT-Bates [4], with a recoil polarization technique, gave an unexpected result. The scaling law, which provides $G_E \simeq G_M$, confirmed until that time by all the measurements [5] (based on the Rosenbluth formula [6]), was disproved by a steep decreasing behaviour of these polarization data, that, in addition, seems to prelude to a space-like zero lying somewhere around $q^2 \sim 15\ GeV^2$ [7].

Recently, new data for the modulus $|R(q^2)|$ became available (see fig.1). Two points have been obtained by performing a reanalysis of old angular distributions measured by the experiments FENICE [8], DM2 [9] and E835 [10]. While, a brand new measurement has been made by BaBar [11]. We consider also a measurement of $|R(q^2)|$ performed by Lear [12] in 1994. Since these last two sets of data, obtained respectively through the processes $e^+e^- \to p\overline{p}$ and $p\overline{p} \to e^+e^-$, are incompatible, in the following we will consider them separately. Each result will be given in two cases, the first by using, in the time-like region, only the Lear data and the second by using all the other time-like data.

The surprising space-like Jlab data have brought a great deal of interest in the nucleon ff's field. Many models [13–16] have been built in order to fit these data and to justify [17] the incompatibility with the previous measurements and then, with the scaling law.

The fits that we have considered to make the comparison (Fig.4b) have been chosen either because they are the simplest models that one can use to fit the space-like data [13–15], or, as in case of the more complicated, but without space-like free parameters, Iachello-Jackson-Landé model (IJL) [16], because it represents a surprisingly accurate prediction of the "non-scaling", low energy, space-like behaviour.

But, while in the space-like region, over the interval of the data $q^2 \sim (-6 \div 0)\ GeV^2$, all these models agree, their analytic continuations in the time-like region are far each other.

The reason for this time-like incompatibility must be sought in the functional structure of these fit functions that are not rigorously analytic in q^2.

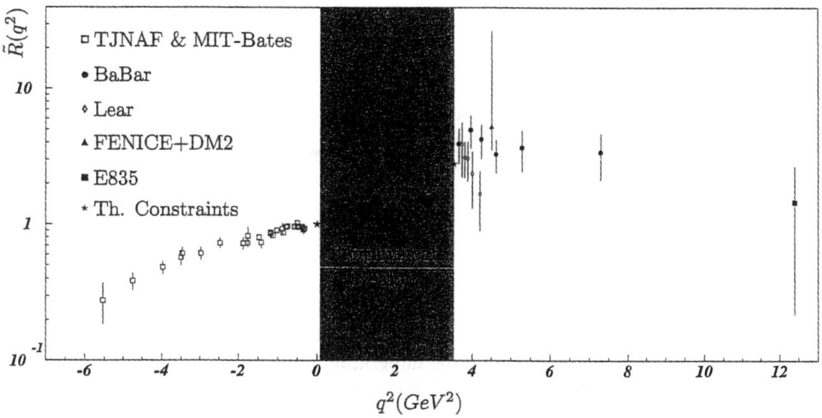

Figure 1: Compilation of all the used data in the space-like (TJNAF and MIT-Bates) and time-like (BaBar, Lear, FENICE+DM2 and E835) regions. The stars are the theoretical constraints and the shadowed area is the unphysical region. The label $\tilde{R}(q^2)$ of the ordinate axis is defined as: $\tilde{R}(q^2) = R(q^2)$ for $q^2 \leq s_0$ and $\tilde{R}(q^2) = |R(q^2)|$ for $q^2 > s_0$.

2.2 Dispersion Relations

Due to the unitarity, the ff's are analytic functions on the q^2-complex plane with a cut along the real axis, starting from the so called theoretical threshold $s_0 = 4m_\pi^2$ up to infinity. Such a threshold represents the energy of the first and then the lightest channel that opens, according to the conservation laws.

By taking advantage from other properties of the ff's such as the asymptotic behaviour, and by applying the Cauchy theorem [18] to the path of integration C in fig.2, when the radius R diverges, we get:

$$f(q^2) = \frac{1}{\pi} \int_{s_0}^{\infty} \frac{\mathrm{Im}[f(s)]}{s - q^2} ds, \tag{1}$$

with $q^2 \leq 0$. This is the DR for the imaginary part and it connects the space-like, real values of the ff $f(q^2)$ to the integral over the cut of its imaginary part.

We want to use the relation of eq.(1), not for a single ff, but for the ratio between the electric and magnetic proton ff's, the function $R(q^2)$. Such a function inherits all the properties of both the functions $G_E^p(q^2)$ and $G_M^p(q^2)$, except for the asymptotic behaviour and mainly the analyticity itself.

This because a zero for $G_M^p(q^2)$ represents a pole for $R(q^2)$, which would not be more an analytic function. In principle, the Cauchy theorem and then relation of eq.(1), can be generalized to the case of a function with a set of isolated poles only if the positions of these poles are known. But, since, as we will see later (§3.4), we have verified that $G_M^p(q^2)$ has no zeros in the q^2-complex plane [19], we consider $R(q^2)$ as an analytic function. Concerning the asymptotic behaviour, the pQCD [2] predicts that both the electric and magnetic ff's should vanish as $(1/q^2)^2$ when $q^2 \to \infty$, as a consequence the ratio $R(q^2)$ should have a constant limit or, in any case, it should behave as a power of $\log(q^2)$. In order to account for this possibility we use the DR

$$R(q^2) = R(0) + \frac{q^2}{\pi} \int_{s_0}^{\infty} \frac{\mathrm{Im}[R(s)]}{s(s - q^2)} ds \equiv 1 + \frac{q^2}{\pi} \int_{s_0}^{\infty} \frac{\mathrm{Im}[R(s)]}{s(s - q^2)} ds, \tag{2}$$

which is subtracted at $q^2 = 0$, where $R(q^2)$ is real and equal to 1. Once the imaginary part is known, we can use the so-called DR for the real part

$$\mathrm{Re}[R(s)] = 1 + \frac{s}{\pi}\mathcal{P}\int_{s_0}^{\infty} \frac{\mathrm{Im}[R(s')]}{s'(s'-s)} ds', \tag{3}$$

to compute the real part of $R(s)$ at each space-like and time-like value of s.

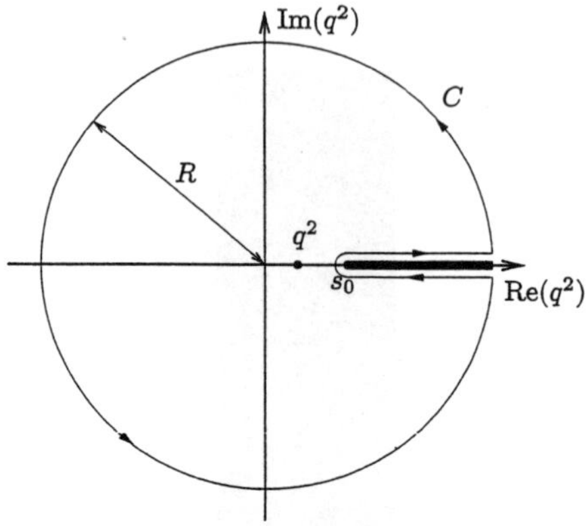

Figure 2: Path of integration C on the complex plane of q^2, with a cut (solid line) along the real axis, from the threshold s_0 up to ∞.

2.3 The dispersive approach

Contrary to the previous models, that have been constructed starting from the data in the space-like region, we start from the imaginary part of $R(q^2)$, which is defined only in a portion of the time-like region, i.e. over the cut $[s_0, \infty[$. We parametrize this function in quite general and model-independent way, that is by means of two series of orthogonal polynomials:

$$\mathrm{Im}[R(s)] \equiv \mathcal{I}(\boldsymbol{C},\boldsymbol{D},s) = \begin{cases} \sum_i^M C_i T_i(x) & x = \frac{2s-s_1-s_0}{s_1-s_0} \quad s_0 \le s \le s_1 \\ \sum_J^N D_j T_j(x) & x = \frac{s_1}{s} - 1 \quad s > s1, \end{cases} \tag{4}$$

where $s_1 = 4M_p^2$ is the physical threshold.

The two series, with coefficients $\boldsymbol{C}= (C_1, C_2, \ldots, C_M)$ and $\boldsymbol{D}= (D_1, D_2, \ldots, D_N)$, cover two naturally distinguished intervals: the unphysical region $[s_0, s_1]$, where the resonances contributing to both G_E^p and G_M^p are smoothed out in the ratio, and the remaining part of the cut, that is the time-like experimentally accessible region: $[s_1, \infty[$.

There are at least two advantages in using this procedure:

- some important theoretical constraints, lying exclusively in the time-like region, can be directly imposed on $\mathrm{Im}[R(q^2)]$;
- with this parameterization of the imaginary part we can apply a rigorous analytic continuation technique, which consists in using the DR's of eqs.(2,3) to reconstruct the ratio $R(q^2)$ for all values of q^2 and we obtain:

$$R(q^2) = 1 + \frac{q^2}{\pi}\mathcal{P}\int_{s_0}^{\infty} \frac{\mathcal{I}(\boldsymbol{C},\boldsymbol{D},s)}{s(s-q^2)} ds + i\mathcal{I}(\boldsymbol{C},\boldsymbol{D},q^2) \tag{5}$$

where, in the right-hand side the dependence on the parameters $\boldsymbol{C}$ and $\boldsymbol{D}$ is shown explicitly through the function $\mathcal{I}(\boldsymbol{C},\boldsymbol{D},q^2)$, which represents the parameterization of $\mathrm{Im}[R(q^2)]$, as defined in eq.(4).

Once the vectors of the coefficients are fixed, the function $R(q^2)$ becomes completely known, in fact we know its real and imaginary part in the whole q^2-complex plane.

In order to select the best values for the free parameters $\boldsymbol{C}$ and $\boldsymbol{D}$ we define a χ^2 which has both, experimental and theoretical contributions. While the experimental constraints shape the function $R(q^2)$ in regions where there are data, the theoretical conditions provide a correct analytic structure that allows to extend this definition of $R(q^2)$ to all the space-like and time-like values of q^2.

2.4 Results and predictions

In the pictures of fig.3 is shown the obtained dispersive descriptions for the ratio $R(q^2)$. As we have already stated in §2.1, due to the incompatibility of the BaBar and Lear time-like data, the dispersive procedure has been applied by considering separately these two sets of data. More precisely, in the so-called *Lear case* (dark band), we use in the time-like region **only** the Lear data, while in the *BaBar case* (gray band) we consider in the time-like region all the other measurements, that is, the BaBar data, the point of FENICE+DM2 and that of E835.

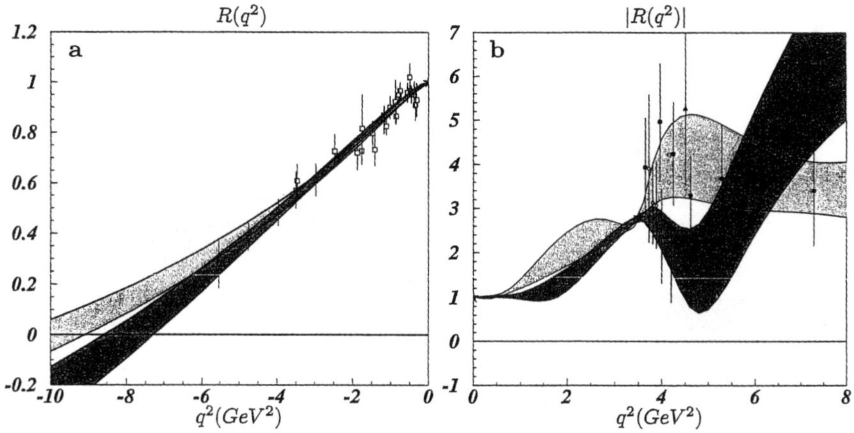

Figure 3: **a**) Dispersive description (the solid bands represent the error) in the space-like region, both the *Lear case* (dark band) and the *BaBar case* (gray band) predict a zero. **b**) Dispersive description in the time-like region, here the two cases have different behaviours, according to the incompatibility of the corresponding data sets.

The picture in fig.3a, shows the results in the space-like region. Here, the descriptions obtained with the BaBar and Lear data, are compatible, in fact in both the cases we obtain a space-like zero at:

$$t_0^{\rm BaBar} = (-10 \pm 1)\ GeV^2 \qquad t_0^{\rm Lear} = (-7.9 \pm 0.7)\ GeV^2. \tag{6}$$

This means that, the unique set of space-like data constraints the two incompatible sets of time-like data to be analytically consistent with the hypothesis of a space-like zero.

The picture of fig.3b, shows the behaviour of $|R(q^2)|$ in the time-like region, where $R(q^2)$ is complex. It is just in the time-like region that the difference between the two cases is more evident and it will be more and more in the asymptotic regime.

Another outcome of this dispersive analysis is the asymptotic behaviour that can be computed in both, the space-like and time-like limits. According to the Phragmèn-Lindelöff theorem, that we have implemented by imposing a vanishing asymptotic value for $\mathrm{Im}[R(q^2)]$, we obtain a real time-like limit for the ratio $R(q^2)$, as $q^2 \to +\infty$. As shown in the picture of fig.4a, the space-like and time-like limits have, in modulus the same value, but they have opposite sign, i.e.:

$$\lim_{q^2\to\pm\infty} \frac{G_E^p(q^2)}{G_M^p(q^2)} = \left[\frac{G_E^p}{G_M^p}\right]_\infty^{\rm Exp}, \qquad \left[\frac{G_E^p}{G_M^p}\right]_\infty^{\rm Exp} = \begin{cases} 0.95 \pm 0.20 & \text{Exp=BaBar} \\ 2.3 \pm 0.7 & \text{Exp=Lear} \end{cases} \tag{7}$$

where $\left[\frac{G_E^p}{G_M^p}\right]_\infty^{\rm Exp}$ is the real and positive time-like limit for $\frac{G_E^p(q^2)}{G_M^p(q^2)}$. By considering the result of eq.(7), the scaling law, providing $G_E^p \simeq G_M^p$, that, in light of the new space-like data, is no more valid at low $|q^2|$, could be restored at least in particular conditions. In fact, only in the *BaBar case*, in the time-like asymptotic regime we get: $G_E^p \simeq G_M^p$, while, in the space-like region, as $q^2 \to -\infty$, the two ff's have opposite limits, i.e. $G_E^p \simeq -G_M^p$. This opposite sign is due to the zero. Moreover, since the Phragmèn-Lindelöff theorem predicts for the phase of a function, with Z zeros and P poles, the asymptotic value: $\Phi(\infty) = (Z - P)\pi$, the fact that we have:

$$R(-\infty) = |R(\infty)|e^{i\Phi(\infty)} = -R(\infty), \tag{8}$$

being $P = 0$, is a further confirmation of the presence of a zero (or, in any case, of an odd number of zeros).

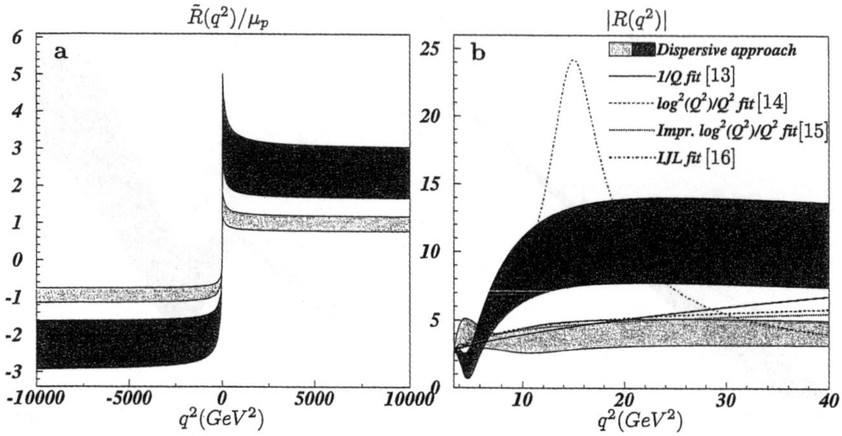

Figure 4: **a**) Ratio $G_E^p(q^2)/G_M^p(q^2)$ [$\equiv R(q^2)/\mu_p$] over a wide range of q^2. **b**) Ratio $R(q^2)$ in the time-like region with $q^2 \geq s_1$ and comparison with the time-like analytic continuations of various models ($Q^2 = -q^2$). $\tilde{R}(q^2)$ is defined as in fig.1.

As a direct consequence of the limits reported in eq.(7), we can also give a prediction for the asymptotic behaviour of the ratio between the Pauli and Dirac ff's as:

$$\lim_{q^2\to\infty} \frac{q^2}{4M_p^2}\frac{F_2^p(q^2)}{F_1^p(q^2)} = \begin{cases} -0.05 \pm 0.20 & \text{Exp=BaBar} \\ 1.3 \pm 0.7 & \text{Exp=Lear} \end{cases} . \tag{9}$$

This means that the ratio $F_2^p(q^2)/F_1^p(q^2)$ scales like $(1/q^2)$ or faster, in the *BaBar case*, as q^2 diverges. The complete knowledge of the complex structure of $R(q^2)$ allows to give predictions concerning also quantities, such as the polarization observables, which depend explicitly on the real and imaginary parts. In fact, in the time-like region, when we consider the annihilation $e^+e^- \to p\bar{p}$, the complex phase of the ff's gives rise to special polarization effects, that is, contrary to what happens in the space-like region, the outgoing proton may experience a polarization even if there are no polarized initial leptons. More precisely, this polarization is determined by the relative phase between $G_E^p(q^2)$ and $G_M^p(q^2)$, which is just the phase $\Phi(q^2)$ of the ratio $R(q^2)$. The time-like polarization vector $\vec{\mathcal{P}}$ has the components [20]:

$$\mathcal{P}_y = -\frac{\sin(2\theta)|R|\sin(\Phi)}{D\sqrt{\tau}} \qquad \mathcal{P}_x = -P_e\frac{2\sin(\theta)|R|\cos(\Phi)}{D\sqrt{\tau}} \qquad \mathcal{P}_z = P_e\frac{2\cos(\theta)}{D}, \tag{10}$$

where $\tau = \frac{q^2}{4M_p^2}$ and

$$D = \frac{1+\cos^2(\theta) + \frac{1}{\tau}|\tau|^2\sin^2(\theta)}{\mu_p}, \tag{11}$$

z is the direction of the outgoing proton in the center of mass system and y is orthogonal to the scattering plane, P_e is the longitudinal polarization of the initial lepton and θ is the scattering angle. As already said, the y-polarization $\mathcal{P}_y$ does not depend on P_e, while the longitudinal polarization does not depend on the phase Φ.

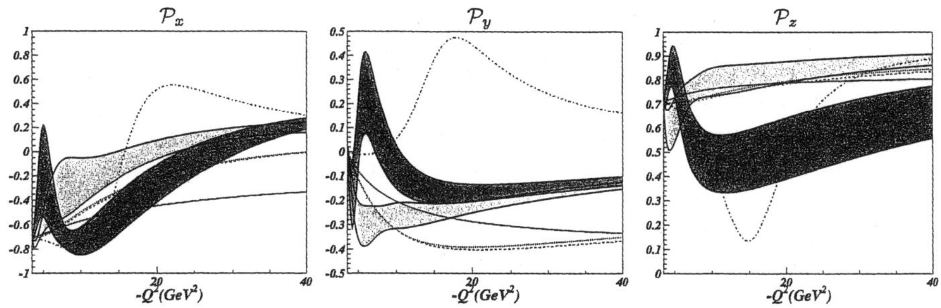

Figure 5: Predictions (solid bands) for polarizations $\mathcal{P}_x$, $\mathcal{P}_y$ and $\mathcal{P}_z$ in the time-like region compared with those of the other models (same references of fig.4b). The plots are for the scattering angle $\theta = 45^o$ and electron polarization $P_e = 1$.

2.5 G_E^p/G_M^p: Conclusions and perspectives

We have used a dispersive approach to construct an expression for the ratio $R(q^2)$, defined in the whole q^2 complex plane, which verifies all the constraints imposed by the theory (analyticity, asymptotic behaviour, etc.) and by the data available at this moment. The knowledge of the complex structure of $R(q^2)$ allows to formulate a wide range of predictions as, for instance, the low energy and asymptotic behaviour of the ratio between Dirac and Pauli ff's, the single and double nucleon polarizations, etc. The other prediction that should be confirmed or refuted soon, is the presence of the space-like zero, that we estimate at $q^2 = (-7.9 \pm 0.7)\ GeV^2$ in the *Lear case* and $q^2 = (-10 \pm 1)\ GeV^2$ in the *BaBar case*, the latter is in agreement with ref. [7]. New polarization measurements are scheduled at Jlab to push down the space-like limit at $q^2 \simeq -10\ GeV^2$ [21].

Anyway, the measurement of $|R(q^2)|$ and of the polarizations [eq.(10)] in the time-like region would have a crucial significance not only to disentangle among the models, but also to gain a rather complete experimental knowledge of the proton ff's.

3 The "inverse" problem: extracting unphysical nucleon magnetic ff's from time-like and space-like data

3.1 Motivations

As we have already stated in §2.2, the nucleon ff's are analytic functions on the q^2-complex plane, then they are well defined at each space-like and time-like value of q^2. In spite of this, while the space-like region is completely covered by the elastic scattering process $ep \to ep$, the time-like region, where the ff's can be measured only through the annihilation $e^+e^- \to p\bar{p}$, is not fully experimentally accessible. The interval which goes from $q^2 = 0$ up to $q^2 = s_1$, the physical threshold for the process $e^+e^- \to p\bar{p}$, represents the so-called unphysical region (fig.1), that is, for those values of q^2 the ff's cannot be measured at all.

Nevertheless, by means of the DR's (§2.2) we can relate values of the ff's in different regions of q^2. Then, in principle, it is possible to extract the values of a ff on a certain interval of q^2, by knowing its values in other regions.

In order to use directly the space-like and time-like data for the ff's $G_M^{p,n}(q^2)$, we consider a special kind of DR, that is, the DR for the logarithm [1, 19]:

$$\ln[G(q^2)] = \frac{\sqrt{s_0 - q^2}}{\pi} \int_{s_0}^{\infty} \frac{\ln|G(s)|ds}{(s-q^2)\sqrt{s-s_0}} \tag{12}$$

with $q^2 \leq s_0$.

Contrary to the usual form of DR's, reported in eq.(1), this DR does not contain the imaginary part, but the modulus of the ff over the cut, this is just the quantity that can be measured in the time-like region where the ff is complex. By means of these ingredients: space-like and time-like data, analyticity of the ff's and DR's, we define a practical strategy to determine ff's in the unphysical region in a model independent way, without any bias towards expected resonances.

Uncovering the nucleon unphysical region is an interesting task by many points of view:

- the time-like data for both $|G_M^p(q^2)|$ and $|G_M^n(q^2)|$, in the region just above the physical threshold s_1, show a steep enhancement, that, in agreement with the multi-hadronic cross section, could be interpreted as a resonance (or a superposition of more structures) lying near, but below, the physical threshold;
- all the vector meson contributions to the nucleon ff's should be in the nucleon unphysical region;
- the presence of the $\phi(1020)$ resonance would be a clear indication of a sizeable $s\bar{s}$ content in the nucleons.

3.2 The integral equation

Consider the DR of eq.(12) for a normalized nucleon ff $G(q^2) \equiv G_M^{p,n}(q^2)/\mu_{p,n}$ with a subtraction at $q^2 = 0$:

$$\ln[G(q^2)] = \frac{q^2\sqrt{s_0 - q^2}}{\pi} \int_{s_0}^{\infty} \frac{\ln|G(s)|ds}{s(s-q^2)\sqrt{s-s_0}}. \tag{13}$$

The subtraction, that makes the integral less dependent on the ff asymptotic beahviour, does not involve any additional term, being $\ln[G(0)] = 0$. For $q^2 \geq s_0$, over the cut, the phase of the ff, defined as $G(q^2) = |G(q^2)|e^{\delta(q^2)}$, is given by the principal value integral

$$\delta(q^2) = -\frac{q^2\sqrt{q^2 - s_0}}{\pi} \mathcal{P}\!\int_{s_0}^{\infty} \frac{\ln|G(s)|ds}{s(s-q^2)\sqrt{s-s_0}}. \tag{14}$$

So in principle the method consists in using the DR of eq.(13) to define an integral equation, where the modulus of the ff's in the unphysical region is the unknown and, data and asymptotic behaviour, as predicted by the pQCD, represent the inputs.

More precisely we split the integral of eq.(13) into 2 parts:

- ⋆ the unknown unphysical region $[s_0, s_1]$ where the ff's are to be determined.
- ⋆ the remaining part of the cut $]s_1, \infty[$ for which data exists and, as previously stated, quite accurate asymptotic estimates are available.

We obtain an integral equation of the first kind and linear in the unknown $\ln|G(q^2)|$

$$\ln[G(q^2)] - I(q^2) = \frac{q^2\sqrt{s_0 - q^2}}{\pi} \int_{s_0}^{s_1} \frac{\ln|G(s)|ds}{s(s-q^2)\sqrt{s-s_0}}, \tag{15}$$

where

$$I(q^2) = \frac{q^2\sqrt{s_0 - q^2}}{\pi} \int_{s_1}^{\infty} \frac{\ln|G(s)|ds}{s(s-q^2)\sqrt{s-s_0}} \tag{16}$$

represents a known quantity, since it can be computed by means of the time-like data and a pQCD recipe to extend them to very high q^2 values.

To avoid instability, usually met in solving integral equations of the first kind, we use some prescriptions devoted to improving smoothness in the solutions and also a regularization scheme, to sum up:

- $I(t)$ [eq.(16)] is calculated using time-like data through a rational, smooth function with the expected asymptotic behaviour. The subtraction at $q^2 = 0$ helps in diminishing the impact which the asymptotic behaviour has on the results.
- There is a steep spike near the $N\bar{N}$ threshold. To avoid any ensuing instabilities, the upper limit of the unphysical region has been raised to $s_2 = s_1 + \Delta$ where $\Delta \approx 0.5 GeV^2$ and continuity is imposed there. A new DR is constructed for the region $[s_1, s_2]$ [19].
- Our regularization consists in requiring the local curvature of the ff in the unphysical region

$$R_2 = \int_{s_0}^{s_2} \left[\frac{d^2|G(q^2)|}{(dq^2)^2}\right]^2 dq^2 \tag{17}$$

to be limited. Instead of the second derivative of $\ln|G(q^2)|$, as is standard, we employ the second derivative of $|G(q^2)|$ for this purpose. The reason is that fluctuations in $|G(q^2)|$ are important only when $|G|$ is large, while $\ln|G(q^2)|$ fluctuations would be large also when $|G(q^2)|$ is small.

- Eq.(15) is then linearized by transforming the integrals into sums over $M = 50$ suitable sub intervals in q^2, with their widths increasing with q^2. This introduces further smoothness, by effectively integrating over any structure with a narrower half width.
- The quantity to be minimized, without the "smoothness improving terms" is:

$$R_0 = \sum_{i=1}^{L} \left[\sum_{j=1}^{M} F_j \frac{t_i\sqrt{s_0 - t_i}}{\pi} \int_{s_j}^{s_{j+1}} \frac{ds}{s(s-t_i)\sqrt{s-s_0}} + I(t_i) - \ln G(t_i) \right]^2 , \tag{18}$$

where $F_j = \ln|G[(s_{j+1} + s_j)/2]|$, with $j = 1, 2, \ldots, M$, is calculated in the middle of the j-th sub interval and the values t_i, with $i = 1, 2, \ldots, L$, correspond to the experimental points available in the space-like region.

- Finally, we minimize

$$R_{\text{total}} = R_0 + \tau^6 R_2 + C. \tag{19}$$

where R_0 is just the integral equation [eq.(18)], R_2 is the regularization term and C contains all the continuity constraints, which are imposed across the theoretical threshold s_0 and the shifted physical one s_2. The "dumping parameter" τ, defined in unit of energy, has to be chosen by trial (see §3.3) and error: if it is set too large, it will not respond to sharp structures, while unstable solutions will result if it is set too low.

- The uncertainties in the solution of eqs.(15) and (14), due to experimental errors, were estimated by simulating new space- and time-like data according to the quoted errors and then solving the DR for each simulated set.

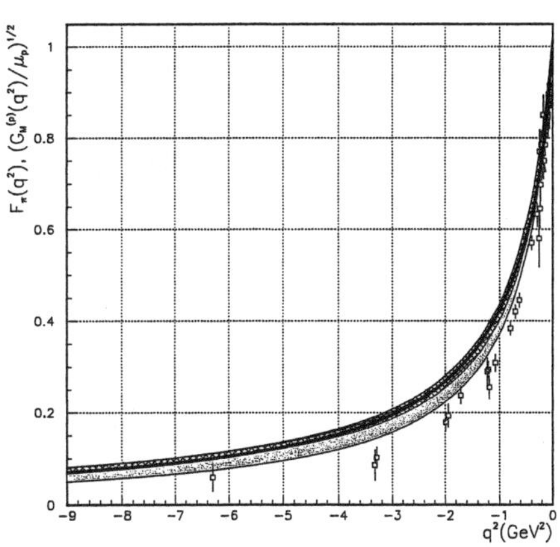

Figure 6: Pion space-like ff computed via DR without (grid band) and with (light band) subtraction, compared with pion ff space-like data and square route of proton space-like ff data fit (dark band).

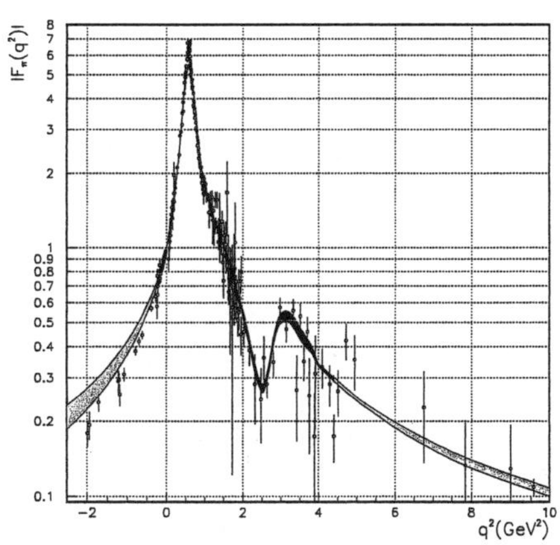

Figure 7: Pion ff, the dark band is the solution of eq.(19), while the gray bands, in the space-like region and in the time-like region above the shifted nucleon physical threshold, represent the inputs.

3.3 Test of the regularization method

To test the entire procedure and also to get a suitable range for the parameter τ, we computed the space-like pion FF using time-like data. In the time-like region, this FF is known up to the J/Ψ mass and at higher q^2, it was extrapolated using first order pQCD [2]. In fig.(6), comparison is made with the measured (low q^2) space-like data. (Higher space like Q^2 data points are through extrapolations from pion electroproduction data and thus may have systematic errors). We have also made other tests obtaining good agreement with the ρ peak, the ρ width and also a dip at 1.6 GeV^2 for $\tau \approx m_\pi$ fig.(7). The phase of the pion ff approaches just above 2 GeV to its expected asymptotic value of 180 degrees fig.(8).

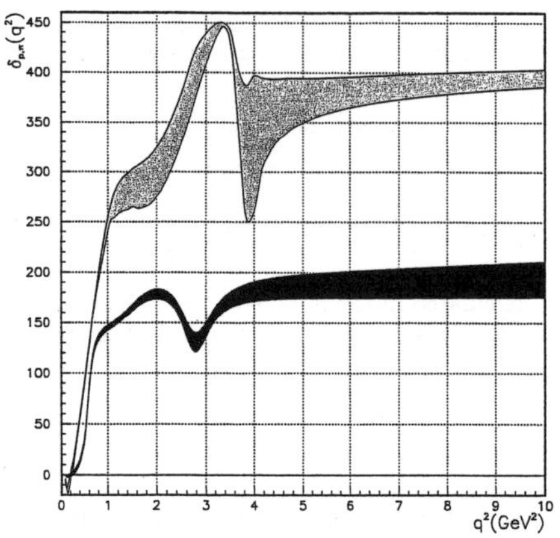

Figure 8: Phase of proton (light band) and pion (dark band) ff computed via DR of eq.(14).

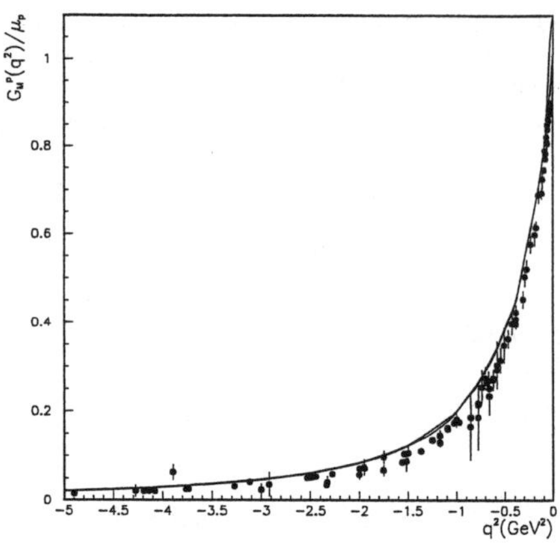

Figure 9: Reconstructed magnetic proton ff (light band) compared with the data.

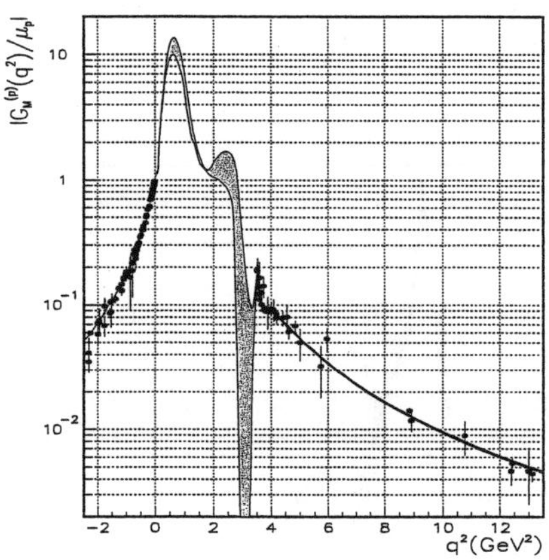

Figure 10: Proton magnetic ff computed via DR (light band). The dark bands are the inputs.

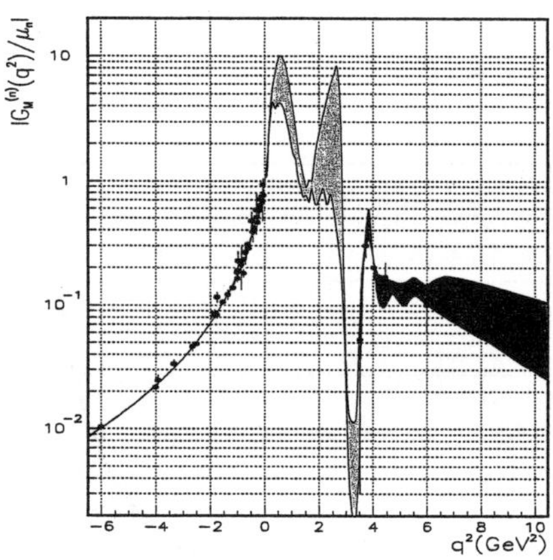

Figure 11: Neutron magnetic ff computed via DR (light band). The dark bands are the inputs.

3.4 Results for the unphysical nucleon ff's

Below we summarize some salient features of our findings:

- For the first time, resonant structures have been generated from "smooth" inputs figs.(10,11). The method is stable and reliable.
- The combined ($\rho + \omega$) peaks and the $\rho'(1600)$ are generated at the right mass. However, the ρ peak is much broader. Earlier analyses [22] had also found a similar discrepancy.
- No $\Phi(1020)$ signal is visible thus signaling a very small $s\bar{s}$ content in the nucleon.
- Phases for the nucleon are consistent with expectations: $\delta_N \rightarrow 360^o$ within the error bands fig.(8).
- There is an interference pattern near threshold ($M \approx 1.88\ GeV$) which may be related to baryonium figs.(10,11).
- Once the modulus and phase of the proton magnetic ff are obtained, under the hypothesis of no zeros for $G_M^p(q^2)$, the imaginary part can be used to recomputed the space-like value of this ff, via the usual DR of eq.(1). The agreement between data and recoputed space-like ff (see fig.9) represents a positive check for the initial hypothesis of no zeros for $G_M^p(q^2)$ in the q^2 complex plane.

3.5 The "inverse problem": conclusions

Nucleon time-like magnetic ff's have been obtained in an almost model independent way by means of DR for $\ln G(q^2)$, using a regularization method in conjunction with space and time-like data.

Resonances have been found consistent with the $\rho(770)$ and $\rho'(1600)$ masses. However, a very large ρ width is obtained as in previous DR analyses. Further work is in progress to understand the sources of discrepancies as well as the relationship of our results with other DR analyses [23]. New data on proton magnetic ff are now available [11], these data are in agreement with the old ones, however we are upgrading the analysis by including them.

Acknowledgments

We thank R. Baldini, C. Bini, P. Gauzzi, M. Mirazita, M. Negrini and Y. Srivastava for discussions on the subject of this talk.

References

[1] M. Gourdin, Phys. Rep. **11** (1974) 29 and other references therein.

[2] V.A. Matveev, R.M. Muradyan, A.N.Tavkhelidze, Lett. Nuovo Cim. **7** (1973) 719; S.J. Brodsky, G.R. Farrar, Phys. Rev. Lett. **31** (1973) 1153; S.J. Brodsky, G.R. Farrar, Phys. Rev. **D11** (1975) 1309; S.J. Brodsky, P.G. Lepage, Phys. Rev. **D22** (1980) 2157; T. Gousset, B. Pire, Phys. Rev. **D51** (1995) 15.

[3] M.K. Jones *et al.*, Phys. Rev. Lett. **84** (2000) 1398; O. Gayou *et al.*, Phys. Rev. **C64** (2001) 038202; O. Gayou *et al.*, Phys. Rev. Lett. **88** (2001) 092301.

[4] B.D. Milbrath *et al.*, Phys. Rev. Lett. **80** (1998) 452; Phys. Rev. Lett. **82** (1999) 2221(E).

[5] R.C. Walker *et al.*, Phys. Rev. **D49** (1994) 5671; L. Andivahis *et al.*, Phys. Rev. **D50** 5491 (1994) 5491; J. Litt *et al.*, Phys. Lett. **B31** (1970) 40; C. Berger *et al.*, Phys. Lett. **B35** (1971) 87; L.E. Price *et al.*, Phys. Rev. **D4** (1971) 45; W. Bartel *et al.*, Nucl. Phys. **B58** (1973) 429.

[6] M.N. Rosenbluth, Phys. Rev. **79** (1950) 615.

[7] S. Dubnicka and A.Z. Dubnickova, Fizika **B13** (2004) 287 [hep-th/0401081].

[8] A. Antonelli *et al.*, Nucl. Phys. **B517** (1998) 3.

[9] D. Bisello *et al.*, Nucl. Phys. **B224** (1983) 379.

[10] G. Stancari, PhD Thesis and M. Andreotti, Laurea Thesis Ferrara University (2001).

[11] R. Baldini (BaBar Coll.), "Preliminary BaBar results on Proton Time-like Form Factors by I.S.R." preseted at "HEP2005", July 21st-27th 2005, in Lisbona, Portugal.

[12] G. Bardin *et al.*, Nucl. Phys. **B411** (1994) 3.

[13] J. P. Ralston, P. Jain, hep-ph/0302043; J.P. Ralston, P. Jain, R. V. Buniy, AIP Conf. Proc. **549** (2000) 302; G.A. Miller, M.R. Frank, Phys. Rev. **C65** (2002) 065205 ; M.R. Frank, B.K. Jennings, G.A. Miller, Phys. Rev. **C54** (1996) 920; G. Holzwarth, Z. Phys. **A356** (1996) 339; F. Cardarelli, S. Simula, Phys. Rev. **C62** (2000) 065201; R.F. Wagenbrunn *et al.*, Eur. Phys. J. **A14** (2002) 17.

[14] A.V. Belitsky, X. Ji, F. Yuan, Phys. Rev. Lett. **91** (2003) 092003.

[15] S.J. Brodsky *et al.*, Phys. Rev. **D69** (2004) 054022.

[16] F. Iachello, A.D. Jackson, A. Lande, Phys. Lett. **B43** (1973) 191; M. Gari, W. Krmpelmann, Z. Phys. **A322** (1985) 689; E.L. Lomon, Phys. Rev. **C66** (2002) 045501; **64** (2001) 035204.

[17] P.A. Guichon, M. Vanderhaeghen, Phys. Rev. Lett. **91** (2003) 142303; P.G. Blunden, W. Melnitchouk, J.A. Tjon, Phys. Rev. Lett. **91** (2003) 142304; J. Arrington, Phys. Rev. **C69** (2004) 032201(R); M.P. Rekalo, E. Tomasi-Gustafsson, Eur. Phys. J. **A22** (2004) 331; A.V. Afanasev *et al.*, hep-ph/0502013.

[18] E.C. Tichmarsch, *The theory of functions*, London, Oxford University Press, 1939.

[19] R. Baldini *et al.*, Eur. Phys. J. **C11** (1999) 709.

[20] A.Z. Dubnickova, S. Dubnicka, P.M. Rekalo, Nuovo Cim. **A109** (1966) 241.

[21] C.F. Perdrisat *et al.*, Jefferson Lab. experiment E01-109.

[22] See for example, B. Orman, Phys. Rev. **145** (1966) 1140 and other references therein.

[23] P. Mergell, Ulf-G. Meßner and D. Drechsel, Nucl. Phys. **A596** (1996) 367;
H. Forkel, Prog. Part. Nucl. Phys. **36** (1996) 229;
Ulf-G. Meßner, Nucl. Phys. **A623** (1997) 340.

Reprinted with permission from the electronic only supplement "Young Scientist" (DOI: 10.1140/epjcd/s2006-03-004-4) of The European Physical Journal (EPJ) C

Semiotic Dynamics in Online Social Communities

C. Cattuto[1,2]

[1] Museo Storico della Fisica e Centro Studi e Ricerche "Enrico Fermi"
Compendio Viminale, 00184 Roma, Italy
e-mail: ciro.cattuto@roma1.infn.it

[2] Dipartimento di Fisica, Università di Roma "La Sapienza"
P.le A. Moro, 2, 00185 Roma, Italy

Abstract. A distributed classification paradigm known as *collaborative tagging* has been successfully deployed in large-scale web applications designed to manage and share diverse online resources. Users of these applications organize resources by associating with them freely chosen text labels, or *tags*. Here we regard tags as basic dynamical entities and study the *semiotic dynamics* underlying collaborative tagging. We collect data from a popular system and focus on tags associated with a given resource. We find that the frequencies of tags obey to a generalized Zipf's law and show that a Yule-Simon process with memory can be used to explain the observed frequency distributions in terms of a simple model of user behavior.

1 Collaborative Tagging

A new paradigm has been quickly gaining ground in information systems on the World Wide Web: Collaborative Tagging [1,2]. In web-based applications like *Del.icio.us*[1], *Flickr*[2], *CiteULike*[3], users enrich diverse resources – ranging from photographs to scientific references and web pages – with semantically meaningful information in the form of text labels, or "tags". Tags are freely chosen and users associate resources with them in a totally uncoordinated fashion, for their own use.

The tagging activity of each user is globally visible to the user community (see Fig. 1) and the tagging process develops genuine social aspects and complex interactions. Remarkably, despite the selfish nature of users' behavior, tagging systems exhibit cooperative dynamics that eventually lead to a bottom-up categorization of resources, shared throughout the user community. The open-ended set of tags used within the system – commonly referred to as "folksonomy" – can be used as a sort of semantic map to navigate the contents of the system itself. It has been argued that the surging popularity of collaborative tagging is due to its comparatively small cognitive overhead with respect to taxonomic categorization [3,1], so that tagging is a very natural activity for web users. Collaborative tagging systems leverage this aspect, recruiting simple and robust behaviors of individual users in order to create cooperation and foster the emergence of shared conventions at the system level.

Focusing on tags as basic dynamical entities, collaborative tagging falls within the scope of Semiotic Dynamics [4,5], a new field that studies how populations of humans or agents can establish and share semiotic systems, typically driven by their use in communication. New web applications hinged on collaborative tagging fall precisely in this perspective and can be regarded as cases of Semiotic Dynamics at play: folksonomies, in fact, do exhibit dynamical aspects also observed in human languages [6, 7], such as the emergence of naming conventions, competition between terms, takeovers by neologisms, and more.

In the following we briefly describe the structure of collaborative tagging systems and discuss the experimental procedures that we employ to collect data from *Del. icio.us*, one of the most popular and paradigmatic social bookmarking systems. We adopt a resource-centric view of the system and focus on the tags that have been associated with a given resource. We report on the distribution of tag frequencies and show that a stochastic Yule-Simon process with memory can be used to successfully describe the observed frequency distributions. We close by casting our work in a more general research perspective.

2 Experimental Data

The activity of users interacting with a collaborative tagging system consists of either navigating the existing body of resources by using tags, or of adding new resources. In order to add a new resource into the system the user is prompted for a reference to the resource and a set of tags to associate with it. Thus, the basic unit of information in a collaborative tagging system is a (`user, resource, {tags}`) triple, here referred to as "post" (see Fig. 2). Tagging events build a tripartite graph with partitions corresponding to users, resources and tags, respectively. A post

[1] http://del.icio.us/

[2] http://flickr.com/

[3] http://citeulike.org/

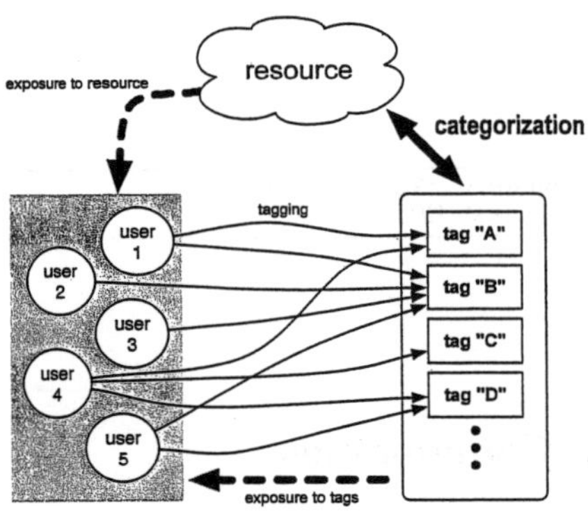

Fig. 1. Schematic depiction of the collaborative tagging process: web users (circles on the left) are exposed to a resource and freely associate tags with it (rectangles on the right). In their interaction with the system users are also exposed to tags previously entered by themselves and by other users. The collective tagging activity creates a dynamical correspondence between a resource and a set of tags, i.e. an emergent categorization in terms of tags shared by a community.

typically contains a temporal marker recording the (physical) time of the tagging event, so that temporal ordering can be preserved in storing and retrieving posts.

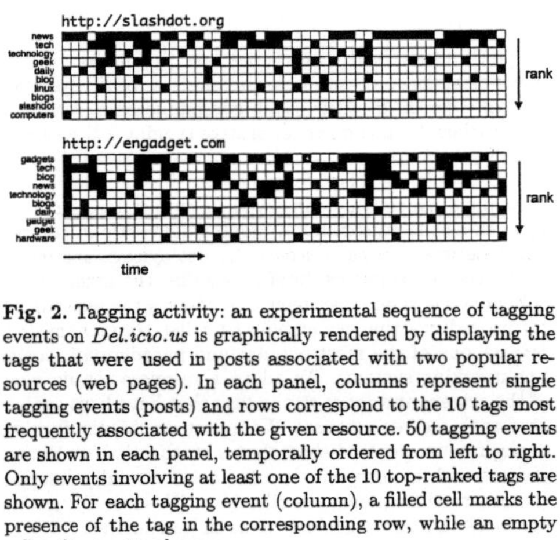

Fig. 2. Tagging activity: an experimental sequence of tagging events on *Del.icio.us* is graphically rendered by displaying the tags that were used in posts associated with two popular resources (web pages). In each panel, columns represent single tagging events (posts) and rows correspond to the 10 tags most frequently associated with the given resource. 50 tagging events are shown in each panel, temporally ordered from left to right. Only events involving at least one of the 10 top-ranked tags are shown. For each tagging event (column), a filled cell marks the presence of the tag in the corresponding row, while an empty cell indicates its absence.

Our analysis will focus on *Del.icio.us* for several reasons: i) it was the first system to deploy the ideas and technologies of collaborative tagging, so it has acquired a paradigmatic character and it is the natural starting point for any quantitative study. ii) because of its popularity, it has a large community of active users and comprises a precious body of raw data on the static and dynamical properties of a folksonomy. iii) it is a *broad folksonomy* [8], i.e. single tagging events (posts) retain their identity and can be individually retrieved. This affords unimpeded access to the "microscopic" dynamics of collaborative tagging, providing the opportunity to make contact between emergent behaviors and low-level dynamics. It also allows us to define and measure the multiplicity (or frequency) of tags in the context of a single resource. Contrary to this, popular sites falling in the *narrow folksonomy* class (*Flickr*, for example) foster a different model of user interaction, where tags are mostly applied by the content creator, no notion of tag multiplicity is possible in the context of a resource, and no access is given to the raw sequence of tagging events. On studying *Del.icio.us* we adopt a resource-centric view of the system, that is we investigate the dynamical correspondence between a given resource and the tags that users associate with it. In line with our focus on semiotic dynamics, we factor out the detailed identity of the users involved in the process and only deal with streams of tagging events and their statistical properties.

To perform automated data collection of raw data we use a custom web (HTTP) client that connects to *Del.icio.us* and navigates the system's interface as an ordinary user would do, extracting the relevant metadata and storing it for further post-processing. *Del.icio.us* allows the user to browse its content by resource: our client requests the web page associated with the resource under study and uses an HTML parser to extract the post information (user, tags and time stamp). Fig. 2 graphically depicts the raw data we gather for the case of two popular resources on *Del.icio.us.* The data used for the present analysis were retrieved in November 2005.

3 Temporal Evolution

Fig. 3 displays the amount of tagging data associated with a single popular resource (the same shown in the top panel of Fig. 2) as a function of time. The time stamp attached to each post is used to build a time series of tagging events and cumulated values are shown for the number of entries associated with the resource and the total number of tags used. The data shown in Fig. 3 span a time interval of about 2 years, during which the popularity of *Del.icio.us* surged and its user base increased enormously. Correspondingly, the amount of tagging data associated with the selected resource increased by several orders of magnitude. In striking contrast with this, the relative proportions of tags associated with a given resource quickly approach a quasi-stationary condition: in Fig. 4 we plot the fraction of occurrence of the 10 top-ranked tags of Fig. 2 (top panel) as a function of the number of posts associated with the resource. After an initial transient, during which the fractions fluctuate significantly, the relative proportion of each tag settles towards an approxi-

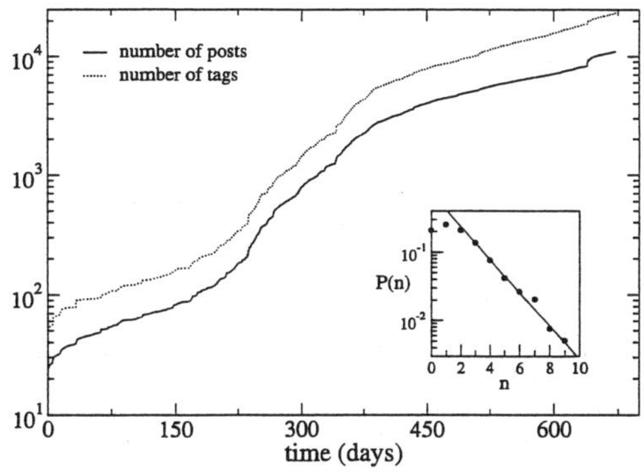

Fig. 3. Amount of metadata associated with a popular resource in *Del.icio.us*, as a function of time. Data are shown for the same resource of Fig. 2, top panel. Over a time interval of about 2 years, the number of posts (solid lines) and the number of total tags (dotted line) increased by several orders of magnitude. Inset: the fraction $P(n)$ of posts containing n tags displays an exponential tail for high values of n. The average value $\bar{n}$ corresponds to the vertical offset between the solid line and the dotted line.

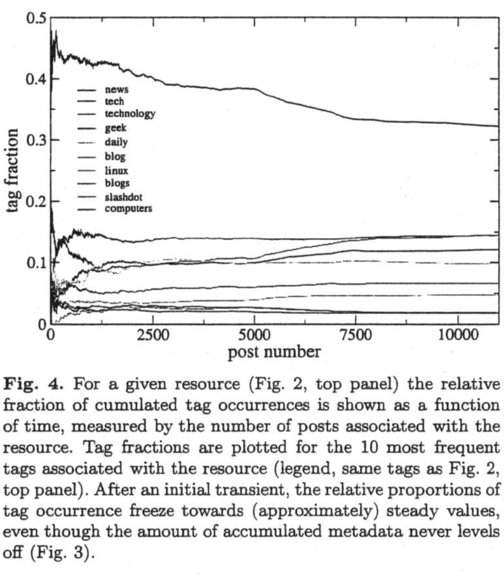

Fig. 4. For a given resource (Fig. 2, top panel) the relative fraction of cumulated tag occurrences is shown as a function of time, measured by the number of posts associated with the resource. Tag fractions are plotted for the 10 most frequent tags associated with the resource (legend, same tags as Fig. 2, top panel). After an initial transient, the relative proportions of tag occurrence freeze towards (approximately) steady values, even though the amount of accumulated metadata never levels off (Fig. 3).

mately constant value. This is reminiscent of the behavior observed in Polya's urn problems [9] and suggests the existence of a multiplicative process underlying the tagging activity of users [10], where the usage pattern of tags is subjected to some kind of frequency-bias. Once the number of posts associated with a resource is sufficiently large, single tagging events have a negligible effect on the global distribution of tags and the existing distribution is reinforced, generally becoming more and more stable. This kind of robustness is a very important property of collaborative tagging: on the one hand, the fact that tag fractions stabilize quickly allows the emergence of a clearly defined categorization of the resource in terms of tags, with a few top-ranked tags defining a semantic "fingerprint" of the resource they refer to (see also Fig. 2). On the other hand, the long-term stability of tag proportions makes the emergent categorization robust against noise. Both aspects contribute greatly to the actual usability of collaborative tagging systems. Occasionally, interesting non-stationary behaviors and transitions can be observed, where new tags are invented and become socially adopted by the user community [10].

4 Tag Frequencies

To better probe the emergent categorization of a resource in terms of tags, we compute the standard frequency-rank distributions for the tags associated with the resources of Fig. 2, i.e. we compute the number of occurrences of tags and rank them: our results are shown in Fig. 5 (black dots). The high-rank tails of the experimental curves display a power-law behavior reminiscent of Zipf's law[11] which is characteristic of self-organized communication systems and is commonly observed in natural languages and written text [12]. The observed exponent of the power law is greater than 1 (the lines in Fig. 5 have slope 4/3) and its explanation requires more complex microscopic mechanisms than those usually invoked to explain Zipf's law [13]. Moreover, the low-rank part of the distributions of Fig. 5 displays a flattening behavior typically not observed in systems strictly obeying Zipf's law, related to the co-existence (and possibly the competition) of the few low-rank tags which characterize the resource in the statistically strongest way.

In order to model the observed frequency-rank behavior, we use a Yule-Simon stochastic process [14–16] with long-term memory, introduced in Ref. [17] (Fig. 6). We move from the observation that actual users are in principle exposed to all the tags stored in the system (like in the original Yule-Simon model), but the way in which they choose among them, when tagging a new post, is far from being a simple uniform distribution (see also [18]). It seems more realistic to assume that users tend to use recently added tags more frequently than old ones, according to a skewed memory kernel. Our modification of the Yule-Simon's model consists in weighting the probability of choosing an existing tag according to a power-law memory kernel which controls the visibility of the past history of the resource. This hypothesis about the functional form of the memory kernel is supported by findings in Cognitive Psychology [19] (where power laws of latency and frequency have been shown to model human memory) as well as by recent analysis on patterns of human activity [20]. Specifically, previous work on tag co-occurrence [17] shows that a hyperbolic memory kernel is needed to match the

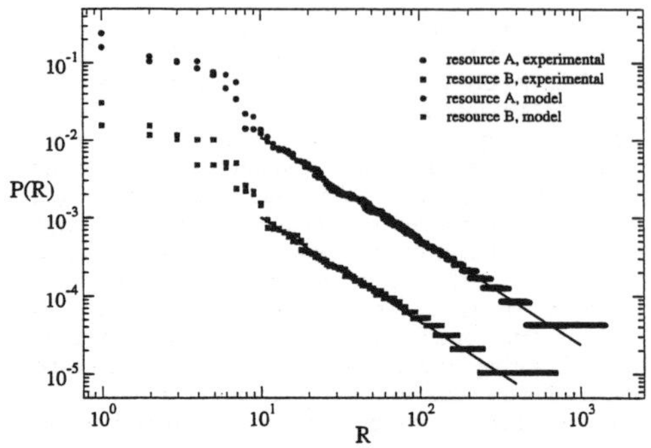

Fig. 5. Frequency-rank plots for the tags associated with a given resource. Experimental data (black symbols) are shown for the resources of Fig. 2 (top panel and bottom panel correspond to resource A and B, respectively). For the sake of clarity the data for resource B were shifted down by one decade. The high-rank tail of the curves displays a power-law behavior corresponding to a generalized Zipf's law. For reference, the blue lines correspond to the power law $R^{-4/3}$. Red symbols are theoretical data obtained by computer simulation of the stochastic model described in the text (Fig. 6). The parameters of the model, i.e. the probability p, the memory parameter τ and the initial number of words n_0 were adjusted to match the experimental data, yielding approximately $p = 0.035$, $\tau = 175$ and $n_0 = 10$ for resource A, and $p = 0.04$, $\tau = 170$ and $n_0 = 10$ for resource B.

experimental data. Thus, our model (Fig. 6) can be stated as follows: the process by which users of a collaborative tagging system associate tags to resources can be regarded as the construction of a "text", built one step at a time by adding "words" (i.e. tags) to a text initially comprised of n_0 words. This process is meant to model the behavior of an effective average user in the context of a single resource. At a generic time step t, a new word may be invented with probability p and appended to the text, while with probability $1 - p$ one word is copied from the existing text, going back in time by i steps with a probability that decays with the offset i as a power law, $Q(i) = C(t)/(\tau + i)$. $C(t)$ is a time-dependent normalization factor and τ is a characteristic time-scale over which recently added words have comparable probabilities. By simulating the stochastic process described above and adjusting the values of the model parameters in order to match the experimental data, we obtain a very good agreement between experimental data (black dots) and model (red dots), as shows in Fig. 5. This proves that the simple Yule-Simon process with fat-tailed memory can be successfully applied to model the frequency-rank distribution of tags associated with a given resource, providing insights into the average behavior of users in the context of that resource.

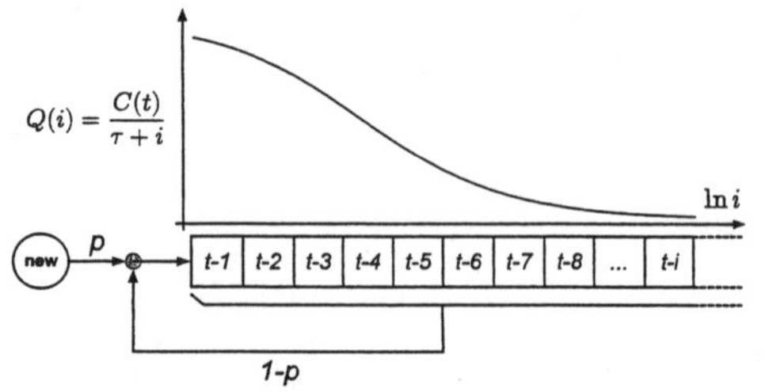

Fig. 6. A Yule-Simon's process with memory. A stream of tags is generated by iterating the following step: with probability p a new tag is created and appended to the tag stream, while with probability $1 - p$ a tag is extracted from the past history of the system and appended to the text, going back in time by i steps with a probability $Q(i) \sim 1/(\tau + i)$.

5 Conclusions and Perspective

Information systems on the World Wide Web have been increasing in size and complexity to the point that they presently exhibit features typically attributed to *bona fide* complex systems. They display rich high-level behaviors that are causally connected in non-trivial ways to the dynamics of their interacting parts. Because of this, concepts and formal tools from the science of complex systems can play a potentially important role in understanding and designing the behavior of such systems. Here we made a first step in this direction and focused on the new paradigm of collaborative tagging. Collaborative tagging systems involve the computer-mediated interaction of a large number of human agents. They have an emergent semiotic system (the folksonomy) at their core and can be regarded as a sort of "laboratory" of semiotic dynamics. In addition to this, their evolution can be easily monitored both at the global level (folksonomy) and at the agent-level (single tagging events of users), providing an interesting opportunity to connect the two levels by means of a suitable theoretical framework. We grounded our work on actual tagging data and made use of statistical tools and stochastic models to gain insights into the dynamical properties of collaborative tagging. We studied the frequency distributions of tags and found that a simple stochastic process, a long-term memory version of the classic Yule-Simon process, is able to describe very accurately the experimental data, allowing us to develop a model of user behavior and link it to the statistical properties of the folksonomy. Overall, our findings suggest that users of collaborative tagging systems share universal behaviors which, despite the intricacies of personal categorization, tagging procedures and user interactions, appear to obey to simple activity patterns.

The author wishes to thank V. Loreto, A. Baldassarri, A. Baronchelli, V. Servedio and L. Steels for stimulating discussion and suggestions. This research has been partly supported by the ECAgents project funded by the Future and Emerging Technologies program (IST-FET) of the European Commission under the EU RD contract IST-1940. The information provided is the sole responsibility of the authors and does not reflect the Commission's opinion. The Commission is not responsible for any use that may be made of data appearing in this publication.

References

1. A. Mates, *Folksonomies - Cooperative Classification and Communication Through Shared Metadata*, Computer Mediated Communication, LIS590CMC, Graduate School of Library and Information Science, University of Illinois Urbana-Champaign (2004).
2. T. Hammond, T. Hannay, B. Lund and J. Scott, *Social Bookmarking Tools (I): A General Review*, D-Lib Magazine **11**(4), (2005).
3. R. Sinha, `http://rashmisinha.com/archives/05_09/tagging-cognitive.html` (2005).
4. L. Steels and F. Kaplan, *Collective learning and semiotic dynamics.* In D. Floreano and J.-D. Nicoud and F. Mondada, editors, Advances in Artificial Life: 5th European Conference (ECAL 99), Lecture Notes in Artificial Intelligence **1674**, 679-688, Berlin, (1999).
5. J. Ke, J. W. Minett, A. Ching-Pong, W.S.-Y, Wang, *Self-organization and selection in the emergence of vocabulary*, Complexity **7**, 41-54 (2002).
6. M. A. Nowak, N. L. Komarova and P. Niyogy, *Computational and evolutionary aspects of language*, Nature **417**, 611-617 (2002).
7. S. Kirby, *Natural language and artificial life*, Artificial Life **8**, 182-215 (2002).
8. T. Vander Wal, *Explaning and Showing Broad and Narrow Folksonomies*, `http://www.personalinfocloud.com/2005/02/explaining_and_.html`, (2005).
9. F. Eggenberger and G. Polya, *Uber die Statistik verketter vorgage*, Zeit. Angew. Math. Mech. **1**, 279 (1923).
10. S. Golder and B. A. Huberman, *Usage patterns of collaborative tagging systems*, Journal of Information Science **32**, 198 (2006); e-print `cs/0508082`.
11. G. K. Zipf, *Human Behavior and the Principle of Least Effort*, Addison-Wesley, Cambridge, MA, (1949).
12. `http://linkage.rockefeller.edu/wli/zipf/index_ru.html`
13. R. Ferrer Cancho and V. D. P. Servedio, *Can simple models explain the Zipf's law for all exponents?*, Glottometrics **11**, 1 (2005).
14. M. E. J. Newman, *Power laws, Pareto distributions and Zipf's law*, Contemporary Physics **46**(5), 323-351 (2005).
15. G. U. Yule, *A Mathematical theory of evolution based on the conclusions of Dr. J.C. Willis*, Philos. Trans. R. Soc. London B **213**, 21-87 (1925).
16. H. A. Simon, *On a class of skew distribution functions*, Biometrika **42**, 425 (1955).
17. C. Cattuto, V. Loreto and L. Pietronero, *Semiotic Dynamics and Collaborative Tagging*, submitted (2006); e-print cs/0605015.
18. D. H. Zanette and M. A. Montemurro, *Dynamics of text generation with realistic Zipf's distribution*, Journal of Quantitative Linguistics **12**, 29-40 (2005).
19. J. R. Anderson, *Cognitive Psychology and its implications*, fifth edition, Worth Publisher, New York (2000).
20. A.-L. Barabasi, *The origin of bursts and heavy tails in human dynamics*, Nature **435**, 207 (2005).

Reprinted with permission from the electronic-only supplement "Young Scientist" (DOI: 10.1140/epjcd/s2006-03-001-7) of The European Physical Journal (EPJ) C

Development of a method for fast, easy and optimized tuning of selection cuts

Modification of the Fisher Linear Discriminant Analysis for use in the low signal-to-noise environments

J. Faivre

INFN sezione di Padova, Padova – ITALY
e-mail: julien.faivre@pd.infn.it

Abstract. In situations where the signal of the analysed particle is tangled up with orders of magnitude more background, its analysis may benefit from the use of a pattern classification method to discriminate the signal out of the background candidates. We present and explain the basic Linear Discriminant Analysis and the modifications brought – the use of cascaded cuts and of a locally optimized criterion – to adapt it to the conditions encountered in the field of heavy ion Physics. We show that this optimized multicut Linear Discriminant Analysis has a higher performance than classical selection cuts and provides a very fast and easy selection cut optimization.

1 Introduction

Particle search in a collision event consists in discriminating the signal (what is wanted) and the background, or noise (fake candidates). This *pattern classification* is achieved through the measurement of several characteristics for each candidate, herein called *cut variables* or *observables*, and the discrimination relies on the fact that the probability distributions of these characteristics are different for the signal and the background.

The simplest method consists in applying a "straight selection cut" on each of these variables separately, as shown in the top of figure 4. We will refer to this approach as *classical analysis* or *classical cuts*. Because these – numerous – cut values are considered as independent parameters while those variables are generally correlated, this method may be very long to tune and usually provides an improvable discrimination. In this article, we will describe the adaptation of Fisher Linear Discriminant Analysis (Fisher-LDA), a pattern classification method widely used in data processing, to the extreme signal-to-noise conditions of the relativistic heavy ion collisions, and show its advantages over the "classical analysis".

The first section explains why such methods are needed for heavy ion Physics, and gives a examples. The second section is a short introduction to pattern classification. Fisher-LDA and its modifications are presented in the third and fourth sections. Finally, the last section explains how the final multi-variable cut is tuned and used in practice, and shows some results.

[1,2] have helped writing paragraphs 3.1 and 4.2.

The method described in this paper has been implemented as a plug-and-play C^{++} class. Its source code and documentation are available upon request to the author.

2 Low signal-to-noise environments

As an example of low signal-to-noise environment, we introduce here briefly the context which led us to use the LDA method. More information about (ultra-)relativistic heavy-ion collisions can be found elsewhere, e.g. [3–5].

2.1 Relativistic heavy-ion collisions

Lattice-QCD predicts that when the energy density of a strongly interacting system of hadrons is large enough, matter should undergo a phase transition from a hadronic state to a quark-gluon plasma (QGP), in which the degrees of freedom are partonic. Parton deconfinement is made possible in such a medium by the strong force screening resulting from the high density of colour charges, similarly to the Debye screening in an electromagnetic plasma.

In this purpose, (ultra-)relativistic heavy ion (Pb-Pb, Au-Au) collisions are made, with $\sqrt{S_{NN}}$ currently ranging from a few GeV (AGS) to 200 GeV (RHIC), and up to 5.5 TeV when LHC starts to run.

Some of the probes of the QGP are based on strangeness and charm, absent in the initial state of the collision. Hadrons containing those quarks decay weakly, which enables an identification up to transverse momenta in the pQCD domain.

Those particles are best studied by reconstructing topologically their secondary decay vertex. This can be achieved in the detectors STAR at RHIC [6], and ALICE at LHC [7], thanks to the tracking subdetectors : in both cases, a time projection chamber (TPC) surrounding several layers of silicon detectors makes possible the reconstruction of the trajectory of the charged particles and their extrapolation towards the primary vertex.

2.2 Cut variables

This paragraph gives examples of cut variables which can be used for discriminating between the signal and the background (fortuitous associations of tracks) in the case of a $\Lambda^0 = uds \rightarrow p\pi^-$ analysis by topological reconstruction.

A weak decay is characterized by a sizeable decay length ($c\tau$ of a hundred microns for charm decays, of a couple of centimeters for strange decays). The reconstruction of a neutral particle decay (V0 vertex) is made by examining all combinations of pairs of opposite charge tracks, and filtering out those (background) which have a geometry incompatible with that of a real particle (signal).

In reality, because of the finite resolution of the detectors, the reconstruction is not perfect and real particles and a significant fraction of the background have a similar geometry. This makes the discrimination challenging, and achievable only statistically : the candidates selected as signal are *mostly* signal, those which are filtered out are *mostly* background. The proportion of signal kept or rejected by the selection process can be estimated by simulation studies for instance.

The projection in two dimensions of the geometry of a V0 vertex is shown in Fig. 1. The charged tracks are bent by the magnetic field (here perpendicular to the figure plane), and because the reconstruction is imperfect, the tracks of the two decay daughters do not cross and the trajectory of the reconstructed parent particle does not meet the primary vertex.

The decay length, the distances of closest approach between the tracks, or between a track and the primary vertex, constitute geometrical variables which can be used to discriminate the background and the signal. Most of these variables are correlated, e.g. the distance of closest approach between a daughter track and the primary vertex is correlated with the Λ decay length.

The cosine of the decay angle ($\cos\theta^*$) is also often used to eliminate the background : the distribution of this variable shows strong peaks at -1 and $+1$ for the background. Examples of other cut variables may be found in [8], and in [9,10] for other analyses.

2.3 Examples of signal-to-noise ratios

Heavy-ion collisions make topological reconstruction of the weak decays a challenging task, because of the high charged track density (multiplicity) in the detectors. The amount of background for a 2-particle decay scales with

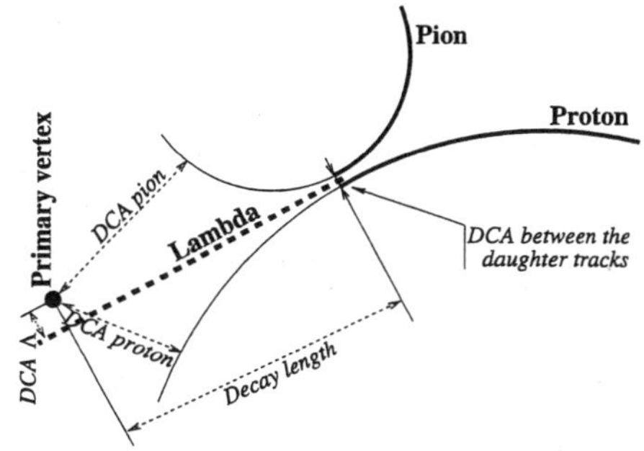

Fig. 1. 2-dimensional geometry of a V0 vertex. The trajectory of each of the daughter particles is a thick solid line, the extrapolations towards the primary vertex are thin solid lines. The trajectory of the reconstructed parent particle is the thick dashed line. DCA stands for "distance of closest approach".

the square of this multiplicity, while for of a 3-particle decay it scales with its cube.

For the case of the $\Omega^- = sss \rightarrow \Lambda^0 K^-$ in STAR's central collisions, the yield of about 0.6 $\Omega + \overline{\Omega}$ per event [11] and the multiplicity of more than 3 000 tracks give an initial signal-to-noise ratio[1] only slightly above 10^{-10}. At the reconstruction stage[2], loose cuts are applied to reduce the computing time and the disk space taken by the storage. While these cuts remove 99.99 % of the combinatorics, the signal-to-noise ratio is still as low as 10^{-6}.

For the $D^0 = c\bar{u} \rightarrow K^-\pi^+$ in ALICE, the initial signal-to-noise ratio is of the order of 10^{-8} [12]. Although this is higher in value than for the Ω, the fact that the signal and background distributions of the geometrical variables differ more in the case of the Ω than in that of the D^0 makes the latter more difficult to reconstruct than the Ω.

Analyses in such extreme conditions, also encountered in the fields of top quark analysis or Higgs search, benefit from the advantages brought by the pattern classification methods. Other fields – industry, health, image processing in general – do not deal with such situations, but rather with poor training statistics and high numbers of observables and/or classes. The methods created for their needs therefore do not meet ours, which made necessary the development of a method adapted to our conditions.

3 Pattern classification

3.1 Short introduction and general procedure

Pattern classification consists in classifying an object (a candidate) in a category (class). The input data are gen-

[1] Here, not calculated in an invariant mass window selecting the signal peak. It can therefore not be compared with the numbers given in Fig. 6.

[2] Reconstruction of the secondary decay candidates from the tracks, themselves reconstructed from the hits in the detectors.

erally : p classes of objects (e.g. signal, noise), n observables defined for all the classes (n is thus the dimension of the space with which we will work), and, for each of the p classes, a sample of N_k objects for training and test, k being the class index. These notations will be kept in the rest of the article. The observables used may be chosen amongst the parameters which are directly measured (see § 2.2 for an example), or may be calculated from these.

The aim is the creation of an algorithm which is able to classify an object into one of the classes defined. A training (or *learning*) phase first optimizes (tunes) the method's parameters until a maximum of candidates whose class is known are classified correctly. A new object, the class membership of which is unknown, can then be presented to the algorithm for classification.

Figure 2 describes the data classification process. The phase involving the detectors is the data collection. In our case, it is the collection of the hits in the subdetectors for example. The phases of segmentation and feature extraction transform this low-level information into mid-level information, smaller in size but more suited to distinguish various classes. For us, segmentation corresponds for example to the track and vertex reconstruction, and feature extraction is the calculation of the various cut variables of the candidates.

Sorting is the phase in which pattern classification methods are involved. It consists in calculating, from the previously mentioned mid-level information, high-level information : only a handful of variables – or even just one – but which contain the relevant information to distinguish signal and background. They are those used for the discrimination between the classes. At this stage, two objects can be compared. Yet, the final decision – classifying the candidate into one of the classes – can be taken only after the post-treatment phase, which takes into account an efficiency and a background rejection, via the minimization of a cost, in the calculation of the decision.

In our case, the number p of classes is two, hereafter called *signal* and *background* (or *noise*) and indexed by $k \in \{1; 2\}$. The signal is made of the real particle, while the background is made of all the other candidates (e.g. combinatorial association of tracks, in the case of particles which decay into two or three daughters).

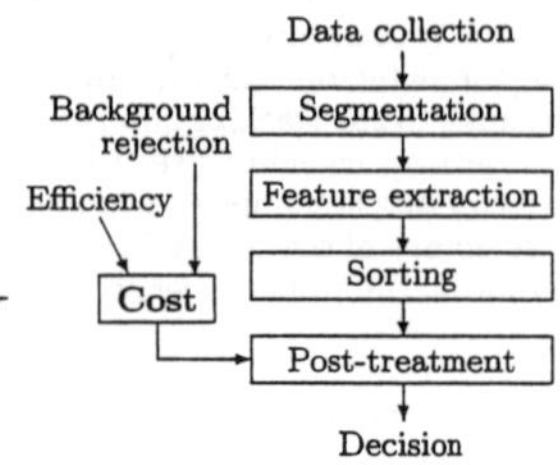

Fig. 2. Data classification process.

3.2 Comparing the performance of various methods

In this paragraph, S and N will refer respectively to an amount of signal and of noise left when cuts are applied.

Here are some variables that can be used as indicators of cuts' performance :

- *Amount of signal S*
- *Efficiency* or *sensitivity* or *detection probability* $\varepsilon_S = \frac{S_{\text{post-cuts}}}{S_{\text{pre-cuts}}}$: proportion of signal that is kept by the cuts
- *Background rejection* $1 - \varepsilon_N = \frac{N_{\text{removed by cuts}}}{N_{\text{pre-cuts}}}$: proportion of background that is rejected by the cuts
- *Signal-to-noise ratio* S/N
- *Purity* or *specificity* $\pi_S = \frac{S}{S+N}$: proportion of kept candidates that actually *are* signal
- *False alarms rate* $\frac{N}{S+N}$: proportion of kept candidates which are actually background
- *Significance* $S/\sqrt{N}$ or $S/\sqrt{S+N}$
- *Relative uncertainty* σ_S/S, where σ_S is the error on S

Our cost function will be the relative uncertainty, as it is the indicator that directly guarantees the smallest possible statistical error on the result[3]. To determine the performance of a method or to compare various methods, it is common to show other indicators as well, usually by pairs :

- Signal with respect to the signal-to-noise ratio
- Signal with respect to purity : this diagram is strictly equivalent to an efficiency-purity diagram, and also strictly equivalent to the diagram mentioned below
- Efficiency with respect to the false alarms rate : this diagram is called "ROC curve" (Receiver Operating Characteristic)
- Relative uncertainty with respect to signal

In such diagrams, all the points that are reachable with a given method, by changing the cuts, define a region, which may be a surface (case of the classical cuts) or a curve (case of most of the other methods). In a signal-S/N or an efficiency-purity diagram, a movement along the curve (or along the border of the surface) inducing an improvement of one of the variables results in a deterioration of the other one. In a diagram showing the relative uncertainty versus the efficiency, the curve is a decreasing, then increasing function which has a global minimum. The latter corresponds to the searched optimal cut.

The performance of a method can then be defined as the minimal relative uncertainty achievable. In other fields though, using the relative uncertainty as the cost function may be irrelevant. In the case of tumor detections for example, one certainly wants to use the tumor detection probability, and set the cut value to a high probability : this selects more background but is safer.

3.3 Some pattern classification methods

Many types of pattern classification methods exist, each of them having several subtypes. We can e.g. mention the

[3] Thus no discrimination criterion will be defined here to compare various pattern classification methods.

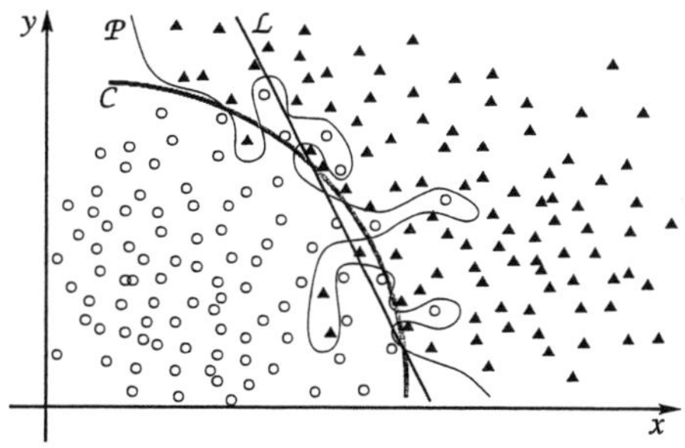

Fig. 3. *Training* samples and shape of various boundaries.

Markov fields, the nearest neighbours methods, the trees, the Parzen windows, the discriminant analyses or the neural networks, on top of unsupervised learning methods which are able to determine themselves how many classes are dealt with. Details can be found in [1]. Some methods have already been applied to particle Physics [9,13–17]. Discriminant analyses themselves can be subdivided into Linear, Quadratic and higher orders Discriminant Analyses, each with several criteria usable for the training, and each with additional possibilities of non-linear data transformations, space dimensionality expansion or reduction, entangled with the basic method.

A common property of all of them is their ability to better discriminate between the signal and the background with respect to the classical method, and therefore to provide better results.

Many of the methods share a second essential advantage over the classical cuts : they provide a transformation of the n-dimensional space of the cut variables to a single output value, and can be seen as functions defined from $\mathbb{R}^n$ to $\mathbb{R}$. While using the classical cuts consists in minimizing a function (the relative uncertainty) of n variables, most pattern classification methods reduce the problem of cut-tuning to the minimization of an only 1-dimensional function.

We decided to give a try to Linear Discriminant Analysis (LDA). Table 1 compares several characteristics of the classical cuts, of the LDA method developed in this article (multicut-LDA, *cf.* § 5), and of the artificial neural networks (ANN), the latter being probably the most used pattern classification method in particle Physics. The positive characteristics are emphasized in bold.

The choice of LDA over a higher order discriminant analysis or over a pattern classification method like the neural networks is justified by its extreme simplicity, which has direct consequences like a better control of how data are handled and selected, as well as a large gain in the amount of time spent on setting up and tuning the method [17–19]. Moreover, although the ANN should reach a higher performance than LDA in theory, choosing the right configuration, the right values of the free parameters and the training method is far from trivial, and in fact may rapidly result in lower performances than what could be expected.

Neural networks also suffer from the huge background statistics : they focus on removing its overwhelming part and leave untouched the comparatively small amount of background which is close to the signal area. This problem can be avoided by cascading several neural networks (it can be seen as an equivalent of multicut-LDA, which naturally solves the problem), each stepping up the S/N ratio by an order of magnitude, but at the cost of an exploding number of parameters of the method : it scales with n^2, while that of LDA scales with n.

The performances of cascaded neural networks, multicut-LDA and classical cuts have been compared for $\Lambda^0 \rightarrow p\pi^-$ topological decays in ALICE. Although almost an order of magnitude more time had been dedicated to the ANN, LDA achieved a better performance [18,19].

For all methods, a test phase is essential to obtain an algorithm which works well, as its performance is not the same if calculated on the training sample or on an independent test sample. The latter performance is always worse than the former, which is biased since the cuts have been optimized for the training sample.

This is illustrated by Fig. 3, in which the distributions of two classes are shown for the *training* sample, as well as three examples of border : a straight line $\mathcal{L}$, a simple curve $\mathcal{C}$ which describes a bit better the boundary between both classes, and a complex parametrization $\mathcal{P}$ that describes the samples almost candidate-by-candidate.

The result of those boundaries on a test sample will be very different : the line will have a fair performance and the simple curve will have a better one, but the performance of the complex curve will be bad, because, while two samples are globally close to identical, they have yet significant local differences. This phenomenon is called *overtraining* when $\mathcal{L}$, $\mathcal{C}$ and $\mathcal{P}$ are actually obtained with the same pattern classification method.

The performance therefore always has to be evaluated on a sample that is independent of the training sample. Because they are able to select very sophisticated shapes in the variables space, fancier, non-linear methods are intrinsically prone to overtraining, while basic LDA is not. We will show that the LDA method developed here is also not, provided a simple condition is respected.

4 Fisher Linear Discriminant Analysis

4.1 Basic principle of LDA

The principle of the LDA method is illustrated by the two drawings of Fig. 4. It has been supposed that two observables, x and y, were accessible to the observer, and the signal and background candidates have respectively been attributed opened circles and closed triangles.

The first drawing shows the behavior of the classical cuts, i.e. straight cuts on one or several of the observables. It has to be kept in mind that we are interested in applications where the number of background candidates is much

Table 1. Comparison of the characteristics of classical cuts, ANN, and the multicut-LDA method presented in this article.

	Classical cuts	Multicut-LDA	Neural networks
Setting up of the method	**Trivial, fast**	**Easy, fast**	Complex, long
Nb. of free parameters to be chosen	Several (or **none**[a])	**One**	Several
Training	**None** (or long and complex[a])	**Simple**	Complex Overtraining
Nb. of parameters tuned during the training	**None (or few**[a]**)**	**Few**	Many
Clarity of the analysis and data treatment	**Under control**	**Under control**	"Black box"
Linear treatment of the observables	**Yes**	**Yes**	No
Gives optimized cuts (in the method's scope)	No[a]	**Yes**	**Yes**
Final tuning to minimize the cost function	Complex, long	**Simple, fast**	**Simple, fast**
Shape of the boundary signal/background	Linear	**Linear, but multicut ⇒ OK**	**Non linear**
Volume selected as being signal	Convex	Convex	**Non connex**

[a] A maximization algorithm of the n-dimensional function can always be set up, but the complexity and processing time is prohibitive as soon as more than half a dozen variables are used [18].

higher than that of signal candidates. When the cuts are chosen loose for the efficiency to be high (solid thick lines, the eliminated region is in grey), the contamination of the signal by the background is high. Tighter cuts (dashed thick lines, the additional region cut is hatched in black) drastically reduce the background, but the price to pay is a small efficiency.

LDA consists in cutting along a linear combination of all the observables, rather than along each of the observables. This linear combination is defined by an LDA direction (or axis). The result, shown in the bottom plot, is a better discrimination between the classes. In an efficiency-purity diagram for example, this translates into a more interesting position than any of the positions accessible with classical cuts.

The algorithm consists in calculating the direction of the axis that gives an optimal discrimination between the classes according to a given criterion. After this training phase, a cut on the axis's coordinate minimizing the cost function is then chosen, which defines as border between both classes a hyperplane perpendicular to the axis.

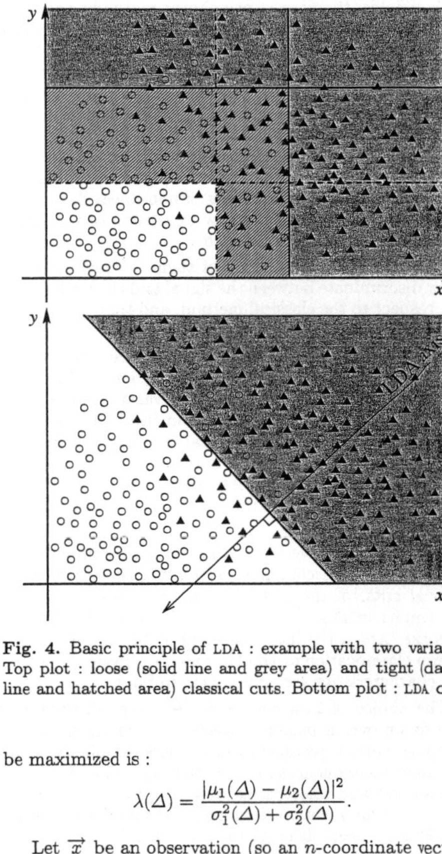

Fig. 4. Basic principle of LDA : example with two variables. Top plot : loose (solid line and grey area) and tight (dashed line and hatched area) classical cuts. Bottom plot : LDA cut.

4.2 Fisher criterion

The most frequently used criterion for the calculation of the axis's direction is the Fisher criterion, which results in what is called Fisher-LDA, introduced by Ronald Fisher in 1936 [20]. The advantage of the Fisher criterion is that, on top of being easy to settle, it gives the exact expression of the direction of the LDA vector, without need for a numerical optimization. There is indeed maximization, but the solution is analytical.

Let's call $\mathcal{D}_k$ the training samples (with e.g. $k = 1$ for the signal and $k = 2$ for the background). The Fisher criterion consists in requiring the best separation of the projections of the classes on an axis Δ defined by $\vec{u}$, i.e. the averages μ_k of those projected distributions should be as far as possible from one another relatively to their squared widths $\sigma_k^2 = \sum_{\vec{x}\in\mathcal{D}_k}(\vec{u}.\vec{x}-\mu_k)^2$, for the overlap between the distributions to be minimal. The criterion to be maximized is :

$$\lambda(\Delta) = \frac{|\mu_1(\Delta) - \mu_2(\Delta)|^2}{\sigma_1^2(\Delta) + \sigma_2^2(\Delta)}. \quad (1)$$

Let $\vec{x}$ be an observation (so an n-coordinate vector). The normalized n-coordinate vector $\vec{u}$ which drives the line Δ characterizes, together with the cut value defined on the axis's coordinate – that is to say, on the scalar product $\vec{x}.\vec{u}$ –, the hyperplane which plays the role of a border between both classes.

Let's call $\vec{m_k}$ the n-dimensional averages of the distributions and write ${}^tM$ for the transposed matrix of a generic matrix M. With $S_B = (\vec{m_1} - \vec{m_2}).{}^t(\vec{m_1} - \vec{m_2})$ the between-class scatter matrix, $S_k = \sum_{\vec{x} \in \mathcal{D}_k} (\vec{x} - \vec{m_k}).{}^t(\vec{x} - \vec{m_k})$ and $S_W = S_1 + S_2$ the within-class scatter matrix, we can write the Fisher criterion matricially :

$$\lambda(\Delta) = \lambda(\vec{u}) = \frac{|\mu_1 - \mu_2|^2}{\sigma_1^2 + \sigma_2^2} = \frac{{}^t\vec{u}\, S_B\, \vec{u}}{{}^t\vec{u}\, S_W\, \vec{u}}. \qquad (2)$$

Maximizing this expression can be done analytically by using the Lagrange multiplier method. A vector $\vec{u}$ maximizing expression (2) obeys : $\exists\, \omega \in \mathbb{R} \,/\, S_W^{-1} S_B \vec{u} = \omega \vec{u}$. ${}^t(\vec{m_1} - \vec{m_2}).\vec{u}$ being a scalar, $S_B \vec{u}$ is always collinear to $\vec{m_1} - \vec{m_2}$, and the expression giving $\vec{u}$ becomes :

$$\exists\, \xi \in \mathbb{R} \quad / \quad S_W^{-1}(\vec{m_1} - \vec{m_2}) = \xi \vec{u}. \qquad (3)$$

The problem is therefore reduced to a matrix inversion.

4.3 Problems with Fisher-LDA

Using the Fisher criterion, even though it is satisfactory for many applications, raises problems in our case. The fact that Fisher relies only on the mean and width of the distributions makes it a "global" criterion, hardly sensitive to the local features of the distributions. The Fisher approach can not succeed in our situations, where the initial S/N is extremely small and the background populates the whole space, including all the signal area. A better discrimination between signal and background requires a local description of the zones where the signal lies and where the background has to be cut. The next section describes how LDA can be improved to meet this requirement without resorting to non-linearity.

5 Optimized multicut-LDA

5.1 Multicut-LDA

A first modification brought to LDA to better cope with the low S/N environments is the application of several successive LDA cuts. This also allows for a finer description of the non-linear boundary between the classes, the same way as a circle can be approximated by an n-sided polygon, all the better as n is high, and yet keeping linear properties to some extent.

The mechanism of this method, which will be called *multicut*-LDA, is depicted in Fig. 5. The first LDA direction is determined by a learning using all the candidates of both samples. A cut value is then determined according to a criterion which will be described in paragraph 5.2, with an efficiency on the signal close to 100 %. This first cut is applied to the learning samples, and a second

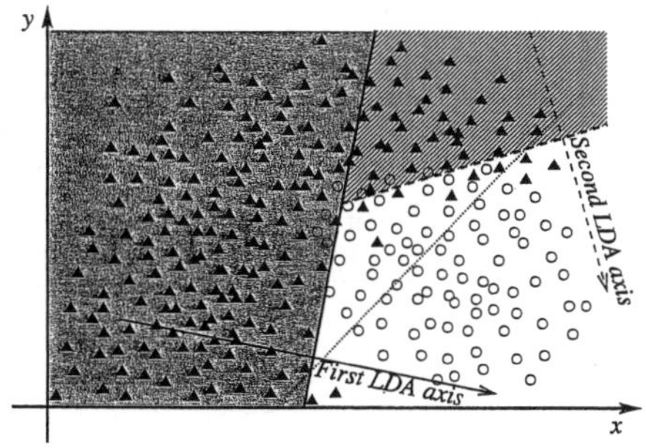

Fig. 5. Mechanism of the multicut-LDA method. The first LDA cut is the thick solid line and eliminates the grey area. The second LDA cut is the thick dashed line and filters out the hatched area. If only one cut had to be used to reach a similar background rejection, its efficiency would be much lower, as shown by the dotted line.

LDA direction is calculated with the remaining candidates. The process is then repeated until not enough candidates remain in the training samples to calculate more LDA axis.

Multicut-LDA therefore provides a set of LDA directions, each being a vector $\vec{u_i}$ of the space of the observables (n coordinates). It also provides a cut value c_i associated to the direction $\vec{u_i}$. The value of c_i depends on $\vec{u_i}$, and the direction $\vec{u_i}$ is a function of $\vec{u_{i-1}}$ and c_{i-1}. Each pair (direction , cut) defines a hyperplane, and this set of hyperplanes demarcates a connex and even convex shape, by construction, in which the candidates are considered as being signal by the algorithm. Further studies can estimate the amount of background selected as signal.

5.2 Optimized criterion

Multicut-LDA can be improved by replacing the Fisher criterion by another one, which takes the local – and not global – behaviour of the distributions into account, therefore more adapted to the multicut method.

Here are two such criteria :

- Optimized criterion I : given an efficiency of the LDA cut on the signal, maximization of the amount of background removed ;
- Optimized criterion II : given a background rejection of the LDA cut, minimization of the amount of signal removed.

Their formulation is antisymmetric for the signal and the background, but, as the criterion II requires a prohibitive computing time to be run, we have not tested if these two criteria give a similar performance. We therefore used only the criterion I.

5.3 Function to maximize

Contrarily to the Fisher criterion, using an optimized criterion requires the implementation of a maximization algorithm, here expressed for the criterion I.

Let ε_{S_i} be the given efficiency (chosen) of the ith cut on the signal and $\mathcal{D}_{S_i}$ the set of candidates of the signal sample after the $i-1$ first cuts. If 1 is assigned to *true* and 0 to *false* in the sum in (4), the number of signal candidates removed by cutting at value c_i along the axis $\overrightarrow{u_i}$ is :

$$S_i - S_{i+1} = (1-\varepsilon_{S_i})S_i = \sum_{\vec{x}\in\mathcal{D}_{S_i}} (\vec{x}.\overrightarrow{u_i} < c_i), \quad (4)$$

with S_i the number of signal candidates (in the training sample) used to determine the ith direction. The value of c_i can also be determined so as to obey the following equality :

$$1 - \frac{\sum_{\vec{x}\in\mathcal{D}_{S_i}} (\vec{x}.\overrightarrow{u_i} < c_i)}{S_i} = \varepsilon_{S_i}. \quad (5)$$

Using directly the number $S_i - S_{i+1}$ to calculate c_i is however more judicious than using the efficiency, because it allows to control the "locality degree" of the optimized criterion. This "locality degree" is determined by a comparison of the numbers of candidates which are removed with two numbers. 1°) The number of signal candidates removed has to be larger than the typical size of the statistical fluctuations for a sample of size S_i, for the algorithm not to trigger on one of those fluctuations. The number of candidates removed should not be too much above this threshold though, as the efficiency of the cut should be kept close to 100 % to take all advantage of the multicut method. 2°) The numbers of signal and background candidates which are removed have to be higher than a fixed absolute number which ensures that the candidates that are removed are numerous enough to be really representative of the actual shape of the distributions in the area that is cut.

In our studies, in 25 dimensions, values of 500 signal candidates removed and above were satisfactory, for samples of 10 000 to 100 000 signal candidates.

Respecting this simple condition guarantees that there is no overtraining, so this problem can basically be considered as absent from the optimized multicut-LDA method.

The function f that is maximized is of course the number of background candidates that are removed by the cut ; hence we can write :

$$\begin{aligned} f : \mathbb{R}^n &\longrightarrow \mathbb{N} \\ \overrightarrow{u_i} &\longmapsto \sum_{\vec{x}\in\mathcal{D}_{B_i}} (\vec{x}.\overrightarrow{u_i} < c_i), \end{aligned} \quad (6)$$

where $\mathcal{D}_{B_i}$ is the set of candidates of the background sample after the $i-1$ first cuts.

The efficiency ε_{S_i} being fixed (it is a chosen parameter), the optimization consists in maximizing f as a function of $\overrightarrow{u_i}$. As the value of c_i depends on $\overrightarrow{u_i}$, it needs to be recalculated at each step.

For the results presented in the next section, the algorithm chosen to maximize f consists in varying each coordinate of the vector $\vec{u}$ at a time, moving the vector by a given angle α (for example 8°). The first coordinate is changed until f has reached a maximum, then the second coordinate is changed, etc... When all n coordinates have been changed, the process is repeated until the vector does not move anymore. Then α is divided by two and the whole algorithm is repeated. One keeps dividing α by two until the gain in number of background candidates removed becomes null or unsignificant.

A common problem to many maximization algorithms is the possibility of being trapped in a local maximum. In our case, the problem is partially resolved by the initial condition : the natural start vector for this algorithm is the direction found with the Fisher criterion, which guarantees that the final result will necessarily be better than with the Fisher criterion. We observed an average improvement with respect to it of around 50 % more background candidates cut per LDA cut, although with strong variations. One can also implement a genetic algorithm, which in principle converges to the global maximum [21].

5.4 Size of the training samples

Determining the minimal statistics needed for the calculation of the LDA directions can be done by calculating the performance for various sizes of the training samples. The performance should rise, with possible oscillations, when the size of the training samples is increased, and saturate when the latter reaches the minimal size necessary for a good determination of the LDA directions.

A lazier way to do, but a priori as reliable, is to check that the performance of the ith LDA cut is better than that of the $i-1^{\text{th}}$ cut tightened beyond the cut value at which the ith cut should begin to be applied. If it is not the case, it is a strong indication that the statistics used to determine the ith direction was not sufficient.

Preliminary studies done with the "lazy method" indicate that 2 000 candidates in each sample, possibly even less, are already enough. This was done with 10 dimensions and is not expected to change with the number of dimensions.

Unlike for Fisher-LDA (see Eq. (1)), when the optimized criterium is used the relative proportion of signal and background candidates in the training samples has no importance.

5.5 Variables used

Because multicut-LDA selects a convex (and therefore connex) signal area, the signal's distributions have to show only one main peak whenever possible, for the method to be efficient.

It is also preferable to use reasonably well shaped distributions, e.g. an angle value is better than using its cosine, as the cosine function will flatten everything towards

1, which may cause the algorithm to fail using that variable in the optimization (the peak would be extremely narrow). Normalizing the distributions' shapes eliminates this constraint, but is a non trivial issue.

The number of variables to use is a study by itself, although not very time-consuming because it should simply be as high as possible, so as to reach the highest possible discrimination. Non-linear combinations of the initial cut variables can be added. The number of variables to use can yet be limited by statistics and processing time reasons – numbers of a few tens are not critical. Methods exist to reduce this number of variables while avoiding a drop in discrimination (see [8]) : under-optimal LDA, Principal Component Analysis (much faster, but unsupervised, matrix-based analysis which reduces the dimension of the working space by taking out the least informative dimensions), or fractional-step LDA [22] (performs dimensionality reduction while keeping a better discrimination than Principal Component Analysis). Principal Component Analysis provides a partial solution to the normalization problem by giving a new base of the space.

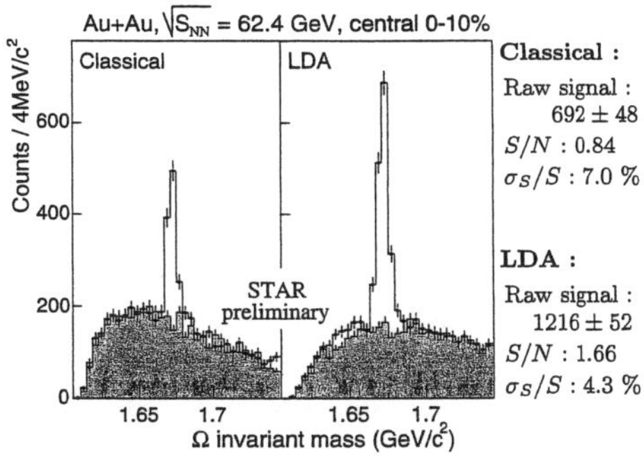

Fig. 6. STAR preliminary $\Omega + \overline{\Omega}$ invariant mass distributions for central Au–Au collisions at $\sqrt{S_{NN}} = 62$ GeV, with classical cuts (left) and LDA cuts (right) [24]. The grey histograms show the amount of background estimated by rotating.

6 LDA practical guide and results

6.1 Results on real data

Optimized multicut-LDA has been tested with 25 cut variables on the Ξ and Ω multi-strange baryons in the 200 GeV Au–Au collisions in STAR and has shown to provide more precise results than the classical cuts [11] : the statistical error on the production yields was 25 % (Ξ) to 40 % (Ω) larger with the classical cuts than with LDA. Furthermore, the transverse mass spectra obtained with LDA had small enough error bars to rule out the formula that was then commonly used to fit it.

The method has then succesfully been applied to STAR's 62 GeV Au–Au collisions [23], and has proven one of its advantages : while classical cuts had to be tediously re-tuned to be adapted to the new – lower than at 200 GeV – track multiplicity, a simple loosening of the LDA cuts calculated in [11] provided, with respect to the re-tuned classical set of cuts, 75 % more signal and a relative uncertainty on the raw amount of signal 1.6 times lower, as shown in Fig. 6. This loosening of the LDA cuts is made by plotting the new valley-shaped curve (relative uncertainty versus efficiency) and determining its minimum. One could also have re-calculated LDA directions with 62 GeV simulation to have possibly more optimized LDA cuts.

6.2 How to do the final tuning

The final tuning consists in finding how many LDA directions have to be applied, and how much is the cut value of the last one, to reach the minimum of the relative uncertainty. In this process, those of the last directions which have not been calculated with enough candidates in the training samples should be discarded. The way the final tuning is done can be illustrated by the example of the $D^0 \to K^-\pi^+$ study in ALICE, in central Pb–Pb collisions at $\sqrt{S_{NN}} = 5.5$ TeV. As no data have been taken yet, the results shown come from a simulation sample independent of the training sample, and the relative and absolute amounts of signal and background have been rescaled to match the amounts that are expected to be found in 10^7 real events. The number of cut variables used is 10, and the number of signal candidates of the training sample to be removed by each LDA cut has been set to 499 (the candidate with $\overrightarrow{x}.\overrightarrow{u_i} = c_i$ is not removed), for an initial sample size of 15 967. One may also choose to remove 500 D^0 for the two first cuts, and then 1 000 or more for the following ones, which have a lower background rejection factor. This strategy was used in [11], as the initial signal training sample was larger.

Figure 7 shows our cost function : with respect to the amount of signal that passes the cuts, the relative uncertainty on that variable. The valley is clearly visible, and finding its minimum is a trivial task. In the zoom presented in the inset, black squares have been put when an additional LDA cut was applied : the rise in performance is very visible, in the form of a steeper slope with an additional direction (left of a point) than without (right).

The corresponding efficiency-purity plot is shown in Fig. 8. Such plots can be used when a different cost function is wanted. One may for example wish to have a higher background rejection to avoid problems due to the background estimation : here, the proportion of background falls down by a factor of more than 3.5 for the same efficiency as the classical cuts. Conversely, a higher efficiency may be desired to reduce possible bias due to tight cuts : here, LDA multiplies the efficiency by 2 for a purity equal to that given by the classical cuts. It can then be checked on the relative uncertainty curve that the statistical error obtained is not too much higher than the minimal one.

A zoom on Fig. 7 is presented in Fig. 9 to illustrate the multicut process. The black curves (LDA) and the black point (classical) have been obtained by assuming a back-

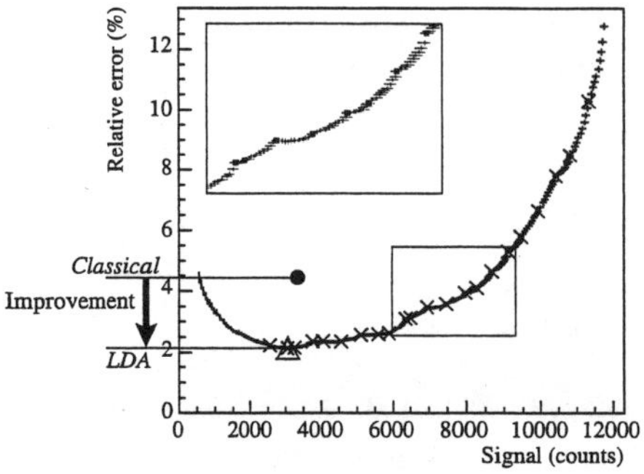

Fig. 7. σ_S/S of the D^0 as a function of the amount of signal, for central Pb–Pb collisions at $\sqrt{S_{NN}} = 5.5$ TeV in ALICE. The closed circle shows the performance reached with the current classical cuts [25], and the valley-shaped curve is the locus of the points described when the LDA cut is gradually tightened (right to left). The crosses (×) mark each addition of a new LDA direction ; the open triangle shows the minimal value reached. A zoom on the boxed area is shown in the inset.

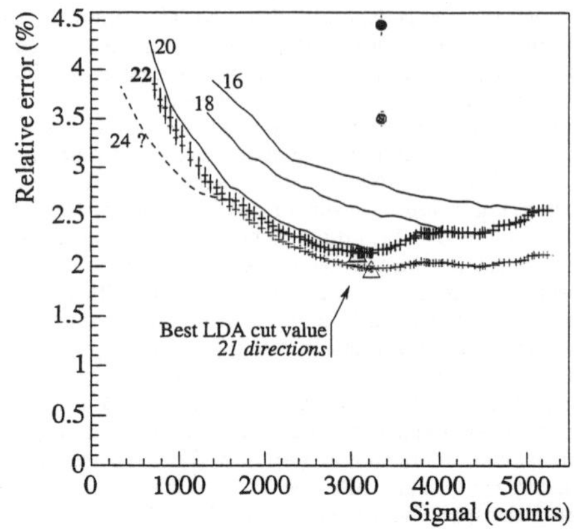

Fig. 9. Zoom on Fig. 7. Black and grey curves and symbols are for background estimation respectively with rotating and event-mixing. The lines show the curves that are obtained with 16, 18 and 20 LDA cuts instead of 22 (with rotating), the dashed line shows how could be this curve when 24 cuts are applied.

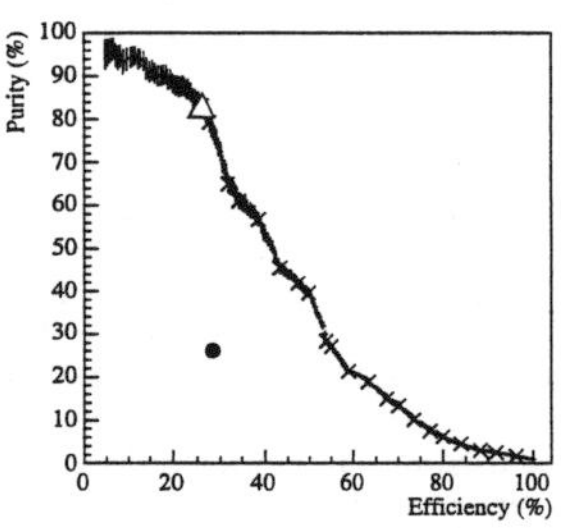

Fig. 8. D^0 efficiency-purity plot corresponding to Fig. 7. The efficiency is that of the cuts only.

ground estimation via rotating, i.e. the V0 vertices are reconstructed with one of the two tracks rotated by 180°. This preserves the combinatorial background and destroys the signal (it can not be reconstructed). The grey curve and point will be explained below. In this relative uncertainty versus efficiency diagram, as already said, each LDA cut results in a valley-shaped curve. A cut along the first direction is applied and progressively tightened until the cut value c_1 is reached (*cf.* § 5.3). At this point, the algorithm has calculated a second LDA direction, which has a better performance than the first one. The second LDA cut is therefore applied and progressively tightened, till the cut value c_2, and so on. The plot shows the end of the process, and the envelope of all the valley-shaped curves is the locus drawn when the LDA cut is gradually tightened (and directions progressively added).

The minimum of this locus, indicated by the opened triangle, is obtained with the searched number of directions, here 21. The optimal number of cuts to use in this case is therefore 21, and the 21st cut value corresponding to the minimal relative uncertainty can easily be determined by the program. It obviously belongs to $]-\infty; c_{21}]$.

The curve obtained with a 22nd direction is rather similar to that obtained by keeping tightening the 21st cut beyond c_{21}. The 22nd direction has been determined with 5 488 signal and 1 120 background D^0 in the training samples, and was the last one calculated for low statistics reasons. If the statistics were enough, further calculated directions should behave like the dashed curve labelled "24 ?" : an improvement is brought, but the relative uncertainty does not fall below the previously found minimal point, as this point is the absolute minimum of the envelope.

Yet, the final value of the LDA cut is not determined at this stage, because the final error bar depends on other factors. There is usually further "manipulation" of the data, such as an efficiency correction and a fit to some distribution. The minimal final statistical error may then be obtained for another value of the LDA cut. The process is identical though, therefore simple and instantaneous, and one may wish to use various LDA cut values, depending on the physics observable looked at.

The grey curve of Fig. 9 illustrates how fast is a retuning : this curve results from the assumption that the estimated background is obtained from event-mixing rather than rotating. Event-mixing consists in reconstructing V0 vertices with one track taken in an event, and the second track taken in another event of similar multiplicity and primary vertex location. Like rotating, this destroys the signal candidates, but preserves the combinatorial background. As many events can be mixed together, event-mixing provides an estimation of the amount of background with a much smaller error bar than rotating (other characteristics have to be taken into account though, but a comparison of the background estimation techniques will not be addressed in this article). This results in a different cost function curve. Again, the new minimum is easily found, as well as the corresponding LDA cut.

As a summary, these preliminary results show that the relative uncertainty on the raw number of D^0 is divided by 2 for a similar efficiency when LDA is used, compared with the classical cuts. When rotating is replaced by event-mixing to estimate the amount of background, the bottom of the valley is not much lower, but is much flatter and extends up to 5 500 signal counts. The cut efficiency can thus be multiplied by 1.8 while the relative uncertainty is left untouched.

Conclusions

Because it uses linear combinations of the n observables, LDA transforms $\mathbb{R}^n$ into a set of segments that is equivalent to $\mathbb{R}$. Cut-tuning is therefore obvious, as it consists in a simple minimization of a one-dimensional function (e.g. the relative uncertainty). Moreover, as the n-dimensional information of the distributions is taken into account, rather than the n projections as the classical cuts do, LDA also provides an improvement of the statistical error bars. While Fisher-LDA can not deal with low initial signal-to-noise ratios such as those considered above, optimized multicut-LDA significantly reduces the statistical uncertainty in the analyses presented. A drawback of this method is that it can be used only with two classes : it is not able to distinguish e.g. several sorts of backgrounds, and removes all of them as if it was a single contribution. Moreover, the selected area is always convex.

The method has fewer parameters to be tuned during the training phase than pattern classification methods like the neural networks. The minimal size of the samples needed to train the method correctly is therefore lower for LDA. Furthermore, LDA has only one parameter to be set by the user – namely, the efficiency of each cut –, which makes the method fast to set up.

LDA can also be used to calculate a systematic error due to the cuts : the LDA cut can be tightened and loosened while permanently staying on the optimized curve shown in Fig. 7. As a set of LDA cuts is easy and fast to determine, results can be obtained with several of them and compared together. Finally, classical cuts can be rapidly derived from the projected distributions of the candidates cut with LDA, and they can for instance also be used to estimate a systematic error.

When an analysis is done under other conditions (e.g. other collision centrality, other range of transverse momentum $p_\perp$), the relative proportion of background may be different, and hence tighter or looser cuts may be needed. While classical cuts require another n-dimensional minimization, previously calculated LDA cuts can simply be adapted to the new environment by a tightening or a loosening, until the new minimum in relative uncertainty is reached. LDA can also be trained with specific candidates (e.g. low-$p_\perp$, mid-multiplicity events, ...) and therefore be optimized for such characteristics.

Acknowledgements

I thank the organizers and lecturers of the Erice school of subnuclear Physics, as well as the juniors for the nice discussions and lively atmosphere. All junior's talks really deserved attention, the proceedings of some of them can be read in [26].

I also warmly thank S. Faisan for the numerous fruitful discussions about pattern classification and related topics.

References

1. R. Duda, P. Hart, D. Stork, *Pattern classification* (John Wiley & Sons, 2001)
2. S. Faisan. Private communication
3. J. Collins, M. Perry, Phys. Rev. Lett. **34**, 1353 (1975)
4. N. Cabibbo, G. Parisi, Phys. Lett. **B59**, 67 (1975)
5. J. Bjorken, Phys.Rev. **D27**, 140 (1983)
6. STAR collaboration, Nucl. Instrum. Meth. **A499**, 624 (2003)
7. ALICE collaboration, J. Phys. G: Nucl. Part. Phys. **30**, 1517 (2004)
8. J. Faivre. Development of a pattern classification method (LDA) to improve signal selection and cuts optimization. STAR note 04xx (2005). To be released at the following URL : `www.star.bnl.gov/STAR/sno/sno.html`
9. T. Carli, B. Koblitz, Nucl. Instrum. Meth. **A501**, 576 (2003)
10. D0 collaboration, Phys. Lett. **B606**, 25 (2005)
11. J. Faivre, Reconstruction and study of the multi-strange baryons in ultra-relativistic heavy-ion collisions at $\sqrt{S_{NN}} = 200\ GeV$ with the STAR experiment at RHIC. Ph.D. thesis, Université Louis Pasteur, Strasbourg (2004). In French
12. A. Dainese, Charm production and in-medium QCD energy loss in nucleus-nucleus collisions with ALICE. Ph.D. thesis, Università degli studi di Padova (2003)
13. B. Knuteson, H. Miettinen, L. Holmström, Comput. Phys. Commun. **145**, 351 (2002)
14. M. Mjahed, Nucl. Instrum. Meth. **A481**, 601 (2002)
15. B. Roe, H.J. Yang, J. Zhu, Y. Liu, I. Stancu, G. McGregor, Nucl. Instrum. Meth. **A543**, 577 (2005)
16. D0 collaboration, Phys. Lett. **B517**, 282 (2001)
17. M. Mjahed, Nucl. Instrum. Meth. **A449**, 602 (2000)
18. L. Gaudichet, Optimization of the reconstruction of strange hadrons in ALICE (2005). Proceedings of the Hadron Collider Physics Symposium, Les Diablerets, Switzerland, July 2005. To be published in Eur. Phys. J
19. L. Gaudichet. Private communication
20. R. Fisher, Ann. Eugen. **7**, 179 (1936)
21. D. Goldberg, *Genetic Algorithms in Search, Optimization and Machine Learning* (Kluwer Academic Publishers, 1989)
22. R. Lotlikar, R. Kothari, IEEE trans. pattern analysis and machine intelligence **22**, 623 (2000)
23. J. Speltz, for the STAR collaboration, J. Phys. G: Nucl. Part. Phys. **31**, S1025 (2005)
24. J. Speltz. Private communication
25. ALICE collaboration, ALICE physics performance report, vol. 2. CERN/LHCC 2005-030, to be published
26. Proceedings of the Erice school of subnuclear Physics (2005)

Reprinted with permission from the electronic-only supplement "Young Scientist" (DOI: 10.1140/epjcd/s2006-03-002-6) of The European Physical Journal (EPJ) C

N=2 Superparticles, RR Fields, and Noncommutative Structures of (super)-Spacetime

P. A. Grassi[1,2,3],

[1] Centro Studi e Ricerche E. Fermi, Compendio Viminale, I-00184, Roma, Italy,
[2] CERN, Theory Unit, Physics Department, 1211-CH Geneva 23, Switzerland,
[3] DISTA, Università del Piemonte Orientale, via Bellini 25/g, I-15100, Alessandria, Italy.

Abstract. The recent developments in superstring theory prompted the study of non-commutative structures in superspace. Considering bosonic and fermionic strings in a constant antisymmetric background yields a non-vanishing commutator between the bosonic coordinates of the spacetime. Likewise, the presence of constant Ramond-Ramond (RR) background leads to a non-vanishing anti-commutator for the Grassmann coordinates of the superspace. The non-vanishing commutation relation between bosonic coordinates can also be derived using a particle moving in a magnetic background, we use superparticle to show how the non-commutative structures emerge in superspace. The derivation is original and it is shown that only a $D0$-brane in supergravity background reproduces the results obtained in string theory.

PACS. PACS-key discribing text of that key – PACS-key discribing text of that key

1 Introduction

During the last years, several new ideas emerged from the marriage of non-commutative geometry to quantum field theory and string theory. This is due to the discovery that the spacetime generated by strings propagating on a non-trivial background [1–5] is non-commutative. Furthermore, string theory provides a meaningful way to construct quantum field theories on non-commutative spaces.

Nevertheless the history of non-commutative geometry and non-commutative structures of spacetime is definitely longer, and it has its roots in quantum mechanics. Indeed, it was first realized by Peierls using non-relativistic quantum mechanics that the motion of a charged particle in presence of non-trivial external magnetic fields can be described by a free Hamiltonian assuming non-vanishing commutation relations for the coordinates [6–8].

The simplest example of non-commutative spacetime [9,10] is represented by the Heisenberg algebra of the coordinates

$$[x^m, x^n] = i\,\theta^{mn}, \qquad (1)$$

where θ^{mn} is a constant antisymmetric tensor. Based on this prototype, it has been developed an enormous amount of new mathematics (which we are not going to review here, and we refer to [11] for the most recent work on a wide class of non-commutative structures used in quantum field theory and to their collection of references). The commutation relations (1) can be recovered from the Hamiltonian of a charged particle moving in a background of magnetic field in the massless limit $m \to 0$. In that limit the particle is confined in the lowest Landau level and this limit can consistently be taken if some constraints on the momenta are imposed. These are second-class constraints that have to be treated using the Dirac brackets and this yields the commutation relations (1). We briefly review this model in Sec. 2. Then, we move to the supersymmetric version.

As is well-known, bosonic particles and bosonic string theories are not sufficient to provide a complete description of particle physics since they do not include fermionic degrees of freedom (except maybe only for some unphysical ghost fields). Thus, we have to extend the bosonic theory to a fermionic one. There are essentially two ways to do it: adding some fermionic (anticommuting) *worldline* spinors ψ^m (or worldsheet spinor in the case of superstrings), or adding some fermionic *target-space* spinors θ^α [12,13]. In the former case supersymmetry on the worldsheet has to be imposed for a consistent formulation of the model, whereas for the latter case one has to impose a new gauge symmetry, known as κ-symmetry [14], and this leads to supersymmetry in the target space. We recall the basic ingredients of this superparticle model in Sec. 3.

The quantization of superparticle is unfortunately very problematic. The action is obtained from the bosonic one by replacing the momentum Π^m with its supersymmetric version $\Pi^m = \dot{x}^m + \theta\gamma^m\dot{\theta}$, and this leads to fermionic constraints since the momentum p_α is algebraically related to its conjugated variable θ^α. However, these constraints mix first-class constraints – which generate the κ-symmetry – with second-class constraints and there is no Lorentz-covariant way to separate the twos. Several procedures were conceived to covariantly quantize these models (see for example [15] and the references therein), but most of them were nonpractical for computations and were abandoned.[1]

On the other hand, the recent work by N. Berkovits [17] provides a new technique to handle the quantization of the superparticle and the superstring theory. In this new framework, the action of the superparticle is replaced by a free action and the physical states are constructed using a BRST charge acting on the Hilbert space of free fields. To be more precise, some of fields are not really free. Indeed, to define a BRST charge Q, one has to introduce new degrees of freedom which play the role of ghosts here denoted by λ^α, and the nilpotency of Q implies the quadratic constraints

$$\lambda^\alpha \gamma^m_{\alpha\beta} \lambda^\beta = 0 \,, \tag{2}$$

where $\gamma^m_{\alpha\beta}$ are the Dirac matrices in the Majorana basis and they are symmetric (in 10 dimensions). The spinors satisfying eq. (2) are known as *pure spinors* and the formalism is now denoted as Pure-Spinor Formulation. Here, we consider only the Pure Spinor formulation of superparticle and $D0$-branes and we refer to [18–20]. The possibility to remove the constraint by adding new ghost fields has also been considered [23–25], and in the specific case of superparticle this was explored in [26]. However, for the purposes of the present work we will use the Pure-Spinor formulation, whose basic ingredients will be reviewed in Sec. 3.2. Thus, given a consistent way to quantize the superparticle we can study the spectrum and the interactions.

At the massless (lowest) level string theory can be described by an effective theory of supergravity and the spectrum consists of a bosonic sector with the graviton G_{mn}, the NS-NS antisymmetric tensor B_{mn} , the dilaton ϕ and a set of p-forms F_p[2], and a fermionic sector, a.k.a. NS-R or R-NS sector, which contains the gravitinos Ψ^α_m (see [27] for a complete reference). The interest of superparticles in this context is due to the fact that they can be viewed as truncations of string theory to the massless sector. Therefore these models are useful to deduce some general aspects of string theory such as the spectrum of the massless modes, their equations of motion, and some radiative corrections, even if they can be used only a limited amount of amplitude computations [28,29].

To be more precise, the N=1 d=10 superparticle describe the multiplet of N=1 super-Yang-Mills theory. The spectrum is characterized by the gluon (8 on-shell dofs) and the gluino (8 fermionic dofs). It is formulated in the superspace, but there are no auxiliary fields since the multiplet is on-shell. An N=2 superparticle in 10 dimensions describes the on-shell modes of N=2 supergravity, namely 64 bosonic and 64 fermionic degrees of freedom.

Whereas string theory can be consistently formulated only in 10 dimensions, superparticle models can be formulated also in lower dimensions.[3] These models in lower dimensions are easier to be used since the BRST conditions for physical states does not put the theory on shell and there is a wider range of consistent backgrounds (vacua).

For what concerns the interactions we have to recall that the superparticle as well as superstrings couple to their own background. This means, for instance, that N=2 d=4 superparticle couples to N=2 d=4 supergravity. In Sec. 3.3 this is described in detail. More important, we have to underline that N=2 d=4 supergravity is characterized by a graviton, two gravitinos and a RR field (known as graviphoton in the literature) and the coupling with N=2 d=4 superparticle is dictated by the BRST symmetry. In addition, since for the d=4 model the supergravity does not need to be on-shell we can choose to set to zero all background fields except the RR field.

We will show later how the RR fields lead to deformations of anticommutative structure of superspace. But before describing this result, it is worth to say few words about superspace.

Let us remind the reader that *superspace* [30] is a powerful technique to handle supersymetric theories, it is characterized by the bosonic coordinates of manifold and a set of Grassmann coordinates in the spinor rapresentation of the Lorentz group. The superspace technique provides *1)* a very compact way to write the equations of motion for the entire supersymmetric multiplet, *2)* an extremely economic way to compute Feynamn diagrams taking into account supersymmetry and, *3)* a guideline to construct effective actions of supersymmetric theories. Finally, superspace is naturally embedded in the Pure Spinor formulation of string theory.

Now, we are finally in the position to study the supersymmetric analog of noncommutative geometry of bosonic theory (1). At this time, we study the deformation of the anticommutator between fermionic coordinates θ^α [36,37]. We have to recall that there are several studies in that direction [38–40] where the second-class constrained were used to show that there is a fundamental non-commutative superspace in the quantization of superparticle. This is reviewed in this new formulation, for two reasons: *1)* the pure spinor quantization method is the only consistent way to quantize superparticle without losing the super-Poincaré invariance; *2)* in Sections 4 and 4.1 the role of the RR fields in the present analysis is shown.

However, only recently [41–43], using the pure spinor formulation of string theory, the RR fields have been shown to deform the anticommutation relations as follows

$$\{\theta^\alpha, \theta^\beta\} = \alpha'^2 F^{\alpha\beta} \,, \tag{3}$$

[1] It is important to mention that the superparticle can be quantized using the light-cone gauge. In that case the spectrum can be easily computed and tree level computations can be performed [16]. However, there are several limitations to go beyond this point because of lacking of Lorentz covariance.

[2] In the case of N=2 d=10 supergravity there are two possibilities: type IIA with F_2, F_4 and type IIB with F_1, F_3, F_5^+ (where the last form is selfdual).

[3] Recently, it has been discovered that also superstrings can be formulated in lower dimensions [31–34] and these models can be viewed either as the uncompactified part of 10 dimensional superstrings or as the non-Liouville sector of non-critical superstrings [35].

where $F^{\alpha\beta}$ are the RR fields of N=2 supergravity. It is discussed the implications of these new anticommutator relations on quantum field theory in paper [44] and in the several papers that followed it. The relation (3) can also be derived in the context of quantum mechanics, or better in the context of superparticle and this is the purpose of this note. We show that in order to reproduce the relation (3) we have to use a peculiar type of superparticle known as $D0$-brane. The pure spinor formulation is used again, and it is shown that the quantization of the D0 brane leads to non-commutativity in the superspace.

In the paper, we also discuss the perturbation theory and the limit where the RR fields can be taken either very weak or very strong. Some interesting results emerge from this preliminary analysis and further developments will be discussed elsewhere.

2 A Particle in a Magnetic Field

We briefly review the formulation of a non-relativistic massive and charged particle in presence of magnetic field. We will do it in a generic dimension and the position of the particle is described by its coordinates x^m (with $m = 1, \dots, d$). We introduce a constant background $B_{mn} = -B_{nm}$, thus, the action reads ($\dot{A} = \partial_\tau A$)

$$S = \int d\tau \Big(\Pi_m \dot{x}^m - \frac{1}{2m} \Pi_m \Pi^m + B_{mn} \dot{x}^m x^n \Big) . \quad (4)$$

The indices are raised and lowered with the flat metric η_{mn}. The conjugate momentum P_m can be easily computed and it gives $P_m = \Pi_m + B_{mn} x^n$. Now, if we impose the quantization rules $[P_m, x^n] = i\delta_m^{\ n}$, we have that

$$[x^m, x^n] = 0 \,, \quad [\Pi_m, x^n] = i\delta_m^{\ n} \,, \quad [\Pi_m, \Pi_n] = iB_{mn} \,. \quad (5)$$

The other equations of motion are

$$m\dot{x}^m = \Pi^m \,, \qquad \dot{\Pi}^m + B^m_{\ n}\dot{x}^n = 0 \,. \quad (6)$$

For this equations, it follows that $\dot{P}_m = 0$. Therefore, in order to take the limit $m \to 0$, we have to impose the constraints

$$\Pi^m \approx 0 \,, \quad (7)$$

which are second-class constraints. They have to be treated using the Dirac brackets (see [46] for the definition of Dirac brackets) and this leads to the non-commutation relations for the coordinates

$$[x^m, x^n]_D = (B^{-1})^{mn} \,, \quad (8)$$

where the subscript denotes the Dirac brackets.

This model is interesting for three aspects: *1)* it represents a simple solvable model of a particle moving in a non-trivial background; *2)* it gives the non-commutative relations between the coordinates of the spacetime and, finally, *3)* it requires Dirac brackets for its quantization. All these ingredients will be found again in the subsequent sections.

3 Superparticles and the BRST symmetry

3.1 Action and κ-symmetry

We use Dirac basis for gamma matrices, and the spacetime is taken to be 4 dimensional. The field content is represented by the bosonic coordinates x^m where $m = 0, \dots, 3$, two anticommuting Dirac spinors $\theta^\alpha_L, \theta^\alpha_R$ with $\alpha = 1, \dots, 4$ and their conjugate momenta $P_m, p_{L\alpha}$ and $p_{R\alpha}$. Since we are considering an N=2 model, we have introduced the notation L/R to distinguish between the two flavours of the spinors. In the case of d=4, there is no distinction between type IIA/B since the theory is not chiral in the present case. The Dirac matrices $\gamma^m_{\alpha\beta}$ are the usual 4×4 matrices and satisfy the Fierz identities $\gamma_{m,(\alpha\beta}\gamma^m_{\gamma)\delta} = 0$.

Let us consider the superparticle action [12–14]

$$S = \int d\tau (P_m \Pi^m - \frac{e}{2} P_m P^m) \quad (9)$$

in the first order formalism with

$$\Pi^m = \dot{x} + \frac{i}{2}\theta_L \gamma^m \dot{\theta}_L + \frac{i}{2}\theta_R \gamma^m \dot{\theta}_R \,. \quad (10)$$

This action is invariant under the κ-symmetry and under the reparametrization of the worldline

$$\delta\theta^\alpha_L = (\not{P}\kappa_L)^\alpha \,, \quad \delta\theta^\alpha_R = (\not{P}\kappa_R)^\alpha \,, \quad \delta P^m = 0$$

$$\delta x^m = \zeta P^m + \frac{i}{2}\Big(\theta_L \gamma^m \not{P}\kappa_L + \theta_R \gamma^m \not{P}\kappa_R\Big) \,, \quad (11)$$

$$\delta e = \dot{\zeta} + 2i\Big(\dot{\theta}^\alpha_L \kappa_{L\alpha} + \dot{\theta}^\alpha_R \kappa_{R\alpha}\Big) \,.$$

where $\kappa_{L/R}$ are the infinitesimal gauge parameters of κ-symmetry and ζ is the parameter for diffeomorphisms.

From the action (9), we deduce the fermionic constraints

$$d_{L\alpha} = p_{L\alpha} + \frac{i}{2} P_m (\gamma^m \theta_L)_\alpha \approx 0 \,, \quad (12)$$

$$d_{R\alpha} = p_{R\alpha} + \frac{i}{2} P_m (\gamma^m \theta_R)_\alpha \approx 0 \,,$$

which satisfy

$$\{d_{L\alpha}, d_{L\beta}\} = P_m \gamma^m_{\alpha\beta} \,, \qquad \{d_{R\alpha}, d_{R\beta}\} = P_m \gamma^m_{\alpha\beta} \,, \quad (13)$$

$$\{d_{L\alpha}, d_{R\beta}\} = 0 \,.$$

The last equations are obtained using the canonical commutation relations $[P_m, x^n] = i\eta_m^{\ n}$, $\{p_{L\alpha}, \theta^\alpha_L\} = -i\delta_\alpha^{\ \beta}$, $\{p_{R\alpha}, \theta^\beta_R\} = -i\delta_\alpha^{\ \beta}$.

We have to notice the following facts: *1)* there are first- and second-class constraints generated by the operators $d_{L\alpha}$ and $d_{R\alpha}$, and they cannot be disentangled without breaking Lorentz covariance, so the technique of Dirac brackets cannot be used here, *2)* the second-class constraints of the superparticle have been used in [38–40] to derive non-(anti)commutation relations among the fermionic coordinates of the superspace. However, since as it stands the superparticle model cannot be quantized we use the Pure-Spinor Formulation.

3.2 Quantization

We briefly review the Pure-Spinor formulation of superparticle [18–20].

We introduce the commuting spinors λ_L^α and λ_R^α, which satisfy the pure spinor conditions

$$\lambda_L \gamma^m \lambda_L = 0\,, \qquad \lambda_R \gamma^m \lambda_R = 0\,, \tag{14}$$

and their conjugate momenta $w_{L\alpha}, w_{R\alpha}$.

We define the BRST operators

$$Q_L = \lambda_L^\alpha d_{L\alpha}\,, \qquad Q_R = \lambda_R^\alpha d_{R\alpha}\,. \tag{15}$$

They have the usual form *ghost* × *constraint*. Due to pure spinor constraints (14), they are nilpotent up to the gauge transformations of $w_{L\alpha}, w_{R\alpha}$ with the local parameters Λ_L and Λ_R given by

$$\delta w_{L\alpha} = \Lambda_{Lm}(\gamma^m \lambda_L)_\alpha\,, \qquad \delta w_{R\alpha} = \Lambda_{Rm}(\gamma^m \lambda_R)_\alpha\,. \tag{16}$$

These gauge transformations remove the degrees of freedom from the spinors $w_{L\alpha}$ and $w_{R\alpha}$ to match those of the pure spinors λ_L^α and λ_R^α. Following the usual prescription of the BRST quantization rules, we can define the quantum action as follows [47]

$$S_0 = \int d\tau (P_m \Pi^m - \frac{1}{2} P_m P^m) + \tag{17}$$

$$+ Q_L \int d\tau w_{L\alpha} \dot\theta_L^\alpha - Q_R \int d\tau w_{R\alpha} \dot\theta_R^\alpha\,.$$

Even if it seems the usual BRST procedure, we have to notice that the BRST operators Q_L and Q_R are nilpotent only up to gauge transformations (16). This compensates the fact that the Brink-Schwarz superparticle action (9) is not invariant under the BRST transformations. In addition, we can always add to the action BRST invariant terms. The reparametrization is fixed by the gauge condition $e = 1$, and we have to add the corresponding ghosts $\int d\tau b\dot c$. However, there is no procedure to get (17) from an honest gauge fixing of the action (9) (a suggestion how this might work is given in [47,48]).

By exploiting the different contributions in (17), we obtain

$$S_0 = \int d\tau \Big(\dot\theta_L^\alpha p_{L\alpha} + \dot\theta_R^{\bar\alpha} p_{R\bar\alpha} + P_m \dot x^m - \frac{1}{2} P_m P^m +$$

$$- w_{L\alpha} \dot\lambda_L^\alpha - \dot\lambda_R^\alpha w_{R\alpha} \Big)\,, \tag{18}$$

which is BRST invariant and invariant under the gauge transformation (16) if the spinors $\lambda_L^\alpha, \lambda_R^\alpha$ are pure. The action is also invariant under supersymmetry transformations generated by $Q_\epsilon = \epsilon_L^\alpha q_{L\alpha} + \epsilon_R^\alpha q_{R\alpha}$ where

$$q_{L\alpha} = p_{L\alpha} - \frac{i}{2} P_m (\gamma^m \theta_L)_\alpha\,, \tag{19}$$

$$q_{R\alpha} = p_{R\alpha} - \frac{i}{2} P_m (\gamma^m \theta_R)_\alpha\,,$$

which anticommute with the BRST operators Q_L and Q_R.

The physical states are identified with the BRST cohomology at ghost number 1 and the cohomology is computed by the following equations

$$Q_L|\psi> = 0\,, Q_R|\psi> = 0\,, \tag{20}$$

$$|\psi> \neq Q_L|\Omega_L> + Q_R|\Omega_R>$$

with $Q_R|\Omega_L> = Q_L|\Omega_R> = 0$. The physical state $|\psi>$ has ghost number one and the parameters of the gauge transformations $|\Omega_{L/R}>$ have ghost number zero. The states $|\psi>$ are obtained by acting with normal-ordered combinations of operators $x^m, \theta^\alpha, \ldots\ldots$ on the vacuum $|0>$. The complete analysis of eqs. (20) in d=10 N=2 case has been given in [49,50], and based on those results it can be shown that the solution of these equations yields the off-shell multiplet of N=2 d=4 supergravity.

3.3 Coupling the superparticle to the background

As illustrated in [20] the superparticle $N = 2$ can be coupled to a $N = 2$ supergravity background. The deformation of the action $S + \int d\tau V$ has to be BRST invariant in order to define gauge invariant correlation functions. For constant backgrounds, the BRST invariant action is given by

$$S_R = S_0 + \int d\tau \Big(P^m g_{mn} P^n + B_{mn} L^{mn} + \Psi_{Lm}^\alpha q_{L\alpha} P^m + \tag{21}$$

$$+ q_{R\bar\alpha} P^m \Psi_{Lm}^\alpha + i\, q_{L\alpha} F^{\alpha\beta} q_{R\beta} \Big)\,,$$

where g_{mn} is the usual metric deformation, B_{mn} is the NS-NS two form, Ψ_{Lm}^α and Ψ_{Rm}^α are the gravitinos and $F^{\alpha\beta}$ are the R-R field strengths. The fields $P_m, q_{L\alpha}, q_{R\bar\alpha}$ and

$$L^{mn} = P^{[m} x^{n]} + \frac{1}{2} p_L \gamma^{mn} \theta_L + \frac{1}{2} p_R \gamma^{mn} \theta_R + \tag{22}$$

$$\frac{1}{2} w_L \gamma^{mn} \lambda_L + \frac{1}{2} w_R \gamma^{mn} \theta_R\,,$$

are BRST invariant. As a consequence, the action (21) is invariant if the backgrounds are constant. Given that, we can obtain the action (21) by an equation similar to (17) (see [47,20]). In the following we will set all background fields to zero except for the RR graviphoton and the metric G_{mn} and we will take them to be constant.

The advantage of working in 4 dimensions is due to weaker constraints to which the background has to satisfy and the absence of backreaction. Indeed, as is been shown in [32], in d=4 the BRST cohomology implies only that the background fields belong to off-shell supermultiplets and no equations of motion are necessary (see also [21]). In the case of closed superstrings [22], it can be shown that the BRST conditions implies only some kinematical restrictions on the background. For that reason, one can choose suitable background enforcing the absence of the backreaction.

4 RR fields and Non-commutative Superspace: Act I

Setting the background fields B_{mn} and the gravitinos Ψ^{α}_{Lm}, $\Psi^{\hat\alpha}_{Rm}$ to zero and we assume that $F^{\alpha\beta} = F^{\beta\alpha}$, we obtain the new action

$$S_R = \int d\tau \Big[\dot\theta^{\alpha}_L p_{L\alpha} + \dot\theta^{\alpha}_R p_{R\alpha} +$$

$$+ i\left(p_{L\alpha} - \frac{i}{2}P_m(\gamma^m\theta_L)_\alpha\right)F^{\alpha\beta}\left(p_{R\beta} - \frac{i}{2}P_m(\gamma^m\theta_R)_\beta\right)\Big]$$

$$+ \int d\tau \Big[P_m\dot x^m - \frac{1}{2}G_{mn}P^mP^n - w_{L\alpha}\dot\lambda^{\alpha}_L - \dot\lambda^{\alpha}_R w_{R\alpha}\Big], \quad (23)$$

where $G_{mn} = \eta_{mn} + g_{mn}$. The presence of RR fields breaks the supersymmetry. The amount of supersymmetry preserved in this background is given by the equations

$$P_m F^{\alpha\beta}\gamma^m_{\beta\gamma}\epsilon^{\gamma}_R = 0, \qquad P_m F^{\alpha\beta}\gamma^m_{\alpha\gamma}\epsilon^{\gamma}_L = 0, \qquad (24)$$

which are the usual Killing equations for spinors if one redefines the supersymmetry parameters with $\epsilon'_L = \not P\epsilon_L$ and $\epsilon'_R = \not P\epsilon_R$, for off-shell momentum P_m. No contribution is added to the ghost action and this simplifies the analysis.

From the action (23) we can derive the equations of motion for θ^{α}_L and θ^{α}_R

$$-\dot\theta^{\alpha}_L + i\,F^{\alpha\beta}p_{R\bar\beta} + \frac{1}{2}(F\not P)^{\alpha}{}_{\beta}\theta^{\beta}_R = 0, \qquad (25)$$

$$-\dot\theta^{\alpha}_R - iF^{\beta\alpha}p_{L\beta} - \frac{1}{2}(F\not P)_{\beta}{}^{\alpha}\theta^{\beta}_L = 0,$$

which can be solved in terms of $p_{L\alpha}$ and $p_{R\bar\alpha}$. We assume for the time being that $F^{\alpha\beta}$ is an invertible matrix. On the contrary, if F is not invertible on the spinor space, one can decompose any spinor into a part belonging to $\ker(F)$ and to $\ker(F)^{\perp}$. The spinors belonging to the kernel of F do not enter the coupling term in (23) and therefore can be treated separately, in that case there is a residual supersymmetry.

Notice that the RR field $F^{\alpha\beta}$ plays the role of the mass in the case of non-relativistic charged particle in Sec. 2. Therefore, we are interested in studying the limit $||F|| \to \infty$, where $||\cdot||$ means a measure of the intensity of the RR field strength.[4] Moreover, the RR fields does not seem to play the role of the magnetic field B_{mn} of Sec. 2. Indeed, the anticommutation relations among the fermionic constraints $d_{\alpha L}$ and $d_{\alpha R}$ do not contain the field $F^{\alpha\beta}$ in contrast to the corresponding constraints Π^m of the bosonic case.

To exploit the analogy between the mass term of (4) and the RR-dependent terms of (23), we further manipulate the action. Substituting (25) in the action (23), one obtains

$$S_R = \int d\tau \Big[i\dot\theta^{\alpha}_L F^{-1}_{\alpha\beta}\dot\theta^{\beta}_R + \frac{i}{2}\theta_L\not P\dot\theta_L + \frac{i}{2}\dot\theta_R\not P\theta_R\Big] + \qquad (26)$$

$$\int d\tau \Big[P^m\dot x^m - \frac{1}{2}P^mP_m - w_{L\alpha}\dot\lambda^{\alpha}_L - \dot\lambda^{\alpha}_R w_{R\alpha}\Big].$$

The first term is a kinetic term for the fermions which is quadratic in the derivatives. This is an usual term for spinors, but it is always present in string models in superspace. The second and the third term in the action resemble the spinorial part of the Brink-Schwarz action.

This action suggests that this is a superparticle moving on a supergroup manifold with coordinates x^m, θ^{α}_L and $\theta^{\hat\alpha}_R$. Denoting by $\mathcal{Q}_{L\alpha}$, $\mathcal{Q}_{R\bar\alpha}$ and by $\mathcal{P}_m$ the abstract generators of the algebra $\{\mathcal{Q}_{L\alpha}, \mathcal{Q}_{L\beta}\} = -\gamma^m_{\alpha\beta}\mathcal{P}_m$ and $\{\mathcal{Q}_{R\alpha}, \mathcal{Q}_{R\beta}\} = -\gamma^m_{\alpha\beta}\mathcal{P}_m$, we find the following MC forms

$$g^{-1}dg = (dx^m + d\theta_L\gamma^m\theta_L + d\theta_R\gamma^m\theta_R)\mathcal{P}_m + \qquad (27)$$

$$+ d\theta^{\alpha}_L\mathcal{Q}_{L\alpha} + d\theta^{\alpha}_R\mathcal{Q}_{R\alpha},$$

and the metric for the algebra have the following nonvanishing entries

$$(\mathcal{P}_m, \mathcal{P}_n) = g_{mn}, \qquad (\mathcal{Q}_{R\alpha}, \mathcal{Q}_{L\beta}) = F^{-1}_{\alpha\beta}. \qquad (28)$$

Notice that the MC forms are not supersymmetric invariant since the supersymmetry is broken by the presence of the RR fields. Eliminating P_m from the action (26) and using the metric given in (28), the action (26) can be written as

$$S_R = \int d\tau \Big[(g^{-1}\dot g, g^{-1}\dot g) - w_{L\alpha}\dot\lambda^{\alpha}_L - \dot\lambda^{\bar\alpha}_R w_{R\bar\alpha}\Big]. \qquad (29)$$

This is similar to the result found in [52], where the authors showed that in the case of $AdS_3 \times S^3$, one finds a sigma model on a supergroup. It is interesting that the same situation is reproduced in the present context (it should also be possible to do it for string theory in 10 dimension with constant RR fluxes as studied in [53,54,41], but nobody found a convenient set of variables yet).

After the conjugated momenta $p_{\alpha L/R}$ are removed, the equations of motion for the spinors read

$$\ddot\theta^{\alpha}_L + \frac{1}{2}F^{\alpha\bar\beta}\,\not P_{\bar\beta\bar\gamma}\dot\theta^{\bar\gamma}_R = 0, \qquad (30)$$

$$\ddot\theta^{\bar\alpha}_R - \frac{1}{2}\dot\theta^{\gamma}_L\,\not P_{\gamma\beta}F^{\beta\bar\alpha} = 0.$$

Since on-shell we have $\dot P_m = 0$ and since we choose constant RR field strengths, we can integrate once the above equations to get

$$\dot\theta^{\alpha}_L + \frac{1}{2}F^{\alpha\bar\beta}\,\not P_{\bar\beta\bar\gamma}\theta^{\bar\gamma}_R = C^{\alpha}_L, \qquad (31)$$

$$\dot\theta^{\bar\alpha}_R - \frac{1}{2}F^{\bar\alpha\beta}\,\not P_{\beta\gamma}\theta^{\gamma}_L = C^{\bar\alpha}_R.$$

where $C_{L/R}$ are integration constants to be fixed by boundary conditions (in order to avoid any new constant noncovariant quantity we choose to set them to zero). Inserting (31) into (30), we arrive at the decoupled equations for θ^{α}_L and $\theta^{\bar\alpha}_R$

$$\ddot\theta^{\alpha}_L - \frac{1}{4}(F\not PF^T\not P)^{\alpha}{}_{\beta}\theta^{\beta}_L = 0, \qquad (32)$$

[4] The relation between lowest Landau levels and RR fields is explored in [51]

$$\ddot{\theta}^{\bar{\alpha}}_R - \frac{1}{4}\theta^{\bar{\beta}}_R(\not{P}F^T \not{P}F)^{\bar{\alpha}}_{\bar{\beta}} = 0 .$$

These equations show that the RR fields play the role of a mass term for the fundamental fields θ^α and the matrix $(F\not{P}F^T\not{P})$ is constant. This is a well-known phenomena in *pp*-waves background [55,56]: the spectrum of worldsheet theory becomes massive. In addition, we can see that the theory is not supersymmetric: the bosonic partner is not massive.[5]

Nevertheless, if the matrix $(F\not{P}F^T\not{P})$ has some zero eigenvalues, along those directions we recover a partial supersymmetry.

Let us study two interesting limits: $||F|| \to 0$ and $||F|| \to \infty$. The first limit is the regime where perturbation theory can be used to perform worldline computations. The second limit is certainly more interesting due to the fact that very little is know about string theory in the presence of strong background fields.

In the limit $||F|| \to \infty$, where the norm $||\cdot||$ is properly defined, the first term can be neglected, and the action is invariant under a new κ-symmetry (notice that we have replaced the classical κ-symmetry of the action (9) with the BRST symmetry given in eq. (15))

$$\delta_k x^m = \theta_L\gamma^m \not{P}\kappa_L + \theta_R\gamma^m \not{P}\kappa_R + \zeta P^m , \qquad (33)$$

$$\delta_k\theta^\alpha_L = -(\not{P}\kappa)^\alpha_L + \zeta\dot{\theta}^\alpha_L , \qquad \delta_k\theta^{\bar{\alpha}}_R = -(\not{P}\kappa)^{\bar{\alpha}}_R + \zeta\dot{\theta}^{\bar{\alpha}}_R ,$$

where the diffeomorphism ghost ζ compensates the gauge choice $e = 1$, and it is $\zeta = -\int^\tau d\tau'(\dot{\theta}^\alpha_L\kappa_{L\alpha} + \dot{\theta}^{\bar{\alpha}}_R\kappa_{R\bar{\alpha}})$.

Notice also that from eqs. (25) it turns out that to take properly the limit of $||F|| \to \infty$ one has to impose the constraints

$$q_{\alpha L} \approx 0 , \qquad q_{\alpha R} \approx 0 . \qquad (34)$$

These constraints generate the κ-symmetry (33) with the opposite sign is front of the transformation rules for the spinors. In addition, they include also second-class constraints. The theory is quantized and therefore we can use the technique discussed in [13,38,39]. This yields the same non-(anti) commutative superspace which is a consequence of the structure of second-class constraints of $q_{\alpha L/R}$, and it does not depend upon the RR field. Applying the Dirac procedure, one finds that

$$\left\{\theta^\alpha_R, \theta^\beta_R\right\}_D \sim \gamma^{\alpha\beta}_m x^m , \qquad \left\{\theta^\alpha_L, \theta^\beta_L\right\}_D \sim \gamma^{\alpha\beta}_m x^m , \qquad (35)$$

$$\left\{\theta^\alpha_L, \theta^\beta_R\right\}_D = 0 .$$

This result for the superparticle is quite different from the result of superstrings [41,42,44] given in (3); for the superparticle, the anticommutation relations of the Grassmann coordinates (35) are related to the bosonic coordinates and not the RR field. This is due to the fact that we have only one set of free parameters, namely $F^{\alpha\beta}$, which have to be interpreted as a mass matrix and not as a "magnetic field".

Moreover, for highly curved space one has to take into account the radiative corrections to the action (26) before taking the limit $||F|| \to \infty$. In fact, as we shall show below, at one loop there are new pieces generated by radiative correction at one-loop in the worldline.

On the other hand, in the limit $||F|| \to 0$, the first term becomes dominant over the other fermionic terms (for the bosonic terms in (26), one can also add a background metric $-\frac{1}{2}g^{mn}P_mP_n$ and therefore they cannot be neglected) and the action (26) with background metric g_{mn} reduces to

$$S_R = \int d\tau\Big[-\dot{\theta}^{\bar{\alpha}}_R F^{-1}_{\bar{\alpha}\beta}\dot{\theta}^\beta_L + \frac{1}{2}g_{mn}\dot{x}^n\dot{x}^m + w_{L\alpha}\dot{\lambda}^\alpha_L + \dot{\lambda}^{\bar{\alpha}}_R w_{R\bar{\alpha}}\Big] . \qquad (36)$$

We conclude that we can use the RR background to set up a perturbation theory around weak RR backgrounds. Namely, we can consider the action (36) as the quadratic part of the action from which the propagators can be computed, and the rest has to be considered as a perturbation.

4.1 Radiative Corrections

It is easy to compute the radiative corrections to the bosonic inverse propagator $\langle x^m(\tau)x^n(0)\rangle$. By computing the free propagators of x^m and the off-diagonal propagator of θ_L and θ_R, one can obtain the one loop contribution

$$G^{mn}(\tau) = -\frac{1}{16}\gamma^m_{\alpha\beta}F^{\beta\bar{\gamma}}\gamma^n_{\bar{\gamma}\bar{\delta}}F^{\alpha\bar{\delta}}\int_0^1 dt|\tau - t|\,|t| = \qquad (37)$$

$$= -\frac{1}{16}\mathrm{Tr}(\gamma^m F\gamma^n F)\mathcal{P}(\tau) ,$$

where $\mathcal{P}(\tau)$ is a polynomial of third order in τ. In the same way the one-loop corrections to the off-diagonal inverse propagator $\langle\theta^\alpha_L\theta^{\bar{\beta}}_R\rangle$

$$G_{\alpha\bar{\beta}}(\tau) = -\frac{1}{16}\gamma^m_{\alpha\delta}F^{\delta\bar{\gamma}}\gamma^n_{\bar{\gamma}\bar{\beta}}g_{mn}\mathcal{P}(\tau) . \qquad (38)$$

Form this computation, we see how the presence of RR backgrounds in the $\langle\theta^\alpha_L\theta^{\bar{\beta}}_R\rangle$ leads to modifications in all the couplings, and therefore the analysis at string coupling $||F|| \to \infty$ cannot be performed without taking into account the radiative corrections. It may be possible to re-sum all the contributions. Notice for example that the spinors θ_L and θ_R appear only quadratically in (36), therefore one can integrate over those fields obtaining the following determinant

$$\det\begin{pmatrix} \frac{i}{2}\not{P}_{\alpha\beta}\partial_\tau & \frac{1}{2}F^{-1}_{\alpha\bar{\beta}}\partial^2_\tau \\ \frac{1}{2}F^{-1}_{\bar{\alpha}\beta}\partial^2_\tau & \frac{i}{2}\not{P}_{\bar{\alpha}\bar{\beta}}\partial_\tau \end{pmatrix} . \qquad (39)$$

The next step will be to integrate over the field P_m. We will not pursue this analysis here and we refer to a subsequent publication.

[5] It has to be recalled that the mass parameter in a curved space does not carefully measure the masslessness of the field. The best way to reveal a supersymmetry breaking is to analyze the Killing spinor equations in a curved background.

5 D0 branes and the BRST symmetry

Since we have seen that the introduction of the RR fields in the case of N=2 d=4 (or d=10) superparticle does not really help for deriving the anticommutation relation (3) for the Grassmann coordinates, we need to use a different type of superparticle to do it.

As is known, in string theory there are solitonic degrees of freedom which are known as Dp-branes [27]. A given Dp-brane has a worldvolume which has $p+1$ dimensions. They are characterized by the coupling to the RR fields of the superstrings, and they can be described by a low energy effective action which is the sum of a Born-Infeld action and a Wess-Zumino term. Among the Dp-branes, we can consider the D0-brane which is a particle (the worldvolume is 1-dimensional) described by the effective action

$$S = \int d\tau (P_m \Pi^m - \frac{e}{2} P_m P^m) + \int d\tau f_{\alpha\beta} \theta_L^\alpha \dot{\theta}_R^\beta \quad (40)$$

in the first order formalism with

$$\Pi^m = \dot{x} + \frac{i}{2}\theta_L \gamma^m \dot{\theta}_L + \frac{i}{2}\theta_R \gamma^m \dot{\theta}_R . \quad (41)$$

and $f_{\alpha\beta}$ is constant. The first term is the Born-Infeld term written in the first order formalism. The second term is a Wess-Zumino term and the coefficient $f_{\alpha\beta}$ is related to the brane tension [57]. Indeed it can be viewed as the mass of the $D0$-brane

The action is invariant under κ-symmetry which yields the supersymmetry of the $D0$-brane and it can be quantized using the BRST technique discussed above in Sec. 3.2. An objection to this might be: the $D0$-brane is not a fundamental degree of freedom and there is no need of quantizing it! However it is shown in [45] that the BRST based on Pure Spinor formulation replaces the κ-symmetry and provides a guideline how to coupled the $D0$-brane to a supergravity background as in the case of the N=2 superparticle.

In analogy with eqs. (19), we derive the supersymmetry generators $q_{\alpha,L}$ and $q_{\alpha,R}$ for the $D0$-brane.

$$q_{L\alpha} = p_{L\alpha} - \frac{i}{2} P_m (\gamma^m \theta_L)_\alpha + f_{\alpha\beta} \theta_R^\beta , \quad (42)$$

$$q_{R\alpha} = p_{R\alpha} - \frac{i}{2} P_m (\gamma^m \theta_R)_\alpha + f_{\alpha\beta} \theta_L^\alpha .$$

They couple to the RR field of the supergravity background and the supersymmetry is broken because of the presence of RR fields, or equivalently to the presence of $D0$-branes.

6 RR fields and Non-commutative Superspace: Act II

Finally, we can couple the $D0$-brane to RR field of the supergravity background and this introduces a new term of the form

$$\int d\tau F^{\alpha\beta} q_{\alpha L} q_{\beta R} \quad (43)$$

such as in the case of a superparticle coupled to RR fields. To simplify as much as possible the analysis, we can set the momentum of the particle to zero and we find that in the limit of $||F|| \to \infty$, we need the constraints

$$q_{\alpha L} \approx 0 , \qquad q_{\alpha R} \approx 0 . \quad (44)$$

where $q_{\alpha L}$ and $q_{\alpha R}$ are given by (42) and they depend on the RR field $f_{\alpha\beta}$ generated by the $D0$-brane. Applying the Dirac procedure and using the canonical brackets $[p_{\alpha L/R}, \theta^\beta_{L/R}] = i\delta_\alpha^{\ \beta}$, we end up with the commutation relations

$$\{\theta_L^\alpha, \theta_R^\beta\}_D = (f^{-1})^{\alpha\beta} , \quad (45)$$

which finally gives the non-(anti)commutation relations between the fermionic coordinates.

To conclude, we have shown that the coupling of the RR fields and the Wess-Zumino term, in the case of superparticle and $D0$-brane, are fundamental to generate a deformation of anti-commutation relations among Grassmann coordinates of the superspace. It is worth pointing out that the RR fields of the supergravity background cannot generate the wanted commutation relations, but replace the role of the mass as in the case of the non-relativistic particle moving in a magnetic field. Furthermore, it is shown that the brane tension together with a supergravity background yields the wanted commutation relations (3) obtained also in string theory. The role of the mass and of the RR fields is inverted: in the case of massive bosonic charge particle, we take the limit $m \to 0$ and we derive (1) deformed by B_{mn}, in the case of superparticles, we take the limit $||F|| \to \infty$ and we derive (3) deformed by the brane tension f.

The author would like to thank the organizers and the directors of the 43rd International School on Subnuclear Physics held in Erice, Sicily in Sept. 2005 where a lecture based on the present paper has been delivered. We also thank P. Aschieri, I. Bars, L. Castellani, A. Lerda, F. Morales, Y. Oz, L. Tamassia and N. Wyllard for useful discussions on related subjects and comments. A special thank to L. Anguelova for a careful reading of the manuscript and for useful suggestions.

References

1. C. S. Chu and P. M. Ho, "Noncommutative open string and D-brane," Nucl. Phys. B **550**, 151 (1999) [arXiv:hep-th/9812219].
2. C. S. Chu and P. M. Ho, "Constrained quantization of open string in background B field and noncommutative D-brane," Nucl. Phys. B **568**, 447 (2000) [arXiv:hep-th/9906192].
3. F. Ardalan, H. Arfaei and M. M. Sheikh-Jabbari, "Noncommutative geometry from strings and branes," JHEP **9902**, 016 (1999) [arXiv:hep-th/9810072].
4. N. Seiberg and E. Witten, "String theory and noncommutative geometry," JHEP **9909**, 032 (1999) [arXiv:hep-th/9908142].

5. D. Bigatti and L. Susskind, "Magnetic fields, branes and noncommutative geometry," Phys. Rev. D **62**, 066004 (2000) [arXiv:hep-th/9908056].
6. R. Peierls, "Zur Theorie des Diamagnetismus von Leitungselektronen," Z. Phys. **80**, 763 (1933).
7. G. V. Dunne, R. Jackiw and C. A. Trugenberger, "'Topological' (Chern-Simons) Quantum Mechanics," Phys. Rev. D **41**, 661 (1990).
8. R. Jackiw, "Physical instances of noncommuting coordinates," Nucl. Phys. Proc. Suppl. **108**, 30 (2002) [Phys. Part. Nucl. **33**, S6 (2002 LNPHA,616,294-304.2003)] [arXiv:hep-th/0110057].
9. H. S. Snyder, "Quantized Space-Time," Phys. Rev. **71**, 38 (1947).
10. C. N. Yang, "On Quantized Space-Time," Phys. Rev. **72**, 874 (1947).
11. P. Aschieri, M. Dimitrijevic, F. Meyer and J. Wess, "Noncommutative geometry and gravity," arXiv:hep-th/0510059.
12. R. Casalbuoni, "The Classical Mechanics For Bose-Fermi Systems," Nuovo Cim. A **33**, 389 (1976).
13. L. Brink and J. H. Schwarz, "Clifford Algebra Superspace," CALT-68-813
14. W. Siegel, "Hidden Local Supersymmetry In The Supersymmetric Particle Action," Phys. Lett. B **128**, 397 (1983).
15. P. A. Grassi, G. Policastro and M. Porrati, "Covariant quantization of the Brink-Schwarz superparticle," Nucl. Phys. B **606**, 380 (2001) [arXiv:hep-th/0009239].
16. M. B. Green, J. H. Schwarz and E. Witten, "Superstring Theory. Vol. 1: Introduction," and "Superstring Theory. Vol. 2: Loop Amplitudes, Anomalies And Phenomenology,"
17. N. Berkovits, "Super–Poincaré covariant quantization of the superstring," *JHEP* **04** (2000) 018, hep-th/0001035.
18. N. Berkovits, "Covariant quantization of the superparticle using pure spinors," JHEP **0109**, 016 (2001) [arXiv:hep-th/0105050].
19. N. Berkovits, "ICTP lectures on covariant quantization of the superstring," hep-th/0209059.
20. N. Berkovits, "Covariant quantization of the supermembrane," JHEP **0209**, 051 (2002) [arXiv:hep-th/0201151].
21. P. A. Grassi and J. F. Morales, "Partition functions of pure spinors," arXiv:hep-th/0510215.
22. P. A. Grassi and L. Tamassia, in preparation.
23. P. A. Grassi, G. Policastro, M. Porrati, and P. van Nieuwenhuizen, "Covariant quantization of superstrings without pure spinor constraints," *JHEP* **10** (2002) 054, hep-th/0112162; P. A. Grassi, G. Policastro, and P. van Nieuwenhuizen, "The massless spectrum of covariant superstrings," *JHEP* **11** (2002) 001, hep-th/0202123; P. A. Grassi, G. Policastro, and P. van Nieuwenhuizen, "On the BRST cohomology of superstrings with/without pure spinors," *Adv. Theor. Math. Phys.* **7** (2003) 499, hep-th/0206216;
24. M. Chesterman, "Ghost constraints and the covariant quantization of the superparticle in ten dimensions," *JHEP* **02** (2004) 011, hep-th/0212261;
25. Y. Aisaka and Y. Kazama, "A new first class algebra, homological perturbation and extension of pure spinor formalism for superstring," JHEP **02** (2003) 017, hep-th/0212316.
26. P. A. Grassi, G. Policastro and P. van Nieuwenhuizen, "The covariant quantum superstring and superparticle from their classical actions," Phys. Lett. B **553**, 96 (2003) [arXiv:hep-th/0209026].
27. J. Polchinski, "String theory. Vol. 1: An introduction to the bosonic string," J. Polchinski, "String theory. Vol. 2: Superstring theory and beyond,"
28. M. B. Green, M. Gutperle and P. Vanhove, "One loop in eleven dimensions," Phys. Lett. B **409**, 177 (1997) [arXiv:hep-th/9706175]. M. B. Green, H. h. Kwon and P. Vanhove, "Two loops in eleven dimensions," Phys. Rev. D **61**, 104010 (2000) [arXiv:hep-th/9910055].
29. L. Anguelova, P. A. Grassi and P. Vanhove, Nucl. Phys. B **702**, 269 (2004) [arXiv:hep-th/0408171].
30. S. J. Gates, M. T. Grisaru, M. Rocek and W. Siegel, "Superspace, Or One Thousand And One Lessons In Supersymmetry," Front. Phys. **58**, 1 (1983) [arXiv:hep-th/0108200].
31. N. Berkovits, "Pure spinor formalism as an $\mathcal{N} = 2$ topological string," hep-th/0509120.
32. P. A. Grassi and N. Wyllard, "Lower-dimensional pure-spinor superstrings," arXiv:hep-th/0509140.
33. N. Wyllard, "Pure-spinor superstrings in d = 2, 4, 6," arXiv:hep-th/0509165.
34. O. Chandia, "D = 4 pure spinor superstring and N = 2 strings," arXiv:hep-th/0509185.
35. P. A. Grassi and Y. Oz, "Non-critical covariant superstrings," arXiv:hep-th/0507168.
36. S. Ferrara and M. A. Lledo, "Some aspects of deformations of supersymmetric field theories," JHEP **0005**, 008 (2000) [arXiv:hep-th/0002084].
37. D. Klemm, S. Penati and L. Tamassia, "Non(anti)commutative superspace," Class. Quant. Grav. **20**, 2905 (2003) [arXiv:hep-th/0104190].
38. L. Brink and J. H. Schwarz, "Quantum Superspace," Phys. Lett. B **100**, 310 (1981).
39. J. H. Schwarz and P. Van Nieuwenhuizen, "Speculations Concerning A Fermionic Substructure Of Space-Time," Lett. Nuovo Cim. **34** (1982) 21.
40. I. Bars, C. Deliduman, A. Pasqua and B. Zumino, "Superstar in noncommutative superspace via covariant quantization of the superparticle," Phys. Rev. D **68**, 106006 (2003) [arXiv:hep-th/0308107].
41. J. de Boer, P. A. Grassi and P. van Nieuwenhuizen, "Noncommutative superspace from string theory," Phys. Lett. B **574**, 98 (2003) [arXiv:hep-th/0302078].
42. H. Ooguri and C. Vafa, "The C-deformation of gluino and non-planar diagrams," Adv. Theor. Math. Phys. **7**, 53 (2003) [arXiv:hep-th/0302109].
43. N. Berkovits and N. Seiberg, "Superstrings in graviphoton background and N = 1/2 + 3/2 supersymmetry," JHEP **0307**, 010 (2003) [arXiv:hep-th/0306226].
44. N. Seiberg, "Noncommutative superspace, N = 1/2 supersymmetry, field theory and string theory," JHEP **0306**, 010 (2003) [arXiv:hep-th/0305248].
45. L. Anguelova and P. A. Grassi, "Super D-branes from BRST symmetry," JHEP **0311**, 010 (2003) [arXiv:hep-th/0307260].
46. M. Henneaux and C. Teitelboim, "Quantization of gauge systems,", Princeton University Press, Princeton, New Jersey.
47. I. Oda and M. Tonin, "On the Berkovits covariant quantization of GS superstring," Phys. Lett. B **520** (2001) 398 [arXiv:hep-th/0109051]. M. Matone, L. Mazzucato, I. Oda, D. Sorokin and M. Tonin, "The superembedding origin of the Berkovits pure spinor covariant quantization of superstrings," Nucl. Phys. B **639**, 182 (2002) [arXiv:hep-th/0206104].
48. Y. Aisaka and Y. Kazama, "Origin of pure spinor superstring," JHEP **0505**, 046 (2005) [arXiv:hep-th/0502208].

49. N. Berkovits and P. S. Howe, "Ten-dimensional supergravity constraints from the pure spinor formalism for the superstring," Nucl. Phys. B **635**, 75 (2002) [arXiv:hep-th/0112160].
50. P. A. Grassi and L. Tamassia, "Vertex operators for closed superstrings," JHEP **0407**, 071 (2004) [arXiv:hep-th/0405072].
51. M. Hatsuda, S. Iso and H. Umetsu, "Noncommutative superspace, supermatrix and lowest Landau level," Nucl. Phys. B **671**, 217 (2003) [arXiv:hep-th/0306251].
52. N. Berkovits, C. Vafa and E. Witten, "Conformal field theory of AdS background with Ramond-Ramond flux," JHEP **9903**, 018 (1999) [arXiv:hep-th/9902098].
53. N. Berkovits and O. Chandia, "Superstring vertex operators in an AdS(5) x S(5) background," Nucl. Phys. B **596**, 185 (2001) [arXiv:hep-th/0009168].
54. N. Berkovits, "Conformal field theory for the superstring in a Ramond-Ramond plane wave background," JHEP **0204**, 037 (2002) [arXiv:hep-th/0203248].
55. R. R. Metsaev, "Type IIB Green-Schwarz superstring in plane wave Ramond-Ramond background," Nucl. Phys. B **625**, 70 (2002) [arXiv:hep-th/0112044].
56. R. R. Metsaev and A. A. Tseytlin, "Exactly solvable model of superstring in plane wave Ramond-Ramond background," Phys. Rev. D **65**, 126004 (2002) [arXiv:hep-th/0202109].
57. M. Aganagic, C. Popescu and J. H. Schwarz, "Gauge-invariant and gauge-fixed D-brane actions," Nucl. Phys. B **495**, 99 (1997) [arXiv:hep-th/9612080].

Reprinted with permission from the electronic-only supplement "Young Scientist"
(DOI: 10.1140/epjcd/s2006-03-003-5) of The European Physical Journal (EPJ) C

Searches for New Physics in Photon Final States

A. Loginov[a]

SSC RF ITEP, 117218 Moscow, Russia

Abstract. The Run I results on the searches for new physics in photon final states were intriguing. The rare $ee\gamma\gamma \not{E}_T$ candidate event and the measured event rate for the signature $\ell + \gamma + \not{E}_T$, which was 2.7 sigma above the Standard Model predictions, sparked signature-based searches in the $\gamma\gamma + X$ and $\ell\gamma + X$ channels. With more data in Run II we should be able to answer a simple question: was it an anomaly or were the Run I results the first evidence for new physics? We present searches for New Physics in Photon Final States at CDF Run II, Fermilab, with substantially more data and a higher $\bar{p}p$ collision energy, 1.96 TeV, and the upgraded CDF-II detector.

PACS. 13.85.Rm Limits on production of particles – 12.60.Jv Supersymmetric models – 13.85.Qk Inclusive production with identified leptons, photons, or other nonhadronic particles – 14.80.Ly Supersymmetric partners of known particles – 14.80.-j Other particles (including hypothetical)

1 Introduction

The Standard Model (SM) [1] is an effective field theory that has so far described the fundamental interactions of elementary particles remarkably well. However the model breaks down at energies of a few TeV, in that the cross-section for scattering of longitudinal W bosons would otherwise violate unitarity. The Fermilab Tevatron has the highest center-of-mass energy collisions of any present accelerator, with $\sqrt{s} = 1.96$ TeV, and thus has the potential to discover new physics. As of September, 2005, the CDF experiment at Fermilab has recorded 1 fb^{-1} of data. Physics results using 202 pb^{-1} to 345 pb^{-1} are presented in this paper.

1.1 Motivation

Why do we consider the photon final states a good signature for observing new physics?

- Well Motivated Theories
  - Most importantly Supersymmetry
- History
  - Follow up on some of the anomalies from CDF in Run I [2–6]
- From the experimentalists' point of view, just because...
  - The photon is coupled to electric charge, and thus is radiated by all charged particles, including the incoming states (important for searching for invisible final states)
  - The photon is massless and thus kinematically easier to produce than the W or Z
  - The photon is stable, which implies a high acceptance, as there are no branching ratios to 'pay'
  - The photon is a boson and could be produced by a fermiphobic parent
  - And if we then require
    - Additional Lepton(s) ⇒ high-E_T [1] photon + high-P_T lepton + X signature is rare in SM, backgrounds are low for searches
    - Additional Photon(s) ⇒ the photons have moderate signal-to-noise but good efficiency and mass peak resolution

1.2 Run I Results

1.2.1 $ee\gamma\gamma \not{E}_T$ Candidate Event

In 1995 the CDF experiment, measuring $\bar{p}p$ collisions at a center-of-mass energy of 1.8 TeV at the Fermilab Tevatron, observed an event [2,3,7] consistent with the production of two energetic photons, two energetic electrons, and large missing transverse energy [2], $\not{E}_T$ (Figure 1).

This signature is predicted to be very rare in the Standard Model of particle physics, with the dominant contribution coming from the WW$\gamma\gamma$ production:
$WW\gamma\gamma \rightarrow (e\nu)(e\nu)\gamma\gamma \rightarrow ee\gamma\gamma \not{E}_T$,
from which we expect 8×10^{-7} events. All other sources

[a] *email: loginov@fnal.gov, for the CDF Collaboration*

[1] Transverse momentum and energy are defined as $P_T = p \sin\theta$ and $E_T = E \sin\theta$, respectively. The CDF coordinate system of r, φ, and z is cylindrical, with the z-axis along the proton beam. The pseudorapidity is $\eta = -\ln(\tan(\theta/2))$.

[2] Missing E_T ($\not{E}_T$) is defined by $\not{\vec{E}}_T = -\sum_i E_T^i \hat{n}_i$, where i is the calorimeter tower number for $|\eta| < 3.6$, and $\hat{n}_i$ is a unit vector perpendicular to the beam axis and pointing at the ith calorimeter tower. We define the magnitude $\not{E}_T = |\not{\vec{E}}_T|$.

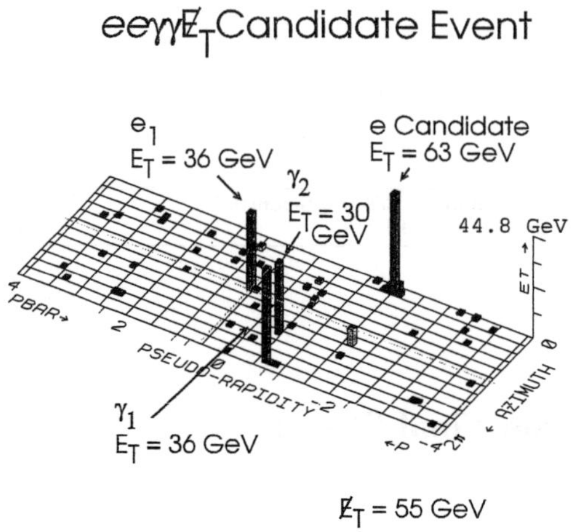

Fig. 1. The Run I $ee\gamma\gamma\not{E}_T$ Candidate Event.

(mostly detector misidentification) lead to 5×10^{-7} events. Therefore, we expect $(1 \pm 1) \times 10^{-6}$ events, which would give us one $ee\gamma\gamma\not{E}_T$ candidate event if we had taken million times more data than we actually had in Run I.

The event raised theoretical interest, however, as the two-lepton two-photon signature is expected in some models of physics 'beyond the Standard Model' such as gauge-mediated models of supersymmetry [8]. For example, possible interpretation will be:

$p\bar{p} \to \tilde{e}^+\tilde{e}^-(+X)$, $\tilde{e} \to \tilde{\chi}_2^0 + e$, $\tilde{\chi}_2^0 \to \tilde{\chi}_1^0\gamma$,

where $\tilde{e}$ is the selectron (the bosonic partner of the electron), and $\tilde{\chi}_1^0$ and $\tilde{\chi}_2^0$ are the lightest and next-to-lightest neutralinos.

1.2.2 $\gamma\gamma$+X Search

The detection of this single event led to the development of 'signature-based' inclusive searches to cast a wider net: in this case one searches for two photons + X [2,3,7], where X stands for anything, with the idea that if pairs of new particles were being created these inclusive signatures would be sensitive to a range of decay modes or to the creation and decay of different particle types.

In Run I Searches for $\gamma\gamma$**+X** all results were consistent with the SM background expectations with no other exceptions other than the observation of the $ee\gamma\gamma\not{E}_T$ candidate event(Table 1) [3].

1.2.3 From $\gamma\gamma$ to $\ell\gamma$: $\ell\gamma+X$ Search

Another 'signature-based' inclusive search, motivated by $ee\gamma\gamma\not{E}_T$ event was for $\ell\gamma+X$ [5,6,9].

Table 1. Number of observed and expected $\gamma\gamma$ events with additional objects in 86 pb^{-1}[3].

Signature (Object)	Obs.	Expected
$\not{E}_T > 35$ GeV, $\lvert\Delta\phi_{\not{E}_T-\text{jet}}\rvert > 10°$	1	0.5 ± 0.1
$N_{\text{jet}} \geq 4$, $E_T^{\text{jet}} > 10$ GeV, $\lvert\eta^{\text{jet}}\rvert < 2.0$	2	1.6 ± 0.4
b-tag, $E_T^b > 25$ GeV	2	1.3 ± 0.7
Central γ, $E_T^{\gamma_3} > 25$ GeV	0	0.1 ± 0.1
Central e or μ, $E_T^{e \text{ or } \mu} > 25$ GeV	3	0.3 ± 0.1
Central τ, $E_T^{\tau} > 25$ GeV	1	0.2 ± 0.1

In general the data agree with expectations, with the exception for the $\ell\gamma\not{E}_T$ category. We have observed 16 $\ell\gamma\not{E}_T$ events on a background of 7.6 ± 0.7 expected. The 16 $\ell\gamma\not{E}_T$ events consist of 11 $\mu\gamma\not{E}_T$ events and 5 $e\gamma\not{E}_T$ events, versus expectations of 4.2±0.5 and 3.4±0.3 events, respectively. The SM prediction yields the observed rate of $\ell\gamma\not{E}_T$ with **0.7%** probability (which is equivalent to **2.7** standard deviations for a Gaussian distribution).

One of the first SUSY interpretation of the CDF $\mu\gamma\not{E}_T$ events [10] was resonant smuon $\tilde{\mu}$ production with a single dominant R-parity violating coupling (Figure 2).

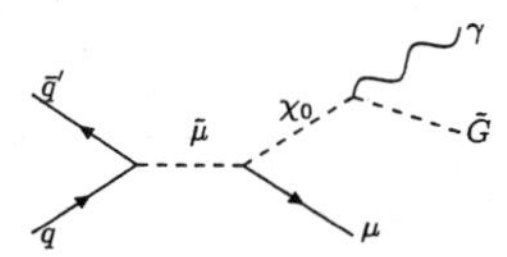

Fig. 2. Resonant smuon production and subsequent decay, producing the $\mu\gamma\not{E}_T$ signature.

The Run I search was initiated by an anomaly in the data itself, and as such the 2.7 sigma excess above the SM expectations must be viewed taking into account the number of such channels a fluctuation could have occurred in.

2 From Run I to Run II

Having many different hints from the signature-based searches for new physics in photon final states in Run I, the

Table 2. Run I Photon-Lepton Results: Number of observed and expected $\ell\gamma$ events with additional objects in 86 pb^{-1}[6].

Category	μ_{SM}	N_0	$P(N \geq N_0\lvert\mu_{SM})$,%
All $\ell\gamma+X$	–	**77**	–
Z-like $e\gamma$	–	17	–
Two-Body $\ell\gamma X$	24.9±2.4	33	9.3
Multi-Body $\ell\gamma X$	20.2±1.7	27	10.0
Multi-Body $\ell\ell\gamma X$	5.8 ± 0.6	5	68.0
Multi-Body $\ell\gamma\gamma X$	0.02±0.02	1	1.5
Multi-Body $\ell\gamma\not{E}_T X$	**7.6 ± 0.7**	**16**	**0.7**

strategy for Run II was straightforward: take more data. The main points were:

- Increase the Collision Energy: 1.80 → 1.96 TeV
- Increase the rate at which we take data: 3500 → 396 ns (timing between bunches)
- Upgrade the Detectors

2.1 CDF Run II Detector

The CDF-II detector [11] is a cylindrically symmetric spectrometer designed to study $\bar{p}p$ collisions at the Fermilab Tevatron, that uses the same solenoidal magnet and central calorimeters as the CDF-I detector [12] from which it was upgraded. Because the analyses described here have been motivated by the Run I searches, we note especially the differences from the Run I detector relevant to the detection of photons, leptons, and $\not\!\!E_{\rm T}$.

The central calorimeters are physically unchanged; however, the readout electronics has been replaced to accommodate the smaller proton and anti-proton bunch spacing of the Tevatron in Run II. The end-cap (plug) and forward calorimeters have been replaced with a more compact scintillator-based design, retaining the projective geometry [13].

The tracking system used to measure the momenta of charged particles has been replaced, with the central outer tracker upgraded to have smaller drift cells [14], and the inner tracking chamber and silicon system replaced by a system of silicon strip chambers with more layers, now in 2-dimensions [15]. The new inner tracking system has substantially more material, resulting in more bremsstrahlung (photons) produced by high-$P_{\rm T}$ electrons.

The central CMU, CMP, and CMX muon systems[3] are also physically unchanged in design, but the coverage of the CMP and CMX muon systems [16] has been extended by filling in gaps in φ [11].

3 Run II: Searches for New Physics in Photon Final States

The Run I results on the searches for new physics in photon final states were intriguing [2,3,5,6]. The rare $ee\gamma\gamma\not\!\!E_{\rm T}$ candidate event and the measured event rate for the signature $\ell+\gamma+\not\!\!E_{\rm T}$, which was 2.7 sigma above the SM predictions, sparked signature-based searches in the $\gamma\gamma+X$ and $\ell\gamma+X$ channels.

With more data in Run II we should be able to answer a simple question: was it an anomaly or were the Run I results the first evidence for the new physics?

There are lots of searches involving photon final states at CDF in Run II. Some of the analyses are presented in this paper:

- Search for High-Mass Diphoton State and Limits on Randall-Sundrum Gravitons (Section 3.1)
- Search for Anomalous Production of Diphoton Events with $\not\!\!E_{\rm T}$ and Limits on GMSB Models (Section 3.2)
- Search for Lepton-Photon-X Events (Section 3.3)

3.1 Search for High-Mass Diphoton State and Limits on Randall-Sundrum Gravitons

Searches for new particles decaying into two identical particles are broad, inclusive and sensitive. The production of the new particle may be direct or in association with other particles, or in a decay chain. The discovery of a sharp mass peak over background would be a compelling evidence for the production of a new particle. The diphoton final state is important because the photons are bosons and the parent may be fermiphobic. The photons have moderate signal-to-noise but good efficiency and mass peak resolution.

One model producing a diphoton mass peak is Randall-Sundrum gravitons [17]. Current string theory proposes that as many as seven new dimensions may exist and the geometry of these extra dimensions is responsible for gravity being so weak. The Randall-Sundrum model has the property that a parameter, the warp factor, determines the curvature of the extra dimensions and therefore the mass of the Kaluza-Klein graviton resonances, which decay to two bodies including photons.

Details on this analysis are reported in [18].

3.1.1 Data Sample

The sample corresponds to 345 pb^{-1} of data taken between February 2002 and July, 2004. We require that the data were taken under good detector conditions for a reliable photon identification. We apply selection cuts as follows:

- Photons in Central Calorimeter
- $E_{\rm T}^{\gamma} > 15$ GeV
- $M(\gamma,\gamma) > 30$ GeV

To select a photon in a central calorimeter (approximately $0.05 < |\eta| < 1.0$), we require a central electromagnetic cluster that: a) is not near the boundary in ϕ of a calorimeter tower [4] b) have the ratio of hadronic to electromagnetic energy, Had/EM, $< 0.055 + 0.00045 \times E^{\gamma}(\rm GeV)$; c) have no tracks, or only one track with $p_T < 1$ GeV/c, extrapolating to the towers of the cluster; d) is isolated in the calorimeter and tracking chamber [5] e) have a shower

[3] The CMU (Central Muon Chambers) system consists of gas proportional chambers in the region $|\eta| < 0.6$; the CMP (Central Muon Upgrade) system consists of chambers after an additional meter of steel, also for $|\eta| < 0.6$. The CMX (Central Muon Extension) chambers cover $0.6 < |\eta| < 1.0$.

[4] The fiducial region has ~87% coverage in the central region.

[5] To reject hadronic backgrounds that fake prompt photons, candidates are required to be isolated in the calorimeter and tracking chamber. In the calorimeter the isolation is defined as the energy in a cone of 0.4 in $\eta-\phi$ space, minus the photon cluster energy, and corrected for energy loss into cracks as well as the number of reconstructed $\bar{p}p$ interactions in the event. We require isolation $< 0.1 \times E_{\rm T}^{\gamma}$ for $E_{\rm T}^{gamma} < 20$ GeV, and

shape in the CES [6] consistent with a single photon; f) have no other significant energy deposited nearby in the CES.

The final dataset consists of 3339 events, for which the data histogrammed with bins equivalent to one σ of invariant mass resolution are shown in Figure 3. The highest mass events occur at masses of 207, 247, 304, 329, and 405 GeV/c^2 (Figure 4).

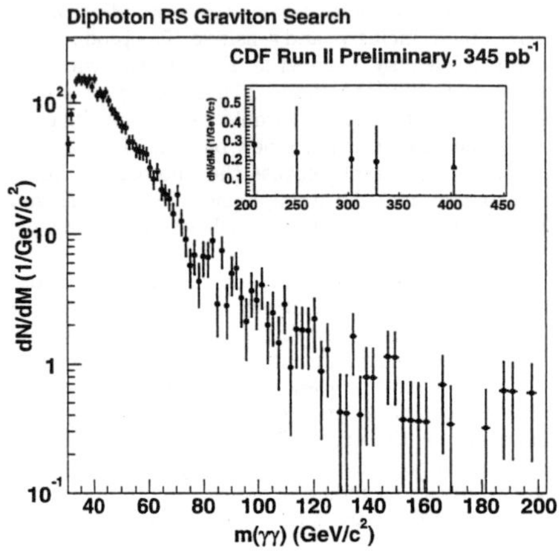

Fig. 3. The diphoton invariant mass distribution histogrammed in bins of approximately one σ of mass resolution.

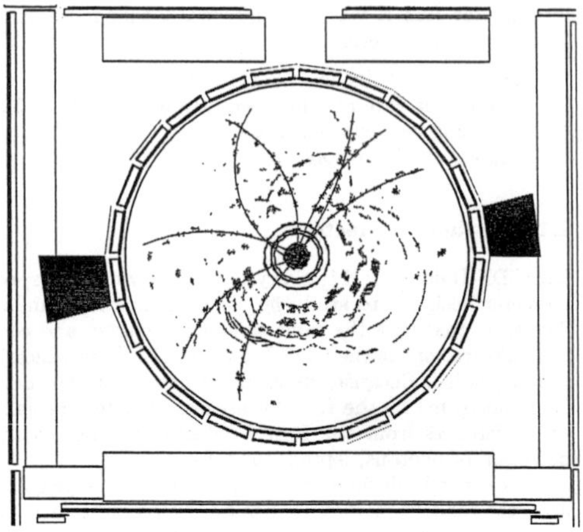

Fig. 4. $\gamma\gamma$ Highest Mass Event. M ($\gamma\gamma$) = 405 GeV/c^2, $E_T^{\gamma 1}$ = 172 GeV, $E_T^{\gamma 2}$ = 175 GeV.

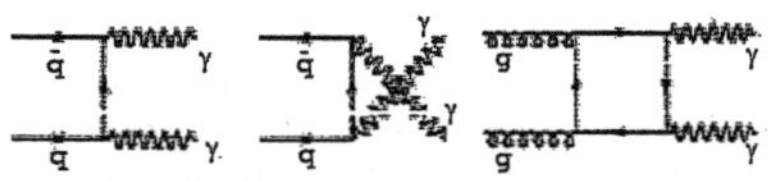

Fig. 5. Standard Model diphoton production diagrams.

Figure 6 shows the data mass spectrum compared to the prediction.

3.1.2 Backgrounds

There are two significant backgrounds to the $\gamma\gamma$ sample. The first is SM diphoton production which accounts for 30% of the events (Figure 5). This background is estimated using a NLO Monte Carlo, diphox [19], which we normalize to $\mathcal{L}$=345 pb^{-1}.

The second background comes from high-E_T π^0's from jets. To create a control sample, we loosen several cuts (including relaxing the isolation cuts by 50%), and we get 9891 events, from which we then reject events in the signal sample and are left with 6552 events in the "photon sideband" sample. We then derive the shape of the mass distribution by fitting this sample to a sum of several exponentials. We then subtract the estimate from the SM contribution and normalize the fakes background to the low mass ($m_{\gamma\gamma}$ between 30 and 100 GeV).

< 2.0 GeV+0.02 × (E_T^{γ} − 20 GeV) for E_T^{γ} > 20 GeV. In the tracking chamber we require the scalar sum of the p_T of all tracks in a cone of 0.4 to be < 2.0 GeV+0.005 × E_T^{γ}, where all values of E_T^{γ} are in GeV.

[6] CES: Central EM Strip Chambers.

3.1.3 Limits on Randall-Sundrum Gravitons

Since the data are consistent with the SM prediction, we place upper limits on the cross sections times branching ratio of Randall-Sundrum graviton production and decay to diphotons (Figure 7).

Figure 8 shows the combined 95% confidence level RS graviton mass limits of the di-photon ($\mathcal{L}$=345 pb^{-1}) and di-lepton ($\mathcal{L}$=200 pb^{-1}) searches [20] in the graviton mass versus coupling, k/M_{Planck}, plane. Note, that $\gamma\gamma$ has a larger Branching Ratio (Br(G→ $\gamma\gamma$) = 2×Br(G→ee)) and the $\gamma\gamma$ spin factors improve the acceptance.

3.2 Search for Anomalous Production of Diphoton Events with $\not\!\!E_T$ and Limits on GMSB Models

For theoretical reasons [21,22], and because of the $ee\gamma\gamma\not\!\!E_T$ candidate event (Figure 1) recorded by the CDF detector in Run I [2,3], we want to search for the production of heavy new particles that decay producing the signature $\gamma\gamma + \not\!\!E_T$. Of particular theoretical interest are supersymmetric (SUSY) models with gauge-mediated SUSY-breaking (GMSB). Characteristically, the effective SUSY-breaking scale (Λ) can be as low as 100 TeV, the lightest

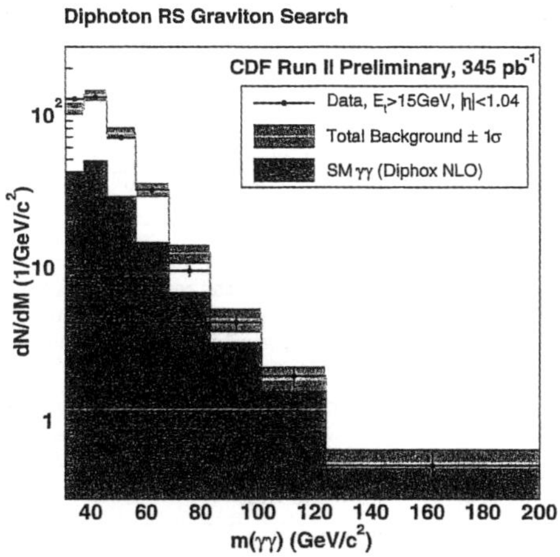

Fig. 6. Comparison of the SM di-photon contribution plus misidentified jets with the observed diphoton mass spectrum. Variable bins are used for statistical comparison to the background prediction.

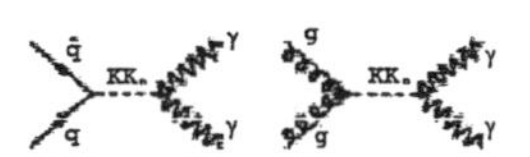

Fig. 7. Randall-Sundrum graviton production and decay to diphotons.

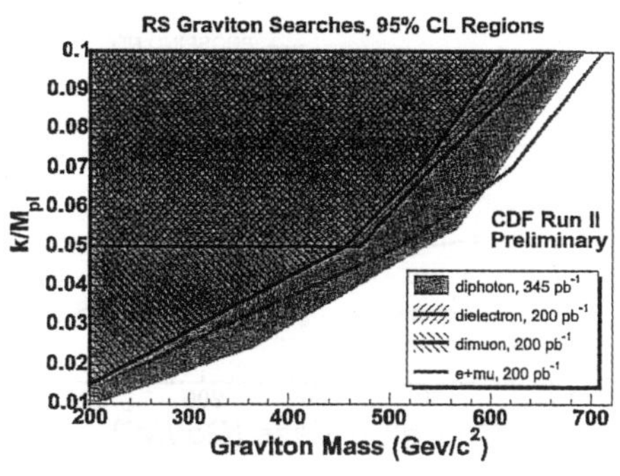

Fig. 8. Combined 95% confidence level Randall-Sundrum graviton mass limits of the di-photon and di-lepton searches.

SUSY particle is a light gravitino ($\tilde{G}$) that is assumed to be stable, and the SUSY particles have masses in a range that may make them accessible at Tevatron energies [21]. In these models the visible signatures are determined by the properties of the next–to–lightest SUSY particle (NLSP) that may be, for example, a slepton or the lightest neutralino ($\tilde{\chi}_1^0$). In the GMSB model investigated here, the NLSP is a $\tilde{\chi}_1^0$ decaying almost exclusively to a photon and a $\tilde{G}$ that penetrates the detector without interacting, producing $\not\!\!E_T$. SUSY particle production at the Tevatron is predicted to be dominated by pairs of the lightest chargino ($\tilde{\chi}_1^{\pm}$) and by associated production of a $\tilde{\chi}_1^{\pm}$ and the next–to–lightest neutralino ($\tilde{\chi}_2^0$). Each gaugino pair cascades down to two $\tilde{\chi}_1^0$'s, leading to a final state of $\gamma\gamma + \not\!\!E_T + X$, where X represents any other final state particles.

Details on this analysis can be found in [23, 24].

3.2.1 Data Sample

The analysis selection criteria have been optimized to maximize, *a priori*, the expected sensitivity to GMSB SUSY based only on the background expectations and the predictions of the model. Event selection requirements for the diphoton candidate sample are designed to reduce electron and jet/π^0 backgrounds while accepting well-measured diphoton candidates.

We require two central (approximately $0.05 < |\eta| < 1.0$) electromagnetic clusters that should pass standard photon selection cuts (Section 3.1.1). For this analysis we require $E_T^\gamma > 13$ GeV.

3.2.2 Backgrounds

Backgrounds for the $\gamma\gamma$+X analysis are:

- QCD background: fake photon (jj, jγ)
- QCD background: $\gamma\gamma$
- $e\gamma$
- Non-Collision: beam-related, cosmic rays

Before the $\not\!\!E_T$ requirement, the diphoton candidate sample is dominated by QCD interactions producing combinations of photons and jets faking photons. In each case only small measured $\not\!\!E_T$ is expected, due mostly to energy measurement resolution effects.

Events with an electron and a photon candidate ($W\gamma \to e\nu\gamma$, $Wj \to e\nu\gamma_{fake}$, $Z\gamma \to ee\gamma$, etc.) can contribute to the diphoton candidate sample when the electron track is lost (by tracking inefficiency or bremsstrahlung) to create a fake photon. For W decays large $\not\!\!E_T$ can come from the neutrinos. This background is estimated using $e\gamma$ events from the data.

Beam–related sources and cosmic rays overlapped with a SM event can contribute to the background by producing spurious energy deposits that in turn affect the measured $\not\!\!E_T$. While the rate at which these events contribute to the diphoton candidate sample is low, most contain large $\not\!\!E_T$. The spurious clusters can pass photon cuts.

Backgrounds and observed number of events are summarized in Table 3.

Table 3. Numbers of events observed and events expected from background sources as a function of the $\not{E}_T$ requirement. Here "QCD" includes the $\gamma\gamma$, γj and jj processes. The first uncertainty is statistical, the second is systematic.

$\not{E}_T$ Requirement	Expected				Observed
	QCD	$e\gamma$	Non-Collision	Total	
25 GeV	$4.01 \pm 3.21 \pm 3.76$	$1.40 \pm 0.52 \pm 0.45$	$0.54 \pm 0.06 \pm 0.42$	$5.95 \pm 3.25 \pm 3.81$	3
35 GeV	$0.30 \pm 0.24 \pm 0.22$	$0.84 \pm 0.32 \pm 0.27$	$0.25 \pm 0.04 \pm 0.19$	$1.39 \pm 0.40 \pm 0.40$	2
45 GeV	$0.01 \pm 0.01 \pm 0.01$	$0.14 \pm 0.06 \pm 0.05$	$0.12 \pm 0.03 \pm 0.09$	$0.27 \pm 0.07 \pm 0.10$	0
55 GeV	(negligible)	$0.05 \pm 0.03 \pm 0.02$	$0.07 \pm 0.02 \pm 0.05$	$0.12 \pm 0.04 \pm 0.05$	0

3.2.3 Limits on GMSB Models

The $\not{E}_T$ spectrum for events with two isolated central photons with $E_T^\gamma > 13$ GeV is shown in Figure 9, along with the predictions from the GMSB model. No excess is observed in two photons + energy imbalance events.

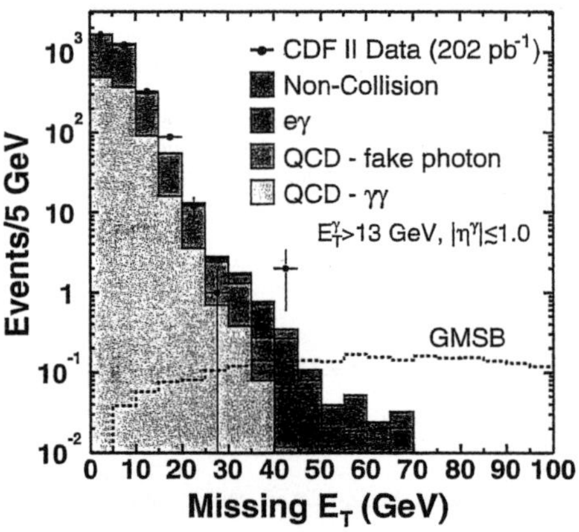

Fig. 9. The $\not{E}_T$ spectrum for events with two isolated central photons with $E_T^\gamma > 13$ GeV and $|\eta| \lesssim 1.0$ along with the predictions from the GMSB model with a $\tilde{\chi}_1^\pm$ mass of 175 GeV/c^2, normalized to 202 pb^{-1}. The diphoton candidate sample data are in good agreement with the background predictions. There are no events above the $\not{E}_T > 45$ GeV threshold. The properties of the two candidates above 40 GeV appear to be consistent with the expected backgrounds.

Since there is no evidence for events with anomalous $\not{E}_T$ in the diphoton candidate sample, we set limits on new particle production from GMSB using the parameters suggested in Ref. [25]. Using the NLO predictions we set a limit of $M_{\tilde{\chi}_1^\pm}$, and then from mass relations in the model, we equivalently set limits on $M_{\tilde{\chi}_1^0}$ and Λ:

$M_{\tilde{\chi}_1^\pm} > 167$ GeV/c^2, $M_{\tilde{\chi}_1^0} > 93$ GeV/c^2, $\Lambda > 69$ GeV/c^2.

The combined CDF+DØ limit [24] is significantly larger (i.e. more stringent) than either experiment alone [23,26].

The details on the combination of the results on the CDF and DØ searches for chargino and neutralino production in GMSB SUSY using the two-photon and missing E_T channel are explained in [24].

Figure 10 shows the combined CDF and DØ result for the observed cross section [24] as a function of $M_{\tilde{\chi}_1^\pm}$ and $M_{\tilde{\chi}_1^0}$ along with the theoretical LO and NLO production cross sections.

The combined CDF+DØ limits are:
$M_{\tilde{\chi}_1^\pm} > 209$ GeV/c^2, $M_{\tilde{\chi}_1^0} > 114$ GeV/c^2, $\Lambda > 84.6$ GeV/c^2 at 95% C.L. in GMSB Model. This is a first combined Run II result and it sets the world's most stringent limits on the GMSB SUSY.

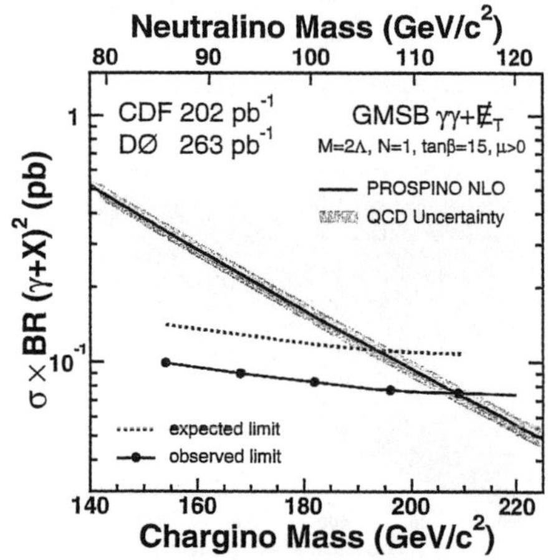

Fig. 10. The 95% C.L. upper limits on the total production cross section times branching ratio versus $M_{\tilde{\chi}_1^\pm}$ and $M_{\tilde{\chi}_1^0}$ for the light gravitino scenario using the parameters proposed in [25]. The lines show the experimental combined CDF+DØ limit and the LO and NLO theoretically predicted cross sections. We set limits of $M_{\tilde{\chi}_1^\pm} > 209$ GeV/c^2, $M_{\tilde{\chi}_1^0} > 114$ GeV/c^2, $\Lambda > 84.6$ GeV/c^2 at 95% C.L. in GMSB Model.

3.3 Search for Lepton-Photon-X Events

In Run I lepton+photon+X search the results were consistent with SM expectations in a number of channels with "the possible exception of photon-lepton events with large $\not\!\!E_T$, for which the observed total was 16 events and the SM expectation was 7.6 ± 0.7 events, corresponding in likelihood to a 2.7 sigma effect." [6]). We concluded "However, an excess of events with 0.7% likelihood (equivalent to 2.7 standard deviations for a Gaussian distribution) in one subsample among the five studied is an interesting result, but it is not a compelling observation of new physics. We look forward to more data in the upcoming run of the Fermilab Tevatron." [6]. In this section we report the results [27] of repeating the $\ell\gamma+X$ search with the same kinematic selection criteria in a substantially larger data set, $\mathcal{L}$=305 pb^{-1}, a higher $\bar{p}p$ collision energy, 1.96 TeV, and the CDF II detector.

3.3.1 Data Sample

The data presented here were taken between March 21, 2002, and August 22, 2004 and represent 305 pb^{-1} for which the silicon detector and all three central muon systems (CMP, CMU and CMX) were operational.

A 3-level trigger [11] system selects events with a high transverse momentum [7] lepton ($P_T > 18$ GeV) or photon ($E_T > 25$ GeV) in the central region, $|\eta| \lesssim 1.0$. Photon and electron candidates are chosen from clusters of energy in adjacent CEM [8] towers; electrons are then further separated from photons by requiring the presence of a COT [9] track pointing at the cluster. Muons are identified by requiring COT tracks to extrapolate to a reconstructed track segment in the muon drift chambers.

We have reused the Run I selection kinematic cuts for Run II analysis, so that they are *a priori*:

- *Tight* Muons: $P_T > 25$ GeV
- *Tight* Central Electrons, Photons: $E_T > 25$ GeV
- *Loose* Muons: $P_T > 20$ GeV
- *Loose* Central Electrons: $E_T > 20$ GeV
- *Loose* Plug Electrons: $E_T > 15$ GeV
- $\not\!\!E_T > 25$ GeV

The identification of photons (see Section 3.1.1) and leptons is essentially the same as in the Run I search [5], with only minor technical differences, mostly due to the changes in the construction of the tracking system and end-plug calorimeters.

A muon passing the 'tight' cuts is required to: a) have a track in the COT that passes quality cuts on the minimum number of hits on the track; b) deposit energy in the electromagnetic and hadronic compartments of the calorimeter consistent with that expected from a muon, c) match a muon 'stub' track in the CMX detector or in both the CMU and CMP detectors; d) not be a cosmic ray (determined from measuring timing with the COT).

'Tight' central electrons are required to have a high-quality track with P_T of at least half the shower energy [10], minimal leakage into the hadronic calorimeter [11], a good profile in the z dimension (the dimension in which the electron track is not bent by the magnetic field) at shower maximum that matches the extrapolated track position, and a lateral sharing of energy in the two calorimeter towers containing the electron shower consistent with that expected.

The additional muons are required to have $P_T > 20$ GeV and to satisfy the same criteria as for "tight" muons but with fewer hits required on the track, or, alternatively, a more stringent cut on track quality but no requirement that there be a matching "stub" in the muon systems. Additional central electrons are required to have $E_T >$ 20 GeV and to satisfy the tight central electron criteria but with a track requirement of only $P_T > 10$ GeV (rather than $0.5\times E_T$), and no requirement on a shower maximum measurement or lateral energy sharing between calorimeter towers. 'Loose' electrons in the end-plug calorimeters are required to have $E_T > 15$ GeV, minimal leakage into the hadron calorimeters, a 'track' containing at least 3 hits in the silicon tracking system, and a shower transverse shape consistent with that expected, with a centroid close to the extrapolated position of the track.

Missing transverse energy $\not\!\!E_T$ is calculated from the calorimeter tower energies in the region $|\eta| < 3.6$. Corrections are then made to the $\not\!\!E_T$ for non-uniform calorimeter response [28] for jets with uncorrected $E_T > 15$ GeV and $\eta < 2.0$, and for muons with $P_T > 20$ GeV.

3.3.2 Control Samples and Backgrounds

We use W and Z^0 production as control samples to ensure that the efficiencies for high-P_T electrons and muons, as well as for $\not\!\!E_T$, are well understood. The photon control sample is constructed from events in which one of the electrons radiates a high-E_T γ such that the $e\gamma$ invariant mass is within 10 GeV of the Z^0 mass.

The dominant source of photon-lepton events at the Tevatron is electroweak diboson production (Figure 11), in which a W or Z^0 boson decays leptonically ($\ell\nu$ or $\ell\ell$) and a photon is radiated from either an initial-state quark, the W or Z^0, or from a charged final-state lepton. The number of such events is estimated using leading-order (LO) matrix element event generators [30–32]. A correction for higher-order processes (K-factor) has been applied [29].

To simulate the triboson channels $W\gamma\gamma$ and $Z\gamma\gamma$ we have used MadGraph [30] and CompHep[32].

[7] We use the convention that "momentum" refers to pc and "mass" to mc^2.

[8] CEM: Central EM Calorimeter.

[9] COT: Central Outer Tracker.

[10] The P_T threshold is set to 25 GeV for $E_T > 100$ GeV.

[11] The fraction of electromagnetic energy E_{em} allowed to leak into the hadronic compartment is $0.055 + 0.00045E_{em}$ for tight and loose central electrons; for loose plug electrons and for photons the fraction must be less than 0.125.

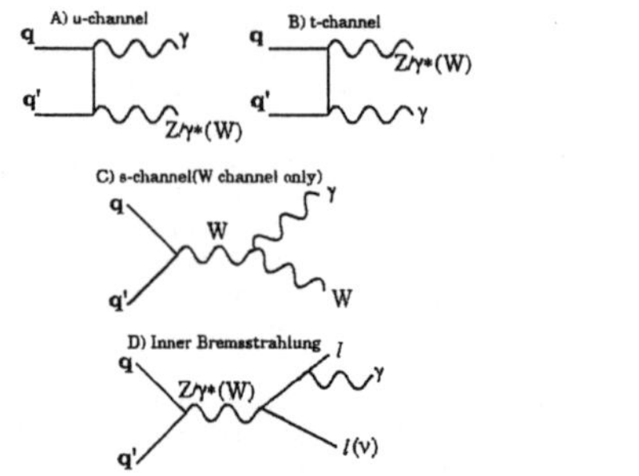

Fig. 11. Standard Model $W\gamma$ and $Z\gamma$ production diagrams.

3.3.3 Lepton-Photon-X Results

Following the Run I analysis strategy, we define the $\ell\gamma\not\!E_T$ subsample by requiring that an event contains, in addition to the central lepton and central photon, $\not\!E_T > 25$ GeV. A second signal subsample, the $\ell\ell\gamma$ sample, is constructed by requiring, in addition to the central lepton and central photon, a second 'loose' lepton with $E_T > 25$ GeV. These two subsamples were selected as the search regions of interest from the Run I results with the same kinematic selections; these two searches in the Run II data are thus *a priori*. Both sample selections are 'inclusive', in that there are no requirements on the presence or absence of other objects.

In addition to the expectations from real SM processes that produce real lepton-photon events, there are backgrounds due to misidentified leptons and photons, and also incorrectly calculated $\not\!E_T$.

We consider two sources of fake photons: QCD jets in which a π^0 or a photon from hadron decay mimics a direct photon, and electron bremsstrahlung, in which an energetic photon is radiated off of an electron which then has much lower energy and curls away from the photon.

Backgrounds from fake leptons and/or fake missing E_T ('QCD') we estimate from a sample, in which we expect to have very little real lepton content [33] by selecting on loose leptons and rejecting events from the W or Z.

The predicted and observed totals for both the $\ell\gamma\not\!E_T$ and $\ell\ell\gamma$ searches are shown in Table 4. We observe 42 $\ell\gamma\not\!E_T$ events, versus the expectation of 37.3 ± 5.4 events. If the Run I ratio of observed to expected, which was 16/7.6, had held up, the 2.7 σ excess observed in Run I would have resulted in an observation of 79 ± 11 events when applying the same analysis to the Run II data, versus the 42 events observed. In the $\ell\ell\gamma$ channel, we observe 31 events, versus an expectation of 23.0 ± 2.7 events. No $e\mu\gamma$ events are observed.

While the number of events observed is somewhat larger than expectations(Table 4), there is not a significant excess in either signature, and the kinematic distributions are in reasonable agreement with the SM predicted shapes.

The distributions for events in the $\ell\gamma\not\!E_T$ sample are shown in Figure 12 for the electron channel and in Figure 13 for the muon channel. The dominant contribution for $\ell\gamma\not\!E_T$ is SM $Z\gamma$ and $W\gamma$ production.

The distributions for events in the $\ell\ell\gamma$ sample are shown at Figure 14 for electron channel and Figure 15 for muon channel. The dominant contribution for $\ell\ell\gamma$ is SM $Z\gamma$ production.

Table 4. A comparison of the numbers of events predicted by the SM and the observations for the $\ell\gamma\not\!E_T$ and $\ell\ell\gamma$ searches. The SM predictions are dominated by $W\gamma$ and $Z\gamma$ production [30–32]. Other contributions come from $W\gamma\gamma$ and $Z\gamma\gamma$, leptonic τ decays, and misidentified leptons, photons, or $\not\!E_T$.

Lepton+Photon+$\not\!E_T$ Events			
SM Source	$e\gamma\not\!E_T$	$\mu\gamma\not\!E_T$	$(e+\mu)\gamma\not\!E_T$
$W^{\pm}\gamma$	13.70±1.89	8.84±1.35	22.54±2.80
$Z^0/\gamma^* + \gamma$	1.16±0.40	4.49±0.64	5.65±1.03
$W^{\pm}\gamma\gamma, Z^0/\gamma^* + \gamma\gamma$	0.14±0.02	0.18±0.02	0.32±0.03
$W^{\pm}\gamma, Z^0/\gamma^* + \gamma \rightarrow \tau\gamma$	0.71±0.18	0.26±0.08	0.97±0.22
$W^{\pm}$+Jet faking γ	2.8±2.8	1.6±1.6	4.4±4.4
$Z^0/\gamma^* \rightarrow e^+e^-, e \rightarrow \gamma$	2.45±0.33	-	2.45±0.33
Jets faking $\ell + \not\!E_T$	0.7±0.7	0.3±0.3	1.0±0.8
Total	**21.7±3.4**	**15.7±2.2**	**37.3±5.4**
Observed	**25**	**17**	**42**
Multi-Lepton+Photon Events			
SM Source	$ee\gamma$	$\mu\mu\gamma$	$ll\gamma$
$Z^0/\gamma^* + \gamma$	12.50±1.53	7.81±0.88	20.31±2.40
$Z^0/\gamma^* + \gamma\gamma$	0.24±0.03	0.12±0.02	0.36±0.04
Z^0/γ^*+Jet faking γ	0.3±0.3	0.2±0.2	0.5±0.5
$Z^0/\gamma^* \rightarrow e^+e^-, e \rightarrow \gamma$	0.23±0.09	-	0.23±0.09
Jets faking $\ell + \not\!E_T$	0.6±0.6	1.0±1.0	1.6±1.2
Total	**13.9±1.7**	**9.1±1.4**	**23.0±2.7**
Observed	**19**	**12**	**31**

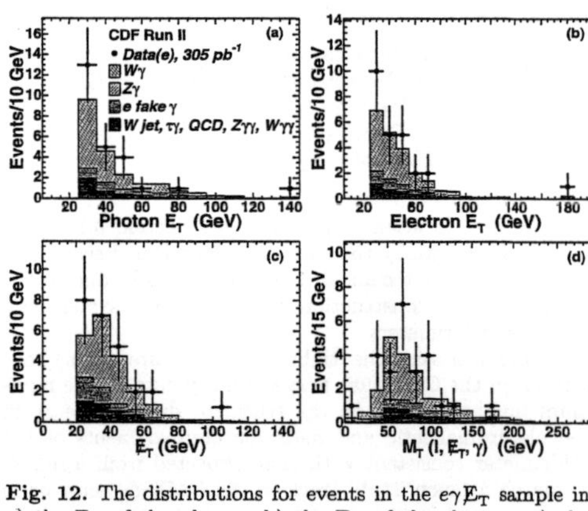

Fig. 12. The distributions for events in the $e\gamma\not\!E_T$ sample in a) the E_T of the photon; b) the E_T of the electron, c) the missing transverse energy, $\not\!E_T$, and d) the transverse mass of the electron-photon-$\not\!E_T$ system. The histograms show the expected SM contributions, including estimated backgrounds from misidentified photons and leptons.

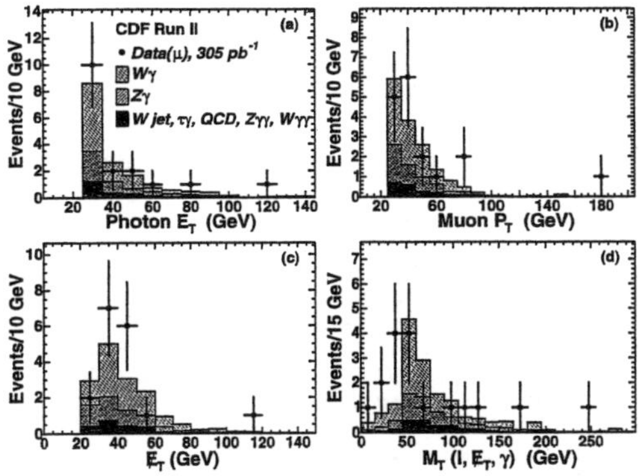

Fig. 13. The distributions for events in the $\mu\gamma\not\!\!E_{\rm T}$ sample in a) the $\rm E_T$ of the photon; b) the $\rm P_T$ of the muon, c) the missing transverse energy, $\not\!\!E_{\rm T}$, and d) the transverse mass of the muon-photon-$\not\!\!E_{\rm T}$ system. The histograms show the expected SM contributions, including estimated backgrounds from misidentified photons and leptons.

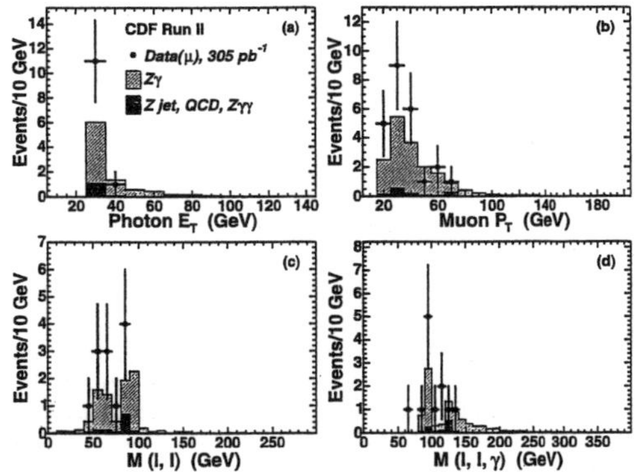

Fig. 15. The distributions in a) the $\rm E_T$ of the photon; b) the $\rm P_T$ of the muon, c) the 2-body mass of the dimuon system, and d) the 3-body invariant mass $m_{\mu\mu\gamma}$.

For the $Z\gamma$ process occurring via initial state radiation, the dilepton invariant mass distribution will be peaked around the Z^0-pole. For the final state radiation, the three body invariant mass (m(l, l, γ)) distribution will be peaked around the Z^0-pole (Figures 14, 15, (c) and (d)).

We do not expect missing $\rm E_T$ in the events in the $\ell\ell\gamma$ sample based on the SM backgrounds; the $ee\gamma\gamma\not\!\!E_{\rm T}$ event was of special interest due to the large value of $\not\!\!E_{\rm T}$. Figure 16 shows the distributions in $\not\!\!E_{\rm T}$ for the $ee\gamma$ and $\mu\mu\gamma$ subsamples of the $\ell\ell\gamma$ sample. No events are observed with $\not\!\!E_{\rm T} > 25$ GeV.

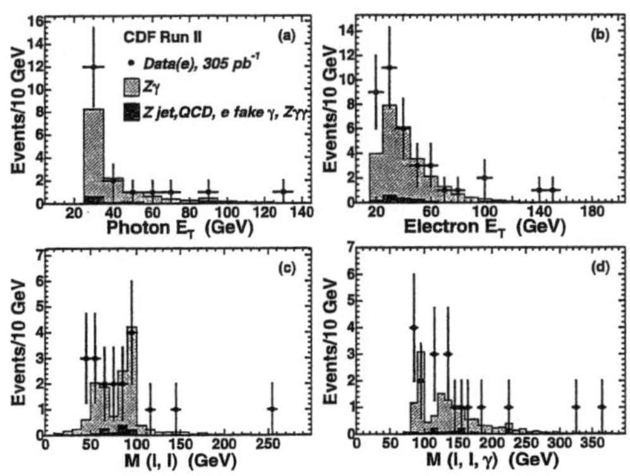

Fig. 14. The distributions in a) the $\rm E_T$ of the photon; b) the $\rm E_T$ of the electron, c) the 2-body mass of the dielectron system, and d) the 3-body invariant mass $m_{ee\gamma}$.

In conclusion, we have repeated the search for inclusive lepton + photon production with the same kinematic requirements as the Run I search, but with a significantly larger data sample and a higher collision energy. We find that the numbers of events in the $\ell\gamma\not\!\!E_{\rm T}$ and $\ell\ell\gamma$ subsamples of the $\ell\gamma$+X sample agree with SM predictions. We observe no $\ell\ell\gamma$ events with anomalous large $\not\!\!E_{\rm T}$ or with multiple photons and so find no events like the $ee\gamma\gamma\not\!\!E_{\rm T}$ event of Run I.

In summary, while we are disappointed that we found no more $ee\gamma\gamma\not\!\!E_{\rm T}$ events in a much larger sample than in Run I, and the Run I excess in $\ell\gamma\not\!\!E_{\rm T}$ became less significant rather than more, we have conclusively settled a question that generated much interest in the theoretical community. The channels we have investigated will remain interesting, and the techniques we have developed and the knowledge gained will be useful for similar searches at the Tevatron and at the LHC.

4 Summary and Outlook

To summarize, we will list the main points for the Run II results presented in this paper:

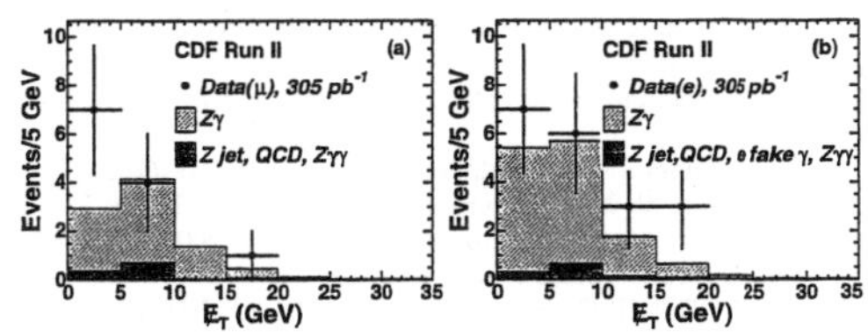

Fig. 16. The distributions in missing transverse energy $\not\!\!E_{\rm T}$ observed in the inclusive search for a) $\mu\mu\gamma$ events and b) $ee\gamma$ events. The histograms show the expected SM contributions.

- Search for $\ell\gamma$+X: the Run I 2.7 sigma excess in $\ell\gamma\not{E}_T$ is not confirmed when repeating the analysis with much more data. We observe no $\ell\ell\gamma$ events with anomalous large $\not{E}_T$ or with multiple photons.
- Search for $\gamma\gamma\not{E}_T+X$: no excess is observed in the two photons + energy imbalance channel. The combined CDF and DØ Result provides world's most stringent limits on GMSB SUSY. No new $ee\gamma\gamma\not{E}_T$ (or similar) candidate events have been found.
- Search for high-mass diphotons: the data agree with predictions.

The Fermilab plan is to have a factor of 10-20 more data than presented here by the end of Run II of the Tevatron. A recent upgrade, the EM Timing system [34], provides a vitally important handle that could confirm (or disprove) that all the photons in unusual events are from the primary collision.

Currently, CDF is actively pursuing topics and analyzing up to 1 fb^{-1} of delivered luminosity. New and exciting results are coming out quickly. Further information regarding the analyses presented in this paper and new results can be found in [35].

References

1. S. L. Glashow, Nucl. Phys. **22** 588, (1961); S. Weinberg, Phys. Rev. Lett. **19** 1264, (1967); A. Salam, Proc. 8th Nobel Symposium, Stockholm, (1979).
2. F. Abe *et al.*, Phys. Rev. D 59, 092002 (1999); hep-ex/9806034.
3. F. Abe *et al.*, Phys.Rev.Lett 81, 1791 (1998); hep-ex/9801019.
4. D. Affolder *et al.*, Phys.Rev.D 65, 052006 (2002); hep-ex/0106012.
5. D. Acosta *et al.*, Phys.Rev.D 66, 012004 (2002); hep-ex/0110015.
6. D. Acosta *et al.*, Phys.Rev.Lett 89, 041802 (2002); hep-ex/0202004.
7. D. Toback, Ph.D. thesis, University of Chicago, 1997.
8. S. Ambrosanio, G. L. Kane, G. D. Kribs, S. P. Martin, and S. Mrenna, Phys. Rev. D **55**, 1372 (1997); B.C. Allanach, S. Lola, K. Sridhar, Phys.Rev.Lett.89:011801, 2002, hep-ph/0111014.
9. J. Berryhill, Ph.D. thesis, University of Chicago, 2000.
10. B.C.Allanach, S.Lola, K.Sridhar, Phys.Rev.Lett. 89, 011801 (2002), hep-ph/0111014.
11. D. Acosta *et al.* (CDF Collaboration), Phys. Rev. D **71**, 032001 (2005).
12. F. Abe *et al.*, Nucl. Instrum. Methods Phys. Res., Sect. A **271**, 387 (1988).
13. S. Kuhlmann et al., Nucl. Instrum. Meth. A518:39-41, 2004.
14. T. Affolder et al., Nucl. Instr. Meth A526,249 (2004).
15. A. Sill et al. Nucl. Instrum. Meth. A447, 1 (2000); T. Affolder et al., Nucl. Instrum. Meth. A453, 84 (2000); C.S. Hill, Nucl. Instrum. Meth. A530, 1 (2000).
16. The CDF II Collaboration, FERMILAB-PUB-96/390-E (1996), Chapter 10.
17. L. Randall and R. Sundrum, Phys. Rev. Lett **83**, 3370 (1999), hep-ph/9905221.
18. CDF Collaboration, D. Acosta *et al.*, http://www-cdf.fnal.gov/internal/physics/exotic/physics/r2a/20040805.diphoton_rsgrav/public/index.html.
19. See wwwlapp.in2p3.fr/lapth/PHOX_FAMILY/readme_diphox.html.
20. High Mass Dielectrons: http://fcdfhome.fnal.gov/usr/ikado/Bless_Fall_2003/Bless_fall2003.html
High Mass Dimuons: http://www-cdf.fnal.gov/physics/exotic/run2/highmass-mumu-2004/blessed_plots_win04.html.
21. S. Dimopoulos, S. Thomas, J. D. Wells, Nucl. Phys. B **488**, 39 (1997); S. Ambrosanio, G.D. Kribs and S.P. Martin, Phys. Rev. D **56**, 1761 (1997); G. F. Giudice and R. Rattazzi, Phys. Rept. **322**, 419 (1999); and S. Ambrosanio, G. L. Kane, G. D. Kribs, S. P. Martin, and S. Mrenna, Phys. Rev. D **55**, 1372 (1997).
22. R. Culbertson *et al.*, hep-ph/0008070.
23. CDF Collaboration, D. Acosta *et al.*, Phys.Rev.D. 71, 031104(R)(2005); hep-ex/0410053.
24. V. Buescher, R.Culbertson *et al.*, for the CDF and DØ collaborations hep/ex 0504004.
25. B. C. Allanach *et al.*, Eur. Phys. J. **C25** 113 (2002). We take the messenger mass scale $M_M = 2\Lambda$, $\tan(\beta) = 15$, $\text{sign}(\mu) = 1$, the number of messenger fields $N_M = 1$, and negligibly short $\tilde{\chi}_1^0$ lifetimes.
26. DØ Collaboration, V.M. Abazov *et al.*, Phys.Rev.Lett. 94, 041801(2005); hep-ex/0408146.
27. CDF Collaboration, A. Abulencia *et al.*, accepted to Phys.Rev.Lett; hep-ex/0605097.
28. A. Bhatti *et al.*, submitted to Nucl. Instrum. Methods, Oct. 2005; hep-ex/0510047.
29. U. Baur, T. Han and J. Ohnemus, Phys. Rev. D **48**, 5140 (1993); U. Baur, T. Han and J. Ohnemus, Phys. Rev. D **57**, 2823 (1998); hep-ph/9710416. Both the $W\gamma$ and $Z\gamma$ K-factors are fixed at 1.36 for generated $\ell\nu$ masses below 76 GeV and for generated $\ell^+\ell^-$ masses below 86 GeV. Above the poles the K-factors grow with E_T^γ to be 1.62 and 1.53 at $E_T^\gamma = 100$ GeV for $W\gamma$ and $Z\gamma$, respectively.
30. T. Stelzer and W. F. Long, Comput.Phys.Commun. 81 (1994) 357-371; F. Maltoni and T. Stelzer, JHEP 0302:027 (2003), hep-ph/0208156.
31. U. Baur, T. Han and J. Ohnemus, Phys. Rev. D **57**, 2823 (1998); hep-ph/9710416; U. Baur, T. Han and J. Ohnemus, Phys. Rev. D **48**, 5140 (1993); hep-ph/9305314.
32. E. Boos *et al.*, INP-MSU-98-41-542. hep-ph/9908288; E. Boos *et al.*, INP MSU 94-36/358 and SNUTP-94-116.hep-ph/9503280.
33. S. Kopp, Ph.D. thesis, University of Chicago, 1994.
34. "The Timing System for the CDF Electromagnetic Calorimeters", M. Goncharov *et al.*, to be submitted to NIM.
35. See http://www-cdf.fnal.gov/physics/exotic/exotic.html.

DIPLOMAS

Nineteen Diplomas were open for competition among the participants. They have been awarded as follows:

- John S. BELL Diploma to:

  Ben O'LEARY

  University of Edinburgh, UK

- Patrick M.S. BLACKETT Diploma to:

  Kazunori HANAGAKI

  Fermilab, Batavia, USA

- Carlo CASTAGNOLI Diploma to:

  Fabio CARDARELLI

  Enrico Fermi Centre, Rome, Italy

- James CHADWICK Diploma to:

  Jorge MERCADO-PÉREZ

  Ruprecht-Karls-Universität Heidelberg, Germany

- Paul A.M. DIRAC Diploma to:

  Lars NILSE

  University of Manchester, UK

- Vladimir N. GRIBOV Diploma to:

  Mate CSANAD

  Eotvos University, Budapest, Hungary

- Robert HOFSTADTER Diploma to:

  Simone PACETTI

  INFN - Laboratori Nazionali di Frascati, Italy

- Gunnar KÄLLEN Diploma to:

  Shoaib MUNIR

  University of Southampton, UK

- Giuseppe P.S. OCCHIALINI Diploma to:

  Marco GARBINI

  Enrico Fermi Centre, Rome, Italy

- Oreste PICCIONI Diploma to:

  Jun-Ichi NOAKI

  University of Southampton, UK

- Bruno PONTECORVO Diploma to:

  Isabella MASINA

  Enrico Fermi Centre, Rome, Italy

- Isidor I. RABI Diploma to:

  Annalisa MASTROSERIO

  University of Bari, Italy

- Giulio RACAH Diploma to:

  Dominika KONIKOWSKA

  Warsaw University, Poland

- Bruno ROSSI Diploma to:

  Marzia NARDI

  Enrico Fermi Centre, Rome, Italy

- Andreij D. SAKHAROV Diploma to:

  Luigi INGROSSO

  Laboratori Nazionali di Frascati, Italy

- Victor F. WEISSKOPF Diploma to:

  Leonardo GUALTIERI

  University of Rome "Tor Vergata", Italy

- Eugene P. WIGNER Diploma to:

  Viacheslav DIDENKO

  Lebedev Physical Institute, Moscow, Russia

- Bjorn H. WIIK Diploma to:

  Roberto PREGHENELLA

  University of Bologna, Italy

- Chien Shiung WU Diploma to:

  Rita GENOVESE

  Enrico Fermi Centre, Rome, Italy

AWARDS

- The Awards for the best "New Talents-Speaker" is divided into four parts:

i) for an original work in experimental physics:

Andrey LOGINOV

CERN, Geneva, Switzerland

ii) for an original work in theoretical physics:

Pietro Antonio GRASSI

Enrico Fermi Centre, Rome, Italy

iii) for an original work in data analysis:

Julien FAIVRE

INFN, Sezione di Padova, Italy

iv) for an original work in new fields:

Ciro CATTUTO

Enrico Fermi Centre, Rome, Italy

- Award for BEST STUDENT given to:

Alessio MARRANI

INFN - Laboratori Nazionali di Frascati, Italy

- Awards for BEST SCIENTIFIC SECRETARY given to:

i) **Laura ANDRIANOPOLI**

Enrico Fermi Centre, Rome, Italy

ii) **Giovanni IMPONENTE**

Enrico Fermi Centre, Rome, Italy

iii) **Mattia JONA-LASINIO**

Laboratoires Kastler - Brossel, Paris, France

PARTICIPANTS

Laura ANDRIANOPOLI Italy	Enrico Fermi Centre Via Panisperna, 89/A – 00184 ROMA, Italy e-mail: *laura.andrianopoli@cern.ch*
Ignatios ANTONIADIS Greece	CERN - Theory Division CH-1211 GENEVA 23, Switzerland e-mail: *Ignatios.Antoniadis@cern.ch*
Thomas APPELQUIST USA	Department of Physics Yale University P.O. Box 108121 - Prospect Street 217 NEW HAVEN, CT 06520-8121, USA e-mail: *thomas.appelquist@yale.edu*
Silvia ARCELLI Italy	Enrico Fermi Centre Via Panisperna, 89/A – 00184 ROMA, Italy e-mail: *silvia.arcelli@bo.infn.it*
Rinaldo BALDINI FERROLI Italy	Enrico Fermi Centre Via Panisperna, 89/A – 00184 ROMA, Italy e-mail: *baldini@centrofermi.it*
Victor BEGUN Ukraine	Bogolyubov Institute for Theoretical Physics 14 b Metrologicheskaya Str. 03143 KIEV, Ukraine e-mail: *vic1@univ.kiev.ua*
Marcello BINDI Italy	Dipartimento di Fisica Università degli Studi Via Irnerio 46 - 40100 BOLOGNA, Italy e-mail: *bindim@bo.infn.it*
Paolo BOLZONI Italy	Dipartimento di Fisica Università degli Studi Via Celoria 16 - 20133 MILANO, Italy e-mail: *paolo.bolzoni@mi.infn.it*
Valerio BOZZA Italy	Enrico Fermi Centre Via Panisperna, 89/A – 00184 ROMA, Italy e-mail: *valboz@sa.infn.it*
Elisa BRESSAN Italy	Dipartimento di Fisica Università degli Studi Via Irnerio 46 - 40100 BOLOGNA, Italy e-mail: *elisa.bressan@fastwebnet.it*
Francesco Maria CARDANO Italy	INFN - Sezione di Perugia 06123 PERUGIA, Italy e-mail: *cardano@pg.infn.it*
Fabio CARDARELLI Italy	Enrico Fermi Centre Via Panisperna, 89/A – 00184 ROMA, Italy e-mail: *cardafam@virgilio.it*
Ciro CATTUTO Italy	Enrico Fermi Centre Via Panisperna, 89/A – 00184 ROMA, Italy e-mail: *ciro.cattuto@roma1.infn.it*

Cristiano CHIARABELLI
Italy

Enrico Fermi Centre
Via Panisperna, 89/A – 00184 ROMA, Italy
e-mail: *c.chiarabelli@uniroma3.it*

Luisa CIFARELLI
Italy

Dipartimento di Fisica
Università degli Studi
Via Irnerio 46 – 40126 BOLOGNA, Italy
e-mail: *cifarelli@bo.infn.it*

Ignazio CIUFOLINI
Italy

University of Lecce
Viale Gallipoli 49 - 73100 LECCE, Italy
e-mail: *ignazio.ciufolini@unile.it*

Fabrizio COCCETTI
Italy

Enrico Fermi Centre
Via Panisperna, 89/A – 00184 ROMA, Italy
e-mail: *fabrizio.coccetti@roma1.infn.it*

Mate CSANAD
Hungary

Department of Atomic Physics
Eotvos University
Pazmany Peter Setany 1/a - H-1117 BUDAPEST, Hungary
e-mail: *csanad@bnl.gov*

Richard Henry DALITZ
Australia

Department of Theoretical Physics
University of Oxford
1 Keble Road - OXFORD, OX1 3NP, UK
e-mail: *r.dalitz1@physics.ox.ac.uk*

Dmitri DENISOV
Russia

FNAL - Fermi National Accelerator Laboratory
Mail Stop 221 - P.O. Box 500
BATAVIA, IL 60510-0500, USA
e-mail: *denisovd@fnal.gov*

Alvaro DE RUJULA
Spain

CERN - TH/SP Division
CH-1211 GENEVA 23, Switzerland
e-mail: *Alvaro.Derujula@cern.ch*

Angela DI GIUSEPPE
Italy

Dipartimento di Fisica
Università degli Studi
Via Irnerio 46 - 40100 BOLOGNA, Italy
e-mail: *digiuseppe@bo.infn.it*

Viacheslav DIDENKO
Russia

I.E. Tamm Department of Theorerical Physics
Lebedev Physical Institute
RU-119991 MOSCOW, Russia
e-mail: *didenko@lpi.ru*

Heng Tong DING
China

Institute of Particle Physics (IOPP)
Central China (Huszhong) Normal University
WUHUAN, 430079 P.R. China
e-mail: *dinght@@iopp.ccnu.edu.cn*

Serena FAGNOCCHI
Italy

Istituto di Fisica
Università degli Studi
Via Irnerio 46 - 40126 BOLOGNA, Italy
e-mail: *fagnocchi@bo.infn.it*

Julien FAIVRE
France

INFN Sezione di Padova
Via Marzolo 8
35131 PADOVA, Italy
e-mail: *julien.faivre@pd.infn.it*

Sergio FERRARA Italy	CERN - TH Division CH-1211 GENEVE 23, Switzerland e-mail: *sergio.ferrara@cern.ch*
Kenji FUKUSHIMA Japan	RIKEN BNL Research Center Brookhaven National Laboratory - Bldg. 510A P.O. Box 5000 - UPTON, NY 11973, USA e-mail: *fuku@quark.phy.bnl.gov*
Marco GARBINI Italy	Enrico Fermi Centre Via Panisperna, 89/A – 00184 ROMA, Italy e-mail: *garbini@bo.infn.it*
Rita GENOVESE Italy	Enrico Fermi Centre Via Panisperna, 89/A – 00184 ROMA, Italy e-mail: *genovese@studenti.ph.unito.it*
Marcello GIORGI Italy	Università di Pisa & INFN Viale Buonarroti 2 – 56100 PISA, Italy e-mail: *giorgi@slac.stanford.edu - marcello.giorgi@pi.infn.it*
Federico GIOVE Italy	Enrico Fermi Centre Via Panisperna, 89/A – 00184 ROMA, Italy e-mail: *federico.giove@roma1.infn.it*
Pietro Antonio GRASSI Italy	Enrico Fermi Centre Via Panisperna, 89/A – 00184 ROMA, Italy e-mail: *pietro.grassi@cern.ch*
Leonardo GUALTIERI Italy	Dipartimento di Fisica Università degli Studi di Roma "Tor Vergata" Piazzale Aldo Moro 2 - 00185 ROMA, Italy e-mail: *Leonardo.Gualtieri@roma1.infn.it*
Kazunori HANAGAKI Japan	Fermi Lab MS 357 P.O. Box 500 BATAVIA, IL 60510, USA e-mail: *kazu@fnal.gov*
Giovanni IMPONENTE Italy	Enrico Fermi Centre Via Panisperna, 89/A – 00184 ROMA, Italy e-mail: *g.imponente@qmul.ac.uk*
Luigi INGROSSO Italy	Laboratori Nazionali di Frascati Via E. Fermi 40 - 00044 FRASCATI, Italy Liceo Touschek - GROTTAFERRATA, Italy e-mail: *luigi.ingrosso@lnf.infn.it - ingrosso_luigi@yhaoo.it*
Mattia JONA-LASINIO Italy	Laboratoires Kastler - Brossel 24 Rue Lhomond F-75213, PARIS Cedex 05, France e-mail: *mjona@lkb.ens.fr*
Vladimir KASHKAN Belarus	Dipartimento di Fisica Teorica Università di Torino Via P. Giuria 1 - 10125 TORINO, Italy e-mail: *kashkan@to.infn.it*
Andrey KAZANTSEV Russia	Institute of General and Nuclear Physics, O-52 RRC "Kurchatov Institute" Kurchatov sq., 1 - RU-123182 MOSCOW, Russia

e-mail: *stkav@vskur4.mbslab.kiae.ru*

Dominika KONIKOWSKA
Poland

Institute of Theoretical Physics - Faculty of Physics
Warsaw University
ul Hoza 69 - PL 00-681 WARSAW, Poland
e-mail: *dominika@fuw.edu.pl*

Christian P. KORTHALS-ALTES
The Netherlands

Centre de Physique Théorique
CNRS-Luminy
163 avenue de Luminy - Case 907
F-13288 MARSEILLE Cedex 9, France
e-mail: *altes@cpt.univ-mrs.fr*

Anastasia KUTUZOVA
Russia

I.E. Tamm Department of Theorerical Physics
Lebedev Physical Institute
Leninsky Prospect 53 - RU-119991 MOSCOW, Russia
e-mail: *kutuz-off@list.ru*

Seymor J. LINDENBAUM
USA

Physics Department
The City College, The City University of New York
NEW YORK, NY 10031, USA
e-mail: *lindenbaum@bnl.gov*

Andrey LOGINOV
Russia

CERN
CH-1211 GENEVA 23, Switzerland
e-mail: *loginov@flanl.gov*

Mauro LOMBARDI
Italy

Dipartimento di Fisica
Università degli Studi
Via Irnerio 46 - 40100 BOLOGNA, Italy
e-mail: *elisa.bressan@fastwebnet.it*

Alessio MARRANI
Italy

INFN - Laboratori Nazionali di Frascati
Via Enrico Fermi 40 - 00044 FRASCATI, Italy
e-mail: *marrani@lnf.infn.it*

Isabella MASINA
Italy

Enrico Fermi Centre
Via Panisperna, 89/A – 00184 ROMA, Italy
e-mail: *Isabella.Masina@roma1.infn.it*

Annalisa MASTROSERIO
Italy

Dipartimento di Fisica
Università degli Studi
Via E. Orabona 4 - 70126 BARI, Italy
e-mail: *annalisa.mastroserio@cern.ch*

Alexey MATVEEV
Russia

I.E. Tamm Department of Theorerical Physics
Lebedev Physical Institute
RU-119991 MOSCOW, Russia
e-mail: *matveev@lpi.ru*

Helenia MENGHETTI
Italy

INFN - Sezione di Bologna
Via Irnerio 46 - 40100 BOLOGNA, Italy
e-mail: *helenia.menghetti@bo.infn.it*

Jorge MERCADO-PÉREZ
Mexico

Ruprecht-Karls-Universität Heidelberg
Physikalisches Institut
Philosophenweg 12 - D-69120 HEIDELBERG, Germany
e-mail: *mercado@physi.uni-heidelberg.de*

Shoaib MUNIR
Pakistan

Department of Physics and Astronomy
Southampton High Energy Physics Group

University of Southampton
SOUTHAMPTON, SO17 1BJ, UK
e-mail: *shobig@phys.soton.ac.uk*

Giovanni MURTAS
Italy
Enrico Fermi Centre
Via Panisperna, 89/A – 00184 Roma
e-mail: *giovanni.murtas@casaccia.enea.it*

Marzia NARDI
Italy
Enrico Fermi Centre
Via Panisperna, 89/A – 00184 ROMA, Italy
e-mail: *nardi@to.infn.it*

Lars NILSE
Germany
Theory Group - School of Physics and Astronomy
University of Manchester
MANCHESTER, M13 9PL, UK
e-mail: *lars@theory.ph.man.ac.uk*

Jun-Ichi NOAKI
Japan
School of Physics and Astronomy
University of Southampton
Highfield, SOUTHAMPTON, SO17 1BJ, UK
e-mail: *cts@phys.soton.ac.uk*

Ben O'LEARY
Ireland
The University of Edinburgh
James Clerk Maxwell Building, The King's Building
Mayfield Road - EDINBURGH EH9 3JZ, UK
e-mail: *ben.oleary@ph.ed.ac.uk*

Simone PACETTI
Italy
INFN - Laboratori Nazionali di Frascati
Via E. Fermi 40 - 00044 FRASCATI, Italy
e-mail: *Simone.Pacetti@lnf.infn.it*

James PINFOLD
Canada
Department of Physics
Director of the Centre for Subatomic Research
University of Alberta
EDMONTON, Alberta T6G 2N5, Canada
e-mail: *pinfold@phys.ualberta.ca*

Roberto PREGHENELLA
Italy
Dipartimento di Fisica
Università degli Studi
Via Irnerio 46 - 40100 BOLOGNA, Italy
e-mail: *preghenella@bo.infn.it*

Lucio ROSSI
Italy
CERN
CH-1211 GENEVE 23, Switzerland
e-mail: *Lucio.Rossi@cern.ch*

Alberto SANTAMBROGIO
Italy
Dipartimento di Fisica
Università di Milano-Bicocca
Piazza della Scienza 3 - 20126 MILANO, Italy
e-mail: *alberto.santambrogio@mi.infn.it*

Gabriella SARTORELLI
Italy
Istituto Nazionale di Fisica Nucleare - INFN
Sezione di Bologna
Via Irnerio 46 - 40126 BOLOGNA, Italy
e-mail: *sartorelli@bo.infn.it*

Marco SELVI
Italy
Dipartimento di Fisica
Università degli Studi
Via Irnerio 46 - 40100 BOLOGNA, Italy
e-mail: *selvi@bo.infn.it*

Maksyn SERBYN
Ukraine

ITEP
B. Cheremushkinskaja 25
MOSCOW, 117259, Russia
e-mail: *maksyn.serbyn@itep.ru*

Shamil SHAKIROV
Russia

ITEP
B. Cheremushkinskaja 25
MOSCOW, 117259, Russia
e-mail: *darkname@ramble.ru*

Irina SHREYBER
Russia

CERN
CH-1211 GENEVA 23, Switzerland
e-mail: *shreyber@mail.cern.ch - shreyber@fnal.gov*

Pasquale STANO
Italy

Enrico Fermi Centre
Via Panisperna, 89/A – 00184 ROMA, Italy
e-mail: *stano@uniroma3.it*

Peter A. STEINBERG
USA

Brookhaven National Laboratory
UPTON, NY 11973, USA
e-mail: *peter.steinberg@bnl.gov*

Francesco SYLOS-LABINI
Italy

Enrico Fermi Centre
Via Panisperna, 89/A – 00184 ROMA, Italy
e-mail: *sylos@roma1.infn.it*

Frank TECKER
Germany

CERN
CH-1211 GENEVE 23, Switzerland
e-mail: *frank.tecker@cern.ch*

Gerardus 't HOOFT
The Netherlands

Department of Theoretical Physics
Spinoza Institute - Utrecht University
NL-3508 TD UTRECHT, The Netherlands
e-mail: *thooft@phys.uu.nl*

Grigory VARTANOV
Russia

Bogoliubov Theoretical Physics Laboratory (BLTP)
JINR - Joint Institute for Nuclear Research
Joliot Curie 6 – DUBNA 141980, Moscow Region, Russia
e-mail: *vartanov@theor.jinr.ru*

Chiara ZAMPOLLI
Italy

Dipartimento di Fisica
Università degli Studi
Via Irnerio 46 - 40100 BOLOGNA, Italy
e-mail: *chiara.zampolli@bo.infn.it*

Antonino ZICHICHI
Italy

CERN
CH-1211 GENEVA 23, Switzerland

Dipartimento di Fisica - Università di Bologna
40126 BOLOGNA, Italy